AF518369

Textbook of

Agricultural Structures and Environmental Control

The Authors

Dr. Mukesh Kumar Garg completed his B.Tech. (Agricultural Engineering) in 1981, M.Tech. (Processing and Agricultural Structures) in 1984 from Punjab Agricultural University, Ludhiana and Ph.D. (Agricultural Structures and Process Engineering) in 1995 from Indian Agricultural Research Institute, New Delhi. He started his career from Haryana Agricultural University, Hisar as Assistant Professor in 1985. He has been actively engaged in teaching, research, administration and management in the field of Agricultural Engineering with special emphasis on Agricultural Processing and Food Engineering for the last more than 35 years. Worked as Dean of the College of Agricultural Engineering and Technology, Hisar from 2008 to 2012 and Officer-on-Special Duty (OSD) to Vice-chancellor from 2016 to 2020. Also worked as Director, Human Resource Management; Estate Officer-cum-SE; First Appellate Authority on behalf of CCS HAU, Hisar under RTI Act, 2005; and Chairman of Central Purchase Committee, CCS Haryana Agricultural University, Hisar. Worked as Head, Department of Processing and Food Engineering from 2013 to 2020. Developed courses as well as laboratories for undergraduate and post-graduate students. PG courses in the discipline of Processing and Food Engineering were prepared as a member of BSMA committee constituted by ICAR. Dr Garg has been involved in design, development and field evaluation of various post-harvest machinery/processing equipment. He has obtained 2 patents and 4 design registrations from the Patent Office, Government of India. Worked as PI/Co-PI of various research projects. Dr Garg has authored 6 books and also published about 100 research papers, 42 popular articles, 13 practical/training manuals, 16 chapters in books and 16 extension leaflets. Received Merit Certificates from the University/College for academic achievements. He has so far guided 23 M.Tech./M.Sc. students as Major advisor. Honored with various awards and certificates from various organizations and associations. He has served as member, Institute Management Committee of Central Institute of Post Harvest Engineering and Technology, Ludhiana; member of Research Advisory Committee of Central Institute of Agricultural Engineering, Bhopal; member of QRT, member of NBA, member of Bureau of Indian Standards (FAD 20). He also worked as referee for several research journals.

Er. Sunil Kumar works as an Assistant Professor in the field of Processing and Food Engineering at the College of Agricultural Engineering and Technology located at CCS Haryana Agricultural University (CCSHAU), Hisar, Haryana. He completed his B.Tech. in Agricultural Engineering from Baba Sahab Bhim Rao Ambedkar College of Agricultural Engineering and Technology (CAET), Etawah, Uttar Pradesh, which is affiliated with Chandra Shekhar Azad University of Agriculture and Technology (CSAUAT), Kanpur, Uttar Pradesh in 2012. Er. Sunil Kumar pursued his master's degree (M.Tech.) in Processing and Food Engineering from CCS Haryana Agricultural University (CCSHAU), Hisar, Haryana in 2014 through the ICAR-JRF examination. He cleared the ICAR-Indian Agricultural Research Institute (IARI) Examination in 2014 for admission to a doctoral degree program at ICAR-Central Institute of Agricultural Engineering (CIAE), Bhopal. Er. Sunil Kumar also obtained the ASRB NET certificate in 2014 in the discipline of Processing and Food Engineering. He also holds two NET certificates in Agricultural Processing and Structures. Er. Sunil Kumar has conducted research work in collaboration with the University of Illinois (UIUC), Urbana-Champaign, and the Tokyo University of Agriculture (TUA), Japan. He has guided one international and co-guided three PG students in the field of Processing and Food Engineering. He has published approximately 26 international research/review papers, 10 national research papers, 12 book chapters, 8 popular articles, and obtained 3 design patents in this field. Er. Sunil Kumar has received above 1000 citations, with an h-Index of 13 and an i10-Index of 15 in the field of Agricultural Processing and Food Engineering.

Textbook of
Agricultural Structures and Environmental Control

– *Authors* –

M.K. Garg

Former Dean
College of Agricultural Engineering and Technology
CCS Haryana Agricultural University, Hisar (India)

Sunil Kumar

Assistant Professor
Department of Processing and Food Engineering
College of Agricultural Engineering and Technology
CCS Haryana Agricultural University, Hisar (India)

2026

Daya Publishing House®
A Division of
Astral International Pvt. Ltd.
New Delhi – 110 002

ISBN 978-93-5919-345-8

Publisher's Note:

Every possible effort has been made to ensure that the information contained in this book is accurate at the time of going to press, and the publisher and authors cannot accept responsibility for any errors or omissions, however caused. No responsibility for loss or damage occasioned to any person acting, or refraining from action, as a result of the material in this publication can be accepted by the editor, the publisher or the authors. The Publisher is not associated with any product or vendor mentioned in the book. The contents of this work are intended to further general scientific research, understanding and discussion only. Readers should consult with a specialist where appropriate.

Every effort has been made to trace the owners of copyright material used in this book, if any. The author and the publisher will be grateful for any omission brought to their notice for acknowledgement in the future editions of the book.

Published by : **Daya Publishing House®**
A Division of
Astral International Pvt. Ltd.
– ISO 9001:2015 Certified Company –
4736/23, Ground Floor, Ansari Road,
Darya Ganj, New Delhi-110 002
Ph. 011-43549197, 8130496929
e-mail: info@astralint.com
Website: www.astralint.com

Printed at : **Replika Press Pvt Ltd.**

Foreword

On the vast canvas of agriculture, agricultural structures is an often-overlooked object that is crucial to the success and sustainability of farming. These structures provide the necessary infrastructure for farmers to cultivate crops, raise animals, store produce, and process goods. From barns to greenhouses, each structure serves a specific purpose, contributing to the efficiency, productivity, and sustainability of agricultural operations. Through careful design and construction, farmers can create environments that promote optimal growth, health, and welfare of both plants and animals.

The basic purpose of agricultural structures is to be able to create environmental conditions favourable to the intended agriculture. Environmental control highlights the importance of environmental factors such as temperature, humidity, ventilation, *etc.* and their control for creating ideal conditions for farming. By using innovative technologies and sustainable practices, farmers can mitigate the impact of climate variability, enhance resource utilization efficiency, and meet the evolving demands of a growing population.

"Textbook of Agricultural Structures and Environmental Control" emerges as a timely and essential resource for students, professionals, and enthusiasts committed to advancing agricultural practices. This comprehensive guide offers an in-depth exploration of the principles, designs, and technologies that influence modern agricultural infrastructure and environmental management.

The authors of this textbook bring a wealth of knowledge and experience, crafting a narrative that is both informative and engaging. They meticulously cover a broad spectrum of topics, from the fundamentals of building materials and structural design to the intricacies of climate control systems and sustainable practices. Each chapter is designed to equip readers with the theoretical foundations and practical skills necessary to innovate and excel in the field.

As we face the dual challenges of feeding a growing global population and mitigating the impacts of climate change, the knowledge contained within this textbook is invaluable. It serves not only as an academic resource but also as a call to action for all stakeholders in the agricultural sector. By adopting the principles and practices outlined in this book, we can work towards a future where agriculture is both productive and sustainable, meeting the needs of today without compromising the ability of future generations to meet their own.

In closing, I extend my heartfelt congratulations to the authors for their remarkable contribution to agricultural literature. "Textbook of Agricultural Structures and Environmental Control" is poised to become an indispensable teaching aid reference, guiding and inspiring the next generation of agricultural innovators and practitioners. I am confident that readers, both teachers and students, will find this book to be a transformative tool in their educational and professional journeys, fostering a deeper understanding and appreciation of the vital role that agricultural structures and environmental control play in present day agriculture.

Dr. Pitam Chandra

Former Director,

ICAR–Central Institute of Agricultural Engineering,

Bhopal (India)

Preface

Agricultural structures form the backbone of modern farming, providing essential infrastructure and facilities to support diverse agricultural operations. From sheltering livestock to storing harvested crops, from processing agricultural products to managing environmental conditions, these structures play a crucial role in ensuring the efficiency, productivity, and sustainability of agricultural enterprises worldwide. Understanding the intricacies of agricultural structures and their multifaceted functions is paramount for farmers, agricultural engineers, and anyone involved in the agricultural sector.

In this book, the authors delve into the realm of agricultural structures, exploring their design, construction, and utilization in the context of contemporary agricultural practices. Drawing upon a wealth of knowledge and research, we aim to provide readers with comprehensive insights into the planning, development, and management of farm buildings and infrastructure.

The journey begins with an exploration of the fundamental functions of agricultural structures, from providing shelter and housing to facilitating livestock management, storage, processing, and environmental control. We examine the diverse array of structures utilized in modern agriculture and their pivotal roles in supporting various aspects of farming operations.

Moving forward, the authors explore into the classification of agricultural structures based on their purpose, design, and construction materials, offering readers a deeper understanding of the different types of farm buildings and facilities. From farmhouses and livestock structures to crop production and storage facilities, the authors explore the wide-ranging categories of agricultural structures and their unique characteristics.

Environmental control emerges as a critical theme in our discussion, highlighting the importance of managing environmental factors such as temperature, humidity, and ventilation to optimize growing conditions, enhance livestock welfare, and improve overall farm efficiency and productivity in terms of milk, egg, meat, wool, crop, *etc.*

The book also covers topics on the role of livestock in the economy of India, systems of livestock production, and animal psychology. As we embark on this journey through the world of agricultural structures, the authors invite readers to delve deeper into the intricacies of farm buildings and infrastructure, exploring the ways in which these structures shape and support modern agriculture. Whether you are a seasoned farmer, a budding agricultural engineer, or simply curious about the inner workings of farm life, the authors hope this book serves as a valuable resource and guide in your quest for knowledge and understanding.

The authors express their sincere gratitude to the colleagues, professional friends, and students who continuously encouraged us in this endeavour. The feedback from both teachers and students for improvement of this book will be highly encouraged and solicited. We also extend our thanks to Mr. Prateek Mittal and the staff of M/s Astral International Pvt. Ltd., New Delhi, for their editing and publishing efforts.

Mukesh Kumar Garg

Sunil Kumar

Contents

Agricultural Structures

Chapter 1

Introduction

Agricultural (or Farm) structures refer to the physical facilities, buildings, and infrastructure utilized in agricultural operations. They are designed to accommodate diverse farming needs, including crop cultivation, livestock management, storage, processing, and environmental control. The primary objective is to understand the requirements of particular farm in the context of development of farm building and how to construct the building in economic manner with the efficient use of available resources in the production of food, dairy and other kind of businesses. Agricultural structures encompass a diverse range of constructions and facilities that play essential roles in supporting modern farming practices. From providing shelter and storage to enabling environmental control and processing operations, these structures contribute to the efficiency, productivity, and sustainability of agricultural operations worldwide.

Farm structures are also known as farm buildings. While the terms "structures" and "buildings" are often used interchangeably, they denote distinct concepts. "Buildings" typically denote shelters designed to accommodate human beings, livestock, farm produce, and machinery. On the other hand, "structures" encompass a broader range, including both buildings and non-sheltered entities like roads, bridges, fences, towers, and agricultural dams. These structures serve various purposes such as facilitating access to farms, offering security, or serving multiple functions.

1.1 Functions of Agricultural (or Farm) Structures

Farm structures serve various functions essential for agricultural operations, animal husbandry, and farm management. They play a critical role in supporting agricultural production, enhancing efficiency, and promoting sustainability in farming operations. Some key functions of farm structures are:

1. **Shelter and Housing:** Farm structures provide shelter and housing for human beings, animals, crops, and farm equipment. This includes barns, sheds, poultry houses, and greenhouses designed to protect livestock, machinery, and plants from adverse weather conditions such as rain, wind, snow, and extreme temperatures. The combined effect of rain, wind, snow, and extreme temperatures can greatly reduce the farmers' efficiency, productivity of livestock, quality of crops and cause wear and tear or rusting of machinery components.
2. **Livestock Management:** Farm structures are designed to facilitate the management and care of livestock. This includes housing for animals such as cattle, pigs, poultry, and sheep, as well as facilities for feeding, watering, and healthcare. Properly designed structures contribute to the health, comfort, and productivity of livestock.
3. **Storage:** Farm structures offer storage facilities for harvested crops, feed, fertilizers, and equipment. Grain silos, hay barns, and storage sheds help preserve agricultural produce, ensuring it remains viable until it is ready for sale or consumption.
4. **Processing and Handling:** Some farm structures are dedicated to processing and handling agricultural products. These may include facilities for washing, sorting, packaging, and processing crops such as fruits, vegetables, and grains. Processing structures help maintain product quality and prepare agricultural goods for market distribution.
5. **Workshop and Maintenance:** Farm structures often include workshops and maintenance facilities for repairing and servicing farm equipment and machinery. These structures provide a space for maintenance activities, equipment storage, and the fabrication of tools and implements essential for farm operations.
6. **Environmental Control:** Certain farm structures, such as greenhouses and livestock housing, allow for environmental control. Farmers can regulate factors like temperature, humidity, and ventilation to create optimal growing conditions for plants or maintain a comfortable environment for human beings and livestock, thus enhancing productivity and efficiency. Farm structures must regulate temperature and humidity to ensure comfort for humans and livestock. Vapour barriers and insulation materials are essential for regulating these factors in enclosures for both humans and livestock. Farm structures are also constructed for storage of the harvested crops and similarly, temperature and humidity have to be regulated using barriers and insulating materials.
7. **Water Management:** Farm structures may incorporate water management systems such as irrigation infrastructure, drainage

systems, and water storage tanks. These systems ensure efficient water distribution to crops and livestock while preventing soil erosion and waterlogging, thereby optimizing agricultural productivity and sustainability.

8. **Accessibility:** Farm structures such as roads, bridges and culverts provide access from farms to urban areas and vice versa. This is important for effective communication, conveyance of farm inputs which include labour, machinery, seeds/seedlings, agro-chemicals and fertilizers to the farms and the transport of harvested farm produce from farms to the urban areas.
9. **Security and Safety:** Farm structures contribute to the security and safety of farm operations. Fenced enclosures, gates, and security systems help protect livestock, crops, and equipment from theft, vandalism, and unauthorized access. Additionally, well-designed structures reduce the risk of accidents and injuries among farm workers and animals.
10. **Energy Production:** Some modern farm structures integrate renewable energy systems such as solar panels, wind turbines, and biomass digesters. These energy-producing structures help reduce reliance on traditional energy sources, lower operating costs, and minimize the environmental footprint of agricultural activities.

1.2 Classification of Agricultural (or Farm) Structures

Farm structures can be classified into various categories based on their purpose, design, construction materials, and functionality. Farm structures based on their construction material can be categorized into earth, wooden, concrete, plastic, and steel structures. The function of a structure is also a factor, dividing them into farm houses, livestock, crop production, storage, processing, and miscellaneous structures. These structures are grouped based on their utilization or function. The classification of farm structures on the basis of its function is as follows:

1. **Farm Houses:** Farmhouses serve as residences for humans, offering lodging for all individuals who need accommodation on the farm. They offer amenities and conveniences conducive to comfortable living, including sleeping quarters, laundry facilities, cooking spaces, and areas for relaxation. These structures shield residents from harsh weather elements such as rain, sunshine, and cold temperatures, as well as provide security, particularly during night-time, against wildlife.
2. **Livestock Structures:** Livestock structures regulate environmental conditions to increase milk, meat, and egg production. These structures reduce animal drudgery to get food and water. Barns, sheds, and pens are used for housing livestock like cattle, goats, pigs, horses, and sheep. Battery cages and deep litter houses are used for poultry housing, and

ponds and tanks are used for fish. Dairy facilities including milking parlours, milk storage tanks, and cooling facilities for dairy cows are required.

3. **Crop Production Structures:** The structures like greenhouses, screen houses, glasshouses, tunnels, and growth chambers modified the ambient conditions as per the requirements of the crop. The modified ambient conditions increase crop production on farms. These structures also allowed the growing of crops in the off-season for commercial purposes.

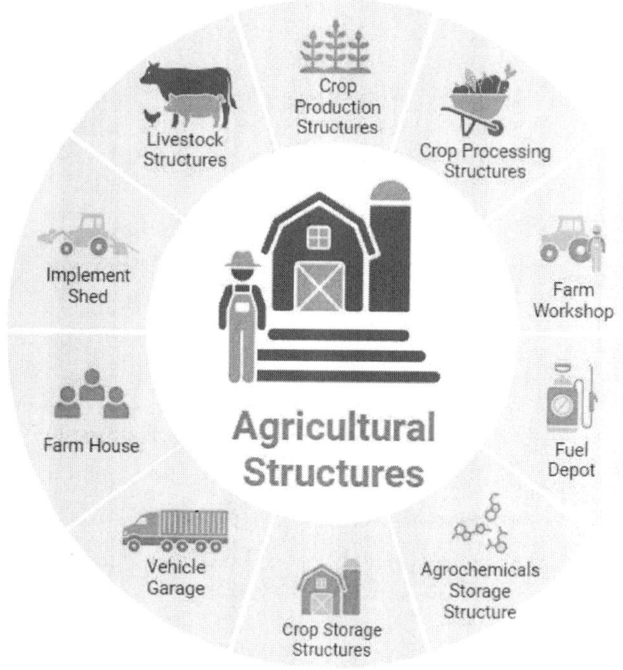

Figure 1.1. Agricultural Structures.

4. **Crop Storage Structures:** The storage structures are used for the storage of agricultural products. The storage can be done for short, medium, and long terms, as per the requirement. The storage shall be strong enough to resist the imposed load of the product, its dead weight, and the live load due to wind, *etc.* The capacity of the storage structure shall meet the volume of agricultural production. Silos, bins, underground pits, and warehouses are designed for storing harvested grains. Refrigerated structures (cold rooms, evaporative coolers) are used for storing perishable agricultural produce such as fruits and vegetables. While feed storage structures are required for storing and distributing animal feed and supplements.

5. **Processing Structures:** It is a space that is used to process agricultural products before marketing or storage. Generally, primary processing like cleaning, grading, *etc.* are carried out at the farm. The processing area shall be designed to promote hygiene conditions with convenient processing lines. Examples are facilities equipped for washing, sorting, packing, and processing agricultural products; structures used for drying crops such as grains, fruits, and vegetables; and facilities equipped for canning, packaging, and labelling agricultural products.
6. **Miscellaneous Structures:** A farm requires various miscellaneous structures such as implement sheds for agro machinery, garages for tractors, farm workshops for repair work, fuel depots to store diesel, water and wastewater treatment plants, composting area, manure storage facilities, dams, fences, gates, irrigation systems, water storage tanks, roads, bridges, culverts, *etc.* to meet the specific needs and optimize agricultural operations.

1.3 Environmental Control

Environmental control in agriculture is essential for creating optimal growing conditions for crops, promoting livestock health and welfare, and improving overall farm productivity and efficiency. By carefully managing environmental factors such as temperature, humidity, light, and air quality, farmers can maximize yields, reduce risks, and sustainably produce high-quality food and agricultural products.

Factors Influencing the Need for Environmental Control

The need for environmental control in agriculture is influenced by a combination of factors related to climate, crop and livestock requirements, pest and disease management, resource efficiency, market demands, regulatory requirements, and technological advancements. By understanding and addressing these factors, farmers can create optimal growing conditions, improve productivity and efficiency, and sustainably produce food and agricultural products to meet the needs of a growing population.

1. **Climate Variability:** Regions with extreme weather conditions, including temperature fluctuations, high humidity, and unpredictable rainfall, may require environmental control measures to mitigate the impact of adverse weather on crop growth and livestock health.
2. **Seasonal Changes:** Seasonal variations in temperature, daylight hours, and humidity levels can affect crop growth cycles and livestock behaviour. Environmental control helps maintain optimal conditions throughout the year, allowing for consistent production and quality.
3. **Crop and Livestock Requirements:** Different crops and livestock species have specific environmental requirements for optimal growth

and performance. For example, some crops may require warmer temperatures and higher humidity levels, while others thrive in cooler and drier conditions. Livestock species also have varying temperature and ventilation needs based on their size, breed, and physiological state.

4. **Pest and Disease Management:** Environmental control measures can help minimize the spread of pests and diseases by creating unfavourable conditions for their development and reproduction. Proper ventilation, temperature regulation, and sanitation practices reduce the risk of infestation and disease outbreaks in crops and livestock.
5. **Resource Efficiency:** Efficient use of resources such as water, energy, and nutrients is essential for sustainable agriculture. Environmental control technologies, such as drip irrigation, energy-efficient lighting, and climate-controlled systems, help optimize resource use and minimize waste, reducing production costs and environmental impact.
6. **Market Demands and Consumer Preferences:** Consumer demand for fresh, high-quality produce year-round drives the need for controlled environment agriculture, such as greenhouse production. Environmental control enables farmers to meet market demands by extending growing seasons, producing out-of-season crops, and ensuring consistent product quality and supply.
7. **Regulatory Requirements:** Environmental regulations and food safety standards may require farmers to implement environmental control measures to ensure compliance with health and safety regulations, minimize environmental impact, and maintain product quality and integrity.
8. **Technological Advancements:** Advances in technology, including sensor-based monitoring systems, automated climate control systems, and precision agriculture techniques, have made it easier and more cost-effective for farmers to manage environmental conditions on the farm. These technologies enable real-time monitoring and adjustment of environmental parameters to optimize crop growth and livestock welfare.

1.4 Livestock Population and Production in India

India has vast resources of livestock and poultry, which play a vital role in improving the socio-economic conditions of rural masses. It is home to one of the largest livestock populations in the world. The country's diverse agro-climatic conditions and cultural practices have contributed to the proliferation of various types of livestock. There are about 303.76 million bovines (cattle, buffalo, mithun and yak), 74.26 million sheep, 148.89 million goats, 9.06 million pigs and about 851.81 million poultry as per 20th Livestock Census in the country (Table 1.1).

Table 1.1. Livestock Population (2019 Livestock census)

Sl.No.	*Species*	*Number (in millions)*
1.	Cattle	193.46
2.	Buffaloes	109.85
	Total bovine (including mithun and yak)	**303.76**
3.	Sheep	74.26
4.	Goats	148.89
5.	Pigs	9.06
6.	Camel	0.25
7.	Others (horse, ponies, mules, donkeys)	0.54
	Total livestock	**536.76**
	Total poultry	**851.81**

India is world's highest livestock owner at about 536.76 million, first in the total buffalo population, second in the population of goats, second largest poultry market, third in the population of sheep, tenth in camel population, second largest producer of fish, and second largest aquaculture nation in the world. India has nearly 57 per cent of the world's buffalo population, 16 per cent of the cattle population, 20 per cent of goat population and 5 per cent of sheep population although India constitutes less than 3 per cent of the world's total land area.

During 2021-22, total production of milk, egg, meat, and wool in India are respectively 221.06 million tonnes, 129.60 billion numbers, 9.29 million metric tonnes, and 33.13 thousand tonnes (Table 1.2). Whereas, the per-capita availability of milk, egg, and poultry meat in India during 2021-22 is respectively 444 g/day, 95 numbers/annum, and 6.82 kg/annum. Per capita availability of meat and eggs in India is far low against the recommendations of Indian Council of Medical Research (ICMR). The council recommended consuming 10.5 kilos of poultry meat and 180 eggs per annum.

Table 1.2. Production of Livestock in India (2021-22)

Sl.No.	*Product*	*Quantity*
1.	Milk in million tonnes	221.06
2.	Eggs in billion numbers	129.60
3.	Meat in million metric tonnes	9.29
4.	Wool in thousand tonnes	33.13

1.5 Role of Livestock in Farmers' Economy in India

Livestock plays a multifaceted role in the economy of farmers in India, contributing to their income, food and nutritional security, agricultural sustainability, and cultural heritage. Livestock, particularly in dairying and mixed farming for small and medium land holdings, is sustainable and yields significant returns. The livestock provide commercially viable food and non-food items. Animal rearing leads to a balanced rural economy and improved economic status for poor people.

About 20.5 million people depend upon livestock for their livelihood. Livestock contribute 16 per cent to the income of small farm households as against an average of 14 per cent for all rural households. Livestock provides livelihood to two-third of rural community. The livestock sector in India employs 8.8 per cent of the population. This area experienced a CAGR of 7.93 per cent from 2014-15 to 2020-21 and contributed to 24.32 to 30.13 per cent (from 2014 to 2021) of the total agriculture and allied sector. In 2020-21, the sector contributed 4.90 per cent of total Gross Value addition (GVA). The development of the livestock sector has improved the per capita availability of milk, eggs, and meat.

The role of livestock in farmers' economy may be summarized as follows:

1. **Source of Livelihood:** Livestock farming is a crucial source of livelihood for millions of farmers across India, particularly in rural areas. Livestock rearing provides employment opportunities and income generation for farmers, especially those with small landholdings. Livestock farmers of cows and buffaloes generate income through milk sales. Sheep and goats provide income during emergencies like marriages, sick treatment, education, and house repairs. Livestock also contributes to the production of wool, hair, hides, leather, and pelts. Some of these products, like leather, wool, *etc.*, are the most significant due to their high export potential.
2. **Diversification of Income:** Livestock farming enables farmers to diversify their sources of income. By engaging in activities such as dairy farming, poultry farming, goat rearing, and animal husbandry, farmers can mitigate risks associated with crop failure or market fluctuations.
3. **Nutritional Security:** Livestock serves as a source of various essential nutrients such as milk, meat, and eggs, contributing to the nutritional security of rural households. Dairy products, in particular, are a significant source of protein and micronutrients for many Indians.
4. **Social Security:** Livestock provide social security to owners and enhance their status in society. Farming and animals are synergistic with each other. Farms provide food to animals, and animals offer energy for ploughing, irrigation, manure, transportation, *etc.*, along with milk and other animal products. The rearing of animals also upgrades the financial status of landless families. The animals provide them with a better place among households without animals.

5. **Agricultural Sustainability:** Livestock farming contributes to agricultural sustainability through the recycling of crop residues and organic waste. Livestock utilize agricultural by-products such as crop residues and leftover grains, converting them into valuable products like milk, meat, and manure.
6. **Manure and Organic Fertilizers:** The dung of dairy animals serves as fuel, fertiliser, and plastering material in rural areas. Livestock manure, a valuable organic fertiliser, improves soil fertility and enhances crop productivity, with its value worth several crores of rupees. Farmyard manure (FYM) is a valuable organic fertiliser used to improve soil fertility. The dried animal dung is used as fuel for regular domestic work like cooking food. The poor people in rural areas use the diluted dung with water for plastering over the floor and walls of houses. Many farmers use animal waste as a natural alternative to chemical fertilizers, reducing their dependency on external inputs and promoting sustainable agricultural practices.
7. **Draft:** Bullocks are crucial for Indian agriculture, especially for small and marginal farmers. They are used for ploughing, carting, and transporting agricultural inputs and outputs. Other pack animals like camels, horses, donkeys, ponies, and mules are also used for transportation. In hilly terrains, mules and ponies are the only alternatives. The army also relies on mules and ponies for transporting items in high-altitude areas.
8. **Cultural Significance:** The Indian culture is deeply rooted in the practice of animal rearing. Livestock holds cultural significance in many rural communities across India. Animals like cows, buffaloes, and goats are revered and play integral roles in religious rituals and ceremonies. During marriage, animal gifting between relatives is a prevalent practice across various regions of the country.
9. **Access to Credit and Insurance:** Livestock are considered as 'moving banks' because of their potentiality to dispose-off during emergencies. Livestock ownership enables farmers to access credit and insurance services. Livestock can serve as collateral for obtaining loans from banks and financial institutions, allowing farmers to invest in their agricultural activities and improve their livelihoods. Additionally, livestock insurance schemes provide financial protection to farmers against the risks of livestock mortality and diseases.
10. **Sports/recreation:** People also use the animals like cocks, rams, bulls *etc.* for competition and sports. Despite ban on these animal competitions the cock fights, ram fights and bull fights (*jalli kattu*) are quite common during festive seasons.

1.6 Systems of Livestock Production

Livestock production systems vary widely depending on factors such as geographic location, climate, cultural practices, available resources, and market demands. Some common systems of livestock production are:

1. **Extensive:** In extensive systems, animals are raised primarily on natural forage such as grasses and shrubs. This is one of the oldest method. This animal farming system is characterised by a low productivity per animal and per surface. This system requires vast land areas and minimal inputs in terms of labour and supplementary feed. Examples include nomadic pastoralism, where herders move their animals seasonally to find fresh grazing lands.
2. **Semi-Intensive:** Semi-intensive systems combine grazing on open pastures (extensive) and feeding of forages in confined spaces (intense). This system is suitable for dairy animals and goats. Although sheep can be reared under this system. The animals are confined under the roof for part of the day or night. The animals are allowed to graze during the day, whereas concentrate feed is provided during confinement.
3. **Intensive:** Intensive livestock farming systems means the raising of animals where their environment in total is provided for them to maximize profit out of their rearing. The environment here includes housing, water, nutrition, temperature, humidity, and management. In this system, animals are confined to shed throughout the year. These systems allow for the direct supervision of a large number of animals. Animals are fed formulated diets, and production practices are highly mechanized. Space requirement is less when compared with extensive or semi-intensive systems. Poultry, pig, and rabbit are ideal for this system. Although, milk dairies in developed countries are also managed through an intensive system.
4. **Mixed Farming:** Mixed farming systems integrate crop production with livestock or poultry rearing on the same farm, with each component of the farm operating somewhat independently of the others. For example, a mixed farm might grow wheat, corn, and soybeans, while also raising chickens, pigs, and cows. The different crops and animals are usually managed separately, with each part of the farm having its own specific set of inputs, management practices, and output markets.
5. **Integrated Farming System:** The integrated farming system addresses the limitations of mixed farming by planning, monitoring, and executing the farm work based on farm size, resources, and agro-climatic factors. Livestock species or poultry enterprises are selected based on feed, fodder, and water availability on the farm. An integrated farm might include crops, livestock, and other elements such as fish ponds, composting systems, and agroforestry plots. The different

components of the farm are designed to work together in a way that maximizes resource use efficiency, reduces waste and environmental impacts, and increases overall productivity. For example, animal manure might be used as fertilizer for crops, while crops can provide feed for livestock, and trees can provide shade and habitat for animals. This integrated approach helps to optimize resource use and enhance farm sustainability.

1.7 Animal Psychology

Animal psychology, also known as comparative psychology or ethology, is the study of the behaviour, cognition, emotions, and mental processes of animals. It explores how animals perceive, learn, interact with their environment, communicate, and adapt to various situations. It is a multidisciplinary field that integrates principles from psychology, biology, neuroscience, ecology, and anthropology to gain a deeper understanding of the minds and behaviours of non-human animals.

Animal Psychology versus Ethology

Animal psychology and ethology are closely related disciplines that study the behaviour of animals and cognition, but they differ in their methodologies, theoretical frameworks, and research goals. Animal psychology focuses on the mental processes and behaviours of animals, often with an emphasis on individual cognition and learning mechanisms. While ethology is the scientific study of animal behaviour, with a focus on understanding behaviour in natural environments and evolutionary contexts. Animal psychologists often study animals in controlled laboratory settings to investigate specific aspects of behaviour, cognition, and learning, using techniques such as operant conditioning, maze learning, and observational studies. While the ethology emphasizes the observation and analysis of behaviour in the natural habitats of animals, including their interactions with conspecifics, predators, prey, and the physical environment. However, both fields complement each other and contribute to a comprehensive understanding of the complexity of animal behaviour.

Cognitive Animal Psychology

Cognitive animal psychology is a branch of psychology that focuses on understanding the cognitive processes and abilities of non-human animals. It explores how animals perceive, learn, remember, solve problems, make decisions, and communicate. By gaining insights into the cognitive mechanisms underlying animal behaviour, cognitive animal psychology contributes to our understanding of the evolution of cognition and the diversity of cognitive abilities across species.

Important Points to Remember

1. Farm structures refer to the physical facilities, buildings, and infrastructure utilized in agricultural operations.
2. Farm structures provide shelter and housing for human beings, animals, crops, and farm equipment.
3. Environmental control includes temperature, humidity, air velocity, radiation, light intensity, and light duration.
4. Environmental control in agriculture is essential for creating optimal growing conditions for crops, promoting livestock health and welfare, and improving overall farm productivity and efficiency.
5. By carefully managing environmental factors such as temperature, humidity, light, and air quality in agricultural structures, farmers can maximize yields, reduce risks, and sustainably produce high-quality food and agricultural products.
6. Livestock plays a multifaceted role in the economy of farmers in India, contributing to their income, food and nutritional security, agricultural sustainability, and cultural heritage.
7. In extensive systems of livestock production, animals are raised primarily on natural forage such as grasses and shrubs.
8. Intensive livestock farming systems means the raising of animals where their environment in total is provided for them to maximize profit out of their rearing. In this system, animals are confined to shed throughout the year.
9. Semi-intensive livestock farming systems combine grazing on open pastures (extensive) and feeding of forages in confined spaces (intense).
10. Mixed farming systems integrate crop production with livestock or poultry rearing on the same farm, with each component of the farm operating somewhat independently of the others.
11. In the integrated farming system, the type of livestock species or poultry enterprises are selected based on the availability of feed, fodder, water resources of the farm.

Exercises/Questions

1. Describe the various functions of agricultural structures?
2. Give the classification of agricultural structures on the basis of their function.
3. What is the importance of environmental control in agricultural structures?
4. Discuss various environmental factors which should be considered while designing an animal shelter/livestock building.

5. What is the role of livestock in farmers' economy in India?
6. What are the different systems of livestock production? Briefly, discuss each of them.
7. Differentiate between:
   (i) Structures and buildings
   (ii) Animal psychology and ethology
   (iii) Extensive and intensive systems of livestock production
   (iv) Mixed farming systems and integrated farming system

Chapter 2

Farmstead Planning and Farm Residence

FARMSTEAD PLANNING

A farmstead forms the nucleus of the farm operation where a wide range of farming activities take place. It encompasses all the buildings, structures, and land that are used for farming activities, including the farm house, barns, sheds, silos, and other outbuildings (Figure 2.1). The term farmstead is often used to describe a self-sustaining farm that produces food and other products for the family or community that lives there. It is the heart of the farming operation.

Planning a farmstead is a complex and important step towards establishing a successful and sustainable agricultural operation. It is the process of designing and organizing the physical layout of a farmstead. It involves determining the optimal location of buildings, roads, fences, and other infrastructure on the farm, with the aim of maximizing efficiency, productivity, and profitability while ensuring the safety of people and the environment. A well-planned farmstead can optimize productivity, reduce waste, and provide a safe and comfortable living and working environment for all.

2.1 Farmstead Planning

Farmstead planning is important for several reasons which are listed below:

1. **Maximizing Land Use**: Proper planning can help farmers make the most efficient use of their land, allowing them to produce more crops or raise more animals on a smaller area of land.

Figure 2.1. A Typical Farmstead with Animals, Farmer, Farm and House (AI generated).

2. **Improving Efficiency**: A well-designed farmstead can reduce labor and equipment needs, minimizing the time and effort required to carry out daily tasks such as feeding, watering, and harvesting.
3. **Ensuring Safety**: Farmstead planning can help minimize safety risks by separating potentially dangerous areas such as livestock corrals and feed storage areas from areas where people and vehicles may be present.
4. **Enhancing Sustainability**: An optimized farmstead can help minimize the impact of farming activities on the environment, by reducing soil erosion, minimizing runoff, and promoting biodiversity.
5. **Enhancing Profitability**: By optimizing the farmstead layout, farmers can reduce costs and increase productivity, leading to higher profits.

The first step in planning a farmstead is to determine the purpose and goals of the farm. This includes considering the type of crops and livestock that will be raised, the size and scale of the operation, the market demand, and the available resources such as water, soil, and equipment. This information will help guide the layout of the farmstead, and ensure that it is tailored to meet the specific needs of the farm and its owners.

The next step is to consider the layout of the farmstead, including the placement of buildings, structures, and infrastructure. This includes identifying the location of the house, barns, sheds, and other buildings, as well as the placement of fencing, roads, and water systems. The layout should be designed to optimize the flow of people, animals, and equipment, and to minimize wasted space and resources.

When planning the placement of buildings, it is important to consider the orientation and placement of the structures in relation to the sun and wind. This can have a significant impact on the energy efficiency of the buildings, as well as the comfort of those living and working in them. Additionally, it is important to consider the safety and accessibility of the buildings, as well as their proximity to important resources such as water, feed, and equipment.

Another key consideration when planning a farmstead is the management of waste and resources. This includes designing systems to manage manure, wastewater, and other waste products, as well as systems for collecting and storing water, feed, and other resources. Efficient management of these resources can reduce waste, improve productivity, and promote sustainable farming practices.

Finally, it is important to consider the aesthetics of the farmstead, including its visual appeal and its impact on the surrounding environment. This includes choosing materials and colours that are visually appealing and harmonious with the surrounding landscape, as well as landscaping and design elements that promote biodiversity and environmental sustainability.

Safety should also be given a top priority when planning a farmstead. The farmstead should be designed in a way that minimizes the risk of accidents and injuries. This may include the use of safety barriers, adequate lighting, and clear signage to direct traffic and people around the farmstead.

2.2 Site Selection

The location of the farmstead is crucial to the success of any farming operation. It should be situated in a place that is easily accessible to vehicles and farm machinery. In large farms, the farmstead shall be located near the centre of the farm as it helps in better management as the farm labourers and other staff also live near the farmstead. On the other hand, location near the road may facilitate better social life and protection. Also, procuring the farm supplies and disposing of farm produce become easier. Thus, the pros and cons of each site shall be considered before finalizing.

The following are some key factors to be considered for site selection of a farmstead:

1. **Topography and Terrain**: The farm buildings should be located on the highest elevation. In hilly areas or places of undulating topography, the buildings should be located with their longitudinal axis running parallel to the contour lines. The topography should be satisfactory for the required storage and drainage of manure and effluent produced at the farmstead.

2. **Soil Quality**: The quality of the soil is one of the most important factors to consider when selecting a site for a farmstead. Good soil should be fertile, well-drained, and have a pH level that is suitable for the types of crops or animals being raised. Buildings should generally be constructed only on poor soil, leaving the fertile soil for farming. For construction of farm buildings, soil or subsoil must have sufficient load-bearing capacity. Otherwise, there will be an additional expenses for the foundation.
3. **Water Supply**: A reliable source of water is essential for any farmstead. Factors to consider include the quality and quantity of the water supply, as well as the cost of accessing and maintaining it. The farmstead should be located near a source of permanent water supply.
4. **Orientation**: Solar radiation may adversely affect the environment within buildings. It is desired to allow maximum solar radiation during winter and minimum solar radiation during summer. An east-west or longer surface facing north and south is best orientation for this purpose.
5. **Wind Direction**: Prevailing wind direction in the locality should also be taken into account while deciding the location of farm house *etc.* Since wind will carry odours, livestock buildings should be placed downwind from the farm house.
6. **Drainage**: A well-drained site will prevent flooding, and ensure all-weather driveways. It is also essential for satisfactory operation of septic tank drainage fields, and for the removal of other wastes.
7. **Waste Management**: The site should be able to handle manure and other waste without any serious problem.
8. **Utilities and Services**: Access to necessary infrastructure, such as roads, electricity, and telecommunications, should also be considered. The cost of building infrastructure, as well as the availability of existing infrastructure, can impact the overall cost of operating the farmstead. The buildings and utilities shall be so arranged as to provide a sequence of operations without too much back-forth movement.
9. **Proximity to Markets**: The location of the farmstead in relation to potential markets should also be considered. If the farm is too far from markets, transportation costs may be too high, and the farm may not be able to compete with other producers.
10. **Expansion**: Adequate provision should be made for possible future extension of facilities.

2.3 Size of Farmstead

The size of a farmstead is typically determined by the amount of land available for farming, the size of the farming operation, and the needs of the farmer. For example, a small family farm may have a smaller farmstead, while a larger commercial farm may require a larger farmstead to accommodate more animals and equipment. The area of farmstead usually varies from 3 to 5 per cent of the farm area.

2.4 Zone Planning

Zone planning of a farmstead involves dividing the farm into different zones or areas, each with its own specific purpose and function. This approach helps to optimize the use of available resources, improve productivity and efficiency, and enhance the overall sustainability of the farm. The specific zones included in a farmstead zone plan will depend on the needs and goals of the farmer and their farm, as well as the available resources and land. Zone planning can be a useful tool, but it is most effective when planning a new farmstead. The farmstead is divided into zones 10 to 30 metres wide by concentric circles as shown in Figure 2.2.

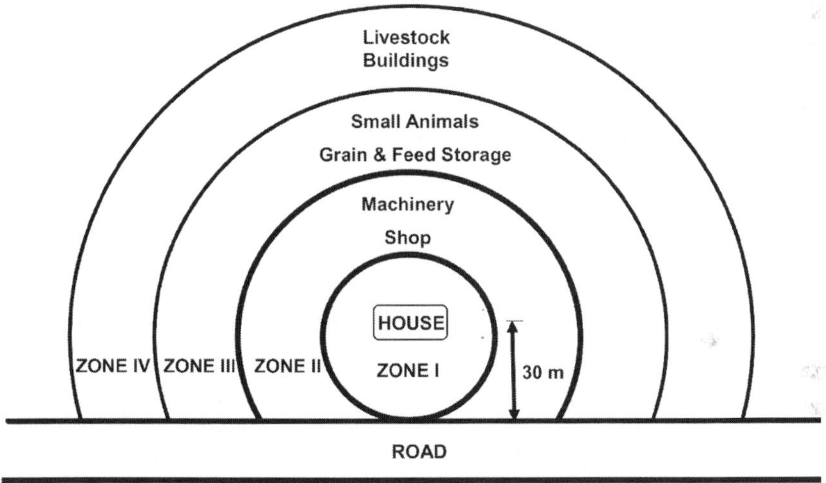

Figure 2.2. Farmstead Planning Zones.

Here are some possible zones that could be included in a farmstead zone plan:

1. **Zone 1 (Residential and Recreational Zone)** is at the centre of the farmstead and encompasses the residential and recreational area. Residential area is where the farmer and their family live. It may also include a guesthouse, staff quarters, or other related structures. Recreational area is where the farmer and their family can relax and enjoy the farm. This area may include a swimming pool, a tennis court, a picnic area, and other related structures. This area should be protected from noise, dust, odour, insect, *etc.*

2. **Zone 2 (Infrastructure Zone)** may be located about 30 metre away from the farm house. In this zone farm equipment, vehicles, and machinery are stored and maintained. This area may include a workshop, a garage, a fuelling station, and other related structures. This area is relatively quiet, dry and odour free, and do not create a nuisance.
3. **Zone 3 (Storage Zone)** is where harvested crops and other farm supplies are stored. This area may include a granary, a silo, a hay loft, pack house and other related structures. Sheep and young animals that need supervision but do not develop heavy odours and waste, can also be located in this zone.
4. **Zone 4 and beyond (Livestock, Production and Processing Zone)** is where the crops and livestock (large animal units) are raised. This area may include fields, pastures, orchards, vegetable gardens, barn, milking parlour, greenhouse, and other related structures. This zone may also include processing facilities where harvested crops and livestock products are processed and prepared for market. This not only reduces the danger of foul smell reaching the farm house but also places the unit which is most likely to be expanded in an area where space will be available.

2.5 Arrangement of Buildings

A farmstead is a collection of buildings and structures that are used for farming activities, such as a farmhouse, barn, stable, storage buildings, and equipment sheds. Arranging various buildings on a farmstead can involve several considerations, including the size and shape of the land, the type of farming operation, the intended purpose of the buildings, and the flow of traffic and resources on the farmstead. Overall, the arrangement of buildings on a farmstead should be done in a way that maximizes efficiency and supports the intended use of the farm while also considering the needs of the land and the surrounding environment. Various buildings shall be so arranged to provide the minimum of walking from one to the other in doing the required work. Here are some general tips that may be helpful:

1. **Determine the Purpose of each Building:** Consider the function of each building and how it will be used on the farm. For example, a barn may be used for storing equipment and housing livestock, while a farmhouse may be used for living quarters.
2. **Plan for Access and Traffic Flow:** Consider the location of roads and pathways on the farmstead, and plan the placement of buildings to minimize traffic congestion and maximize efficiency. Buildings that require frequent access, such as the barn, should be located closer to the main road or entrance.

3. **Consider the Orientation of Buildings:** Orient buildings to take advantage of natural light and wind patterns. For example, a building may be oriented so that the roof ridge or long dimension runs east to west to maximize exposure to the sun. This orientation allows maximum utilization of solar energy in the winter and minimizes solar heating in the summer. Also, the prevailing wind should not blow from the animal houses to the residence.
4. **Allow for Future Expansion:** Plan for future growth and expansion of the farmstead by leaving space for additional buildings or modifications to existing ones.

General Guidelines for Location of Various Buildings

In terms of arrangement, a common design for a farmstead is to have the farmhouse situated near the centre, with other buildings arranged in a functional manner around it. For example, barns and animal housing may be located on one side of the farmhouse, while equipment sheds and storage buildings may be located on another side (Figure 2.3). This type of arrangement helps to optimize workflow and make it easier to move animals and equipment between buildings. The distance between buildings usually is determined by a compromise involving efficiency, fire safety, odour control, disease control, and available space. A distance of 15 to 30 m is usually considered a minimum safe distance between buildings or groups of buildings.

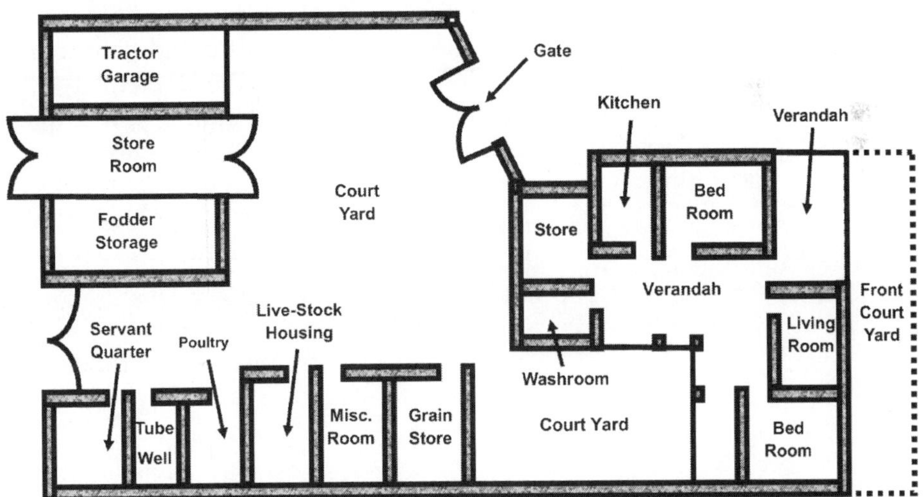

Figure 2.3. Layout of a Typical Farmstead.

The general guidelines for distance between the buildings is as follows:

The farm house should be at least 30 m away from the road. It should be located away from the livestock house and other buildings. This will ensure privacy and reduce the nuisance of flies and smell coming from the livestock house.

Dairy barn of general-purpose barn should be minimum 15 m from machine shed, but not too close to the farm house. It should at least be at least 30 m from the house. The milk house should be near dairy barn. The milk house is generally placed about 6 meter away from the barn.

Pig houses are frequently very malodorous. They should be about 60 m from the residence, and at least 30 m from the cow barn. Sheep, goat sheds and beef cattle shelters may be treated in the same manner as the pig house. For convenience of feeding and egg collection, poultry house should be near the dwelling house, about 30 m away.

The silo pits and feed storage structures should be located near the animal shelters.

Garage is often included in the machine shed and shop. It should be near the house for convenience, but not so close because of noise. A minimum of 20 m from the house is adequate.

2.6 Farmstead Tree-Planting

Farmstead tree-planting is the practice of planting trees around a farm or homestead for various purposes, such as providing shade, improving soil quality, enhancing biodiversity, and increasing the aesthetic value of the property.

Planting trees on a farm can have several benefits. They can provide shade for livestock and reduce heat stress during the hot summer months. Trees can also help to reduce soil erosion, increase soil fertility, and promote healthy soil microbiology. Additionally, trees can act as windbreaks, helping to reduce the impact of strong winds on the farm, which can damage crops, structures, and equipment.

Farmstead tree planting can also enhance biodiversity by providing habitat for wildlife, such as birds, insects, and small mammals. This can lead to a more balanced ecosystem, which can ultimately benefit the farm and surrounding areas.

When planting trees around a farmstead, it's important to consider the type of trees that will thrive in the local climate and soil conditions, as well as the location of the trees in relation to other structures and equipment on the farm. Careful planning and consideration can help to ensure that the tree planting is successful and provides long-term benefits for the farm and the environment.

There are many tree species that are suitable for plantation at a farmstead in India, depending on the specific climate and soil conditions of the region. Some popular options include:

1. Neem (*Azadirachta indica*) – It is a fast-growing, drought-resistant tree that is commonly used in traditional medicine and has insect-repelling properties.
2. Teak (*Tectona grandis*) – It is a hardwood tree that is widely used for furniture and building material, and is tolerant of a range of soil types.
3. Mango (*Mangifera indica*) – It is a fruit-bearing tree that is well-suited to the hot and humid conditions of India, and can provide shade as well as a source of income.
4. Eucalyptus (*Eucalyptus spp.*) – It is a fast-growing tree that is commonly used for paper production and has many other industrial applications.
5. Banyan (*Ficus benghalensis*) – It is a large, spreading tree that can provide shade and shelter for animals, and is also revered in Hindu mythology.
6. Bamboo (*Bambusa spp.*) – It is a fast-growing and versatile plant that can be used for construction, furniture, and handicrafts.
7. Drumstick (*Moringa oleifera*) – It is a tree with multiple uses, including food, medicine, and biofuel production.

When choosing trees for landscaping a farmstead, it is important to consider the aesthetic appeal, as well as the practical benefits they offer. These are just a few examples of trees that can be used for landscaping farmsteads. The factors such as the size, shape, and growth rate of the trees, as well as the overall design of the landscape should be considered when choosing which trees to plant. Here are some popular options for trees suitable for landscaping farmsteads in India:

1. Jacaranda (*Jacaranda mimosifolia*) - A beautiful tree with purple flowers that can provide a stunning focal point in the landscape.
2. Flame of the Forest (*Butea monosperma*) - A deciduous tree with bright red/orange flowers that add a splash of color to the landscape.
3. Ashoka (*Saraca asoca*) - A tree with clusters of bright orange/red flowers that is highly regarded in Hindu mythology and can add a cultural touch to the landscape.
4. Gulmohar (*Delonix regia*) - A tree with brilliant red/orange flowers that can create a dramatic effect in the landscape.
5. Bottlebrush (*Callistemon spp.*) - A tree with unique, brush-like flowers that can add a quirky touch to the landscape.
6. Eucalyptus (*Eucalyptus spp.*) - A tree that has an attractive, aromatic foliage and can provide a natural windbreak.
7. Silk Cotton Tree (*Bombax ceiba*) - A large, deciduous tree with stunning red flowers that can provide shade and shelter for wildlife.

Shading trees are a great addition to any farmstead as they provide protection from the sun and can help to reduce the temperature inside buildings, while also creating a pleasant outdoor environment. Some popular options for shading trees suitable for farmsteads are banyan, *neem*, *peepal*, coconut, rain tree, *jamun*, mahogany, *bakain*, and mulberry. Fruit trees that are suitable for plantation at a farmstead depending on the specific climate and soil conditions of the region are mango, guava, banana, papaya, pomegranate, oranges, lemons, limes, fig, grapes, *bael*, *amla*, *etc.* The fruit trees should be planted to the back portion of the house where they are not visible from the entrance. For a windbreak, timber-trees, *ashoka*, eucalyptus, bamboo are recommended at the farmstead, and also along the canals and farm roads.

FARM RESIDENCE

A farm residence or farm house is a traditional type of dwelling that is found in rural areas where farming is the predominant economic activity. These houses are typically located on or near the farm and serve as both a home for the farmer and their family, as well as a central hub for farm operations.

One of the most significant features of a farm residence is its location. These homes are typically situated on the farm, which allows farmers to be close to their fields and livestock. This proximity makes it easier for farmers to tend to their crops and animals, as well as to monitor their farms for any potential issues or problems.

In terms of architecture, farm residences often have a simple, functional design that reflects their practical purpose. They may be single-story or two-story buildings with a modest footprint, and often have large porches or verandas that offer a comfortable outdoor living space. These spaces are ideal for enjoying the surrounding scenery and for entertaining guests. The materials used to construct these homes vary depending on the region and local resources, but may include brick, stone, wood, or adobe/mud brick.

Another important aspect of farm residences is their role in the local community. These homes serve as a gathering place for neighbours and friends, providing a space for social events and celebrations. In many rural areas, the farm house is the centre of community life, hosting everything from barn dances to community meetings.

Farm residences also play a crucial role in the economic sustainability of rural areas. By living on the farm, farmers are able to reduce their costs and increase their productivity. They can save on transportation costs, as well as reduce their dependence on outside labour. This independence and self-sufficiency are essential to the long-term viability of small-scale farming.

In recent years, there has been a renewed interest in farming and rural living, and many people are rediscovering the benefits of living on a farm. Whether you are a farmer, an aspiring farmer, or simply someone who appreciates the beauty and simplicity of rural life, a farm residence may be the perfect place to call home. With their practical design, strong community ties, and vital role in the local economy, farm residences are an essential part of rural life and culture that will continue to thrive for generations to come.

The landscaping around the farm house should be in keeping with the rural setting. Consider incorporating natural features, such as trees, shrubs, and wildflowers, as well as functional elements, such as vegetable gardens and livestock pastures.

2.7 Layout of Farm House

The layout of a farm house should be functional and efficient. It should be so designed and constructed to provide most comfortable place of stay for the farmer and his family. It should include all the necessary living spaces, such as bedrooms, living areas, kitchen, dining room, and bathrooms, as well as any other spaces that may be needed. The residence should face south or north to take maximum advantage of sunlight. Wherever feasible, the residence shall face onto the main roads or rivers or streams. Consider incorporating open floor plans and large windows to take advantage of natural light and views of the surrounding landscape. The layout and design of each room may vary depending on the size and style of the house, as well as the preferences of the homeowners. The various rooms should be so located as to provide adequate comfort to the family (Figure 2.4).

Bed Room

The bedroom is usually the primary sleeping area in the farmhouse. It may feature a bed, dresser, and wardrobe for storage. Bedrooms can be located on the first or second floor of the house, and may have an attached bathroom. A master bedroom should ideally be around 4.26 m by 4.87 m (14′×16′), while a smaller bedroom should be around 3.05 m by 3.65 m (10′×12′). Cross ventilation with one side exposed to the prevailing breeze is a desirable feature in design.

Drawing Room/Living Room

A drawing room, or living room, is a space for relaxation and socializing. It usually includes comfortable seating such as sofas or armchairs, a coffee table, and sometimes a fireplace. It may also have a television or other entertainment centre. This room is usually one of the larger spaces in a house and should be around 3.65 m by 4.57 m (12′×15′) or larger 4.57 m by 6.01 m (15′×20′), depending on how much space you need for seating and entertaining. The drawing room shall be located on one side of the house, and generally open into the front porch, kitchen, and bed room. The provision of adequate natural light and ventilation in the drawing room should be made.

Kitchen

The kitchen is the heart of the farmhouse, where meals are prepared and shared. It is preferably placed near the living room. The kitchen typically includes appliances such as a stove, refrigerator, sink, cabinets for storage of

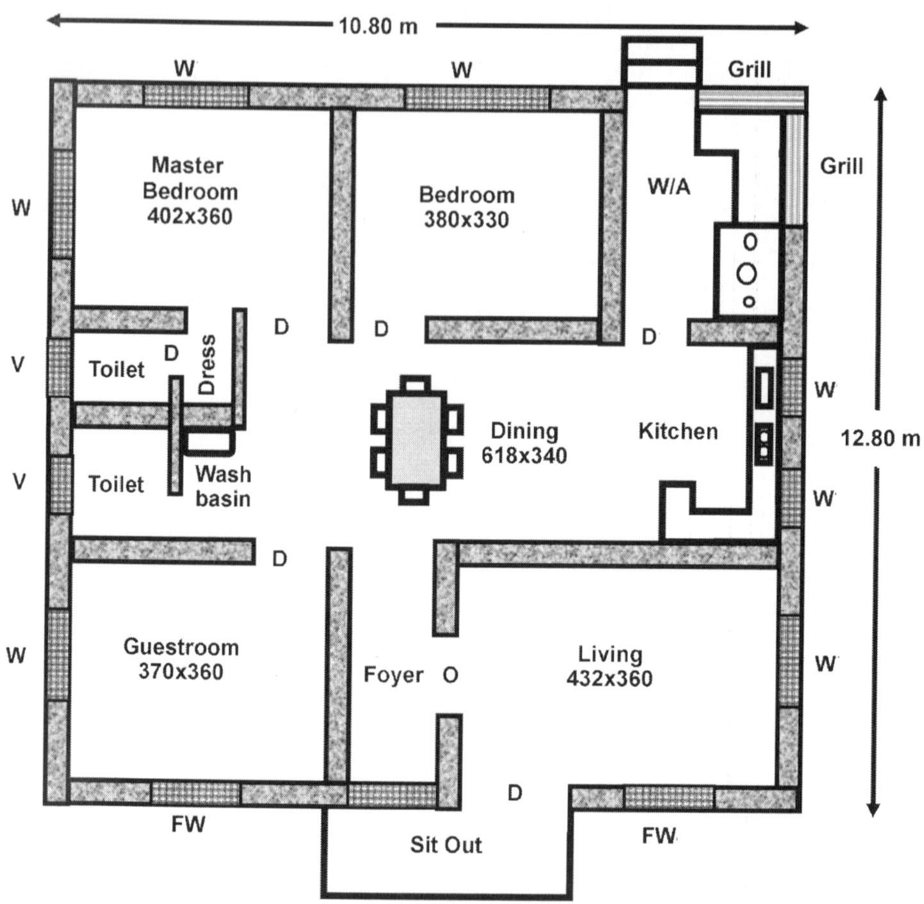

Figure 2.4. Floor Plan of a Typical Farm House.

utensils and supplies, and counter space for food preparation. A functional kitchen should be around 2.44 m by 3.05 m (8′×10′) or larger, depending on how much storage and counter space you need. Store room should be attached to the kitchen. While designing the layout, it should be so positioned such that maximum solar light comes in during the early hours. Now a days, most of the villages in India has the accessibility of LPG gas. Farmers have also installed bio gas plants in their premises for the supply of cooking gas. Due to introduction of gas stoves, design of kitchens has been changed considerably. In areas where gas stoves are not available, improved cook stoves are used. A kitchen typically should have a chimney or ventilation fan to help remove smoke, steam, and cooking odours from the room. It should feature windows for appropriate air circulation. The percentage of window space should be at least 15 per cent of the total floor area.

Toilet Rooms

Toilet rooms, or bathrooms, are essential in any home. They may include a toilet, sink, and bathtub or shower. Bath and toilet facilities may be separate or in the same room. A typical toilet should be around 1.83 m by 3.05 m (6′×10′), but this can vary depending on the layout and features you want to include. In India, there are mainly two types of toilets: Western style toilets and Indian style toilets (squatting type). Western style toilets are commonly used in urban areas and resemble the toilets found in many other parts of the world, while Indian style toilets are more common in rural areas and consist of a hole in the ground with foot rests on either side. Most toilet rooms do not have a direct water supply, but rather rely on a water storage tank to supply water to the toilet bowl. Toilets may have a manual or automatic flushing mechanism that draws water from the tank. Bath and WC (water closet) require enough ventilation and natural light. An exhaust fan may provide enough ventilation. Wash basin with a wall mirror shall be provided either inside or near to the room. Hanger for towels and rack for soap shall also be provided inside the room.

Guest Room

A guest room is a space for visitors to stay in the farmhouse. It may have a bed, dresser, and other amenities similar to a bedroom. It may also have an attached toilet room or be located near a common toilet room. It should open only to the drawing/living room and front veranda. A guest room should be around 3.05 m by 3.65 m (10′×12′) or larger, depending on how many guests you plan to accommodate.

Store Room

A store room is a space for storing various items such as kitchen items, cleaning supplies, or tools. It may have shelves or cabinets for storage and be located in a convenient area of the house. A store room can vary in size, depending on how much storage space you need. A good rule of thumb is to allocate at least 1.52 m by 1.83 m (5′×6′).

Important Points to Remember

1. Farmstead encompasses all the buildings, structures, and land that are used for farming activities, including the farm house, barns, sheds, silos, and other outbuildings.
2. A well-planned farmstead can optimize productivity, reduce waste, and provide a safe and comfortable living and working environment for all.
3. The farm buildings should be located on the highest elevation. In hilly areas or places of undulating topography, the buildings should be located with their longitudinal axes running parallel to the contour

lines. The topography should be satisfactory for the required storage and drainage of manure and effluent produced at the farmstead.

4. Buildings should generally be constructed only on poor soil, leaving the fertile soil for farming.
5. From the management point of view, the farmstead should be located at the centre of the farm.
6. The farmstead should be located near a source of permanent water supply.
7. An east-west or longer surface facing north and south is best orientation for a farmstead.
8. The area of farmstead usually varies from 3 to 5 per cent of the farm area.
9. Zone planning helps to optimize the use of available resources, improve productivity and efficiency, and enhance the overall sustainability of the farm.
10. The arrangement of buildings on a farmstead should be done in a way that maximizes efficiency and supports the intended use of the farm.
11. Barns and animal housing may be located on one side of the farmhouse, while equipment sheds and storage buildings may be located on another side.
12. The residential building on a farm is the heart of the farmstead.
13. The farm house should be at least 30 m away from the road.
14. Dairy Barn of general-purpose barn should be minimum 15 m from machine shed, but not too close to the farm house. It should at least be at least 30 m from the house.
15. Pig house should be about 60 m from the residence, and at least 30 m from the cow barn.
16. For a windbreak, timber-trees, *ashoka*, eucalyptus, bamboo are recommended at the farmstead, and also along the canals and farm roads.
17. The layout of a farm house should be functional and efficient.
18. A master bedroom should ideally be around 4.26 m by 4.87 m (14′×16′), while a smaller bedroom should be around 3.05 m by 3.65 m (10′×12′).
19. The storeroom should preferably be near the kitchen.
20. The common size of guest room should be around 3.05 m by 3.65 m (10′×12′).

Exercises/Questions

1. What are the functional requirements of farm buildings?

2. What is a farmstead? List the objectives of farmstead planning. Explain the principles involved in farmstead planning.
3. List the various factors which influence the location of farmstead on the farm.
4. Discuss zone planning and its importance.
5. Design a farm house. Draw the floor plan of the same.
6. Write short notes on:
   (a) Farmstead planning
   (b) Farmstead tree-planting
   (c) Guidelines about distance between various buildings on a farmstead

Chapter 3

Building Materials

The engineering structures are composed of number of building materials. These are stones, bricks, lime, cement, sand, mortar, aggregate, concrete, steel, timber, *etc.* Availability, cost, permanence and durability are some of the factors in selection of building materials for construction of animal shelters, farm house, and allied structures. The salient characteristics of commonly used construction materials is described in this chapter.

3.1 Stones

The building stones are obtained from rocks. It is an essential and more permanent building material in construction than other natural building materials. They are strong, durable and resistant to weather conditions. It is maintenance free. The different types of building stones are available in India. The most common building stones are granite, *kankar*, limestone, marble, sandstone, slate, quartzite, laterite, *murum*, chalk, *etc.* Each stone has its own properties. Based on the type, stones can be used in buildings for flooring, roofing, masonry, paving roads and also as aggregates for concrete. Marble, granite and sandstone are used for facing work of buildings. Limestone and sandstone are used for general building works. Fine-grained granite, marble, and soft sandstone are used for carvings and ornamental works. Compact limestone and sandstone are used for fire-resistant masonry.

The characteristics of a good building stone are:

1. **Appearance:** Building stones used for the face work of the building should have fine, compact texture. Light coloured stone is usually preferred as dark colours are prone to fade out with time. Marble

and granite are used for face works in buildings due to their uniform colour and excellent appearance when polished.

2. **Durability:** A good building stone should be durable.
3. **Strength:** The stone should be strong enough to withstand the load coming on it. Ordinarily this is not of primary concern since all stones are having good strength. However, in case of a large structure, it may be necessary to test the strength of the stone. Stones with compact fine crystalline texture are stronger. The crushing strength of stones shall be above 100 N/mm^2.
4. **Specific Gravity:** Good building stone has a specific gravity between 2.4 and 2.8, while stones with a specific gravity less than 2.4 are deemed unsuitable for construction.
5. **Structure:** A stone when broken, should not be dull in appearance and should show uniformity of texture. It should be free from cavities, cracks, and patches of loose or soft material. Stratifications, which are usually found in sedimentary rocks should not be visible to naked eye.
6. **Texture:** A good building stone should have fine crystalline structure free from cavities, cracks or patches of soft or loose material. Such stones are usually strong and durable.
7. **Hardness:** The stone used in floors and pavements should be able to resist abrasive forces caused by the movement of men and materials over them. Hence the stones which are to be used in such places should be tested for hardness.
8. **Porosity and Absorption:** Building stones should not be porous. The porous building stones are not suitable in construction especially for exposed surfaces of structures. If it is porous, rainwater enters the pores and reacts with stone and crumbles it. In cold regions, the freezing of water in pores of stones takes place. It results in the disintegration of the stones because of its increase in volume on freezing. Water absorption by stones depends upon their porosity. The more porous the rock is, the more water it will absorb and leads to the damaging of stone. The water absorption test measured by submerge the stones underwater for 24 h and expressed in percentage. A good stone shall have a maximum of 0.60 per cent water absorption.
9. **Toughness:** It is the measure of impact that a stone can withstand. The stones shall withstand the stresses created by vibrations of machinery and moving loads over them.
10. **Ease of Working:** The stones can be easily carved, moulded, cut and dressed. However, the care should be taken so that, this is not at the cost of the required strength and the durability. When selecting stones for building works, it is advisable to focus on ensuring sufficient strength rather than high strength.

11. **Cost:** The cost of building materials plays an important role in their selection for construction. The proximity of a quarry to the building site reduces transportation costs and, thus, the cost of stones.
12. **Fire-Resistant:** To be fire-resistant, building stones should be free from calcium carbonate and oxide of iron and be not composed of minerals with different co-efficient of thermal expansion. The fire resistance of sand-stones is comparatively better. Similarly, argillaceous materials are also offer good fire resistance but are of poor strength.
13. **Seasoning:** The stones should be well seasoned before putting into use.
14. **Weathering:** Stones selected should be capable of resisting adverse effects of natural forces like wind, rain, and heat. Rain and wind can damage the appearance of stones. It is recommended to use weather-resistant stones for face works.

Seasoning

The pores of the fresh stones obtained from quarry have some moisture. This moisture is called quarry sap. Good stones should be free from quarry sap. The presence of this moisture makes the stone soft and weak. Thus, the stones quarried freshly are easy to work. It is, therefore, desirable to do dressing, carving, moulding, *etc.* as early after quarrying as possible. After quarrying, when all the work has been done upon stones, moisture from the stone is removed by cyclic drying and airing. This process is called seasoning. The stones should be dried or seasoned before they are used in structural work. A period of 6 to 12 months is considered sufficient for proper seasoning.

3.2 Bricks

Bricks are rectangular blocks made of clay. The clay is moulded into rectangular units of uniform size, followed by drying and burning. They are commonly used in building work and as a substitute for stone when stone is unavailable.

Properties of Bricks

The basic properties of bricks may be conveniently described under the following four headings: physical, mechanical, thermal and durability properties.

Physical Properties of Bricks

Physical properties of bricks include shape, size, colour, and density of a brick.

1. **Shape:** The shape of an ideal brick is rectangular with well-defined and sharp edges. The surface of the bricks should be regular, even and

free from cracks. For specific purposes, the brick may be manufactured in other shapes also.

2. **Size:** Brick size varies across countries and locations within the same country.
3. **Colour:** Well-burnt bricks typically have a red colour, ranging from deep red to light red. Good bricks have a uniform structure and colour throughout their body. The dark red colour of brick represents over-burnt bricks, whereas yellow is for under-burnt brick. The broken or fractured brick shall have a bright, uniform, compact structure free from voids.
4. **Density:** The density of standard bricks varies from1600 to 1900 kg/m^3. The density of bricks depends upon the type of clay used and the method of brick moulding. The weight of a single brick (19×9×9 cm) shall weigh between 3.2 to 3.5 kg depending upon its density.

Mechanical Properties of Bricks

Under this heading of properties of bricks, compressive strength and flexural strength are included.

1. **Compressive Strength:** The compressive strength of bricks is crucial for load-bearing walls. It is influenced by the clay composition and burning degree. BIS recommends a minimum of 35 kg/cm^2 for ordinary bricks, while first- and second-class bricks must have a minimum of 140 kg/cm^2 and 70 kg/cm^2, respectively. To check strength, the brick shall not break when dropped flat on hard ground from a height of one meter.
2. **Flexural Strength (or Bending Strength):** Bricks should possess sufficient strength against transverse loads to withstand possible bending loads in a building. The flexural strength of a common building brick shall not be less than 10 kg/cm^2. Best grade bricks often possess flexural strength over 20 kg/cm^2.
3. The bricks should produce a distinct metallic ringing sound when struck with each other.
4. The bricks should be sufficiently hard. When scratched with fingernail, no impression should be left on brick surface.

Thermal Properties of Bricks

Ideal bricks should offer adequate insulation against heat, cold, and noise. The bricks should have low thermal conductivity and be soundproof. Dense and heavy bricks have poor thermal and acoustic insulation due to their higher heat and sound conductivity. Although the conductivity of bricks varies significantly with their density and porosity. Thus, bricks should be so designed that they are light and strong and provide adequate insulation.

Durability Properties of Bricks

Durability of bricks means the maximum time for which they remain unaltered and strong when used in construction. Bricks are the most durable man-made building material. The absorption value, frost resistance, and efflorescence are the main properties that determine the durability of brick.

1. **Absorption Value:** The absorption value of bricks is the quantity of water absorbed by bricks in 24 hours. The water absorption depends on the porosity of the bricks. It shall not exceed 20 per cent by weight for first-class bricks and 22 per cent by weight for second-class bricks when soaked in cold water for 24 hours. A higher absorption value indicates deeper water penetration into the brick, which causes dampness. Moreover, bricks of high absorption value are of low strength.
2. **Frost Resistance**: Water on freezing expands by about 10 per cent in volume and exerts a pressure on the order of 140 kg/cm^2. When bricks are used in cold climates, their decay due to this phenomenon of "frost action" may be a common process. This is especially so because bricks are quite porous materials. It is essential to provide proper protection to bricks from rain in the cold climate regions. It minimizes the absorption of water by bricks.
3. **Efflorescence:** Bricks in hot and humid climates often disfigure and deteriorate due to the dissolution of salts. It is visible as white or grey patches on the surface of the brick. The rainwater can dissolve the salts easily. The evaporation starts after the rain, and the salts like sulphates of calcium, magnesium, sodium, and potassium come out of the water. The salt precipitates over the surface and forms thin encrustations. Thus, it is important to check the chemical composition of the clay before using it for brick manufacturing. Good quality clay devoid of salts should be used for brick manufacturing.

Classification of Bricks on the Basis of Quality

Bricks can be classified on the basis of quality, building process, manufacturing method, raw materials *etc.* The classification of bricks on the basis of quality is discussed here.

1. **First Class Bricks**: These bricks shall have sharp, square, smooth, straight edges and surfaces. These bricks shall be well-burnt and uniform in yellow or red. They have a water absorption capacity of less than 10 per cent and a crushing strength of 280 kg/cm^2. The minimum crushing strength of first-class brick shall be 245 kg/cm^2. It shall emit a metallic sound when struck by another brick and be hard enough to resist fingernail expression. They are table-moulded and of standard shape, burnt in kilns, and are used for superior work of a permanent

nature. They do not emit efflorescence and are suitable for use with a thumbnail.

2. **Second Class Bricks**: This type of brick has a uniform yellow or red colour, rough surface, an irregular shape, hair cracks, and uneven edges. The water absorption capacity shall be in the range of 10–15 per cent. The mean crushing strength of second-class brick shall be 175 kg/cm^2. However, a minimum of 154 kg/cm^2 of crushing strength is desired. A metallic sound shall emit when struck two similar bricks with each other. These bricks are commonly used in places where plaster is needed. These bricks are ground-moulded and burnt in kilns. It is hard enough to resist fingernail expression while developing a thumbnail.
3. **Third Class Bricks**: The bricks are irregular in shape and size, with a soft, light red colour and extensive efflorescence. They have a dull or blunt sound when struck together and a water absorption capacity of between 15-20 per cent. The bricks are ground-moulded and burnt in clamps, making them suitable for temporary structures and low-rainfall areas. They are not hard and have a non-uniform texture. It leaves fingernail expressions while developing the thumbnail. The mean crushing strength of these bricks is 140 kg/cm^2, where 105 kg/cm^2 is the minimum value.
4. **Fourth Class Bricks**: These are over burnt bricks with irregular shape and dark colour. These bricks are used as aggregate for concrete in foundations, floors, roads, *etc.* because the over burnt bricks have a compact structure. Sometimes, they are found to be stronger than the first class bricks.

Size and Weight of Bricks

Bricks come in various sizes, although BIS recommends a standard size of 19 cm × 9 cm × 9 cm in India. The nominal size, which includes the mortar thickness in modular bricks, is 20 cm × 10 cm × 10 cm. To complete 1 m^3 of brick masonry, 500 bricks are required. The average weight of a brick is 30 to 35 N. Hence, the weight of 1 m^3 of brick masonry will be about 15 kN. Brick size varies from place to place, and it is recommended to use the correct size for proper burning and placement inside the kiln. Large bricks can be difficult to burn and heavy, while smaller bricks require more mortar.

Testing of Bricks

The tests to be performed for testing of bricks are water absorption test, hardness test, shape and size test, colour test, soundness test, structure test, efflorescence test and crushing strength test. Most of the tests can be performed in the field itself to determine the suitability of bricks for the construction work.

(i) Absorption Test

This test is conducted on brick to find out the amount of moisture absorbed by brick under extreme conditions. In this test, sample dry brick after weighing is immersed in water for a period of 24 hours. Then, note down the weight of wet brick. The difference between wet and dry brick weights will give the quantity of water absorption. For a good quality brick, the quantity of water absorption should not exceed 20 per cent of weight of dry brick.

Absorption in per cent is calculated as:

$$\text{Absorption (per cent)} = \frac{W_2 - W_1}{W_1} \times 100$$

where, W_1 and W_2 are the weight of oven dried and wet brick, respectively.

(ii) Hardness Test

A good quality brick resists scratches against sharp things. Thus for this test, a sharp tool or finger nail is used to make scratch on brick. If there is no scratch impression on brick then it is said to be hard brick.

(iii) Shape and Size Test

Shape and size of bricks are one of the important consideration. The shape of bricks should be rectangular with sharp edges. All bricks used for construction should be of same size. Standard brick size is 19 cm × 9 cm × 9 cm (length × breadth × height). To perform this test, select 20 bricks randomly from a lot and stack them along its length, breadth and height and compare. If all bricks are of similar size, then they are fit for construction work.

(iv) Colour Test

A good brick should possess bright and uniform colour throughout its body. The colour can be judged by visual inspection of the bricks.

(v) Soundness Test

Soundness test of bricks shows the nature of bricks against sudden impact. To perform this test, 2 bricks are chosen randomly and struck with one another. Then sound produced should be clear bell ringing sound and bricks should not break. Then it is said to be good brick.

(vi) Structure Test

In this test, pick one brick randomly from the lot and break it. Observe the inner portion of brick clearly. A good quality brick should be free from lumps and homogeneous.

(vii) Efflorescence Test

A good quality brick should not contain any soluble salts in it. If soluble salts are there, then it will cause efflorescence on brick surfaces. To perform this test, place the brick in a water bath for 24 hours and dry it in shade. After drying, observe the brick surface thoroughly. If there is any white or grey colour deposits, then it contains soluble salts and not suitable for construction.

(viii) Crushing Strength/Compressive Strength

Crushing strength of bricks is determined in the laboratory with the help of compression testing machine. The sample brick shall be placed in compression testing machine for the measurement of crushing strength. Then apply load on it until brick breaks. Note down the value of failure load and find out the crushing strength value of brick. The minimum crushing strength of brick should be 3.50 N/mm^2. If the crushing strength of the brick is less than 3.50 N/mm^2, then it is not useful for construction purpose.

Table 3.1. Comparison of Brickwork and Stonework

Property	*Brickwork*	*Stonework*
Appearance	It does not give solid appearance.	It creates solid appearance.
Architectural effects	It does not provide excellent architectural effects.	Better quality architectural effects can be achieved.
Availability	Plenty of clay is not available in every region of the country.	Only in few areas, stones are easily available.
Cost of material	In regions where clay is available in abundance, the brickwork is cheaper.	In regions where stones are easily available, the stonework is cheaper.
Atmospheric effects	Good resistant against atmospheric variations.	Unable to resist various atmospheric effects.
Construction	Easy to construct openings *etc.*	Difficult to use for construction.
Durability	The structure is more durable due to thin mortar joints.	Less durable.
Fire-resistant	Better fire-resistant.	Stones easily disintegrate in case of fire.
Handling	Easy to handle.	Needs lifting device for handling.
Labour for construction	Less skilled labour is required.	Specialized labour is required for stonework.
Strength	Less strong.	More strong.
Water absorption	Bricks absorb more moisture.	Stones absorb less moisture.

3.3 Lime

The lime is being used as cementing material since time immemorial. It is used in several ancient architectural works. Now a day, cement has replaced lime to a great extent. At places where lime is locally available and when there is acute shortage of cement, lime provides a cheap and a reliable substitute for cement.

Lime is not generally available in nature in free-state. It is obtained by burning one of the material such as limestone from stone hills or boulders of limestone from beds of old rivers or *kankar* found below the ground or shells of sea animals. The main sources of obtaining lime are the limestone which possess clay, soluble silica, magnesium carbonate, alkalis and metallic oxides, sulphates, iron compounds and carbonaceous matter. White chalk is pure limestone and *kankar* is an impure milestone.

As per IS: 712 (1984), limes are classified into following six classes.

Class A	Hydraulic lime	Used for structural purposes
Class B	Semi-hydraulic lime	Used for mortars for masonry work
Class C	Fat lime	Used for finishing coat in plastering, white washing and with suitable admixture such as *surkhi* or any other pozzolanic material to produce artificial hydraulic mortars
Class D	Magnesium or dolomitic lime	Used for finishing coat in plastering, white washing *etc.*
Class E	*Kankar* lime	Used for masonry mortars
Class F	Siliceous dolomitic lime	Used for undercoat and finishing coat of plaster

Uses of Lime

The lime is an important engineering material which has numerous applications. It is used for making mortar for masonry work; for plastering of walls and ceilings; for whitewashing and for serving as a base coat for distemper; for matrix of concrete and mortar *etc.*

Properties of Lime

A good quality lime has the following properties:

1. Good plasticity.
2. Flexible and easily workable.
3. Provide strength to masonry with mortar.
4. Small time for solidification and being hard.
5. It should comprise of excellent binding properties which adhere to brick or stone masonry units perfectly.
6. It should be highly resistant to moisture and can be used for pointing works.

Precautions to be taken in Handling Lime

Slaking of lime is an exothermic reaction and generates a large amount of heat. To avoid accidents, quick lime should not be allowed to come in contact with water before slaking. The workers handling lime should be provided with goggles and respirators as lime dust causes irritation. The lime also causes skin burns, especially if the skin is moist. Thus skin should be protected by rubber gloves, gum boots, skin protective cream, *etc.* The quick lime gives out immense heat while slaking and hence, suitable measures should be taken to avoid chances of any fire hazard. The workers should be instructed to wash the exposed portions of the body with abundant fresh water after handling lime. The workers handling milk of lime, which is hot, should be advised to oil their skin daily to avoid skin burns.

Field Tests

The IS 1624-1986 specifies certain field tests for building lime. These are:

(i) Visual Examination

White colour indicates the pure limestone. The hydraulic lime stones are indicated by bluish grey, brown or some dark colour. The presence of glistering particles on the surface of lime stones indicates the presence of free sand. The presence of lumps gives an indication of quick lime or unburnt limestone.

(ii) Heat Test

Weigh a piece of dry limestone. Heat it in an open fire for few hours. Weigh the sample again. Loss of weight indicates the amount of carbon dioxide. Thus, the amount of calcium carbonate in limestone is worked out.

(iii) Acid Test

A teaspoon of powdered lime is taken in a test tube (the quantity will be almost 10 cc). To this dilute hydrochloric acid is poured. Stir the contents and then the test tube with its contents is kept in the stand for 24 hours.

After 24 hours, if there is vigorous effervescence and less formation of residue, it suggests presence of high content of calcium carbonate. It indicates pure limestone. On the other hand, if content of calcium carbonate is less, there will be less effervescence and more formation of residue. It indicates impure or hydraulic limestone.

If a thick gel is formed which has no kind of flowing capability, it is actually a representation of the presence of Class A lime. If the gel has a flowing capability, it is Class B lime. If there is no gel formation, it indicates lime of Class C.

(iv) Ball Test

A simple test can be performed for finding the type of lime. Form the balls of about 40 mm size of stiff lime by adding enough water and leave them undisturbed for six hours. The balls are then placed in basin of water. If there are signs of slow expansion and slow disintegration within minutes after placing in water, it indicates Class C lime. If there are very little or no expansion and numerous cracks, it indicates Class B lime. If no kind of adverse effect on the ball is found, it indicates Class A lime.

(v) Workability Test

This test is used to determine the sticking quality of the lime-sand mortar mix. Prepare a mix of lime-sand mortar in the ratio of 1:3. Throw this mix over a brick wall with the help of a trowel. If the mix sticks well the workability of the mix is regarded good.

(vi) Impurity Test

Lime of known quantity is taken in a beaker and mixed with water. The solution is then stirred well and allow them to settle for two hours. Transfer the residue to a metal tray with a jet of water. Allow it to settle and decant off the water from the tray. The obtained residue is allowed to dry under the sun for a time period of 8 hours. If the residue is less than 10 per cent in amount, then the lime is good. If it is between 10 to 20 per cent, the lime can be considered fair. Above 20 per cent, the lime is regarded poor or impure.

(vii) Plasticity Test on Blotting Paper

The lime is made into a thick paste with sufficient quantity of water. The sample is left overnight. The next day it is spread over a blotting paper like butter with the help of a knife. If the paste is spreadable with ease without any gritty material and with soft strokes, then it may have good plasticity.

3.4 Cement

Cement is one of the most essential construction material, which is extensively used in the modern-age of construction. It is a binding agent that sets and hardens to stick to building units such as stones, bricks, tiles *etc.* Cement is a prominent ingredient for concrete, mortar, and plaster, *etc.* Cement contains multiple ingredients as its raw material like lime, silica, alumina, iron oxide *etc.* These ingredients interact with one another in the kiln at high temperature during the manufacturing process and make a complex compound known as cement. Cement mixed with water causes a chemical reaction and forms a paste that sets and hardens to bind individual structures of building materials.

Composition of Cement

There are eight major ingredients of cement. The relative proportions of these oxide compositions are responsible for influencing the various properties of cement. The composition of these ingredients in a good ordinary cement is given in Table 3.2.

Table 3.2. Composition of Cement

Sl.No.	*Ingredient*	*Typical Composition (Per cent)*	*Normal Range (Per cent)*
1.	Lime (CaO)	62	60 to 67
2.	Silica (SiO_2)	22	17 to 25
3.	Alumina (Al_2O_3)	5	3 to 8
4.	Calcium sulphate ($CaSO_4$)	4	3 to 5
5.	Iron Oxide (Fe_2O_3)	3	0.5 to 6
6.	Magnesia (MgO)	2	0.1 to 3
7.	Sulphur trioxide (SO_3)	1	1 to 3
8.	Alkaline	1	0 to 1
	Total	100	

Depending upon the specific use, the cement is classified as regular type cement, white cement, and quick hardening cement. The regular type cement sets within 10 hours, and develops its full strength in 28 days if properly cured. White cement is similar to regular type cement in its characteristics except its colour which is white. It is usually used for ornamental work. The quick hardening cement develops its full strength within 3 days after setting. It is used in under water construction. It is also used in rainy and cold weather conditions.

Uses of Cement

Cement is a globally recognised engineering material used in the construction of various structures. It is typically mixed with aggregate to create a strong binding material, *i.e.*, mortar and concrete. There are numerous uses of cement. Some of the possible applications or uses of cement are given below.

1. It can be used in the production of mortar for plastering, masonry work, pointing, *etc.*
2. It can be used in the production of concrete for laying floors, roofs and constructing lintels, beams, stairs, pillars *etc.*
3. It can be used to build fire proof and thermal proof structures in the form of concrete.
4. It can be used to build acid-resistance proof structures.
5. It can be used to build hydrographic and frost resistant structures.
6. It can be used to build waterproof structures.
7. It can be used for making joints for drains and pipes.
8. It can be used to manufacture precast members such as pipes, fencing posts *etc.*
9. It can be used as grout material.

10. It can be used for the construction of wells, water tanks, septic tanks, cement concrete roads, bridges, culverts, dams, tunnels, lighthouses *etc.*
11. Coloured cement can be used for decorating or colouring the structures.
12. It can be used where a hard surface is required for the protection of exposed surfaces of structures against the destructive agents of the weather and certain organic or inorganic chemicals.

Properties of Good Cement

The properties of a good cement primarily depend upon its chemical composition, thoroughness of burning and fineness of grinding. Some of the important properties of a good cement are given below.

1. It should give strength to masonry.
2. It should be an excellent binding material.
3. It should be easily workable.
4. It should offer good resistance to moisture.
5. It should possess a good plasticity.
6. It should harden early.

Cement has a varying range of density depending on the cement composition percentage. The bulk density of cement is usually taken as 1440 kg per cubic meter. Each bag of cement contains 50 kg or 500 N or about 0.035 m^3 of cement. The compressive strength at the end of 3 days should not be less than 115 kg/cm^2 or 11.50 N/mm^2 and that at the end of 7 days should not be less than 175 kg/cm^2 or 17.50 N/mm^2. The tensile stress at the end of 3 days should not be less than 20 kg/cm^2 or 2 N/mm^2 and that at the end of 7 days should not be less than 25 kg/cm^2 or 2.50 N/mm^2. The initial setting time should be about 30 minutes and final setting time should be about 10 hours for ordinary cement.

Field Tests of Cement

Field tests can be conducted within short period of time at the site to check its suitability for construction purpose. Although these tests are not very accurate, yet they provide a fair idea to the engineer regarding the quality of the cement supplied. The some of the basic field tests are:

(i) Colour of Cement

Open the bag and take a look at the cement. The colour of the cement should be grey with a light greenish shade. This colour should be uniform.

(ii) Adulteration Test

Take a pinch of cement and rub between the fingers. It should give a smooth feeling. If it feels gritty, it indicates that the cement is adulterated with sand.

(iii) Presence of Lumps

Lumps are not good to use for construction. It is developed by absorbing the moisture from the atmosphere. Open the bag, look at the cement and thrust your hand into the cement bag. There should not be any lump. The bags which contain such lumps should be rejected.

(iv) Temperature Test

Insert your hand in the bag of cement to check its temperature. It should give you cool feeling. If it's warm inside, then it indicates that the process of hydration is taking place inside the bag.

(v) Float Test

Take some handful of cement from the bag and throw it in a bucket full of water. The cement particles should float for sometimes before it sinks. It should not float on the surface.

(vi) Setting Time/Glass Plate Test

A thick paste is prepared by mixing cement and water, and is placed on a piece of thick glass plate. It is kept under water for 24 hours. After this, it should set and not crack.

(vii) Strength Test

The strength of cement can be roughly ascertained in the following ways:

(i) Briquettes with a lean mortar or weak mortar are made. The size of briquette may be about 75 mm × 25 mm × 12 mm. The proportion of cement and sand may be 1:6. The briquette are immersed in water for a period of 3 days. If cement of sound quality, such briquette will not be broken easily and it will be difficult to convert them into powder form.

(ii) Prepare a block of cement of dimension 25 mm × 25 mm and length of 200 mm. The block is then immersed in water for 7 days. After removing, place it on supports 150 mm apart and load with a weight of 34 kg. If the block shows no sign of failure, then the cement is of good quality.

(viii) Date of Packing

The strength of cement reduces with time. Hence, the manufacturing date play a important role in assessing the strength of cement. Generally, the cement should be used before 90 days from the date of manufacturing.

(ix) Smell of Cement Paste

A thin paste of cement with water should feel sticky between the fingers. If the cement contains too much-pounded clay and silt as an adulterant, the paste will give an earthy smell.

Grade of Cement

The strength of the concrete mix depends on the grade of cement. The cement is available in three grades: 33, 43, and 53. The strength of cement is measured as its compressive strength. It is defined as the strength of cement moulded in a standard cube after 28 days of curing. The compressive strength is measured in Mega-Pascal (MPa) or N/mm^2. As per Indian standards, 33-grade cement has a compressive strength of 33 N/mm^2 after 28 days of curing. Similarly, 43 and 53 grade cement has compressive strength of 43 N/mm^2 and 53 N/mm^2, respectively.

33 Grade cement is suitable for general construction work under normal environmental conditions but may not be suitable for concrete grades above M20. Due to the availability of higher grades, the use of 33-grade cement has declined, and it is now rarely manufactured. 43-grade cement is suitable for plain concrete work and plastering, making concrete mix up to M30. It is also used for precast items like tiles, blocks, and pipes. 53-grade cement has a faster setting time compared to 43-grade cement. 53-grade cement is used for structural purposes, such as reinforced cement concrete, and is suitable for making concrete mix above M25. It can also be used in pre-stressed concrete.

Storage of Cement

Cement is a hygroscopic product. Because of this characteristic, the cement absorbs moisture from the atmosphere very actively and hardens like stone which cannot be used for constructional purpose. An absorption of 1 or 2 per cent of moisture has no appreciable effect on quality of cement. But, if moisture absorption exceeds 5 per cent, the cement becomes totally useless. So, storage of cement should be done with care. This becomes particularly important during the humid season and in coastal regions when atmospheric air contains higher amount of moisture in it.

The cement should be used before its 'best before' date to ensure a durable structure. The shelf life of cement is 3 months. During this 3 months period, it is important to store the cement carefully and prevent it from coming in contact with moisture. Cement should be stored in a dry environment with as little humidity as possible. It should also be stored away from any possible water source. The drainage system should be checked and maintained on regular basis. Doors and windows of the store must be kept closed. Bags must be stacked on a wooden plank or on a strong tarpaulin at 15 cm above the ground. Cement bags should be stacked in a stepwise manner on the wooden plank so that plank is stable and will not topple/overloaded at on one side. In the stacks, the cement bags shall be kept close to each other to reduce the circulation of air as much as possible.

Bags must be stacked at least 30 cm away from the walls. Width of the stack must not be more than length of 4 bags, and height of stack should not be more than 10 bags. A passage of 90 cm shall be provided between the piles of cement. It is advisable to cover the cement with tarpaulins or plastic sheets of at least 175 microns in rainy season or when it is required to be stored for longer time. Never use hooks to lift cement bags. Cement should be removed in order of its storage period, *i.e.* the rule of first in, first out (FIFO) should be followed. Never store the cement bags with fertilizers or other products. The storage space requirement shall be 1 cubic meter for storing of 20 bags or 10 kN (1000 kg) of cement.

Packing of Cement

Cement is usually packed in 50 kg packs. Packaging for cement is often made from paper bags lined with PE to ensure moisture protection. Some of the popular types of cement bags are HDPE cement bags, polypropylene cement bags, perforated cement bags, cement plastic valve bags, PP woven cement bags, recycled polymer cement bags, jute cement bags, *etc.*

Figure 3.1. Stacking of Cement Bags.

Difference between Cement and Lime

The difference between ordinary cement and lime are as follows:

1. The cement can be used under conditions and circumstances which are not favorable for lime. Lime cannot be used for heavy engineering structures.
2. The cement, when converted into paste form, sets quickly while lime sets slowly.

3. No appreciable heat is produced when water is added to cement. Also, there is no slaking action in cement while lime slakes when wetted with water. Process of slaking takes considerable time and the construction work is delayed as compared to cement.
4. Lime is white or greyish in colour while cement is greenish-grey in colour.
5. Lime is best in terms of cost and environment. Lime is cheaper as compared to cement. It doesn't lead to the production of harmful gases and contributes to controlling global warming.
6. It is possible to produce lime in small quantities. Hence, it is also economical to produce.
7. In terms of strength, durability and other qualities, cement is best because it poses high strength and good binding.

3.5 Sand

Sand is a basic raw material used in the construction industries. The sand particles consist of small grains of silica. It is formed by decomposition of sandstones due to various effects of weather. It is finer than gravel and coarser than silt. The particle size of sand ranges from 0.06 mm to 2 mm. Particles which are larger than 0.0078125 mm but smaller than 0.0625 mm are termed silt.

Classification of Sand

Sand is classified according to size of grains.

1. **Fine Sand**: The sand passing through a screen with clear openings of 1.5875 mm is known as fine sand. It is mainly used for plastering.
2. **Coarse Sand**: The sand passing through a screen with clear openings of 3.175 mm is known as coarse sand. It is usually used for masonry work.
3. **Gravelly Sand**: The sand passing through a screen with clear openings of 7.62 mm is known as gravelly sand. It is generally used for concrete work.

Properties of Good Sand

Followings are the desirable properties of good sand:

1. It should be chemically inert.
2. Grains should be sharp, angular, coarse and stable.
3. It should not contain salts (*i.e.*, $CaCl_2$, $MgCl_2$, *etc.*) which attract moisture from the environment.
4. It should not contain any organic and vegetable matter. Generally, 3-4 per cent clay is permitted for practical reasons.

5. It should be free from gritty particles and injurious particles like dust, *kankar*, soft and flaky particles of shale. Sand should be sieved so that organic and gritty particles are removed.
6. It should be well graded and should contain particles of various sizes in suitable proportions. The fineness modulus of sand should be between 2 and 3.

Bulking of Sand

The presence of moisture in sand increases the volume of sand. For a moisture content of about 5 to 8 per cent, the increase in volume may be as much as 20 to 40 per cent, depending upon the grading of sand. Increase in volume shall be more with fine sand and less with coarse sand. In case fine aggregate is damp at the time of proportioning the ingredients for mortar or concrete, its quantity shall be increased suitably to allow for bulkage. Otherwise, it shall result into under-shaded mixes which are harsh and difficult for working and placing.

A simple test is carried out to find out percentage of bulking of sand. Take a container and fill it up to two-third with the sample of sand to be tested. Let the height is 20 cm. Then, take out sand out of container. Care should be taken to see that there is no loss of sand during this transaction. Fill the container with water. Thereafter, slowly drop the sand in the container and stir it thoroughly by means of a rod. Let the height of sand is measured as 16 cm.

$$\text{Then, bulking of sand} = \frac{(20-16)}{16} \times 100 = 25 \text{ per cent}$$

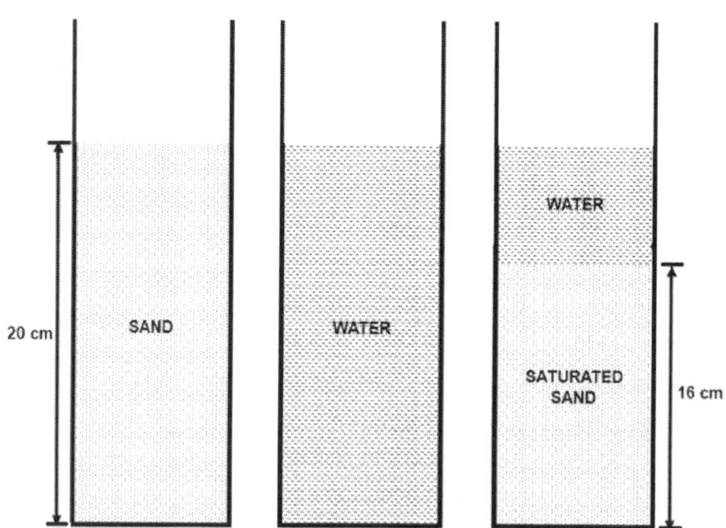

Figure 3.2. Bulking of Sand Test.

Field Tests on Sand

There are some field tests that can be easily carried out in the field in order to check the quality of sand for construction purpose. These are:

1. Take a glass of water and add some quantity of sand in it. Then shake it vigorously and allow it to settle for an hour. Clean sand will settle immediately. If silt or clay particles are present in sand, it will form a distinct layer at the top of sand. The maximum quantity of silt in sand shall not exceed 8 per cent (particle size less than 150 microns). Fine aggregate containing more than allowable percentage of silt shall be washed so as to bring the silt content within allowable limits.
2. In order to detect presence of organic impurities in sand, add sand to the solution of sodium hydroxide or caustic soda and then stir it. If colour of solution changes to brown, it indicates the presence of organic impurities.
3. Take a pinch of sand and taste it. If tasted salty then there exist some salt in sand and it should not be used at all in any construction.
4. Rub the sample of sand with wet palms. Good clean sand will not stick to the hand, whereas sand with clay will stick and change the colour of the palm. OR Take a sample of sand and rub it against the fingers. If fingers are stained, it indicates that sand contains earthy matter.
5. The colour of sand will indicate the purity of sand. The size and sharpness of grains may be examined by touching and observing visually.

3.6 *Surkhi*

Surkhi is an artificial pozzolanic material produced by powdering well burnt bricks or burnt clay balls. It is the popular substitute for sand. It can be used in place of sand during construction. The name *surkhi* is derived due to its reddish colour. *Surkhi* shows pozzolanic properties and is composed of about 40 to 63 per cent of silica by weight.

Surkhi is a complex material whose quality differs depending on the type of clay. *Surkhi* is not standardized product and its properties vary widely. The characteristics of *surkhi* depends upon the composition of soil, degree of burning, and fineness of the powder during grinding. *Surkhi* should be clean and free from any impurities. The maximum quantity of clay, fine silt and fine dust present shall not exceed five per cent by weight.

Surkhi is used as an admixture to lime or cement as it imparts some strength and hydraulic properties to the mortar and concrete. To develop more strength, it should be ground very finely with lime in the mortar grinding mill. *Surkhi* mortar is a mixture of lime, *surkhi* and water. Avoid use of surkhi mortar for external plaster or pointing work due to its disintegration under air and humidity conditions.

Surkhi also makes the concrete more resistant to alkalis and salt solutions. The addition of *surkhi* is accompanied by slight reduction in strength as it attains its full strength after a long period of time. Higher shrinkage than ordinary concrete is another property of *surkhi* concrete. *Surkhi* is also used for reduction of temperature rise during hydration in the mass of concrete and it is also used to reduce cracking in concrete. Concrete which is made by adding *surkhi* is more plastic, bleeds less and segregates less as compared to ordinary concrete. As hydration of cement proceeds, lime is liberated, which reacts with *surkhi* to give dense, compact, and impermeable concrete. *Surkhi* is also used as the main constituent in waterproofing treatment along with lime.

3.7 Mortar

Mortar is being used as a binding material from the dawn of civilization. Water is added to mixture of cement/lime and fine aggregate (sand/*surkhi*/ sawdust, *etc.*) to form mortar. Lime and cement acts as a binding material in the mortar. The durability, quality and strength of mortar mainly depends upon the quality and quantity of the cement or lime. The strength of mortar decreases as the proportion of sand increases. Mortar is used in building construction to bond bricks, stones, tiles, or concrete blocks into a structure. It can also add a decorative pattern in brick or stone masonry. In the order of preference cement mortar, lime mortar or mud mortar may be used. Mud mortar should not be used for the structures where strength and permanence is desired.

Classification of Mortars on the Basis of Kind of Binding Material

On the basis of kind of binding material, the mortars are classified as lime mortar, *surkhi* mortar, cement mortar, gauged mortar or gypsum mortar.

Lime mortar exhibit high plasticity and durability. The high plasticity helps in its easy placement. Although, it takes longer time to stabilize. A good mix is considered to have the ratio of 1:2 (lime: sand). It is usually used for lightly loaded above-ground parts of buildings. *Surkhi* mortar is prepared by using *surkhi* instead of sand. It is used for ordinary masonry work of all kinds in foundation and superstructure. But, *surkhi* mortar cannot be used for plastering or pointing work since *surkhi* is likely to disintegrate after some time.

Cement mortar is the strongest type of mortar and is therefore preferred for use in a different construction of structures subjected to heavy loading. The ratio of cement to sand by volume should normally be 1:3 to 1:6 for the mortar to be workable and strong. The use of mortars of lower cement content is not satisfactory since any notable reduction in cement contents leads to a reduction in workability and less cohesion and will produce porous joints with a tendency for low frost resistance. 1: 8 cement mortar is nearly twice as strong as 1: 3 lime mortar. Cement mortar when used for protective plaster provides a waterproof layer and protects the elements covered from weathering effects. Cement and sand are mixed in a dry state on a watertight platform or steel trough. It is advised to mix two to three times the dry mix. Water is then added, and the ingredients are thoroughly mixed.

Gauged mortar is prepared by adding cement to lime mortar. Addition of cement improves the quality of lime mortar and to achieve early strength. The usual proportion of cement to lime is about 1:6 to 1:8. It shall be used for bedding and for thick brick walls. Gypsum mortar is prepared by using gypsum as binding material.

Table 3.3. Proportions of cement mortar

Nature of Work	*Proportion of Mortar*
Masonry construction	
For ordinary masonry work with brick/stone as a structural unit	1:3 to 1:6
For reinforced brickwork	1:2 to 1:3
For all work in moist situations	1:3
For architectural work	1:6
For load bearing structures	1:3 or 1:4
Damp-proof courses	1:2
Plaster Work	
For external plaster and ceiling plaster	1:4
For internal plaster	1:6
For ceiling	1:3
Flooring Work	
Mortar ratio	1:4 to 1:8
Pointing Work	
Mortar ratio	1:1 to 1:2

Mud Mortar: The soil used for mud mortar should be clayey, free from stones, roots, and other foreign materials. The soil should be soaked in water at least 12 hours before its use.

Properties of a Good Mortar

It is always desirable to use good quality mortar in constructions. Therefore, the properties of a good mortar must be known. Generally, good mortar possesses following properties:

1. Good mortar should provide good adhesion with the building units such as bricks, stones, *etc.*
2. Good mortar should be water resistant.
3. Deformability of mortar should be low.
4. Mortar should be cheap.
5. Mortar should be easily workable in the site condition.

6. The mobility of mortar should be good. It helps the mortar to be paved thinly and evenly.
7. It should possess high durability and keep contact material free of any deterioration.
8. Good mortar shall harden quickly to maintain the speed of construction.
9. Cracks should not be developed in the joints formed by mortar. Mortar should be able to maintain their appearance for long period of time.

Uses of Mortar

Some of the numerous functions of mortar in construction are given below.

1. Mortar is used to bind together the bricks or stones in brick or stone masonry.
2. It is used to give a soft even bed between different layers of brick or stone masonry and equal distribution of pressure over the bed.
3. It is used for pointing and plastering work on exposed surfaces of masonry and to improve the general appearance of structure.
4. It is used in concrete as a matrix to hold coarse aggregates.
5. It is used for moulding and ornamental purpose.
6. It is used to fill up the cracks of the structure.

Precautions Required while Using Mortar Mix

In order to get the best mortar with the available ingredients, some precautions are to be taken. These are:

1. **Consumption of Mortar:** Mortar should be mixed in small quantities so that it can be used conveniently before the mortar starts setting. The mortar which has set already should never be used. If mortar contains lime as an ingredient, it should be used within 36 hours after its preparation, and it should be kept wet or damp. In the case of cement mortar, it should be consumed within 30 minutes after adding water. Thus, it is advisable to prepare cement mortar of one bag at a time. The gauged mortar or composite mortar should be used within 2 hours of the addition of cement.
2. **Soaking of Bricks/Stones:** In masonry construction work, usable bricks or stones should be soaked in water for at least 12 hours before they are used in work with mortar. It will prevent the absorption of moisture from the mortar by the building units before it sets. Otherwise, the mortar will become weak due to lack of water required for hydration.
3. **Workability:** Avoid excess water in mortar to maintain its stiffness. Stiff mortar can be used without any inconvenience. About 22.5 litres of water per bag of cement (50 kg bag) is required for making the mortar suitable consistency.

4. **Quality of Water**: Fresh water should be used in mortar mix preparation. The water should be free from oils, dust, *etc.* Use of seawater in mortar mixture should be avoided. It is not advisable to use sea water in making pure lime mortar or *surkhi* mortar because it will lead to efflorescence.
5. **Sprinkling of Water:** After carrying out the construction work, mortar works should be kept wet by sprinkling water for about 7 to 10 days. It will prevent the rapid drying of mortar, especially in hot weather.
6. **Effect of Frost:** Presence of frost affects the setting of mortar. Else use the cement mortar that will harden or set before the freezing.

3.8 Aggregates

Aggregates are the inert mineral materials used for the manufacturing of mortars and concrete. These are sand, gravel, crushed rock *etc.* Neat cement is very rarely used in construction works as it is liable to shrink too much and cracks on setting. Also it will be costlier to use neat cement. Thus, cement is mixed with some inert, strong and durable hard materials such as sand, gravel, crushed rock, *etc.*, to avert shrinkage and to attain greater crushing strength. These inert materials form approximately 75 per cent volume of the whole mass of concrete.

A good quality aggregate should be chemically inert, clean, sufficiently strong, durable, and hard. It should have limited porosity. The surface of the aggregates should be rough and shape should be preferably spherical or cubical. The aggregates should be completely free from lumps of clay, organic and vegetable matter, fine dust *etc.* The presence of debris in concrete hinders the adhesion of aggregates, thereby reducing its strength.

The aggregates which pass through 75 mm IS sieve and entirely retain on 4.75 mm IS sieve are known as coarse aggregates. Thus, the range of the particle size of coarse aggregate is 4.75 mm to 75 mm. The aggregates which pass through 4.75 mm IS sieve and entirely retain on 0.075 mm (75 micron) IS sieve are known as fine aggregates. Generally, the natural river sand is used as a fine aggregate. The material having particle size, varying from 0.06 to 0.002 mm is termed as silt and that particle size less than 0.002 mm is known as clay. The specific gravity of majority of aggregates varies between 2.6 to 2.9.

The function of using fine aggregate in a concrete mix is to fill up the voids existing in coarse aggregates, and to obtain a dense and strong concrete with less quantity of cement. Use of fine aggregates helps the cement in binding the coarse aggregates and increases the workability of the concrete mix.

The size of aggregates which are commonly used for concrete works is 40 mm, 20 mm and 10 mm. The aggregates of 40 mm size are useful for massive concrete works such as dams, retaining walls, *etc.* For reinforced concrete work, aggregates having a maximum size of 20 mm are usually considered satisfactory. The aggregates of 10 mm size are useful for flooring.

Gravel

Gravel (coarse aggregate) is a loose aggregation of rock fragments. Gravel occurs naturally throughout the world as a result of sedimentary and erosive geologic processes. It is also produced in large quantities commercially as crushed stone. Gravel particle sizes range from 4.75 mm to over 60 mm, and are available in a range of colours, textures and stone types. Gravel or broken stone aggregates with rough and non-glassy texture are the best aggregates because they create a good bond with the cement paste.

Gravel has a wide range of applications in the construction industry. Gravel along with sand, is used for the manufacture of concrete, as well as for mixing with asphalt as part of road construction, and in the production of construction materials like concrete blocks, bricks, and pipes. It can be used as the base layer for roads before being covered with tarmac, and is also commonly used to surface roadways, especially those in rural areas and in icy conditions. It can also be used to as part of roof coverings. Gravel is commonly used for landscaping applications, such as on driveways, walkways, or as a decorative filler over soil instead of grass. It is also used to make roofing shingles, used on icy roads in the winter, for railroad ballast, and water filtration.

3.9 Concrete

Concrete is a structural material composed of sand and coarse aggregate (gravel) bonded together by cement (or lime) and water that hardens over time. Concrete is the second-most-used substance in the world after water, and is the most widely used building material. It is suitable for foundations, pillars, beams, walls, floors, water tanks, *etc.* It is stronger and more durable than brick masonry. Cement concrete is commonly used for building as it has more strength and durability as compared to lime concrete. The cement concrete containing no reinforcement is called plain concrete. Plain concrete is strong in taking compressive stresses but is weak in taking tensile or shear stresses. For taking tensile, excessive compressive or shear stresses, reinforced cement concrete (RCC) is used, in which the reinforcement is embedded.

Properties of Cement Concrete

The salient properties of cement concrete are as follows:

1. It has high compressive strength. The strength of normal concrete varies between 25 and 40 MPa. However, it is low in tensile strength. Thus, the steel reinforcement is placed in cement concrete to take up the tensile stresses.
2. It is extremely resistant to the physico-chemical attack emanating from the environment (frost, rain, atmospheric pollution, *etc.*). It is particularly well-suited for structures exposed to demanding and extreme conditions.

3. The concrete hardens over time, continuing to harden after it has reached sufficient strength.
4. It can be easily moulded into durable structural items of desirable shape and size.
5. It is more economical than steel. Also the maintenance cost is negligible.
6. It has a tendency to be porous. This is due to the presence of voids. It is important to grade and consolidation of aggregates. In addition, maintain a minimum ratio of water and cement.
7. It has a tendency to shrink. This tendency of cement concrete can be minimized by proper curing of concrete.
8. It forms a hard surface, capable of resisting abrasion.
9. Concrete buildings are more resistant to fire than those constructed using steel frames, since concrete has lower heat conductivity than steel.

Materials used in RCC Work

Cement, aggregates (sand and gravel), steel, and water are required for making reinforced cement concrete (RCC). Ordinary portland cement is used for the cement concrete work in building construction. The aggregates should be hard, durable and clean. These should be completely free from lumps of clay, organic and vegetable matter, fine dust, *etc.* as their presence prevents adhesion of aggregates and thus, reduces the strength of concrete. Usually, round bars of mild steel are used for steel reinforcement. These are available in various sizes. Water should be clean and free from harmful impurities such as oil, alkali, acid, *etc.* Impure water can cause problems when setting or in causing premature failure of the structure. Water suitable for drinking should be used for the preparation of concrete.

To make concrete, a paste of portland cement (10-15 per cent) and water (15-20 per cent) is prepared. This paste is then mixed with aggregates (65-75 per cent) such as sand and gravel, or crushed stone. As the cement and water mix, they harden and bind the aggregates into an impenetrable rock-like mass. The cement paste glues the aggregate together, fills voids within it, and makes it flow more freely. The coarse aggregate should be properly washed and cleaned, and the sand sieved.

Grades of Concrete

The proportions of coarse aggregate, fine aggregate, cement and water should be such that the resulting concrete when fresh should have sufficient workability so that it can be placed in the formwork economically. Also, the concrete so formed should be strong and water tight. The ratio of cement, sand and coarse aggregate are fixed depending upon the nature of work for which concrete is required. The general expression for the proportions of cement, sand and coarse aggregate is 1: n: 2n by volume. The recommended proportions are:

1:1:2 and 1:1.2:2.4 - For very high strength concrete

1:1.5:3 and 1:2:4 - For normal work

1:3:6 and 1:4:8 - For foundations and mass concrete work

As per Bureau of Indian Standards, the concrete is designated in seven grades, namely M10, M15, M20, M25, M30, M35 and M40. The letter M refers to concrete mix and number 10, 15 *etc.* denotes the specified compressive strength of 150 mm cube attained at 28 days, expressed in N/mm^2. Thus, M20 signifies mixture of cement, sand and aggregate which are prepared in such a manner that a cement concrete cube of size 15 cm ×15 cm ×15 cm is formed with compressive strength of 20 MPa (N/mm^2) while examining it after being cured for 28 days. The grades of concrete lower than M15 should not be used in RCC work. M20 and M25 are the most common grades of concrete, and higher grades of concrete should be used for severe, very severe and extreme environments. For lean concrete bases and simple foundations for masonry walls, M5 or M7.5 grades of concrete may be used.

The normal mixes correspond approximately to the different grades are as follows:

M5 - 1:5:10 M7.5 - 1:4:8

M10 - 1:3:6 M15 - 1:2:4

M20 - 1:1.5:3 M25 - 1:1:2

Water-Cement Ratio

The water when mixed with cement enters into chemical action. The chemical reaction set and harden the concrete. The water provides lubrication effect to the cement. Therefore, cement fills the voids of aggregates.

Water-cement ratio is the ratio of the amount of water to the amount of cement (by weight). The strength and quality of concrete primarily depend upon this ratio. Water-cement ratio by weight should be 0.50 to 0.60. Thus, one bag of cement (50 kg) will require 30 litres of water for water-cement ratio of 0.6. The strength of concrete is inversely proportional to the water-cement ratio. A lower water-to-cement ratio yields a stronger, more durable concrete, whereas more water gives a free-flowing concrete with higher slump.

Unit Weight of Concrete

The unit weight of concrete depends on percentage of reinforcement, type of aggregate, amount of voids and varies from 23 to 26 kN/m^3. As per IS 456:2000, the unit weight of plain cement concrete is 24 kN/m^3 (2400 kg/m^3); while for reinforced concrete structures, the unit weight of concrete is 25 kN/m^3 (2500 kg/m^3).

Workability

Workability is the ease or difficulty with which the freshly mixed concrete can be placed, consolidated and finished with minimal loss of homogeneity. Workability is a vital property of concrete and related with compaction as well as strength. The desired workability is not same for all types of concrete. Workability depends on water content, water-cement ratio and aggregate-cement ratio. It is also affected by the grading, shape, texture and maximum size of coarse aggregates to be used in the mixture.

Slump Test

Workability of concrete mixture can be measured by the concrete slump test. The concrete slump test measures the workability or consistency of fresh concrete before it sets. It is commonly performed in the field to check the workability of freshly made concrete. Concrete slump test is carried out from batch to batch to check the uniform quality of concrete during construction.

The slump test is very simple and easy to handle. It also demands comparatively less equipment and can be done in a short period of time. These advantages of slump test have made it very popular all over the world. In the slump test, workability of concrete is not measured directly. Instead, consistency of concrete is measured which gives a general idea about the workability condition of concrete mix.

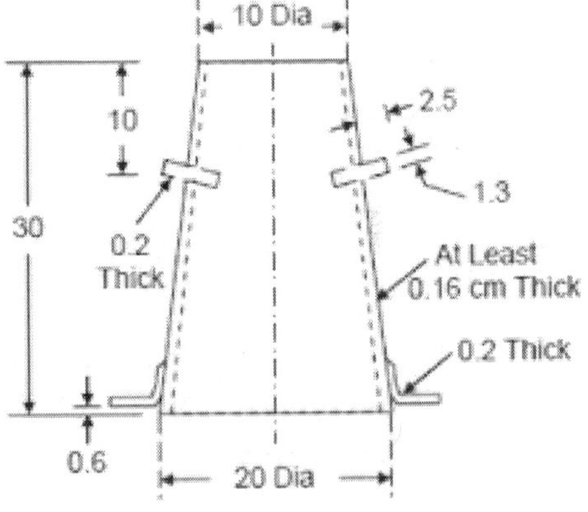

(All dimensions in cm)

Figure 3.3. Slump Cone.

The shape and dimensions of a standard slump cone is shown in Figure 3.3. Firstly, the internal surface of the mould is cleaned carefully. Oil can be applied on the surface. The mould is then placed on a clean, smooth, horizontal and non-porous base plate. The mould is filled with fresh concrete mix in four approximately equal layers. Each layer is tamped 20 to 30 times with a steel rod which is provided with bullet nose at the lower end. The diameter and length of this steel rod are 16 mm and 60 mm, respectively. The tamping should be done uniformly. After filling the mould, excess concrete should be removed and the surface should be levelled with a trowel. When the mould is filled with fresh concrete, the base of the mould is held firmly by handles. The mortar or water leaked out between the mould and the base plate should be cleaned. Then the mould is lifted gradually in the vertical direction and removed. Then unsupported concrete will slump. The decrease in height at the centre point is measured to nearest 5 mm and it is known as 'slump'. Table 3.4 shows the degree of workability on the basis of slump.

Table 3.4. Relation between Workability and Slump Value

Slump	*Degree of Workability*
0 to 25 mm	Very low
25 to 50 mm	Low
50 to 75 mm	Medium
75 to 125 mm	High

Mixing

Concrete materials must be mixed thoroughly, either by hand or using a machine. Hand mixing involves stacking materials on a water-tight platform made of wood, brick, or steel. Mix at least three times in dry conditions before adding water. The prepared mix with water shall be used within 30 minutes. Machine mixing involves feeding all materials, including water, into a revolving drum and rotating it for a certain amount of time. The water should enter the mixer simultaneously or before other materials are placed. The concrete mixer should be thoroughly washed and cleaned after use. The washing and cleaning prevent the formation of hardened concrete cakes inside the rotating drum. Concrete should be used within 30 minutes after discharge from the machine.

Placing of Concrete

The concrete shall be placed on cleaned and well-watered surface. The large quantity of concrete should not be placed at a time. The concrete should be dropped vertically from a reasonable height. For vertical laying, stiff mix should be used. The concrete should be deposited in horizontal layers of about 15 cm height to avoid accumulation of excess water in upper layers. Care should be taken that all edges and corners of concrete surface remain unbroken, sharp and straight in line.

Consolidation of Concrete

Consolidation of concrete means the compaction between aggregate and aggregate; between aggregate and reinforcement; and between aggregate and forms. Consolidation is done to eliminate air voids from the concrete to achieve maximum density. The strength of concrete is reduced by 30 per cent by the presence of 5 per cent of voids. The process of consolidation of concrete can be carried out either with hand or with the help of vibrators. The manual compacting is done by tamping, ramming, spading and slicing with suitable tools such as shovel or steel rod with flattened end. Ramming is stopped when water appears on the upper surface of concrete.

Curing

Concrete must be kept moist for a certain period after placing of concrete in order to achieve optimal strength and durability. Over 90 per cent of a mix's final strength is typically reached within four weeks, with the remaining 10 per cent achieved over years or even decades. The curing protects the concrete surfaces from sun and wind. The presence of water is essential to cause chemical action which accompanies the setting of concrete. Hydration and hardening of concrete during the first three days is critical. Proper curing of concrete leads to increased strength and lower permeability and avoids cracking where the surface dries out prematurely. Care must also be taken to avoid freezing or overheating due to the exothermic setting of cement. Improper curing can cause scaling, reduced strength, poor abrasion resistance and cracking. The period of curing depends on the type of cement and nature of work. For ordinary portland cement, the curing period is about 7 to 14 days. Traditional methods for curing involve spraying or ponding the concrete surface with water. Additional common curing methods include wet gunny bags and plastic sheeting covering the fresh concrete.

The general requirements to produce good quality concrete are:

1. The air bubbles should be removed from the body of the concrete.
2. The cement particles should be of the smallest size.
3. The concrete should be compacted fully in order to remove voids.
4. The concrete should be cured sufficiently and adequately for about 28 days.
5. The water-cement ratio should be kept low.
6. The cubical particles of the aggregates should be used so that good interlocking is obtained.

Yield of Concrete

The quantity of concrete formed from a concrete mix depends on various factors such as water-cement ratio, size of aggregates, compaction, *etc.* The approximate yield of concrete from a concrete mix can be estimated by a rule-of-thumb which is as follows:

$$\text{Yield of concrete} = \frac{2}{3}(a+b+c)$$

where, a, b and c are the parts of cement, sand and coarse aggregate respectively by volume.

Example 3.1

Estimate the yield of concrete per bag of cement for a concrete mix of proportion 1:3:6.

Solution

A 50 kg bag of cement contains 0.035 m^3 of cement.

Thus, **a** = 0.035 m^3

Then volume of sand, **b** = 3×0.035 m^3, and

Volume of coarse aggregate, **c** = 6×0.035 m^3

So, yield of concrete of proportion (1:3:6) = $\frac{2}{3}(1\times 0.035 + 3\times 0.035 + 6\times 0.035)$

$= 0.233 \text{ m}^3$

Estimation of Cement, Sand and Gravel (Fuller's formula)

Fuller's formula is used for the estimation of cement, sand and gravel volumes for concrete proportioning. The volume of cement (in cubic meter) required to prepare a concrete mix of one cubic meter is estimated by the following relationship:

$$C = \frac{1.55}{c + s + g}$$

where,

C = volume of cement per cubic meter of concrete work

c, s, g = cement-sand-gravel ratio (relative amounts of solids by volume in a mixture)

Example 3.2

Calculate the quantities of cement, sand and gravel required to prepare 2 cubic meter of concrete having 1:2:4 as the proportion of aggregates.

Solution

Total quantity of concrete required = 2 m^3

c, s, and g are 1, 2 and 4 respectively.

Therefore, volume of cement required for 1 m^3 of concrete = $C = \frac{1.55}{c + s + g}$

$$= \frac{1.55}{1 + 2 + 4} = 0.22 \text{ m}^3$$

Thus, for 2 m^3 of concrete, material requirement

Quantity of cement required = 0.22 × 2 = 0.44 m^3

Quantity of sand required = 0.44 × 2 = 0.88 m^3

Quantity of gravel required = 0.44 × 4 = 1.76 m^3

One bag of cement (50 kg) has 0.035 m^3 volume

Thus, number of cement bags required = $\frac{0.44}{0.035}$ = 12.57 bags ≈ 13 bags

3.10 Tiles

Tiles are thin slabs of brick which are burnt in kiln. They are thinner than bricks. They are mainly prepared from clay material or any form of inorganic raw material. Tiles are also manufactured from recycled materials, making them eco-friendlier. Different types of tiles used in building construction are available in glazed or unglazed form. They are incombustible in its nature, and unaffected under the light. They are used to cover surfaces like roofs, floors, and walls.

Tiles can be classified into two group *viz.* common tiles and encaustic tiles. Common tiles have different shapes and sizes. They are mainly used for paving, flooring and roofing, while encaustic tiles are used for decorative purposes in floors, walls, ceilings and roofs.

Properties of a Good Tile

1. Well burnt and possess uniform colour.
2. Free from twists, cracks, bends.
3. Regular shape and size.
4. Sound, hard, durable.
5. Clear ringing sound when struck.
6. Even and compact structure of broken tile.

Types of Common Tiles

On the basis of the application or use, there are mainly three types of tiles. These are

1. Roof tiles,
2. Floor or paving tiles, and
3. Drain tiles

Roof Tiles

Roof tiles are used to serve as covering for pitched roof. Various types of roof tiles are available in the country. Most common are Allahabad tiles, corrugated tiles, flat tiles, Guna tiles, Mangalore tiles, *etc.* These tiles are popular in rural and semi-urban areas of the country.

Floor or Paving Tiles

Floor or paving tiles are usually thicker than roofing tiles and vary from 12 mm to 50 mm in thickness. Their shapes may be square, hexagonal or of any other geometrical pattern and may be coloured, if desired. The size of square tiles varies from 150 mm to 300 mm. These tiles should be hard and compact to resist wear and tear in a better way.

Drain Tiles

These tiles retain porous texture after burning, and are used in water-logged areas. In the water-logged areas, these tiles allow subsoil water to pass through them. They are also used to convey irrigation water. The shape of these drains may be circular, semi-circular or segmental. Tiles to be used in the construction of sewage carrying drains should be glazed.

3.11 Steel

Steel is an essential engineering and construction material. It is composed of iron and carbon alloys. The carbon content between 0.25 and 1.5 per cent enhances strength and fracture resistance. Steel also contains 1 per cent manganese and small amounts of silicon, phosphorus, sulphur, and oxygen. It is used widely to build residential buildings, high rise buildings, industrial sheds, parking garages, bridges, *etc.* Even those with a concrete structure employ steel for reinforcing. Steel is used because it binds well to concrete, has a similar thermal expansion coefficient and is strong and relatively cost-effective. The strength-to-weight ratio of steel is more as compared to other materials. Thus, it makes steel buildings stronger and more durable than wood, brick or concrete, despite their light weight. It is more suitable in areas prone to tornadoes, earthquakes, and other natural disasters. In addition to all of these, steel is also safer for the environment as it does not require depletion of woodlands to use for buildings. Structural steel is resistant to external forces such as wind and earthquakes. It can be recycled over and over again without loss of property.

Different Types of Steel

There are four main types of steel, *viz.* carbon steel, alloy steel, tool steel and stainless steel. Carbon steel looks dull, matte-like, and is known to be vulnerable to corrosion. Overall, depending upon the carbon content, there are three subtypes to carbon steel: mild steel, medium carbon steel, and high carbon steel. The carbon content of mild steel is about 0.10 to 0.25 per cent. Mild steel is widely used for building construction. It is also a highly favoured building frame material. While the carbon content of medium and high carbon steel varies from 0.25 to 0.60 per cent, and 0.60 to 1.10 per cent, respectively. High carbon steel is exceptionally strong, which is why it is often used to make things like knives, high-tension wires, automotive parts, and other similar items. With higher carbon content (2 to 4 per cent), the material is defined as cast iron. Alloy steel is a mixture of several different metals, like nickel, copper, and aluminium. These tend to be more resistant to corrosion. Tool steel is famous for being hard and both heat and scrape resistant. They are commonly used to make metal tools, like hammers. These are steels used for tooling activities such as drilling. Commonly made up of molybdenum, vanadium, tungsten, and cobalt, tooling steels are heat-resistant, durable, and strong. Stainless steel is shiny and generally has around 10 to 20 per cent chromium, which is their main alloying element. This combination allows the steel to be resistant to corrosion and very easily moulded into varying shapes. Stainless steel is mostly used in home applications, dairy and food processing equipment, surgical equipment, and as exterior cladding for commercial/industrial buildings.

Market Forms of Steel

1. Angle sections
2. Channel sections
3. Expanded metal
4. Corrugated sheets
5. I-sections
6. T- sections
7. Plates flat bars
8. Ribbed-tor steel bars
9. Round bars
10. Square bars
11. Flat bars
12. Thermo-mechanically treated bars (TMT bars)

1. Angle Sections

The angle sections may be of equal legs or unequal legs (Figures 3.4 and 3.5). The equal angle sections are available in sizes varying from 20 mm × 20 mm × 3 mm to 200 mm × 200 mm × 25 mm. The unequal angle sections are available in sizes varying from 30 mm × 20 mm × 3 mm to 200 mm × 150 mm × 18 mm. The weight per meter length for different sizes of angle sections is given in Appendix B and C. The angle sections are broadly used in the structural steelwork particularly in the construction of steel roof trusses and filler joist floors.

2. Channel Sections

The channel section is a web with two equal flanges (Figure 3.6). The size of the channel is determined by the height of the web and the width of the flange. The size of the channel section ranges from 100 mm × 45 mm to 400 mm × 100 mm. The channel sections are widely used as the structural members of the steel framed structures. The corresponding weights per metre length are given in Appendix D.

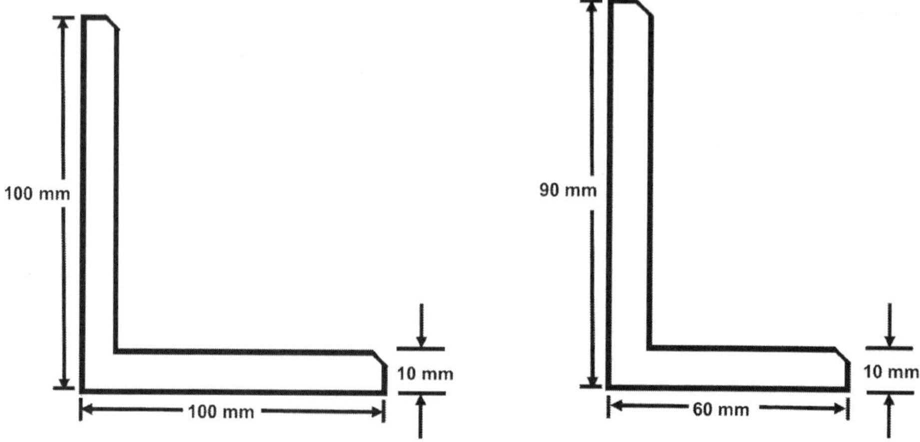

Figure 3.4. Equal Angle Section.

Figure 3.5. Unequal Angle Section.

3. Expanded Metal

Expanded metal is prepared from sheets of mild steel which are machine cut and drawn out or expanded (Figure 3.7). A diamond mesh appearance is thus formed throughout the whole area of the sheet. This is widely used for reinforcing concrete in foundation, roads, bridges *etc.* It is also used as lathing material and for partitions. This is available in different shapes and sizes.

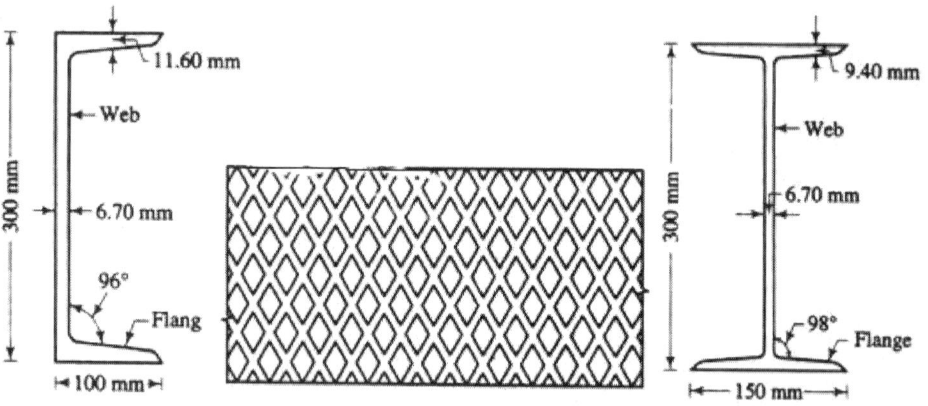

Figure 3.6. Channel Section.

Figure 3.7. Plain Expanded Metal.

Figure 3.8. I-section.

4. Corrugated Sheets

Corrugated sheets are formed by passing steel sheets through grooves. These grooves bend and press steel sheets and corrugations are formed on the sheets. These corrugated sheets are generally galvanized and described as the galvanized sheets or GI sheets. These sheets are used for roof covering.

5. I- Sections

Rolled steel joists, also known as beams, are a type of structural steel used in various applications, such as floor beams, lintels, and columns (Figure 3.8). It consists of two flanges connected by a web and is classified by depth, width, and weight per metre of length. BIS categorises I-section beams into junior, light, medium, wide-flange, and heavy beams (H-section). The size as well as weight per meter of each type of beams are given in Appendix E.

6. T-Sections

The section resembles a letter 'T' and consists of a flange and web (Figure 3.9). The size of the T section is determined by its overall dimensions and thickness. It is available in various sizes. Special T-sections with unequal sides, bulbs at the bottom edge of web are also available in the market. These sections are commonly used in steel roof trusses and built-up sections. The details of size, flange and web thickness along with weight per meter are given in Appendix F.

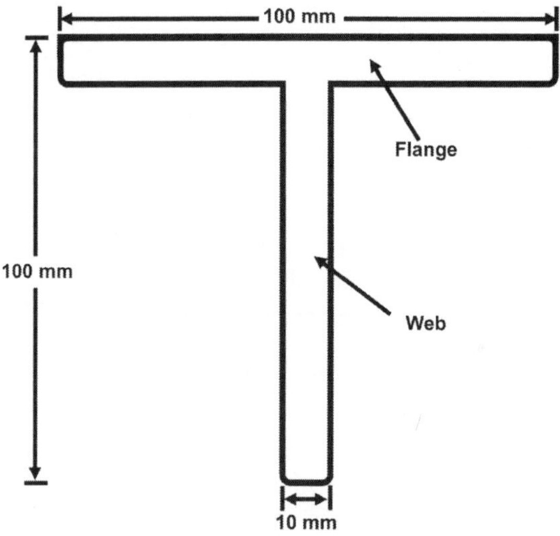

Figure 3.9. T-section.

7. Plates

The plates section of steels is available in different sizes. The thickness of plates varies between 5-63 mm. The standard nominal thickness of plates in mm are 5, 6, 8, 10, 12, 14, 16, 18, 20, 22, 25, 28, 32, 36, 40, 45, 50, 56 and 63. The plates are used mainly for the following purposes in the structural steel works:

1. To connect steel beams for extension of the length;
2. To serve as tension members of steel roof truss; and
3. To form built-up sections of steel.

8. Ribbed-Tor Steel Bars

Ribbed-tor steel, also known as High Yield Strength Deformed (HYSD) bars, is a deformed high-strength steel with ribs or projections on its surface. The ribs, or projection, increase the strength of the bars. It is produced by the controlled cold twisting of hot-rolled bars. These bars, available in sizes ranging from 6 to 50mm in diameter, are widely used as reinforcement in concrete structures like buildings, bridges, roads, irrigation works, pile foundations, and pre-cast concrete works. The weight per metre for 6 mm HYSD bar is 2.22N and for 50 mm is 154.10N. They can be bent through 180° without cracks or fractures and can reduce reinforcement costs by 30–40 per cent. Ribbed bars possess excellent bonding properties and better structural properties than ordinary plain round bars, making them efficient and economical reinforcement. Ribbed mild steel bars, which resemble HYSD bars but have lower allowable stresses, are also available in the market. Avoid the use of ribbed mild steel bars in RCC work.

9. Round Bars

These are available in circular cross-section with diameters varying from 5 mm to 250 mm. The round bars are widely used as reinforcement in concrete structures, construction of steel grill work *etc.* The standard dimensions and weight per meter of round bars are given in Appendix G.

10. Square Bars

These are available in square cross-section with sides varying from 5 mm to 250 mm. The square bars are widely used in the construction of steel grill work for window, gates, *etc.* The standard dimensions and weight per meter of square bars are given in Appendix G.

11. Flat Bars

Flat bars are available in suitable widths varying from 10 mm to 400 mm with thickness varying from 3 mm to 40 mm. They are widely used in the building of steel grill work for windows and gates (Appendix H).

12. Thermo-Mechanically Treated Bars (TMT bars)

Sudden quenching of red hot steel bars by a spray of water can produce steel bars with high strength at the structure with a core of mild steel. The combined strength of these materials raises the yield point of steel with the high percentage of elongation at ultimate failure. TMT bars are also rolled with ribs to increase the bond strength. TMT bars are more corrosion resistance than cold twisted bars.

The rolled steel sections are also available in miscellaneous sections such as acute and obtuse angle sections, rail sections, trough sections, and z-sections, which are not commonly used in the structural steelwork.

Calculation of Weight of Steel

Steel calculation helps in determining the approximate required quantity or weight of steel bars needed for the construction projects. One can calculate weight of different types of steels bars, round/circular solid bars, square shaped solid bars, hexagonal solid bars, metal sheets, H- beam, I- beam *etc.* by using following two methods:

1. By using density of steel
2. By using the standard formula (for round/circular solid steel bars)

1. By using Density of Steel

In this method, the weight of steel is calculated by multiplying the density of the steel by the volume of steel.

i.e., Weight of steel = density of steel × volume of steel

The standard density of steel is taken as 7850 kg/m^3.

Volume of steel depends upon the shape of the sections.

For Rectangular Section (Flat bars)

Volume = length (L) × width (W) × thickness (t)

Weight of steel flat bar = volume of steel bar × density of steel

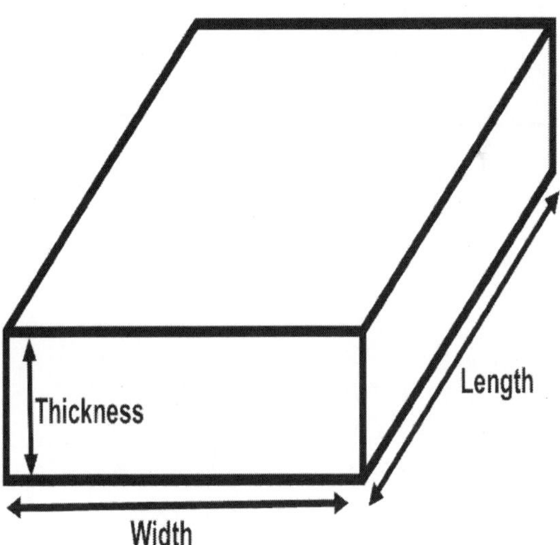

Example 3.3

Determine the unit weight of MS flat bar of width 20 mm and thickness of 10 mm.

Solution

Weight of steel flat bar = volume of steel bar × density of steel

= (length × width × thickness) × 7850

Unit Weight of steel bar = (1 × 0.020 × 0.010) × 7850

= 1.57 kg/m ≈ 1.6 kg/m

Example 3.4

Calculate the theoretical weight of the rectangular steel section having a length of 4 m, a width of 2 m, and the thickness of 40 mm.

Solution

Weight of steel flat bar = volume of steel bar × density of steel

= (length × width × thickness) × 7850

= (4 × 2 × 0.040) × 7850

= 2512 kg

2. By using the Standard Formula (For round/circular solid steel bars)

The weight of round bars and cylindrical section can be calculated by a standard formula.

Weight of round steel bar = volume of steel bar × density of steel

$$= \left(\frac{\pi}{4} \times d^2 \times L\right) \times 7850$$

Where, d and L are respectively the diameter in mm and length of bar in meter.

Thus, weight of L meter length of round bar $= \left(\frac{\pi}{4} \times \frac{d^2}{1000 \times 1000} \times L\right) \times 7850$

$$= \frac{d^2 L}{162.28} \text{(kg)}$$

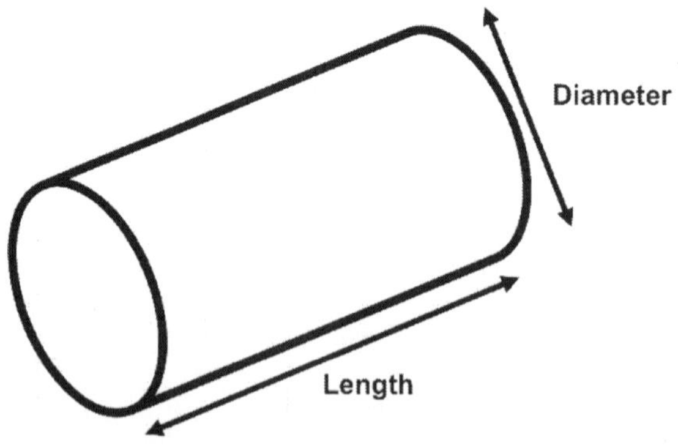

Example 3.5

Calculate the weight of a round steel section having a diameter of 10 mm and a length of 10 meter.

Solution

Diameter (d) = 10 mm

Length (L) = 10 m

By using standard formula,

$$\text{Weight of round steel section} = \frac{d^2L}{162.28} = \frac{(10)^2 \times 10}{162.28}$$

$$= 6.16 \text{ kg} \approx 6.2 \text{ kg}$$

Same problem can be solved by using density of steel.

Weight of round steel bar = volume of steel bar × density of steel

$$\text{Weight of round steel bar} = \left(\frac{\pi}{4} \times d^2 \times L\right) \times 7850$$

$$= \left(\frac{\pi}{4} \times \frac{10^2}{1000 \times 1000} \times 10\right) \times 7850$$

$$= 6.16 \text{ kg} \approx 6.2 \text{ kg}$$

In this way, one can easily calculate the weight of steel having different diameters.

3.12 Timber

The word timber is derived from an old English word, *timbrian* which means to build. The timber denotes wood which is suitable for building or carpentry or various other engineering purposes. It is applied to trees measuring not less than 60 cm in girth (circumference) of the trunk. The salient properties of timber are low thermal conductivity, amenable to mechanical working, small bulk density, and relatively high strength. Timber is categorized amongst the world's most eco-friendly building solution. In addition, timber is renewable as it continually grown in plantations and forests. It has some drawbacks also like fluctuations in properties due to changes in moisture content, susceptibility to decay and inflammability, variations in strength in length and across fibres, *etc.*

Characteristics/Qualities of a Good Timber

The quality of timber depends upon a number of factors. These are environmental conditions of the locality, maturity of the tree, method of seasoning, nature of soil, process of preservation, and time of felling. The good quality timber should have following characteristics:

1. The dark and uniform colour of timber indicates high strength.
2. Straight fibres with uniform structure and closely located regular annual rings are desired.
3. Free from serious defects like dead knots, flaws, and shakes.
4. An unpleasant smell indicates decayed timber. A sweet smell is desired.
5. Heavy weight indicates the soundness and strength of timber.
6. The absence of a clear ringing sound when struck indicates decayed timber.
7. Timber shall be durable and hard and resist the actions of fungi, insects, chemicals, and mechanical wear.
8. The timber should be elastic in nature. It should return to its original shape when load causing its deformation is removed.
9. The timber should be fire resistant. A dense wood offers good resistance to the fire as it requires sufficient heat to cause a flame.
10. Retains shape during conversion or seasoning without bowing, warping, or splitting.
11. It shall offer resistance to vibration shocks.
12. Strong for structural members like joists, beams, and rafters.
13. Timber shall take loads slowly or suddenly, with sufficient strength in both directions.
14. Timber shall show the least surface disintegration due to weathering.
15. The timer shall be workable by machine and manually.

Disadvantages of Timber Construction

The few disadvantages are that they require regular maintenance, fire risk, costly, requires seasoning and treatment with preservatives, and not suitable for cyclonic weather.

Seasoning of Timber

The process of removal or drying out the moisture from timber is known as seasoning. A newly fell tree contains about 50 per cent or more of its dry weight as water. Therefore, it is necessary to remove the water before the timber can be used for any engineering purpose. The timber contains free and bound moisture. Free moisture is present in timber as water vapour while bound moisture is present in cell walls. When it is allowed to seasoning, free moisture evaporated first and this point is called as fibre saturation point. After fibre saturation point, the timber will shrink on drying which is nothing but evaporation of bound moisture. There are two methods of seasoning timber, namely natural (air) and artificial (kiln) drying. Timber is stacked in both methods in such a way that allow full circulation of air. Natural seasoning is the method used with the timber stacked in the open air. Water should be removed at a uniform rate to prevent damage to the wood during drying. Seasoning increases dimensional stability (maintain the size and shape of timber), improve strength, hardness and stiffness of timber. Seasoning reduces or eliminate attack by decay due to fungi. It also makes timber suitable for receiving various treatments like paints, preservatives, varnishes *etc.* Thus, reducing the moisture in the timber used in construction projects allows for a higher quality of finish on the constructions. A well-seasoned timber has 10 to 12 per cent moisture content in it.

Uses of Timber

Timber is a natural and most widely used construction material extracted from trees. Wood has been used for centuries because of its versatile properties, its diversity and aesthetic qualities, which makes it suitable for a wide range of application. Timber as a building material is used for frame and shutters of doors and windows, flooring, roofing, cabinet, cupboards shelves, furniture, scaffolding, *etc.* Timber has good strength and hence also used for making load-bearing members like beams, columns, trusses and piles. Timber is also popularly used in the form of plywood and raw wood. Products like ply blocks and ply boards. It provides good insulation from the cold.

Timber should be kept away from damp earth and damp walls in order to prevent rotting. Timber columns should be supported on concrete or masonry bases to save timber from termite attack.

Indian Timber Trees

Over 150 species of timber are produced in India. Some of the common timbers which are used for the engineering purposes in the country are *teak, deodar, kali, chir, sissoo* (*shisham*), *sal, babul, oak, mango, simul, bamboo, bakul, banyan, siris*, and *mahogany*. The timber commonly used in India for superstructure of permanent buildings are *teak, sal* and *deodar*. However, the best variety available locally, may be used. Salient properties of some of the timbers are given.

Teak (Sagwan): In India, teak is the most commonly used timber. It is commonly found in central and southern India. It is hard, strong and durable. Its colour is deep yellow to dark brown. It is yellow when fresh and brown on seasoning. It is commonly used for various structural works like house and costly furniture. It takes up good polish. It can be easily seasoned and does not corrode the iron fastenings. It is identified by its distinct smell. It is fire resistant and also shrinks a little. It resists the acidic action and action of white ants. It has more weight (760 kg/m^3). Its use is limited to superior work only as it is comparatively very costly.

Sal: Its colour is yellow or light brown or dark brown. It is strong, hard, fibrous and close-grained. It does not take up good polish. It requires slow and careful seasoning. It is also a heavy type timber (800 kg/m^3). It is used for railway sleepers, wagons, ship building, bridges, structural work *etc.*

Deodar (Pine): It is a light coloured timber. Its yellow colour becomes darker when exposed to atmosphere. It gives sweet or pleasant smell. It is hard, strong and compact. It can be easily worked and are more moderately strong. It is very durable and polishes well. Its weight is about 560 kg/m^3.

Sissoo (shisham): Its colour is dark brown for heartwood and light brown for sapwood. The colour darkness on exposure. It is hard, durable and tough. It is a high class timber and has good appearance. It maintains its shape well and can be easily seasoned, but requires a fine polish to be effective. It is also a heavy timber (880 kg/m^3). It is used for high class furniture and a very good material for decorative works and carvings.

Bamboo: Bamboo is flexible, very strong, durable, and fastest-growing renewable natural building material. The compressive strength of bamboo is two times higher than concrete, while the tensile strength is close to steel. Its fibre has a higher shear stress than wood. It is available in most parts of the country and grows abundantly in hot humid tropics. It is cost effective and easy to work. Its enormous elasticity makes it a very useful building material in areas with very high risks of earthquakes. It is used for thatched roofs, walls, scaffolding, rafters, temporary bridges, fences, *etc.* It is also being used as reinforcement in cement concrete. Bamboo mats with a well laid out cement mortar layer provides a good roofing material for humid regions.

Plywood: Plywood is a board made from thin layers of wood or veneers. The three or more veneers were placed one above the other at right angles. Adhesive is applied to maintain their position. This increases the plywood's longitudinal and transverse strengths. Multi-ply plywood, made from more than three piles, can have thicknesses ranging from 6–25 mm or more.

Advantages of Plywood

The advantages of plywood are as follows:

1. The expansion and shrinkage are comparatively very low as plies are placed at right angles to each other.
2. The elasticity of plywood prevents split/cracks due to changes in the atmosphere.
3. Availability in a variety of decorative appearance.
4. Availability in large sizes (widths up to 150 cm and length up to 300 cm) and thickness.
5. Light in weight and also not easily affected by moisture.
6. Easy to work.
7. They do not split in axial direction due to their cross grained nature when nailed.
8. Possess uniform tensile strength in all directions.
9. Uses timber in economical way.

10. A three-ply board is significantly stronger than a solid board of the same thickness.

3.13 Paints, Varnishes and Distempers

Paints are fluid coatings applied to timber and metal surfaces for their protection. Varnish is a transparent or nearly transparent resinous solution used to coat the painted surfaces. The protective power of paints, varnishes, and distempers decreases over time and should be applied periodically at intervals of 2–5 years. Distempers are applied to plastered surfaces.

Paints

Paints shield surfaces from weathering, liquids, fumes, and gases. It prevents wood decay and metal corrosion. They enhance hygiene, cleanliness, and attractiveness while providing a smooth, easy-to-clean surface.

Quality Attributes of an Ideal Paint

The ideal paint should have following characteristics:

1. **Spreading Power**: It should possess a good spreading power. It should cover maximum area of the surface with minimum quantity of the paint. The paint should form a thin film of uniform nature.
2. **Cost**: It should be cheap and economical.
3. **Weathering Action**: It should not be affected by atmosphere.
4. **Ease in Use**: Paint should be easily and freely applied on the surface.
5. **Drying:** The paint should not be dried rapidly. It should be dried in reasonable time.
6. **Durability**: Colour should be maintained for a long time. The paint should form a hard and durable surface. It should possess attractive and pleasing appearance. It should not show cracks when the paint dries.
7. **Side Effects**: The paint should not affect health of workers during its application.

Varnishing

Varnish is applied on wooden surfaces. Varnish is the solution of resins or resinous substances prepared either in alcohol, oil, or turpentine.

Important Points to Remember

1. The crushing strength of stones should be greater than 100 N/mm^2.
2. The specific gravity of good building stone is between 2.4 and 2.8.
3. For a good stone, water absorption should be as small as possible and in no case more than 0.60 per cent.
4. The stones should be well seasoned before putting into use.

5. The colour of the well-burnt bricks is usually red.
6. The weight density of standard bricks varies from1600 to 1900 kg/m^3.
7. As per BIS, the standard size of brick is 19 cm × 9 cm × 9 cm.
8. According to BIS, an ordinary type building brick must possess a minimum compressive strength of 35 kg/cm^2.
9. The first and second class bricks shall have a compressive strength not less than 140 kg/cm^2 and 70 kg/cm^2, respectively.
10. Hollow bricks are used for insulating heat flow.
11. The bricks should give a clear metallic ringing sound when struck with each other.
12. The bricks should not absorb water more than 20 per cent by weight for first class bricks and 22 per cent by weight for second-class bricks, when soaked in cold water for a period of 24 hours.
13. The indentation marks left on bricks during the process of moulding are known as frog.
14. A good quality brick should not contain any soluble salts in it.
15. White chalk is pure limestone and *kankar* is an impure milestone.
16. The bulk density of cement is 1440 kg per cubic meter.
17. The volume of cement in each 50 kg bag is 0.035 cum.
18. The initial setting time should be about 30 minutes and final setting time should be about 10 hours for ordinary cement.
19. The strength of cement reduces with time.
20. 33-grade cement means that the compressive strength of the cement after 28 days is 33 N/mm^2 when tested as per Indian Standards under standard conditions.
21. The storage space requirement shall be 1 cubic meter for storing of 20 bags or 1000 kg of cement.
22. *Surkhi* is an artificial pozzolanic material produced by powdering well burnt bricks or burnt clay balls.
23. About 22.5 litres of water per bag of cement (50 kg bag) is required for making the mortar suitable consistency.
24. After carrying out the construction work, mortar works should be kept wet by sprinkling water for about 7 to 10 days.
25. The maximum percentage of ingredient in cement is that of lime.
26. The minimum percentage of ingredient in cement is that of MgO.
27. The range of the particle size of coarse aggregate is 4.75 mm to 75 mm.
28. The aggregates which pass through 4.75 mm IS sieve and entirely retain on 0.075 mm IS sieve are known as fine aggregates.
29. The strength of normal concrete varies between 25 and 40 MPa.

30. Concrete is low in tensile strength. Thus, the steel reinforcement is placed in cement concrete to take up the tensile stresses.
31. The number of cement bags required per cubic meter of concrete (1:2:4) would be 5 to 6.
32. Full strength of concrete is achieved after 28 days.
33. Curing of concrete is a process of strengthening through hydration.
34. In a 1:3:6 concrete mixtures, the sand is 3/10 of the mixture.
35. M20 grade signifies mixture of cement, sand and aggregate which are prepared in such a manner that a cement concrete cube of size 15 cm ×15 cm ×15 cm is formed with compressive strength of 20 MPa (N/mm^2) while examining it after being cured for 28 days.
36. Slump test for concrete is carried out to measure the workability.
37. Slump cone is 30 cm high having diameter of 20 cm at base and 10 cm at the top.
38. M15 grade corresponds to 1:2:4 concrete mixture.
39. Water-cement ratio by weight should be 0.50 to 0.60.
40. As per IS 456:2000, the unit weight of plain cement concrete is 2400 kg/m^3; while for reinforced concrete structures, the unit weight of concrete is 2500 kg/m^3.
41. The carbon content of mild steel is about 0.10 to 0.25 per cent.
42. Carbon content of medium and high carbon steel varies from 0.25 to 0.60 per cent, and 0.60 to 1.10 per cent, respectively.
43. Standard density of steel is taken as 7850 kg/m^3.
44. The colour of timber should preferably be dark and uniform.
45. The timber should have heavy weight which is considered to be sound and strong.
46. The percentage of moisture in a freshly felled tree may be up to 100.
47. The timber commonly used in India for superstructure of permanent buildings are *teak, sal* and *deodar*.
48. The wood that has maximum resistance against white ants is teak (*sagwan*).
49. The moisture content of a well-seasoned wood is in the range 10 to 12 per cent.
50. An ideal paint should have a good spreading power.

Exercises/Questions

1. What are the main uses of stone in building construction?
2. Briefly discuss the properties of a good building stone.

3. Why seasoning of stone is required?
4. Explain the importance of lime as a building material.
5. Briefly discuss the classification of lime.
6. Explain briefly uses of lime.
7. What are the properties of good cement?
8. Briefly discus the ingredients of cement.
9. What are the uses of cement?
10. What are the different grades of cement and how is it stored?
11. Discuss the difference between cement and lime.
12. What are the constituents of good bricks?
13. What are the qualities of a good brick?
14. Discuss the classification of bricks on the basis of quality.
15. What are the characteristics of a good tile?
16. Define cement concrete. What are the ingredients of concrete? On what factors strength of concrete depends? What is the significance of sand in concrete?
17. What is slump test and explain the factors affecting workability of concrete? What is its significance in building construction?
18. What is water-cement ratio and what is its importance in concrete?
19. How is concrete cured and why is curing required for concrete?
20. Explain Fuller's formula for estimating the quantities of aggregates for concrete preparation.
21. How sand is classified?
22. What are the properties of a good sand?
23. Discuss briefly about the field tests of sand.
24. How can you distinguish between sands with and without organic matter?
25. What are the properties of a good mortar?
26. What are the properties of a good timber?
27. What is meant by seasoning of timber?
28. How is timber seasoned?
29. What are the uses of steel in building works?
30. What are the different market forms of steel?
31. Why is steel used as reinforcement in RCC?
32. What do you mean by paint? Give the characteristics of an ideal paint.

33. Write short notes on:
    (a) Cement and its classification
    (b) Properties of cement concrete
    (c) Grades of concrete
    (d) Compacting of concrete
    (e) Placing of concrete
    (f) Water-cement ratio
34. Calculate the quantities of cement, sand and gravel required to prepare 4 cubic meter of concrete having 1:3:4 as the proportion of aggregates. Use Fuller's formula.

    [Answer: cement – 0.78 cum, sand – 2.34 cum, gravel – 3.12 cum]
35. Estimate the yield of concrete per bag of cement for a concrete mix of proportion 1:2:4.

    [Answer: 0.163 cum]

Chapter 4

Building Construction

Building construction is the process of creating structures, typically designed to provide shelter, workspace, or other functional purposes. It is a fundamental aspect of human civilization and has evolved significantly over the centuries. It is a complex endeavour, involving various parts and components that come together to create a functional structure. The different parts of a building construction are foundation, walls, flooring, roofing, plastering, doors and windows, *etc.* The details of all these parts are discussed in this chapter.

4.1 Foundation

A foundation is a lowest portion of building below ground level that transfers a load of superstructure to the earth through it. It plays an important role to protect the building from overturning during an earthquake. It also provides strength to bear wind load to the building. Depth and width of foundation generally depend upon bearing capacity of soil, load applied from structure to the foundation, and depth of water level below the ground surface. Soil load bearing capacity depends upon the soil type. The soil with poor load-bearing capacity requires a deeper foundation than the soil with good load-bearing capacity. Foundations are categorized into shallow and deep types. Soil with higher load-bearing capacity typically has a shallow foundation, while soil with less capacity has a pile foundation. Deep foundations are used in special cases, like waterlogged areas or poor soil. Weak soils like black cotton and poor sandy soil require a deeper foundation.

The marking at the site shall be done for excavation of foundation. After digging the trench to the proper depth and width, the bottom should be dressed to a uniform horizontal level. Thereafter, the bed should be watered and well rammed. Concrete may then be poured in layers of 15 cm. Every layer of the concrete shall be rammed. Total depth of concrete shall not exceed 30 cm. Construction of the wall in brick or stone masonry should be done on this concrete block, leaving 7.5 cm on each side. The foundation wall is then raised to a height of 60 cm above the ground level. Stepped foundation is mostly preferred if it is laid with the bricks. Stepped foundations should have twice the depth of the projections. Whereas, the base width shall be twice the layer above it. The projections are typically set at 15 cm on either side. Reinforced concrete foundations are used where the soil is very unstable and the ordinary brick masonry or concrete foundation does not provide a strong and dependable base.

Width of Foundation

Width of wall foundations or size of column footing is determined by first calculating the expected load and then dividing that with safe bearing capacity of soil (SBC). SBC of the soil is the capacity of the soil that supports the structural load to the ground without any shear failure or settlement. The values of SBC for various types of soil is given in Table 4.1.

Thus, the width of foundation is calculated by the following equation:

$$w = \frac{W}{SBC}$$

where,

w = width of the foundation in meter;

W = dead load coming on the foundation per unit length (kg/m); and

SBC = safe bearing capacity of the soil (kg/m^2)

Area of column footing is calculated as follows:

$$\text{Area of column footing} = \frac{\textit{load} \text{ carried by column (kg)}}{\textit{safe} \text{ bearing capacity of soil (kg/m}^2\text{)}}$$

Example 4.1

Calculate the width of the masonry foundation having 6000 kg of dead load per running meter of the wall. Assuming that the building has to be constructed in sandy loam soil having the bearing pressure 10×10^3 kg/m^2.

Solution

Dead load coming on the foundation per unit length (W) = 6000 kg

Safe bearing pressure of the sandy loam soil (SBC) = 10×10^3 kg/m^2

Thus, width of foundation (w) = $\frac{W}{SBC} = \frac{6000}{10 \times 10^3} = 0.6$ m

Example 4.2

The total load on foundation for a two-storey building is 8.5 t/m. It is to be constructed over a soil whose properties are as follows:

Angle of repose and density of soil = 30° and1800 kg/m^3

Soil bearing capacity of soil = 10000 kg/m^2

Calculate the width of the foundation.

Solution

Load coming on the foundation per unit length (W) = 8.5 t/m = 8500 kg/m

Angle of repose of soil (ϕ) = 30°

Weight of soil (ω) = 1800 kg/m^3

Safe bearing capacity of soil (SBC) = 10000 kg/m^2

Thus, width required for foundation (w) = $\dfrac{W}{SBC} = \dfrac{8500}{10000} = 0.85$ m

Table 4.1. Safe Bearing Capacity of Various Types of Soil

Sl.No.	*Soil type*	*Safe Bearing Capacity*
1.	Soft, wet clay or muddy clay	50 kN/m^2 (5000 kg/m^2)
2.	Soft clay	100 kN/m^2 (10000 kg/m^2)
3.	Medium clay	250 kN/m^2 (25000 kg/m^2)
4.	Compact clay	450 kN/m^2 (45000 kg/m^2)
5.	Fine, loose and dry sand	100 kN/m^2 (10000 kg/m^2)
6.	Fine sand	150 kN/m^2 (15000 kg/m^2)
7.	Moist clay and sandy clay mixture	150 kN/m^2 (15000 kg/m^2)
8.	Loose gravel or sand gravel	250 kN/m^2 (25000 kg/m^2)
9.	Medium, compact and dry sand	250 kN/m^2 (25000 kg/m^2)
10.	Compact gravel	450 kN/m^2 (45000 kg/m^2)
11.	Black cotton soil	150 kN/m^2 (15000 kg/m^2)
12.	Soft rocks, Coarse sand	450 kN/m^2 (45000 kg/m^2)
13.	Hard rocks	3300 kN/m^2 (330000 kg/m^2)

Minimum Depth of Foundation

As per IS 1080: 1985, all foundations must be at least 50 cm below the natural ground level. The average minimum thickness of the foundation must be at least 200 mm. The strength of the concrete used for unreinforced strip footing must be at least 15MPa or better. For a double storey building reinforced foundations would require 25–30MPa.

The minimum foundation depth for residential buildings and two to three-storey compartments shall be 1.5 m to 2 m from ground level. The depth of the foundation shall be 1.50 times the width of the footing. In cold climates, it should be limited to 1.50 m below the surface due to frost action.

Depth of Foundation Formula (Rankine's Formula)

For all types of foundations, minimum depth required is calculated using Rankine's formula as follows:

$$D_{min} = \frac{P}{\omega}\left(\frac{1-\sin\phi}{1+\sin\phi}\right)^2$$

where,

D_{min} = minimum depth of foundation (m)

P = total load on foundation in (kg/m^2)

ϕ = density of soil (kg/m^3), and

ω =angle of repose or internal friction of soil.

The average values of angle of repose and corresponding unit weight for different types of soil are given in Table 4.2. It is recommended to test the soil to get the accurate values of the angle of repose and unit weight of that particular soil.

Table 4.2. Angle of Repose and Density Values for Various Soil Types

Sl.No.	Soil Type	Angle of Repose (degree)	Density (kg/m^3)
1.	Sand (dry)	30 – 35	1440 – 1700
2.	Sand (wet)	45	1920
3.	Earth	30 – 45	1200
4.	Clay (dry)	30 – 35	1600
5.	Clay (wet)	15	1760
6.	Gravel	40	2000
7.	Mud	0	1600 – 1920
8.	Coal ash	40 – 45	1000

Example 4.3

Calculate the depth of a foundation for a poultry shed over a soil which consists of moist clay mixed with sand with angle of repose 30°, having safe bearing capacity 1.9 kg/cm^2 and weighs 1750 kg/m^3.

Solution

Safe bearing capacity of soil (P) = 1.9 kg/cm^2 = 1.9×10^4 kg/m^2

Density of soil (ω) = 1750 kg/m^3

Angle of repose (ϕ) = 30°

The minimum depth of foundation is

$$D_{min} = \frac{P}{\omega}\left(\frac{1-\sin\phi}{1+\sin\phi}\right)^2$$

$$= \frac{1.9\times10^4}{1750}\left(\frac{1-\sin 30}{1+\sin 30}\right)^2 = 1.2 \text{ m}$$

Example 4.4

Calculate the minimum depth of foundation for a building over a soil with angle of repose 35° and unit weight 17 kN/m³. The load on the foundation is 370 kN/m².

Solution

Load on the foundation (P) = 370 kN/m²

Density of soil (ω) = 17 kN/m³

Angle of repose (ϕ) = 35°

The minimum depth of foundation is

$$D_{min} = \frac{P}{\omega}\left(\frac{1-\sin\phi}{1+\sin\phi}\right)^2$$

$$= \frac{370}{17}\left(\frac{1-\sin 35}{1+\sin 35}\right)^2 \approx 1.6 \text{ m}$$

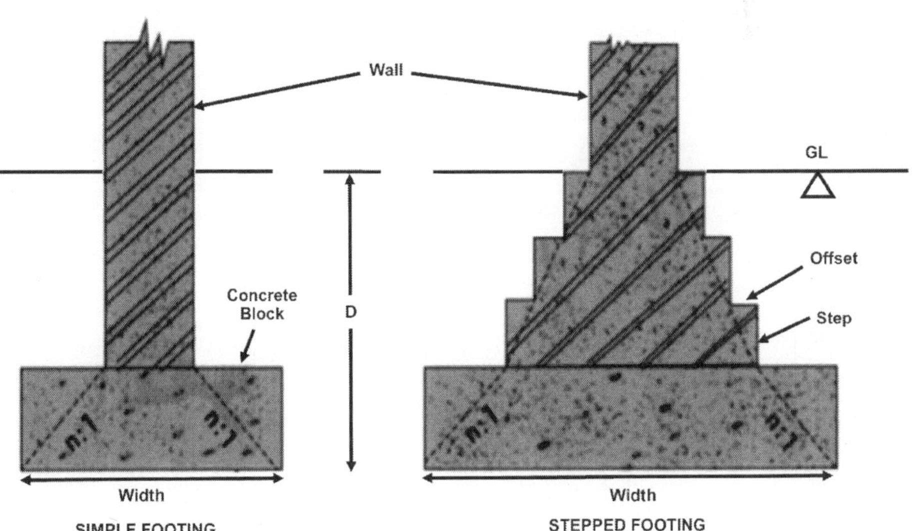

Figure 4.1. Wall Footing.

Example 4.5

Design a square masonry foundation for a 50×50 cm column carrying a load of 60 tons at base. Bearing capacity and angle of repose of soil may be assumed as 25 t/m^2 and 30°, respectively. Weight of soil may be taken as 1800 kg/m^3.

Solution

Load on the foundation = 60 tons

Bearing capacity of soil (P) = 25 t/m^2

Density of soil (ω) = 1800 kg/m^3 = 1.8 t/m^3

Angle of repose (ϕ) = 30°

The minimum depth of foundation is

$$D_{min} = \frac{P}{\omega}\left(\frac{1-\sin\phi}{1+\sin\phi}\right)^2$$

$$= \frac{25}{1.8}\left(\frac{1-\sin 30}{1+\sin 30}\right)^2 \approx 1.5 \text{ m}$$

$$\text{Area of foundation} = \frac{\text{Load on the foundation}}{\text{Bearing capacity of soil}} = \frac{60}{25} = 2.4 \text{ m}^2$$

Considering a square foundation.

Thus, one side of foundation = $\sqrt{2.4} \approx 1.55$ m

Size of foundation = 1.55 × 1.55 m

4.2 Damp-Proof Course (DPC)

Dampness in building may cause unhealthy and difficult working conditions for the occupants. It may result in softening and crumbling of plaster, corrosion of metals, deterioration of timber, growth of termite, *etc.* DPC prevents the moisture rising by capillary action from the ground up the walls. It is applied on the walls rising from the foundation at the plinth level. It consists of a 38-50 mm thick layer of cement concrete at least 150 mm above the ground level. The cement concrete of 1:2:4 proportion shall be mixed with a good quality waterproofing compound over which two coats of melted coal-tar are applied. However, under normal conditions, a layer of cement concrete (1:2:4) shall be used for DPC at plinth level. The DPC should be of uniform thickness. The materials commonly used to check dampness are bitumen felts (which may be hessian based or fibre/glass fibre-based), plastic sheeting (polythene sheets), mastic, asphalt, first-class bricks, stones, slate, cement concrete, *etc.* The choice of material depends upon the climate and atmospheric conditions, nature of the structure, and the situation where DPC is to be provided. Damp proofing material should be impervious, strong, durable, cheap and easily available. Also, it should be capable of withstanding both dead as well as live loads without damage.

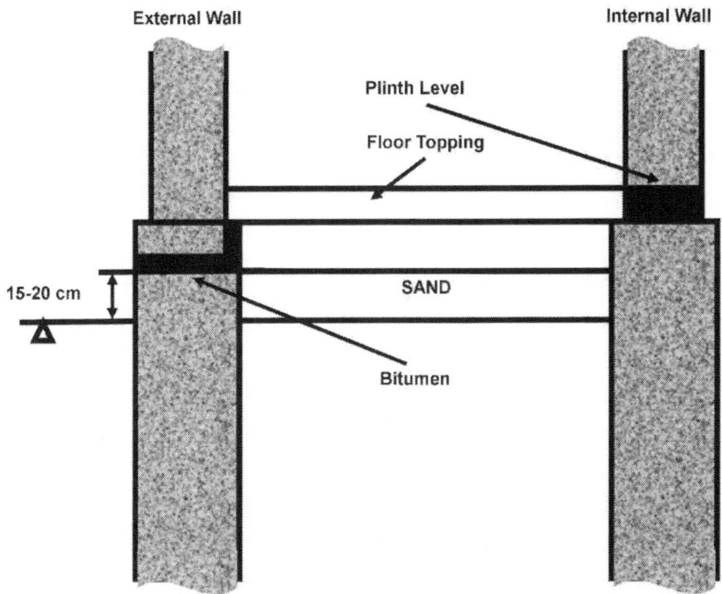

Figure 4.2. Damp-Proof Course.

4.3 Walls

Wall is a structural element which divides the space and also provides safety and shelter. Generally, the walls are differentiated as outer-walls and inner-walls. Outer-walls gives an enclosure to the house for shelter and inner-walls helps to partition the enclosure into the required number of rooms. Inner walls are also called as partition walls.

In the technical point of view, the walls are divided as load-bearing or non-load bearing. In the case of load bearing walls, the entire building construction is supported by walls rather than columns. Strip foundation is adopted for the load-bearing type of wall. Non-load bearing wall does not sustain floor or roof loads above it. In other words, this wall will not carry any of the weight of the structure above it. These walls are typically 100 mm to 125 mm thick. Partition walls inside the building are the best example of it.

Brick, stone, bamboo, wooden plank, and mud are the commonly used materials for the construction of walls in India. The selection of material depends upon utility of structure, availability, climate, and cost. Brick or stone walls are comparatively costly but these walls are durable and hygienic. The brick or stone walls are usually laid in cement mortar or lime mortar. Bamboo and mud walls are economical and useful but are temporary and difficult to keep hygienic. Bamboo walls are constructed with split bamboos which are tied to the posts fixed in a ground about 2.4 meters apart. Mud plaster is usually done over the bamboo structure.

Brick is a versatile and long-lasting building material. Therefore, brick masonry walls have been employed in building construction for thousands of years with minor changes. Walls built with bricks and cement mortar or lime mortar are recommended for permanent structures. When the height of wall is less than 3 meters, 22.5 cm width is sufficient. The width of wall can be increased to 35 cm, if the height of wall exceeds 3 meters. Burnt clay bricks, sand-lime bricks (calcium silicate bricks), concrete bricks, fly ash clay bricks, and firebrick are used in its construction. The brick wall thickness could be 22.5 cm or 11.25 cm. A 22.5 cm wall is used for the outer walls, and 11.25 cm wall is used for inner walls. The length of the brick wall in a single stretch should not exceed more than 4 meter. If it exceeds, a column must be constructed with RCC. Generally, the inner and outer face of walls are plastered.

The lower wall of barns shall be robust due to farm animals. Construction materials like brick, stone, and cement concrete shall be used in lower walls. The wall thickness should be at least 20 cm. Although a wall thickness of 30 cm shall be in bull sheds. Walls and partitions can be made of galvanised corrugated iron or asbestos sheets, fixed to posts 2.5 to 3 metres apart.

4.4 Roof

A roof is the uppermost part of the building that enclose the space. It protects the occupants/space from the effects of rain, sun, wind, heat and snow. The roof should be light, strong, durable, weather-proof, and provide sufficient protection to the inhabitants from heat and cold. Commonly used materials for roofing are clay tiles, corrugated galvanized steel sheets, asbestos cement sheets, asphalt roofing material, aluminium sheets, wood, thatch, *etc.* The selection of roofing material depends upon its availability, cost and climatic conditions.

The advantages of galvanized, asbestos, aluminium sheets are that they are fire-resistant, hygienic, easily fixable, and has reasonably long life. Being light in weight, these do not need heavy roof supporting structures. Thatch roofs are cheaper and act as good insulator. But, these roofs are highly prone to fire hazards, and unhygienic especially during monsoon. These roofs also harbour insects, flies, cobwebs, and vermin. Wood make the most comfortable roof, being a good insulator. But it is liable to fire risk and is quite costly.

The roof shall be gabled type, lean-to-type or flat type. Flat roofs are preferred in low rainfall areas while gabled or lean-to-type roofs are desirable in medium to heavy rainfall areas. The slope of the roof depends upon the rainfall and other climatic conditions of the region. Steep sloping thatched roofs or corrugated metal sheet roofs are generally preferred in high rainfall areas. In moderate rainfall areas where summer days are very hot, tile roofs with a gentle slope are best suited. Country clay tiles are good insulators if placed over bamboo or timber matting. Flat mud roofs are recommended in low rainfall areas. This type of roof keeps the house cool during summer days, but is liable to be eroded during rainy days. These types of roofs need periodically maintenance or require replacement after a certain period of time. Reinforced concrete or reinforced brick roofs are permanent type but they are costly. These types of roofs generally do not need any regular maintenance. In hot climate, asbestos cement or galvanized steel roof shall be overlaid with 8 to 10 cm thick thatch in order to reduce the severity of heat.

For animal housing, a plain stabilized mud roof or a sloping roof consisting either of CGI sheets or cement asbestos sheets is recommended. The attic space in these cases may be covered with bamboo matting with tiles fixed below the sheets.

Single slope roof (shed roof) may be adopted for a span up to 3.6 meters. For span between 3.6 and 4.8 meters, either double sloped roofs with wooden beams or single sloped roofs with steel structural may be adopted. For spans beyond 4.8 meters, trusses are required at an interval of 2.4 to 3 meters. King post roof truss may be used for spans between 16 feet and 26 feet (4.8 to 7.8 meters). The eaves of the roof shall project out sufficiently long in order to provide protection to the animals from direct sun and rain. The eaves should be 2.2 m high from ground level.

The roof is supported by pillars made of stones, brick columns, cast iron pipes, or wooden posts. These pillars can be rectangular or round, with edges rounded and smooth. Pillars are placed at intervals of 2 to 3 meters, depending on the span and roof type. The minimum dimensions of pillars are given below:

Sl.No.	*Material*	*Minimum Dimensions*
1.	Stone	10×10 cm or 8×15 cm
2.	Brick	45×35 cm or 40×30 cm
3.	Iron pipe	10 cm dia
4.	Timber	10×10 cm (rectangular) or 15 cm dia (round)
5.	Mild steel I-section	10×10 cm

Thatch Roof

Thatch roofing is reasonably water-tight, cool and economical. Its main drawbacks are short life and danger of fire. It can be made first by making a split bamboo frame, and fixing it in the roof. The thatch should be at least 15 cm thick and laid in 7.5 cm layers. The support frame of thatch is made of round bamboo rafters spaced at 30 cm and tied using the split bamboos at right angles. The thatch shall be tightly tied to the support framework using tar-dipped ropes or twines. The leakage of rain water can be largely checked by plastering the roof with mud plaster prepared by mixing clayey soil with wheat *bhusa*. The roof may have a slope of about 35°. Thatch roofs are commonly used in sheds, low-cost houses, and village buildings, especially in rural areas due to their affordability and light weight.

A thatch roof is laid over bamboos and *ballies*. It is about 150 mm thick and is done in two ways. Black polyethylene sheet of 700 gauge may be embedded in between the layers to make it water-proof. Treated gunny bags may also be used on the top of the thatch covering to make it water-proof. As per ISI standard IS:11786 – 1986, the treatment consists in the coating of gunny bags with a well-mixed mixture of cement, lime, water, common salt and alum in the volume proportion of 12, 3, 20, 1, and 0.5, respectively. The technology has been developed to make it fire-resistant. For this purpose, thatch is soaked in specially prepared fire-resistant solution.

Shingles

Wood shingles are a popular roof covering material in hilly areas. The wood is cheaper and easily available in hilly areas. It is made from thin, sawn, or split pieces of wood resembling slates or tiles. It is sourced from well-seasoned timber and coated in creosote for preservative properties. These are typically driven on rafters and nailed to their tops. The shingles are available in lengths of 30–50 cm and widths of 5–25 cm.

Galvanized Iron (GI) Sheet or Asbestos Sheet Roof

Galvanized iron sheets and aluminium sheets have the disadvantage of being very good conductors of heat. They get very hot in summer and very cold in winter and thus cause discomfort to the animals and the workers. The aluminium sheets are slightly better because they have reflecting surface and most of the sun's rays are reflected from their surface. Their use is recommended for buildings used for storage of hay, farm implements *etc.* If their use for animal shelters is inevitable, the roof surface should be covered with 30 cm layer of thatch and/or a false ceiling may be provided. Painting the ceiling surface of the metal sheets with black paint and the roof surface with a white reflecting paint is also helpful. The other way is to provide a layer of ordinary tiles at the top of GI sheeting. The sheets should be well nailed to secure against winds. Since the nails are exposed, washers of lead are necessary under the nail head to exclude moisture.

This roofing is suitable for machinery shed, workshops, animal sheds *etc.* The sheets are fixed with hook bolts and nuts to the purlins which run on the principal rafters. 24 BWG (0.63 mm)) thick sheets are normally used. They are available in 1.80 m, 2.20 m, 2.50 m, 2.80 m and 3.20 m lengths. The width is 80 cm and number of corrugations is 10 per sheet. The pitch is 7.6 cm and depth of corrugation is 18 mm. The overlap is 15 cm. Felt washers should be used at all joints. The top surface should be coated with two coats of white aluminium paint to reduce radiation. Sometimes, it is comforting to use thatch on the top of a sheeting. Each lap of 15 cm and side lap of two corrugations are provided at the time of fixing the GI sheets at the top of the roof. GI sheets are lighter and can be fixed easily.

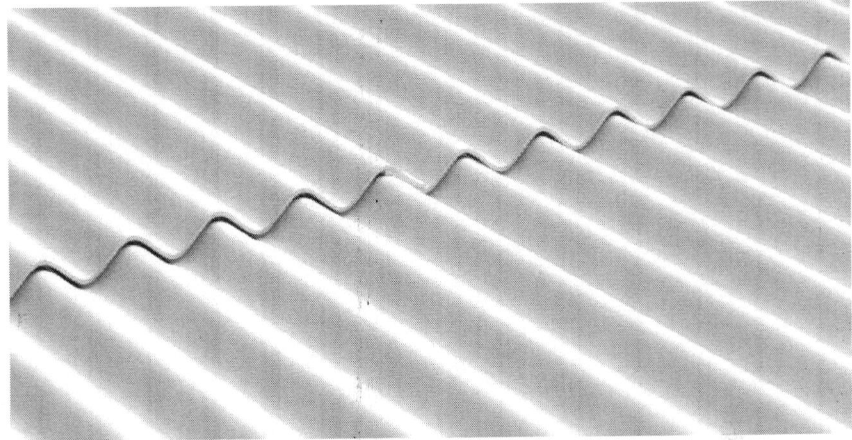

Figure 4.3. Asbestos Sheet.

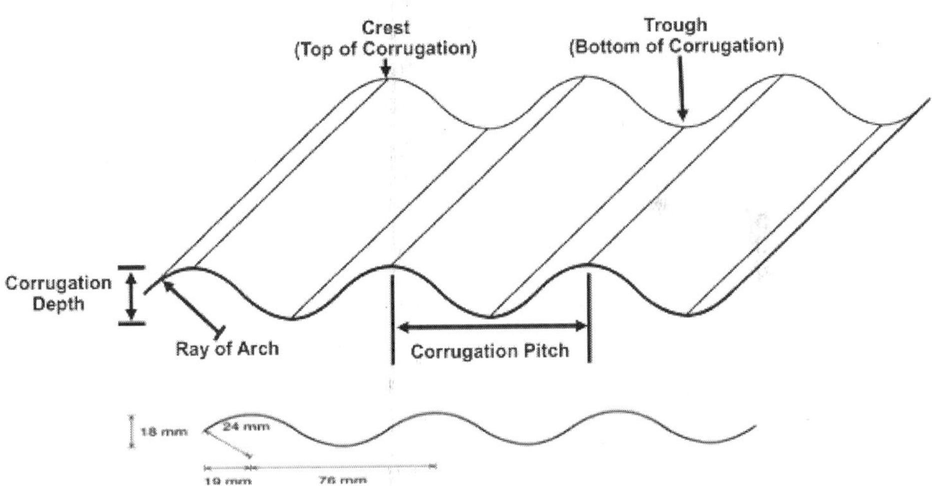

Figure 4.4. GI Sheet.

Asbestos cement is a mixture of portland cement and 15 per cent asbestos fibres. They are comparatively bad conductors of heat, tough, readily available, and easy to install. Also, they are light in weight, cheap, resistant to fire and impervious to water. It is durable without any protective coat. The sheets are secured to the beams with hook bolts. Holes in the sheets should be drilled and never punched as the sheets are very brittle. Sheet roofing may be given a slope of 1 in 5. 10-mm diameter hook bolts are then inserted in 13-mm diameter drilled holes. A bitumen washer must be used under the flat washer.

Various sizes of asbestos cement roof coverings are available in market into two variants *i.e.* flat corrugated and ribbed sheets. Asbestos sheets are available in standard widths of 1.05 m and with lengths of 1.50 m, 1.75 m, 2.00 m, 2.25 m, 2.75 m and 3.00 m. The number of corrugations is 7 per sheet, with pitch as 14.6 cm and depth 4.8 cm. The overlap is 15 cm.

Mud-Roofing

The flat roof consists of 75 mm thick wooden or concrete rafters, topped with brick tiles 300 mm×150 mm×50 mm and covered with 100 mm mud *phuska* laid to slope to 1 in 20. The mud *phuska* can be combined with soil-bitumen mixture containing 1 to 2 per cent bitumen and is finished with dung-plastering.

Roofing Tiles

The most common roofing tiles are flat tiles, pan tiles, pot tiles or half-round country tiles and patent tiles such as Mangalore and Allahabad tiles. Pot tiles and pan tiles are widely used in the rural part of India. These tiles have varying section on both the ends and as such they can be placed on overlapping the other with some kinds of obstruction at the lowest point of the roof to avoid sliding. These tiles are very much susceptible to breakage even by mild jerks. Clay tiles have been widely used as a roof covering material for residential buildings. The use of cement concrete tiles is limited because of high investment and difficulty in their manufacturing.

Clay tiles are durable, strong, pleasing in appearance, economical and non-conductors of heat. Clay tiles offer fire and moisture resistance but are heavy, especially when rafters are placed closer to reduce timber span and to remove rainwater from roof. The average weight of a tile roof is 75 kg/m^2, with further weight increased by reducing timber span.

Flat tiles are rectangular tiles with dimensions ranging from 25×15 cm to 28×18 cm and thicknesses ranging from 9 to 15 mm. They have a slight camber of 5–10 mm. Two projecting nibs and nail holes at one end help in fixing these tiles on roof truss battens.

Curved pan tiles are shorter, less curved, heavier, stronger, and more durable than pot tiles. They are moulded flat and then given the required curvature, measuring 30-35 cm long and 20-25 cm wide.

Pot tiles, also known as Nurria tiles, are a cost-effective roof covering for rural houses. They are laid in pairs of under and over tiles (concave and convex upwards), with a proper overlap of at least 8 cm on all sides. The under tiles are flat with a broad head tapered towards the tail, while the over tiles have a wider tail and a narrower head. The roof is called a double-tiled roof due to the double layer of tiles. Therefore, these roofs require stronger supporting timbers than usual.

Patent tiles are rectangular in shape with face corrugations that helps its interlock with other tiles. These machine made tiles provide a lighter roof with a decent appearance.

Figure 4.5. Roofing Tiles.

Reinforced Cement Concrete (RCC) Roof and Reinforced Brick Concrete (RBC) Roof

RCC is reinforced cement concrete where steel bar is required in slab as per spacing. RBC is reinforced brick concrete used for roof of building where the space between the bars is occupied by the brick placed on its. Steel bars are placed between the bricks in both the direction. Bricks only act as filler material in RBC slab. RBC slab is cheaper than RCC slab. However, strength, load carrying capacity, earthquake resistant and life of RBC slab is lower than RCC slab.

For small rooms up to 3.6 meters in width, a plain flat slab of RCC or RBC type supported only on the side walls can be made. For larger rooms, suitable designed beams have to be used to support the entire load. Flat masonry roofs are best suited for residential buildings, offices, school buildings, and grain storage structures.

4.5 Plastering

Plastering is a process of applying thin layer of mortar over the masonry surface. It covers rough edges and uneven surfaces, and also acts as a damp-proof coat over the brick masonry work. Plastering provides an even, smooth, regular, clean and durable finished surface with improved appearance. It preserves and protect the surface from atmospheric influences. Also, it conceals the defective workmanship and use of inferior quality materials of the masonry work.

Plaster is a mixture of lime or cement concrete, sand, and water. Common types include lime plaster, cement plaster, and mud plaster. Cement plaster develops non-absorbent surfaces. Therefore, it is ideal for damp conditions like bathrooms, floors, reservoirs, water tanks, *etc.* Mud plastering is the cheapest type and is used in village houses and temporary structures. It consists of well-tempered clay, chopped straw, and cow dung and is prepared in a homogenous mass.

Requirements of Good Plaster

1. Adherence to the background in all seasons.
2. Affordable and cost effective.
3. Hard and durable
4. Good workability.
5. Prevent entrance of moisture in masonry

Methods of Plastering

The surface to be plastered should be prepared before the start of plastering work. The entire wall should be roughened. All the joints and surfaces of the wall should be cleaned with a wire brush. There should be no oil or grease *etc.* left on wall surface. Any cavities or holes on the surface should be filled in with appropriate material. Wash the mortar joints and entire wall to be plastered, and keep it wet for at least 6 hours before applying cement plaster.

Plaster should be applied in one or more coats, with a single coat not exceeding 12mm thick. The first coat of plaster mix should be a 1:3 to 1:6 ratios, typically applied by throwing the mix with force against the walls. For cheaper constructions, one coat is typically sufficient, while superior constructions require two or more coats. The surface should be levelled with flat wooden floats and straight edges after the first coat. The final setting coat should be applied after the previous coat is almost dry. The second coat thickness can range from 2–3 mm. The previous surface should be scratched or roughened with a scratching tool and dampened before applying the next coat. As far as possible, the finishing coat should be applied starting from top towards bottom and completed in one operation to eliminate joining marks.

After completion of the plastering work, proper curing should be carried out. Curing shall commence 24 hours after the plaster is placed. The surface is kept wet by sprinkling water for at least 7 days in order to develop sufficient strength and hardness. Gunny bags or other materials is used to keep the plastering works wet in external works. Improper curing may lead to cracks formation or efflorescence in plaster work.

4.6 Floor

The floor shall be hard and warm, impervious to water and easy to clean. Wherever necessary, the flooring may be suitable insulated. The floor may be cement concrete or paved with cement concrete flooring tiles, stone slabs or bricks, and set in cement mortar or lime mortar. If the floor is made of stone slabs or bricks, the joints shall be filled in properly with the binding material. The details of construction shall vary from place to place depending on the climatic conditions and shall be constructed on the recommendations of the local housing experts. The floor shall slope towards the drains.

The details of different types of flooring are as follows:

4.6.1 Stone Slab Flooring

A stone-slab is a big stone which is flat and relatively thin. It is generally used for paving floors and covering walls. The stone slab flooring is hard, durable, long-lasting and wear resistant. This type of flooring is easy to construct and can be repaired easily. Square or rectangular slabs of stones are used for the construction of floor. The most common sizes of stones are 30×30 cm, 45×45 cm, 60×60 cm, 45×60 cm, *etc.* The thickness of stones shall be 20 to 40 mm. The only disadvantage is the effect of climate and environmental changes on stone slabs. Climate and environmental changes do affect the texture of the rock and cause cracking, mildew, and dandruff on the surface. The surface of stones shall be rough or polished. A rough surface is provided usually in godowns, sheds, stores, *etc.* while a polished surface is provided in superior type of works. Polished stone slabs are elegant in appearance, easy to clean but at the same time costs high. Polished stone slabs of granite, marble is used in this type of flooring.

The natural stones such as marble, granite, slate, sandstone, limestone and travertine floors are best suited in Indian homes. Natural stone flooring is one of the best options for homes and commercial spaces. Natural stone is aesthetically pleasing, durable, as well as adds value to house. The term "natural stone" refers to a variety of mineral substances that are born in the mountains.

Natural stones are hard and cool, making them ideal for homes in warmer climates such as India.

(i) Marble Flooring - Marble is highly durable and scratch resistant. It is available in myriad of colours, patterns, and charm. Marble floors add value and sophistication to home. Marble slabs are more expensive.

(ii) Granite Flooring - Granite is a hard-edged stone with a beautiful natural grain. Just like marble, granite is a luxurious natural stone used in several elements of the house for decorative purposes. Granite is available in a wide variety of colors, shades and grains. It has a highly resilient surface, making it suitable for both outdoor and indoor use. It is incredibly durable and has the ability to withstand heavy use, heat, spills, stains and foot traffic. Granite is impenetrable to liquids. But, granite is one of the most expensive natural stone available for flooring, even costlier than marble. Marble slab is required to be polished with carborundum stone, whereas granite slab does not require any polishing.

(iii) Slate Flooring - Slate is a beautiful porous stone which is easier to cut and mould into flat flooring sheets. It is highly durable and resistant to cracks, scratches and dents. It is available in many colour options. Slate is not as luxurious as marble, yet it provides a certain appeal to house with the awe-inspiring essence of natural stone. It is a more affordable choice within the spectrum of natural stone floorings. It is water-resistant, and can be installed in both wet and dry areas of the home like kitchens, dining rooms, bathrooms, living rooms, *etc.* It can be used both outdoors and indoors. It is easy to clean and maintain.

(iv) Limestone Flooring - Limestone is a sedimentary stone consists of carbonate of lime. Limestone creates an earthy appeal to the house and is found in a wide range of colours. This is a soft and porous stone that can be sanded and polished or machine-tumbled to achieve a rough look. It is resistant to mould and bacteria so it works really well as a bathroom option or for an outdoor project. It is the most affordable natural stone for flooring compared to other stones. It can last for longer period of time because of its mineral composition. It could easily get scratched. Kota limestone is one of the most beautiful stones in India. It is quarried from sources in Rajasthan. It is a naturally non-slip and non-porous; suitable for outdoor and high traffic areas. Cuddapah slabs and Shahbad slabs are other examples of limestone available in India.

(v) Sandstone Flooring - Sandstone is good for house flooring. It is a widely popular stone in India with large variations in colour and grain. As the name suggests, its name is derived from the sand as it seems to resemble the look of beach and desert sand. It is highly durable and warm, and lends the room a distinctive natural look with its rocky characteristics. Sandstone is great for outdoor use, too. Sandstone has a high water absorption rate, so it is not suitable for places like kitchens and bathrooms. Even in the outdoors, it is preferable to have the spaces covered so the flooring won't be affected by rain too much. Over time, it will lose its beauty due to weathering and aging. Since it is a soft stone, it is more prone to scratches, dents, and cracks compared to other natural stones.

(vi) **Travertine Flooring -** Although it is commonly referred to as "travertine marble", travertine is actually a type of limestone. It's preferred in both outdoors and indoors in spaces such as paving, walkways, living and common areas. Travertine is an elegant, and character-filled material. When installed correctly, travertine is one of the most durable stones for bathrooms and showers. It is an excellent choice for outdoor areas or around swimming pools as it is both durable and slip resistant. It is comparatively cheaper. Travertine is available in cream, ivory and gold colours. Although not impossible, travertine is hard to damage. Among other natural stones, travertine is one of the most porous since it naturally has many holes in it. Travertine is highly reactive to acidic substances.

Method of Construction

The subgrade is prepared by laying rubbles or bricks aggregates, which is then levelled, compacted and watered. Over this surface, the sub base of lean cement concrete of 1:4:8 or 1:5:10 proportion is laid to a thickness of 10 to 15 cm and consolidated well. When it is sufficiently hard, cement mortar in the ratio of 1:4 or 1:5 is spread to a thickness of 15 to 20 mm, and the stone slabs are fixed and levelled with the adjoining slabs. Stones are dressed on all the edges and washed clean before laying. The joints are flush pointed with cement mortar (1:3). The joints are generally kept as fine as possible. Surplus cement if any on the surface of the slab shall be cleaned off. The flooring shall be cured for a minimum period of 7 days. The surface of the flooring shall be of desired slope (1:40) for proper drainage.

4.6.2 Cement Concrete Floor

Cement concrete flooring is one of the most common types of flooring used in both residential as well as public buildings. Cement concrete floor is durable, non-absorbent, wear resistant, fire-resistant, easy to maintain and economical. It provides a hard, smooth, even and pleasing surface. Owing to its non-absorbent nature, it is very useful for water stores. It is one of the cheapest flooring. There are also some demerits of cement concrete flooring. It is hard underfoot, and has option of only few colours and designs. There are chances of moisture migration from underneath the slab. Also, the defects once developed in cement concrete floor cannot be satisfactorily repaired by patch work.

Method of Construction

First of all, the sub-base is prepared. The earth filling in plinth is consolidated thoroughly so as to ensure that no loose pockets are left in the whole area. A 10 to 15 cm thick layer of clean coarse sand is then spread over the whole area. The sand layer is consolidated and dressed to the required level and slope. Thereafter, a base coat of 7.5 to 10 cm thickness in the proportion of cement concrete 1:8:16 or 1:4:8 is laid on compacted earth under the floor. The base concrete layer is deposited over the whole area, thoroughly tamped and levelled to a rough surface.

When the base concrete layer has fully set and hardened, its surface is thoroughly cleaned and the entire area is divided into rectangular or square panels by use of 4 mm thick glass strips or 5 mm thick plain asbestos strips. The depth of strips should be equal to the thickness of floor. The length of one such panel should not be more than 2 meters and the area of panel should not preferably be more than 2 square meters. The panels should be uniform in size. To ensure a strong bond between the base and topping, dampen the surface of base concrete with water and apply a neat cement slurry coat before laying the topping, ensuring a rough finish. The cement concrete topping, normally in the proportion of 1:2:4 is then laid in required thickness in alternate panels on alternate days. The concrete is spread evenly by using a straight edge and the surface is thoroughly tamped and floated with wooden float till the cream of the mortar comes at top. The surface is then smoothened and finally finished by means of steel trowels. Rough and non-slippery surface is obtained when worked with a wooden float, while a smooth surface is obtained upon steel trowelling. The flooring should be cured for 10 days before use. The floor surface shall be properly grooved in the livestock sheds in order to avoid slipping of animals.

4.6.3 Tile Floor

Clay tiles of different sizes, shapes and colours are available in the market for the construction of floor. They are placed in position on a concrete base with a thin layer of mortar. This type of flooring provides a non-absorbent, decorative and durable surface. They can be quickly placed, and easily repaired when required. The disadvantages of tiled flooring are that tiles are usually expensive, and have high maintenance cost. On becoming wet, they provide a slippery surface.

4.6.4 Terrazzo Floor

Terrazzo is a concrete-based floor made from a mixture of marble chips and cement. The cement can be white or coloured. Marble chips of 3-6 mm are mixed in cement in a ratio of 1:2 to 1:3. The mixture is polished with carborundum stone for a smooth finish. It is a costly flooring option to obtain a clean, attractive, and durable surface.

4.6.5 Wood Flooring

Wooden flooring is highly durable and easy to maintain. It adds aesthetic to a space. Wood floors are more resistant to liquid spills and dirt. Wooden flooring can be used everywhere, except in wet areas like kitchens and bathrooms as hardwood flooring is vulnerable to moisture and humidity. Wood is also prone to termite damage. Before using it, proper termite treatment must be carried out in the space. The best hardwood floors are made with wood species that are readily available. Oak flooring, maple flooring and cherry flooring are all good choices. Other species include bamboo, walnut, ash and mahogany.

4.6.6 Brick-on-Edge Flooring

Brick flooring is one type of floor made of bricks and used for cheap constructions such as *godowns*, sheds, stores, garages, barracks, *etc.* Brick flooring offers a durable, sufficiently hard, economical, easy to construct as well as repair, non-slippery and a fire-resisting floor surface. It is cheaper than terrazzo, mosaic, cement concrete and marble flooring. The only drawback of brick flooring is that it absorbs water. This type of flooring shall be adopted for the areas where frequent washing is not required. Brick-on-edge flooring is developed based on the concept that bricks laid on edge give a better performance than when laid flat. Ordinary bricks are used for flooring low-cost temporary sheds, cattle sheds, courtyard, footpaths, *etc.* Burnt clay bricks are used for floors subject to heavy wear and tear, such as in stores, *godowns*, and platforms.

Figure 4.6. Herringbone Pattern.

Method of Construction

A sub-grade of rubbles or brick aggregates is laid. It is watered and well rammed. Over this, a base course of lean cement concrete mix (1:4:8 or 1:5:10) or lime concrete mix of 10 cm thickness is laid. Slope if required for quick disposal of wash water may be provided in this base course. When the base course is sufficiently hard, 10 mm thick cement mortar (1:4 or 1:5) or lime mortar is spread over the base course and the bricks either flat or on edge are laid in the desired pattern. The bricks may be laid either in herringbone or diagonal or zig-zag pattern. Each brick should be properly bedded on edge and set home by gentle tapping with the handle of trowel or a wooden mallet. The inside faces of the brick should be buttered with mortar before the next brick is laid and pressed against it. Bricks must be sufficiently soaked in water prior to laying. The thickness of vertical joints may vary from 1.5 mm to 6 mm and should not exceed 1.0 cm for any class bricks. The vertical joints formed in the flooring shall be flush finished with the mortar and the entire floor should be cured for 7 days. In the case of fat lime mortar, curing should commence two days after the laying of flooring and should continue for 7 days. In case the joints are to be filled with sand, the bricks need not be soaked. Sometimes, the joints are pointed to obtain a better appearance. Floors in the *verandah*, courtyard kitchens, and baths shall have slopes ranging from 1 in 36 to 1 in 48, depending upon locations. The faces of bricks shall be cleaned, and mortar droppings shall be removed on the same day.

4.6.7 *Moorum* Flooring

Any disintegrated rock is called *Moorum*. Moorum flooring is used in low cost housing, especially in villages. Moorum flooring are cheap and suitable for all climatic conditions. It is easy to construct and repair. It provides smooth, hard and less permeable surface. The maintenance of this flooring is quite difficult. Because, the floor needs wash of cow-dung at least once a week.

Method of Construction

A subgrade of 15 to 25 cm is prepared by laying layer of broken hard brick-bats and then rammed well. On this hard bed, 15 cm thick layer of *moorum* is laid with coarser pieces of brick-bats at bottom and finer at the top. Over this about 3 to 5 cm thick layer of powdery *moorum* is spread. Water is sprinkled on entire surface and rammed well. After ramming, water is sprinkled on the surface until the floor is fully saturated. Then the surface is trampled well till the cream of *moorum* rises to the top. The surface is left in this condition for a day and then rammed again, levelled and well consolidated for about two to three days. When the floor is dry, surface is smeared with a thick paste of cow dung and well rammed once again for two days. Finally, over dried and hard surface, a thin coat of cement and cow-dung slurry in the proportion of 1:4 is applied evenly and wiped clean to prevent cracking and panelling. The cow dung plaster is applied once in a week to keep the floor in good condition.

4.6.8 Mud Flooring

Mud floors are a popular and cost-effective flooring option in rural areas. It is made of locally available material and provide good thermal insulation. Although, mud floors are not fairly good as per the sanitation requirement. It is suitable for areas with high summer temperatures and low winter temperatures. It keeps homes warm in cold weather and cool in hot weather. However, they are not durable, impervious and require periodic maintenance.

Methods of Construction

The mud floor is made by spreading a 25 cm thick layer of broken rubble or brick bats over a bed, then moistening it with water. The layer is then compacted with a rammer to create a 15 cm thick hard-consolidated layer, followed by 2–3 cm thick mud. Ramming should be done frequently to achieve a level surface. Avoid using water while ramming. Moist earth shall be mixed with chopped straw to reinforce the mud floor. The chopped straw prevents crack formation in the mud floor.

4.7 Doors

Doors are fundamental components of architecture, facilitating passage and delineating boundaries within buildings and structures. In a literal sense, doors provide security and privacy, controlling access to spaces and safeguarding occupants. They come in various forms, from simple wooden doors to elaborate, ornate designs crafted from metal, glass, or other materials. The sizes of doors in a farmhouse can vary depending on factors such as the architectural style, building regulations, and specific requirements of the property owner. Standard exterior entry doors typically range in width from 32 inches to 36 inches (81 cm to 91 cm). Heights commonly range from 80 inches to 96 inches (203 cm to 244 cm). These sizes accommodate standard door frames and are suitable for most residential applications, including farmhouse entrances. Interior doors in farmhouses often follow standard residential sizes. These doors are typically narrower than exterior doors, ranging from 61 cm to 91 cm (24 inches to 36 inches) in width and 203 cm to 244 cm (80 inches to 96 inches) in height. Common sizes include 24 inches, 28 inches, 30 inches, and 32 inches in width. Utility doors, such as those leading to kitchen, bathroom, basements, garages, or storage areas, may have different size requirements based on their specific function and location within the farmhouse. These doors often range from 71 cm to 91 cm (28 inches to 36 inches) in width and 203 cm to 244 cm (80 inches to 96 inches) in height. It's important to note that these dimensions are general guidelines, and actual door sizes may vary based on individual preferences, building codes, and architectural considerations. The frames of the doors and windows may be made either of wood or metal. The wooden frame is usually made of 10×7.5 cm cross-section.

4.8 Windows

Windows serve as essential components of architecture, providing openings in structures that allow light, air, and views to penetrate indoor spaces. From the simplest designs to elaborate stained glass masterpieces, windows come in a myriad of forms, reflecting the architectural styles, technological advancements, and cultural preferences of their time. They can vary in size, shape, and placement, from small clerestory windows high on cathedral walls to expansive floor-to-ceiling windows in modern skyscrapers. One of the primary functions of windows is to admit natural light, creating bright, airy interiors that promote well-being and productivity. They also offer ventilation, facilitating air circulation and regulating indoor temperature. Furthermore, windows provide visual connections to the surrounding environment, framing picturesque views of landscapes, cityscapes, or architectural features. Windows also play a role in environmental sustainability and energy efficiency. Advances in window technology, such as double-glazing and low-emissivity coatings, have improved thermal performance, reducing energy consumption and enhancing indoor comfort. Standard window widths typically range from 61 cm to 122 cm (24 inches to 48 inches), although larger windows are also common in modern construction. Heights of standard windows often vary from 91 cm to 183 cm (36 inches to 72 inches). These sizes provide a balance between natural light, ventilation, and architectural proportion. Windows in living rooms, dining rooms, and bedrooms are often larger to maximize natural light and views. In contrast, windows in bathrooms and utility rooms may be smaller to prioritize privacy. Consider the orientation of the room and desired sunlight exposure when determining window sizes. South-facing windows, for example, can maximize natural light and solar heat gain, while north-facing windows may require larger sizes to compensate for less direct sunlight. Windows are fixed in such a manner that the tops of all windows are at the same height as that of the doors. Wire mesh is provided on doors and windows to keep the flies and insects away. Steel grills are also fixed on windows to provide greater safety and security.

4.9 Painting, Distempering and White-Washing

4.9.1 Paints

Paints are fluid materials used as a final finish on surfaces like walls, ceilings, wood, and metal. It protects the surfaces from weathering effects, decay, and corrosion. Paints ensure a durable and long-lasting finish on the surfaces. Painting preserves the interior and exterior surfaces through its protective coat. Painting provides a decorative finish and a clean, hygienic and healthy living atmosphere. It also protects the surface against insects, improves surface's durability and acts as a waterproofing.

There are various types of paints available in the market. These are enamel paints, oil paints, cellulose paints, cement-based paints, aluminium paints, emulsion paints, anti-corrosive paints, plastic paints, water-based paints, bituminous paints, synthetic rubber paints, *etc.* Each paint is specifically created for a variety of uses. Select a right type of paint which provides good quality and enhanced performance at minimal cost. Before the application of paint on the wall, apply two coats of wall putty.

Characteristics of a Good Paint

Good paint should be durable, tough, and resistant to wear. It shall not have any harmful effects on users. It should not crack, fade, or change colour over time. It can spread into thin layers for a smooth and pleasing appearance. The weather should not affect its functionality.

4.9.2 White Washing

Whitewashing is the process of covering a wall with a very thin coat of thin plaster made with water, lime and other ingredients. Fat lime is used for whitewashing as it produces slaked lime (calcium hydroxide, $Ca(OH)_2$) when it reacts with water (H_2O) and absorbs carbon dioxide (CO_2) from the environment, and as a result, it produces a thin layer of calcium carbonate ($CaCO_3$) on the walls. Calcium carbonate is formed after two to three days of whitewashing and gives a shiny finish to the walls. Sometimes various other additives are used.

Whitewash is a type of paint which is widely used for low-cost houses. It revitalizes the sheen and texture of walls in the house. It acts as an insect repellent, protecting the house, building, or barn from different kinds of insects. It prevents animals from getting infected with flies and mites, and creating homes by bees and wasps. Another significant advantage of whitewash is that it prevents mould growth on the surface of walls, ceilings, and floors. Also, it is antibacterial. It makes the surface more hygienic and sanitary. It is non-toxic and safe for animals. Whitewash is used to reduce odours and provide more sanitary conditions in dairy barns, poultry houses, stables, *etc.* The white colour of the coating can also minimize the absorption of heat. Whitewash can be applied on plaster, brick, cement concrete surfaces and woods. The life of this paint is not more. It is required to paint almost every year or once in two years.

Preparation of White Wash: To create a white wash, fresh lime is slaked at the work site and dissolved in clear, potable water. About five litres of water are needed for 1 kg of un-slaked lime. After 24 hours of slaking, the mixture is stirred until it resembles thin cream. The mixture is then screened through a clean, coarse cloth to form a white wash solution. To increase adhesion, one kg of gum is dissolved in hot water and mixed with 10 kg of lime. DDL, SDL, or Fevicol are commonly used as glue in whitewash. To improve adhesive properties, soap and common salt (sodium chloride) can be added. To prevent glaring and increase surface brightness, a small amount (3 g per kg of lime) of ultramarine blue (*neel*) is added to the last two coats of white wash solution.

Preparation of Surface: The surface to be whitewashed must be thoroughly cleaned. All loose scales, dirt, dust, mortar and other foreign matter should be removed from the surface. The holes in patches and plaster should be filled with mortar. The surface be allowed to dry for at least 48 hours before doing whitewash. Putty should be used for small unevenness in walls since it does not require any curing. Cement mortar should be used for big depressions.

Application of White Wash: White wash is applied to new or scrapped surfaces using a *moonj* or brush. Generally, three coats are necessary for new work and scrapped surfaces, and one or two coats for old work. Each coat should dry before applying the next coat. The operation involves one stroke from the top downwards and the other from the bottom upwards, and another stroke from the right and left over the first brush. Ceiling whitewashing should be done before the walls. The finished dry surface should not show cracking or peeling and not easily come off on fingers when rubbed. To check paint quality, rub a finger on the white-washed surface, and it becomes white for the bad-quality white wash.

Colour Washing: A colouring pigment in suitable quantity is added to the screened white wash to prepare a colour wash. Usually, yellow earth, red ochre, and blue vitriol are widely used as pigments. Before mixing these colour pigments, they must be crushed and made into powder. The application of colour wash is similar to the whitewashing procedure. After successfully preparing the surface for colour-washing on new works, the first primary coat shall be of whitewash, and the subsequent coats (minimum two) shall be of colour-wash. The entire surface shall be applied with a smooth and uniform finish.

4.9.3 Distemper

Distempers are water-paints made from a base like chalk, water, a binder like glue or casein, and colouring pigments. Distempers are applied over the plastered surfaces. It is available in a variety of colours. They can be water-bound or oil-bound and are available in powder or paste form. Water-bound distempers are mixed with hot water before use, while oil-bound distempers are washable. Distempers are cheaper than oil paints and provide a neat appearance. They are light in colour and provide a good reflective coating, but they are less durable than oil paints.

4.10 Termite Protection

Termites are grouped on the basis of their habits. These are subterranean or ground nesting termites, and non- subterranean or wood nesting termites. Subterranean termites are most destructive and mainly responsible for the damage to be caused in buildings. Non-subterranean termites frequently construct nests within large dimensional timbers such as doors, windows, *etc.* which they destroy.

Anti-termite treatment is done before the construction completes. In this anti-termite treatment, the soil, building pillars and other joints are treated with chemicals to prevent termites. The application of chemical to the soil in immediate contact with the foundation and floor structure of a building kills or repels termites thus forming a toxic barrier which is lethal to termite entry. The termite proofing treatment should invariably be given in all types of buildings during the construction stage. Pre-construction anti-termite treatment is less expensive as compared to post anti-termite.

For the effective control of termites, the site for the construction of buildings should be cleared of all old tree stumps, dead wood, plant remains *etc.* To prevent termite movement, ensure no bridges are built between building parts and untreated soil. Use concrete or solid foundations to check termite movement from the ground. It is important to treat the superstructure with preservatives to make it termite-proof. Chemicals are used to treat the soil beneath the foundation. The Bureau of Indian Standards (IS 6313 Part-3) has recommended the following chemicals for pre and post construction anti-termite treatments for all types of buildings:

1. Clorpyrifos 20 EC
2. Lindane 20 EC
3. Imidacloprid 30.5 per cent SC

The recommended emulsion concentration by weight for Clorpyrifos 20 EC and Lindane 20 EC is 1.0 per cent which means that 1 litre of these chemical has to be mixed with 19 litres of water. Whereas for recommended concentration by weight for Imidacloprid is 0.075 per cent *i.e.* mix 10.5 ml of Imidacloprid 30.5 SC with 5 litres of water. These chemicals may be used in water emulsion for soil treatment in order to protect a building from termite attack, and in mineral oil or kerosene for treatment of wood to protect from termite attack. These emulsions are applied uniformly at the prescribed rate. During foundation preparation, the emulsion shall be applied on the bottom surface and sides of foundation pits to the height of 30 cm @ 5 litre/square meter of surface area. After completion of foundation, treat the refill earth on both sides of all built up walls (approx. width 30 cm and depth 45 cm) @ 7.5 litre per linear meter. Before laying of floor, the entire levelled surface below ground floor slab be treated with chemical emulsion @ 5 litre per square meter, and after completion of building, treat the earth along the external perimeter of the building @ 7.5 litre of spray per linear meter.

4.11 Maintenance of Buildings

Buildings deteriorate over the time due to age, environmental and other external agents, and improper use. This necessitates proper maintenance and repair of the buildings to serve their purpose throughout their designed life. Thus, timely and proper repair of the buildings is very desirable. A thorough check-up, minor repairs and maintenance jobs should be undertaken at least once a year. However, necessary repairs should be carried out as and when required.

All buildings should be cleaned daily. The washable floors and walls should be scraped, brushed and washed with water. Wooden parts shall be given proper anti-termite treatment and should be painted. All structural steel works should be painted in order to protect them from corrosion. Before painting, rust scales and dirt should be removed by means of iron brushes. Generally, a surface coating of paint, zinc or asphalt is enough. The leakage, if any, from the roof should be repaired immediately to avoid further damage both on the roof and on the items of the buildings. Electric wiring, water pipes and drainage system need regular attention. A few pieces of generally required spare parts should be kept in stock for immediate use when needed. An occasional spray of insecticides is necessary to check the flies *etc.*

Important Points to Remember

1. Foundation is a lowest portion of building below ground level that transfers a load of superstructure to the earth through it.
2. Shallow foundation is provided in the soil which have higher load bearing capacity and pile foundation is provided in the soil which have less load bearing capacity.
3. Safe bearing capacity of the soil is the capacity of the soil that supports the structural load to the ground without any shear failure or settlement.
4. Minimum depth of foundation is kept 1.5 m to 2 m for residential buildings and also for two to three-storey compartments from ground level or at least 1.50 times the width of footing.
5. Damp-proof course (DPC) is provided to prevent the moisture rising by capillary action from the ground up the walls.
6. DPC should be at least 150 mm above the ground level.
7. The wood shingles as a roof covering material is generally used in hilly areas where wood is easily and cheaply available in abundance.
8. 24 BWG (0.63 mm) thick GI sheets are normally used for machinery sheds.
9. Asbestos cement corrugated sheets are available in standard widths of 1.05 m and with lengths of 1.50 m, 1.75 m, 2.00 m, 2.25 m, 2.75 m and 3.00 m.
10. The overlap in case of asbestos cement corrugated sheets is 15 cm.
11. The number of corrugations per sheet in case of asbestos cement corrugated sheets is 7.
12. For asbestos corrugated sheets, the pitch and depth are 14.6 cm and 4.8 cm, respectively.
13. Pot tiles and pan tiles are widely used in the rural part of India.
14. RBC slab is cheaper than RCC slab.
15. Thickness of a single coat of plaster should not be more than 12 mm.
16. Curing shall commence 24 hours after the plaster is placed.
17. Cement concrete floor is durable, non-absorbent, wear resistant, fire-resistant, easy to maintain and economical.

18. Brick flooring is one type of floor made of bricks and used for cheap constructions such as *godowns*, sheds, stores, garages, barracks, *etc.*
19. *Moorum* flooring is used in low cost housing, especially in villages.
20. A good paint should possess high spreading power and should be durable, tough and resistant to wear on drying, and not harmful to the users.
21. Fat lime is used for whitewashing.
22. For the effective control of termites, the site for the construction of buildings should be cleared of all old tree stumps, dead wood, plant remains *etc.*

Exercises/Questions

1. What do you mean by foundation and what are its main objectives?
2. What is the bearing capacity of soil?
3. How width of foundation is calculated? Discuss with the help of relationship.
4. Discuss Rankine's formula for the calculation of minimum depth of foundation.
5. What are the main functions of roofs?
6. What is the main functions of a floor and what are its main components?
7. Briefly discuss the construction details of cement concrete floor.
8. Briefly discuss RCC flooring.
9. Differentiate between load-bearing and non-load bearing walls.
10. Why is plastering required for walls?
11. What are the requirements of a good plaster? Discuss the process of plastering.
12. What are the main characteristics of a good paint?
13. What are the main uses of paints?
14. What is white washing and how is white wash prepared?
15. What is the purpose of damp proofing course (DPC)? How is it laid?
16. Write short notes on:
    (a) Thatch roof (b) Roofing tiles
    (c) Asbestos sheet roof (d) Sandstone flooring
    (e) Stone-slab flooring (f) Moorum flooring
    (g) Brick-on-edge flooring (h) Termite protection of buildings
    (i) Maintenance of buildings

Chapter 5

Water Supply and Sanitation

Water supply and sanitation are fundamental services that underpin the well-being of communities and societies worldwide. Access to clean and safe water is a basic human right and a cornerstone of public health. Likewise, the proper management of wastewater is essential for protecting the environment and preventing the spread of waterborne diseases. Both these aspects are discussed in this chapter.

WATER SUPPLY

Water is a transparent, tasteless, odourless, and nearly colourless chemical substance that is the main constituent of earth's streams, lakes, oceans, and the fluids of most living organisms. It is essential for the survival of all known forms of life on Earth. Water plays a crucial role in many aspects of human life, from agriculture and industry to transportation and recreation. It is used for irrigation, generating electricity, washing and cleaning, cooking and drinking, and many other purposes. In addition, water is also a vital part of many natural ecosystems, providing habitat for aquatic plants and animals and playing a key role in the water cycle. Despite its importance, water is a finite resource, and many regions of the world are currently facing water scarcity due to a combination of factors such as climate change, overconsumption, and pollution. Therefore, it is essential to conserve and manage water resources sustainably to ensure that future generations have access to this vital resource.

Water is essential for a farmstead and plays a crucial role in sustaining agricultural operations. Adequate access to water is critical for sustaining agricultural operations and ensuring the health and well-being of both animals and crops. Water is necessary for crop irrigation. It is needed to supply the crops with the required amount of moisture to grow and develop. Without water, crops can wilt and die, leading to reduced yields and even crop failure. Water is also necessary for livestock watering. Animals need access to clean water to drink, and it is essential for their health and well-being. Lack of water can lead to dehydration, which can result in reduced milk production, weight loss, and even death. Water is essential for cleaning and sanitation purposes. Farmsteads require clean water to wash equipment, animal housing, and other farm structures. This is critical to prevent the spread of diseases and ensure the health and safety of both animals and workers. The water need is met by wells, ponds, streams and rivers, but wells are the main source of the farmstead water supply.

5.1 Sources of Water Supply

Hydrological Cycle

The hydrological cycle (Figure 5.1), also known as the water cycle, refers to the continuous process by which water is circulated between the earth's atmosphere, land, and oceans. It involves the evaporation of water from the surface of the earth, the formation of clouds, precipitation in the form of rain, snow, and hail, and the runoff of water back into rivers, lakes, and oceans.

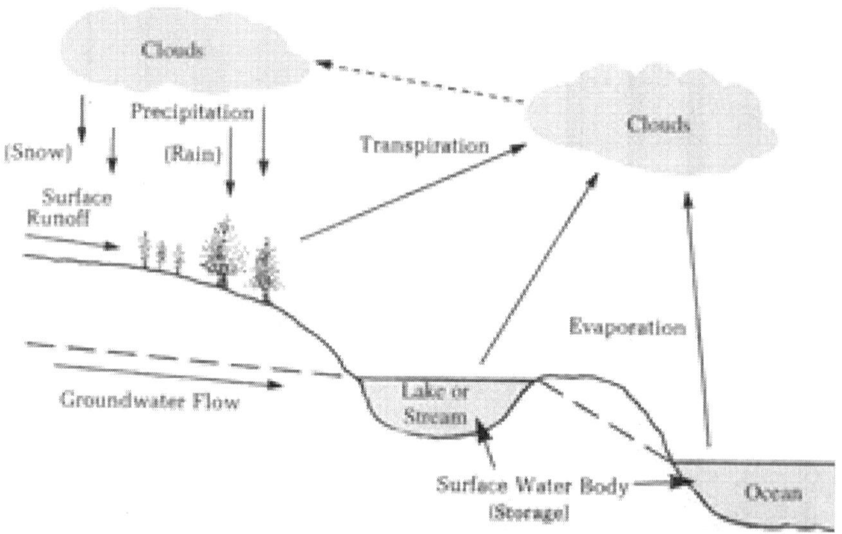

Figure 5.1. Hydrological Cycle.

The hydrological cycle begins with the evaporation of water from bodies of water such as oceans, lakes, and rivers, as well as from the ground and plants through a process called transpiration. This water vapour rises into the atmosphere, where it cools and condenses into clouds. When the clouds become heavy enough, the water droplets fall to the ground as precipitation, which can take many forms, such as rain, snow, sleet, or hail.

Once the precipitation reaches the ground, it can either be absorbed by plants or the soil or runoff into rivers, streams, and oceans, eventually making its way back to the original water source. Some of the water can also seep into the ground, replenishing underground water sources called aquifers. This process of movement of water through various phases and its circulation between different components of the earth is called the hydrological cycle. It plays a vital role in sustaining life on earth by providing freshwater for drinking, irrigation, and other essential needs.

Importance and Necessity of Water Supply System

Water supply systems are critical for ensuring that clean and safe water is available to meet the basic needs of communities, businesses, and industries. Here are some of the reasons why water supply systems are important:

1. **Health**: Access to clean and safe drinking water is essential for maintaining public health. Without reliable water supply systems, people may be forced to rely on contaminated or unsafe sources of water, which can lead to the spread of waterborne diseases.
2. **Sanitation:** Water supply systems are essential for maintaining adequate sanitation, such as flushing toilets and washing hands, which helps prevent the spread of disease.
3. **Agriculture and Industry**: Water is also critical for agricultural and industrial uses. Reliable water supply systems ensure that businesses and industries have access to the water they need to operate, which can help drive economic growth.
4. **Environment:** Water supply systems can also play an important role in protecting the environment by ensuring that water resources are managed in a sustainable way. This includes managing water use, protecting water quality, and preventing water waste.

There are several sources of water supply. These are:

Surface Water: This includes streams, rivers, lakes, ponds, and reservoirs. Surface water is often treated to remove contaminants and made suitable for human consumption. The water yield of the surface sources varies from season to season. In mountainous region, streams are formed by the runoff. Small streams are useful as a source of water only for small villages. Large and perennial stream may be used as source of water supplied by providing storage reservoirs across them. The streams feed their water to the lakes or rivers. Rivers are the most important sources of water for public water supply schemes. Rivers may be perennial or non-perennial. Perennial rivers have water throughout the year. They are fed by rains during rainy season and by snow during summer season.

Groundwater: This is water that is stored underground in aquifers. Groundwater is accessed through wells, and it is often of high quality since it is naturally filtered through soil and rocks. Shallow dug wells (open type) and drilled wells with hand pumps are used if there is only a small water requirement. Where the water requirement is higher than 20,000 litres per hour, high yielding open wells, tube wells, are necessary to meet the need.

Rainwater: Rainwater can be collected and used for non-potable purposes such as irrigation, washing, and cleaning. It can also be treated for potable use in areas where other water sources are limited.

Desalination: This involves the process of removing salt and minerals from seawater to make it suitable for human consumption. Desalination is commonly used in coastal regions where freshwater sources are limited.

Wastewater Treatment: This involves the treatment and reuse of wastewater, also known as recycled water. Recycled water is often used for non-potable purposes such as irrigation and industrial use.

The water need of the farmstead (Figure 5.2) is mainly met by the surface water and the ground water such as wells, ponds, streams and rivers.

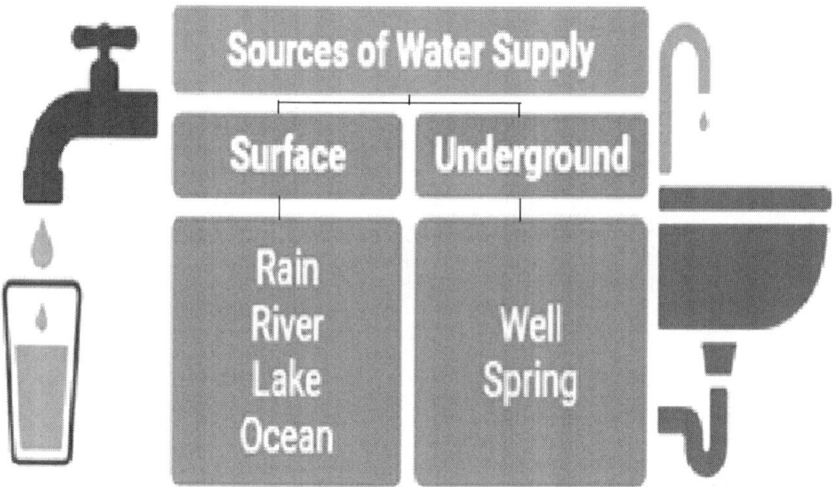

Figure 5.2. Sources of Water Supply.

5.2 Quantity of Water Required

The estimated water demand can be used to determine the size of the water storage and distribution facilities needed to meet the needs of the city or town or farmstead. It is important to consider factors such as population growth, seasonal variations in demand, and potential losses due to leaks and other factors when designing a water supply system.

Farmstead Water Requirement

The water requirement for a farmstead will depend on a variety of factors, including the size of the property, the types of crops or animals being raised, and the climate in the area. For example, in hot and dry regions, more water will be needed to irrigate crops and provide drinking water for animals. Thus, the demand for water is increased during hot summer month and is considerably reduced in winter. In general, water usage on a farmstead can be broken down into following categories:

1. **Household Use**: If there is a residence on the farmstead, water will be needed for drinking, cooking, bathing, and other household uses.
2. **Cleaning and Sanitation**: Water will also be needed for cleaning and sanitation purposes, such as washing equipment, barns, and animal enclosures.
3. **Livestock Watering**: Animals on the farmstead will need a constant supply of fresh water to drink. The amount of water needed will depend on the number and type of animals being raised.
4. **Crop Irrigation**: Depending on the types of crops being grown and the irrigation method used, water usage can vary widely. For example, some crops may require daily watering while others may only need irrigation once or twice a week.

Norms of Water Supply for Human Beings

The water demand for domestic use varies depending on factors such as climate, population density, and lifestyle. It includes the water required for everyday household activities such as drinking, cooking, bathing, washing of clothes and utensils, cleaning of house, flushing of water closets, *etc.* As per IS: 1172 – 1993, water required for domestic purposes for average Indian conditions per head per day may be taken as 135 litres. The demand for kitchen garden is about 1.5 to 2 litres/sqm/day. Table 5.1 shows the details of water requirement for domestic purposes.

Table 5.1. Minimum Domestic Water Consumption for Small Indian Towns and Cities (Annual average)

Sl.No.	*Purpose*	*Consumption (litre/head/day)*
1.	Drinking	5
2.	Cooking	5
3.	Bathing	55
4.	Washing of clothes	20
5.	Washing of utensils	10
6.	Washing and cleaning of houses	10
7.	Flushing of water closets *etc.*	30
	Total	**135**

Norms of Water Supply for Animals

The norms of water supply for livestock can vary depending on the type of animal, age, weight, activity level, feed and environmental conditions. As a general rule of thumb, adult cattle require around 70-100 litres of water per day. Sheep and goats require around 4-14 litres per day, while pigs need around 10-20 litres per day. However, during hot and dry weather conditions or when animals are lactating or pregnant, their water requirements can increase significantly. In such situations, it is essential to provide additional water to ensure the health and productivity of the animals. Table 5.2 presents some general guidelines for water consumption by livestock. It's important to note that these are general guidelines, and specific water requirements may vary depending on the animal's individual needs and environmental conditions.

Table 5.2. Water Requirements for Livestock

Sl.No.	*Livestock*	*Water Requirements (litres/animal/day)*
1.	Calf	30 to 50
2.	Heifer	55 to 100
3.	Cattle (mature cow)	
	Dry	70 to 100
	Lactating	80 to 120
4.	Horses	40 to 50
5.	Sheep and goats	4 to 14
6.	Pigs	10 to 20
7.	Poultry (100 birds)	10 to 30
8.	Dogs	8 to 12

Demand of Water for a Town or City

The water supply for a city or town or village is estimated by considering the various types of water demands. These are:

1. **Domestic Water Demand** - For residential buildings, a minimum water supply of 200 litres per head per day should be assured for a full flushing system. Out of 200 litres per head per day, 45 litres per head per day is meant for flushing arrangement and remaining quantity for other domestic purposes.
2. **Institutional and Commercial Water Demand** – Water required for office buildings, warehouses, hotels, shopping centres, schools, temples, cinema halls, railway stations, bus stands, *etc.*
3. **Industrial Water Demand** – Water required in factories.
4. **Fire Demand** – Water required for fighting a fire outbreak.

5. **Public Use Demand** – Water required for public utility such as washing and sprinkling on roads, watering of parks, gardens, fountains, *etc.*
6. **Water Required to Compensate Losses in Waste and Thefts** – Water loss due to theft, bad plumbing, overflow, *etc.*
7. **Per capita demand** is defined as total yearly consumption of water for a water supply scheme divided by the population of that area and the number of days in a year. It is expressed in litres per capita per day.

$$\text{Per capita demand} = \frac{\text{Total consumption of water in litres}}{\text{Population} \times 365}$$

The break-up of per capita demand for an Indian city is given in Table 5.3.

Table 5.3. Break-up of per Capita Demand of Water

Sl.No.	*Use*	*Demand (Litre/capita/day)*
1.	Domestic use	200
2.	Institutional and commercial use	20
3.	Industrial use	50
4.	Fire demand	15
5.	Public use	10
6.	Waste and thefts	55
	Total	**350**

5.3 Quality of Water

The quality of water can be defined as its suitability for a particular purpose, such as drinking, irrigation, or recreational activities. Water quality is determined by its physical, chemical, and biological characteristics. Physical characteristics of water include its colour, odour, taste, temperature, and turbidity. Chemical characteristics refer to the presence of various substances, such as nutrients, heavy metals, organic compounds, and pesticides. Biological characteristics are related to the presence of microorganisms, such as bacteria and viruses that can cause disease.

Water quality can be affected by various natural and human-made factors, such as climate, geological conditions, land use, wastewater discharges, and agricultural practices. Poor water quality can have adverse effects on human health and the environment, including the contamination of drinking water sources, the spread of waterborne diseases, and the degradation of aquatic ecosystems.

Drinking Water Standards

Drinking water standards are designed to protect public health by ensuring that drinking water is safe and free from harmful contaminants. Drinking water standards set limits on the concentration of various contaminants in public drinking water supplies. The specific standards can vary from country to country and may be established by governmental agencies or international organizations. Compliance with these standards is typically monitored by local water utilities or other regulatory agencies to ensure that the water supply is safe for human consumption.

The drinking water standards in India are set by the Bureau of Indian Standards (BIS) and the Ministry of Drinking Water and Sanitation (MDWS). The primary drinking water standard is the IS 10500:2012 standard, which specifies the maximum permissible limits for various physical, chemical, and microbiological parameters in drinking water. Some of the key parameters and their maximum permissible limits (Table 5.4) under the IS 10500:2012 standard are:

Table 5.4. Maximum Permissible Limit for Various Parameters in Drinking Water

Sl.No.	Parameter	Maximum Permissible Limit
1.	pH	6.5 to 8.5
2.	Total dissolved solids (TDS)	500 mg/L
3.	Total hardness	600 mg/L
4.	Calcium	75 mg/L
5.	Magnesium	30 mg/L
6.	Chloride	250 mg/L
7.	Nitrate	45 mg/L
8.	Fluoride	1 mg/L
9.	Arsenic	0.01 mg/L
10.	Lead	0.01 mg/L
11.	Coliform bacteria	Absent in 100 mL of water

Impurities of Water

Water can contain various impurities, which can affect its quality and suitability for human consumption. Some common impurities of water and their effect is as follows:

1. **Microorganisms**: Water can contain various types of bacteria, viruses, and other microorganisms that can cause illness and disease, such as *E. coli*, *Salmonella*, and *Cryptosporidium*. Waterborne microorganisms such as bacteria and viruses can cause a range of illnesses, including

gastrointestinal diseases, cholera, typhoid fever, and dysentery. These illnesses can lead to symptoms such as diarrhoea, vomiting, and dehydration.

2. **Inorganic Substances**: These are substances that do not contain carbon and are typically found in rocks and minerals. Examples include heavy metals such as lead, arsenic, and mercury, as well as minerals such as calcium, magnesium, and sodium. Heavy metals such as lead, arsenic, and mercury can accumulate in the body and cause long-term health effects such as kidney damage, neurological problems, and cancer. Minerals such as calcium, magnesium, and sodium can affect the taste and hardness of water but are generally not harmful to human health.
3. **Organic Substances**: These are substances that contain carbon and are often associated with human activity. Examples include pesticides, herbicides, and industrial chemicals such as benzene and toluene. Pesticides, herbicides, and industrial chemicals can cause a range of health problems, including cancer, reproductive problems, and developmental issues. These substances can also have a negative impact on the environment and wildlife.
4. **Suspended Solids**: These are small particles that are suspended in water, such as soil, silt, and clay. They can make water appear cloudy and affect its taste and odour. These can cause cloudiness in water, affect its taste and odour, and clog plumbing and appliances.
5. **pH**: The pH of water refers to its acidity or alkalinity. Water that is too acidic or too alkaline can cause skin irritation, and long-term exposure can cause damage to plumbing and appliances. It can also affect its taste.
6. **Hardness**: Hardness refers to the concentration of dissolved minerals such as calcium and magnesium in water. Hard water can cause scaling in plumbing and appliances, affect the taste of water, and reduce the effectiveness of soap and detergent.

The presence of these impurities can affect the taste, odour, and colour of water, and in some cases, can pose a health risk to those who consume it. It is essential to treat water to remove impurities and ensure its safety for human consumption. Water treatment methods may include filtration, disinfection, reverse osmosis, and distillation.

Water and Health

The linkage between water and health is significant and complex. Access to clean and safe water is essential for human health, and inadequate access to water can result in a wide range of health problems. Water is essential for hydration. Water makes up about 60 per cent of the human body, and it is essential for many bodily functions, including regulating body temperature, lubricating joints, and transporting nutrients and oxygen. Lack of access to safe water can lead to waterborne diseases such as cholera, typhoid, and dysentery which are caused by drinking water that has been contaminated with bacteria, viruses, or parasites. These diseases can cause severe diarrhoea, vomiting, and dehydration, and can be life-threatening, especially for young children and people with weakened immune systems. Poor water quality can affect entire communities, leading to widespread health problems such as skin infections, gastrointestinal illnesses, and even chronic diseases like cancer.

The presence of fluoride, arsenic, and nitrate in drinking water can have various effects on human health depending on their concentration and duration of exposure. Excessive intake of fluoride can lead to dental and skeletal fluorosis, which can cause pain, stiffness, and even deformities in bones and joints. Arsenic is a toxic substance that can occur naturally in groundwater and some industrial processes. Long-term exposure to arsenic can cause skin lesions, cardiovascular diseases, and different types of cancers, including skin, bladder, and lung cancer. Nitrate is often found in fertilizers and can enter drinking water sources through agricultural runoff. Excessive levels of nitrate in drinking water can cause methemoglobinemia or "blue baby syndrome," which reduces the blood's ability to carry oxygen and can be fatal to infants. The Bureau of Indian Standards (BIS) has established guidelines for the maximum allowable levels of fluoride, arsenic, and nitrate in drinking water to ensure safe consumption.

Hardness of Water

Water hardness is a measure of the amount of dissolved minerals, particularly calcium and magnesium, present in water. It is usually expressed in terms of the concentration of calcium carbonate ($CaCO_3$) in parts per million (ppm) or milligrams per litre (mg/L). Water with a hardness below 60 mg/L is considered soft, while water with a hardness above 120 mg/L is considered hard. Hard water is not harmful to human health, but it can cause several problems, such as scaling in pipes and appliances, reduced soap lathering, and a rough texture on clothing after washing. Additionally, hard water can affect the taste of beverages such as tea and coffee, and it may cause skin irritation in some people. On the other hand, soft water is more desirable for household use because it is less likely to cause these issues. Soft water may also be more beneficial for the environment, as it typically requires less detergent and energy to heat for washing and cleaning. However, it's worth noting that some minerals in hard water, such as calcium and magnesium, are beneficial for human health, and their absence in soft water may require alternative sources to ensure proper nutrition.

Treatment of Hard Water

Hard water is water that contains high levels of minerals such as calcium and magnesium, and can be a problem in many areas. Here are a few methods to treat hard water:

1. **Water Softeners**: A water softener is an ion exchange system that removes the minerals that cause hardness from the water. This system uses a tank filled with resin beads that attract and remove the calcium and magnesium ions from the water. The water then flows out of the tank, free of the minerals that cause hardness.
2. **Reverse Osmosis**: Reverse osmosis is a water treatment process that removes minerals, particles and impurities from water by forcing it through a semipermeable membrane under high pressure. This process is effective in removing minerals that cause hardness, as well as other impurities in the water.
3. **Distillation**: Distillation is a process that involves boiling water to create steam, which is then condensed back into water. The minerals that cause hardness are left behind during the boiling process, and the resulting water is soft.
4. **Chelating Agents**: Chelating agents are chemicals that can be added to the water to bind with the minerals that cause hardness, making them soluble and preventing them from forming scale or build up. This method is less common than the others, but can be effective for some types of hard water.

5.4 Water Treatment

The raw water which is found in various natural sources cannot be directly used by the public before removing the impurities. The process of removing contaminants and impurities from water to make it safe for human consumption or other purposes is called as water treatment. The water should be free from unpleasant taste, odours, and must have sparkling appearance. It should be free from disease spreading germs.

The main objectives of water treatment are:

1. **To remove or reduce impurities**: The primary objective of water treatment is to remove or reduce impurities from the water to make it safe for human consumption and use. Impurities can include organic matter, bacteria, viruses, minerals, chemicals, and other contaminants.
2. **To improve water quality**: Water treatment aims to improve the overall quality of water by removing unwanted substances that can affect its colour, taste, and odour.
3. **To protect public health**: By removing harmful substances from water, water treatment helps protect public health by reducing the risk of waterborne diseases and illnesses.
4. **To ensure compliance with regulations**: Water treatment also aims to meet regulatory standards for water quality and safety. These regulations are set by government agencies to protect public health and the environment.

5. **To protect the environment**: Water treatment also has a role in protecting the environment by reducing the amount of pollutants discharged into waterways and reducing the demand for new water sources.

Treatment Processes

There are several different water treatment processes, each designed to remove specific contaminants or impurities from water. The specific process used will depend on the source of the water and the contaminants or impurities that need to be removed. Some of the most common water treatment processes include:

1. **Screening:** In this process, a screen is provided in the front of the pumps or the intake works, so as to exclude large sized particles such as leaves, dead animals *etc.*
2. **Plain Sedimentation:** In this process, water is allowed to stand in a tank or basin for a period of time, and the heavier particles settle to the bottom of the tank while the clearer water is removed from the top. Suspended impurities such as silt, clay, sand *etc.* are removed by this process.
3. **Sedimentation with Coagulation:** This process involves adding chemicals such as alum or ferric chloride to water, which causes small particles to clump together into larger particles called flocks. Thereafter, the flocks settle to the bottom of the tank, leaving clear water on top. Fine suspended matter are removed by this process.
4. **Filtration:** Once the water has settled, it is passed through a series of filters, typically made of sand, gravel, and charcoal. These filters remove any remaining particles and impurities from the water such as micro-organism and colloidal matters.
5. **Disinfection:** The water is disinfected using chemicals such as chlorine or ozone to kill pathogenic bacteria and other harmful microorganisms.
6. **Aeration:** This process involves exposing water to air, which helps to remove dissolved gases, odours, and tastes.

There are also additional treatment methods that can be used depending on the specific contaminants present in the water, including reverse osmosis, ion exchange, and activated carbon treatment. The goal of water treatment is to produce clean and safe water that is suitable for human consumption and other uses.

1. **Reverse Osmosis**: In this process, Water is forced through a membrane that filters out impurities and contaminants.
2. **Ion Exchange**: This process involves passing water through a resin bed that exchanges ions with the water to remove contaminants like lead, arsenic, and nitrates.

3. **Membrane Filtration**: This process involves using a membrane to filter out impurities in the water, including bacteria, viruses, and other microorganisms.

5.4.1 Screening

Screening is a commonly used method of water treatment that involves the physical removal of large solids and debris from water. This is achieved through the use of screens or sieves, which can be made from a variety of materials, such as wire mesh or perforated plates. The purpose of screening is to protect downstream treatment processes, such as sedimentation or filtration, from damage or clogging by large particles. Screening is typically used as a pre-treatment step in the water treatment process, and can be used in both industrial and municipal applications.

Screening can be performed using a variety of screen types, including stationary screens, rotating screens, and vibrating screens. Each type of screen has its own advantages and disadvantages, and the selection of the appropriate type of screen will depend on the specific application. Overall, screening is a simple and effective method of water treatment that can be used to remove large solids and debris from water, and is an important component of many water treatment processes.

There are two types of screening

1. Coarse screening
2. Fine screening

Both coarse and fine screening are important steps in the water treatment process as they help to prevent clogging of downstream equipment, protect the environment, and ensure the water is safe for use.

Coarse screening is the initial stage of the screening process and is typically the first step in removing large debris such as branches, leaves, rocks, and other large objects that may be present in the water. This process is typically done using a bar screen, which consists of a series of parallel bars or iron rods placed across the flow of the water. The bars are spaced a certain distance apart (2 to 10 cm centre to centre), depending on the size of the debris being removed, and the water flows through the bars while the debris is caught on the upstream side. Coarse screen bars are generally kept at an angle 18° to 60°. The main purpose of this is to increase the opening area of screen and thus reducing the velocity of flow. The velocity of flow through the screen should not exceed 0.75 to 1.0 m/sec.

Fine screening is the second stage of the screening process and is used to remove smaller particles from the water. This process is typically done using a screen with a finer mesh (less than 10 mm openings) than the coarse screen, and the water is passed through the screen to remove particles such as sand, silt, and other small debris. Fine screening can be done using various types of screens, including micro-screens, drum screens, and disc filters, depending on the size and type of particles being removed. The only disadvantage is that the fine screens gets clogged frequently and require frequent cleaning operations. That's why fine screens are not used these days.

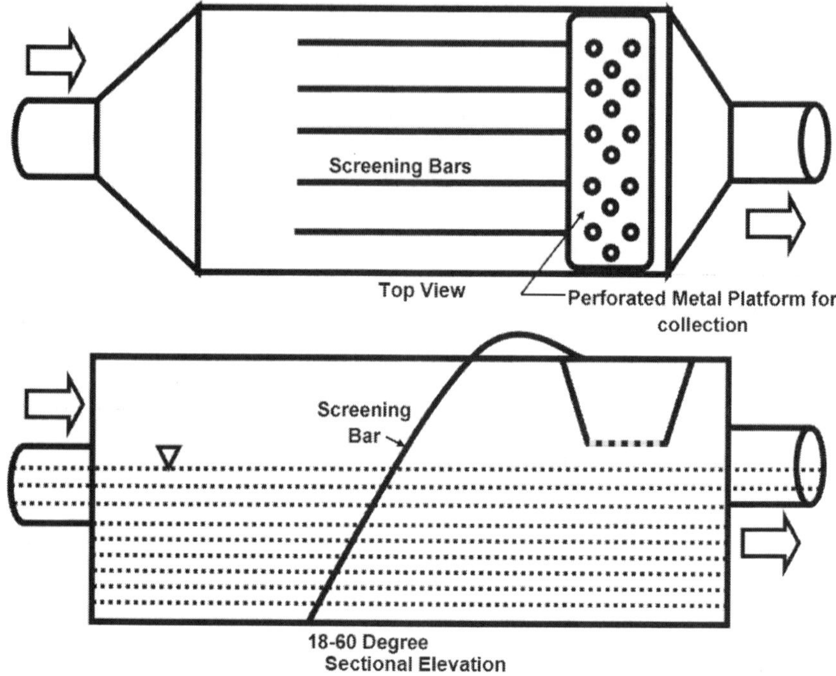

Figure 5.3. Fixed Bar Type Screen.

The debris accumulated on the screens are removed from time to time either manually or mechanically to allow free movement of water.

5.4.2 Plain Sedimentation

Plain sedimentation is an effective and widely used process for the removal of suspended solids from water, and it can be used in combination with other treatment processes for improved water quality. During sedimentation, water is allowed to stand in a tank or basin for a period of time, and the heavier/higher density particles settle to the bottom of the tank due to action of gravity while the clearer water is removed from the top. This process is also known as settling. Sedimentation is typically the first step in the water treatment process and is used to remove large particles, such as sand, clay, and silt, from the water. These particles are removed to prevent them from clogging filters and other equipment downstream in the treatment process.

Following are the advantages of plain sedimentation:

1. This process decreases the load on subsequent processes.
2. Cleaning cost of the tank is less.

3. Very less quantity of chemicals are needed for the subsequent processes.

Factors affecting Plain Sedimentation

Plain sedimentation is the settling of suspended particles in a liquid under the influence of gravity. Several factors can affect the process of plain sedimentation, including:

1. **Particle Size and Shape**: The settling velocity of particles depends on their size and shape. Larger and denser particles settle faster than smaller and lighter particles.
2. **Concentration of Particles**: The settling rate of particles decreases as the concentration of particles in the liquid increases.
3. **Density of the Liquid**: The settling rate of particles is influenced by the density of the liquid. A higher density liquid can support larger and denser particles.
4. **Temperature of the Liquid**: The settling rate of particles is influenced by the temperature of the liquid. A higher temperature can reduce the viscosity of the liquid, which can increase the settling rate.
5. **Turbulence and Agitation**: Turbulence and agitation can disrupt the settling process by suspending particles and preventing them from settling.
6. **Shape and Size of the Sedimentation Vessel**: The shape and size of the sedimentation vessel can affect the settling process by influencing the flow of liquid and the distribution of particles.
7. **Presence of Flocculants**: The addition of flocculants can increase the settling rate by causing particles to agglomerate and settle faster.
8. **pH and Ionic Strength of the Liquid**: The pH and ionic strength of the liquid can influence the surface charge of particles, which can affect their settling behaviour.

There are two methods of plain sedimentation. These are:

1. **Draw and Fill Method** – In this method, the water is filled in the tank and allowed to remain in the tank so that suspended impurities may settle. The water is then taken out from the tank when all the particles has been settled down.
2. **Continuous Flow Type Basin Method** – In this method, water flows continuously at uniform rate through the tank. All the suspended impurities settle in the bottom of the tank and clear water is drawn from the outlet. The design of the tank is such that water has to move a long distance and during this movement suspended particles settle down. (Figure 5.4)

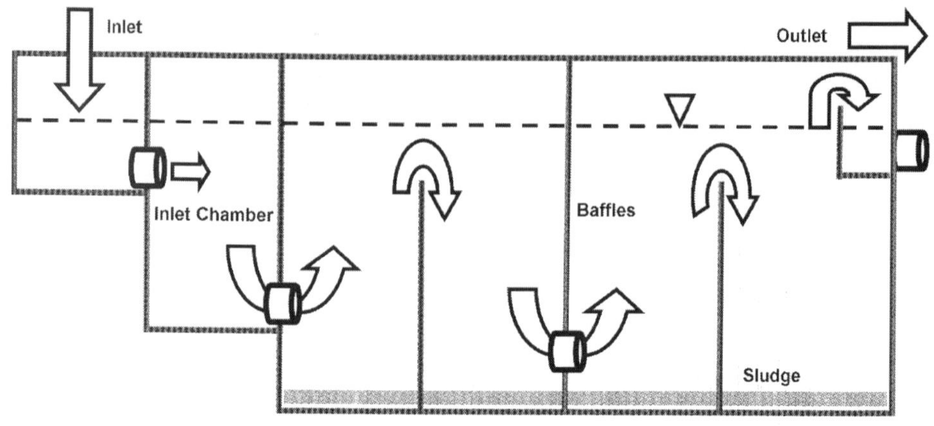

Figure 5.4. Continuous Flow Type.

The sedimentation tanks may be rectangular or circular in plan. Rectangular tanks are rectangular in shape and are used for large-scale water treatment processes. They are commonly used in water treatment plants. While, circular sedimentation tanks are circular in shape and are used for smaller water treatment processes. They are often used in small communities and in industrial processes.

5.4.3 Sedimentation with Coagulation

When water contains large amount of very fine and light colloidal impurities, it is practically difficult to eliminate them by plain sedimentation. However, they can be removed easily by increasing their size by changing them into flocculated particles.

The sedimentation process can be enhanced by adding chemicals, such as coagulants and flocculants, to the water. Coagulants such as alum or ferric chloride cause small particles to clump together, making them easier to settle out of the water. This process is known as coagulation. Flocculants such as polyelectrolytes or starch help the particles to form larger, heavier masses that settle more quickly. This process is known as flocculation.

5.4.4 Filtration

The water after treatment of screening, sedimentation and sedimentation with coagulation may also contain very fine suspended particles and bacteria in it. In order to remove the remaining impurities, the water is passed through a porous medium. The process of passing the water through a porous medium is known as filtration. The porous medium can be made up of sand, gravel, activated carbon, or other materials that have a specific size and shape to capture the impurities. Filtration helps in removing bacteria, colour, taste, odour, and thus producing clear and sparkling water.

The type of filtration system used depends on the quality of water to be treated and the specific impurities that need to be removed. Water treatment plants use a combination of filtration systems to achieve the desired level of water quality.

Filter Materials

Either fine sand or coarse sand is commonly used as filter media in water treatment applications. The layers of sand is supported on gravel and this layer permits the filtered water to move freely to the bottom. Each type of media has its own unique properties that make it well-suited for certain types of water treatment. The choice of filter media depends on the specific water treatment needs and the desired level of filtration. A combination of different media types may be used in multi-layer filtration systems to achieve optimal results.

The salient characteristics of the filter materials are as follows:

Fine Sand

1. Free from dirt and impurities.
2. Effective particle size ranges from 0.35 to 0.60 mm.
3. Has a high surface area to volume ratio, making it effective in removing small suspended particles, turbidity, and coliform bacteria.
4. Has a low porosity, which can lead to a high head loss and frequent backwashing.
5. Can be used alone or as part of a multi-layer filtration system.

Coarse Sand

1. Effective particle size ranges from 0.60 to 2.00 mm.
2. Has a larger pore size than fine sand, allowing for higher flow rates and lower head loss.
3. Less effective in removing small particles than fine sand, but can effectively remove larger suspended solids and some pathogens.
4. Typically used as a support layer for finer media in a multi-layer filtration system.

Gravel

1. Hard, durable, properly rounded and free from impurities.
2. Has a minimum density of 1600 kg/m^3.
3. Typically consists of particles larger than 2.00 mm.
4. Has a very large pore size, allowing for high flow rates and low head loss.
5. Not effective in removing small particles, but can effectively remove larger suspended solids and debris.

6. Typically used as a bottom layer in a multi-layer filtration system to provide support for finer media and prevent clogging.

There are various types of filtration systems used in water treatment, but they all aim to achieve the same goal, which is to remove impurities and produce clean water. Some of the most commonly used filtration systems include:

1. **Slow Sand Filtration**: This type of filtration system is one of the oldest and simplest forms of filtration. It uses a bed of sand as a filter medium, and water is slowly passed through it. The sand bed traps impurities and microorganisms, and the filtered water is collected at the bottom of the filter bed. Slow sand filtration is effective in removing suspended solids and some bacteria, but it is not very effective in removing dissolved solids.
2. **Rapid Sand Filtration**: This type of filtration system is similar to slow sand filtration, but it uses a coarser sand medium, and the water is passed through the sand at a faster rate. Rapid sand filtration is more effective in removing suspended solids, bacteria, and viruses than slow sand filtration, but it does not remove dissolved solids.
3. **Dual Media Filtration**: This type of filtration system uses a combination of two filter media, such as sand and anthracite. The coarser sand layer removes larger particles, while the finer anthracite layer removes smaller particles. Dual media filtration is more effective in removing suspended solids and some dissolved solids than slow or rapid sand filtration.
4. **Activated Carbon Filtration**: This type of filtration system uses activated carbon as the filter medium. Activated carbon has a large surface area that can adsorb impurities such as chlorine, pesticides, and organic compounds. Activated carbon filtration is effective in removing dissolved solids and some chemicals that other filtration systems cannot remove.
5. **Reverse Osmosis**: This type of filtration system uses a semi-permeable membrane to remove impurities from water. The water is forced through the membrane under high pressure, and the impurities are trapped on the other side of the membrane. Reverse osmosis is effective in removing dissolved solids, bacteria, viruses, and some chemicals.

Types of Filters

According to type of filters used for filtration, it is mainly of two types:

1. Slow sand filter
2. Rapid sand filter
3. Gravity type
4. Pressure type

Slow Sand Filter

Slow sand filter was introduced in 1829 and was devised by James Simps who was employed at Chelsea Waterworks Company, London. It uses a bed of sand as a filter medium, and water is slowly passed through it. The sand bed traps impurities and microorganisms, and the filtered water is collected at the bottom of the filter bed. Slow sand filtration is effective in removing suspended solids and some bacteria, but it is not very effective in removing dissolved solids.

A slow sand filter is a type of water treatment system that uses a layer of sand to remove impurities and pathogens from water. The filter is designed to operate at a slow flow rate, typically 0.1 to 0.3 meters per hour, to allow sufficient time for the sand to effectively remove the contaminants from the water.

The slow sand filter (Figure 5.5) works by allowing water to pass through a layer of sand that has been specifically designed to trap and remove impurities. As the water passes through the sand, the impurities become trapped in the spaces between the sand particles. The sand layer also contains a layer of beneficial bacteria, known as the "schmutzdecke," which helps to break down organic matter and further remove impurities.

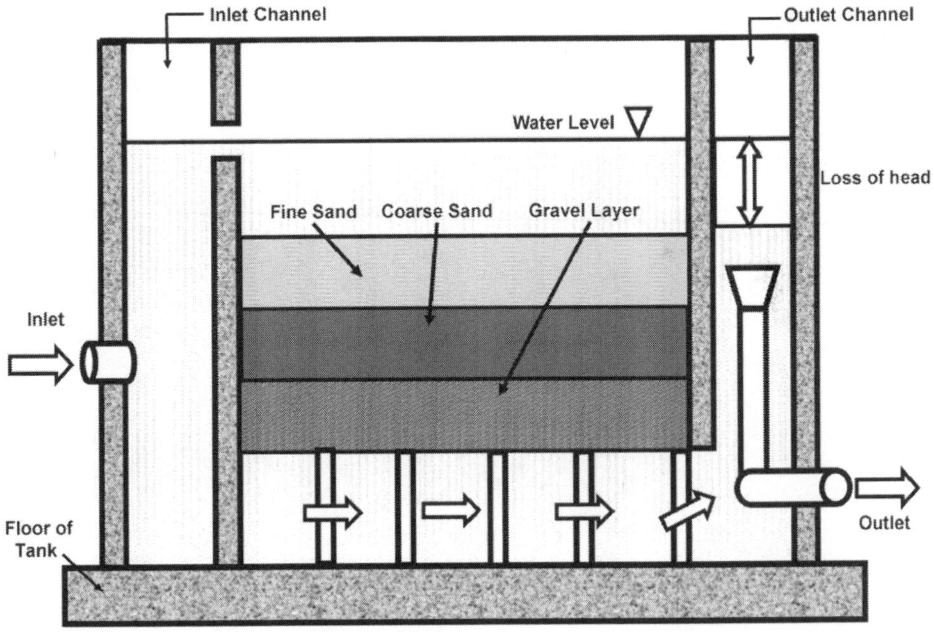

Figure 5.5. Slow Sand Filter (Section elevation).

To maintain the effectiveness of a slow sand filter, it is important to periodically clean the sand layer to remove any accumulated debris and to prevent clogging. The frequency of cleaning will depend on the quality of the incoming water, but typically ranges from several months to several years.

Slow sand filters are commonly used in developing countries as a low-cost, low-maintenance water treatment solution. They can be used to treat water from a variety of sources, including rivers, lakes, and groundwater. However, they may not be suitable for treating water that is heavily contaminated or contains high levels of dissolved minerals.

Rapid Sand Filter

A rapid sand filter is a common type of water treatment filter used to remove impurities and contaminants from water. It consists of a bed of fine sand or other granular material, through which water is passed at a high rate. The filter is designed to remove suspended particles, organic matter, and some microorganisms, as well as to improve the taste, odour, and clarity of the water. These were developed during 1900 – 1910 in USA.

The basic principle of operation of a rapid sand filter (Figure 5.6) is as follows: water is introduced into the top of the filter bed, and flows downward through the sand bed. As the water passes through the sand, suspended particles and other impurities are trapped in the interstices between the sand grains. The filtered water is then collected at the bottom of the filter, and discharged for further treatment or distribution.

To maintain the effectiveness of the filter, regular backwashing is required. During backwashing, water is passed through the filter in the opposite direction, dislodging and removing the accumulated impurities from the sand bed.

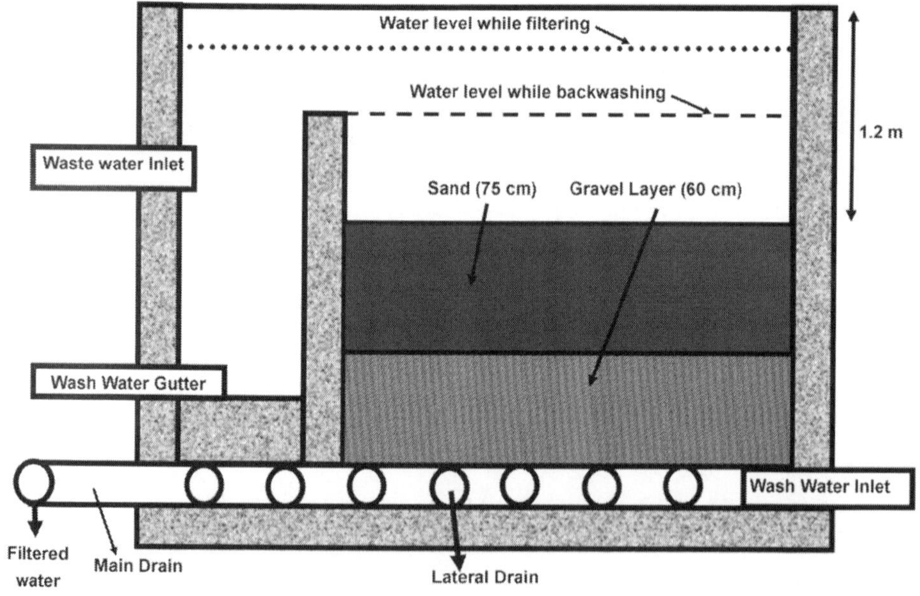

Figure 5.6. Rapid Gravity Filter.

Rapid sand filters are widely used in municipal water treatment plants, as well as in many industrial applications. They are relatively simple and cost-effective to operate, and can remove a wide range of contaminants from water. However, they are not effective at removing dissolved contaminants, such as salts or heavy metals, and may require additional treatment steps for this purpose.

Comparison between Slow Sand and Rapid Gravity Filters

1. **Operating principle**: In slow sand filters, water is passed through a bed of fine sand at a slow rate (0.1-0.3 meters per hour). The sand acts as a physical filter, and biological processes within the filter layer also help to remove suspended particles, organic matter, and some microorganisms. While, in rapid sand gravity filters, water is passed through a bed of coarser sand at a faster rate (2-3 meters per hour), typically under the influence of gravity. The sand bed is cleaned periodically by backwashing with water or air.
2. **Size of one unit**: Slow sand filters requires very large area as compared to rapid gravity filters. The size of one unit of slow sand filter is 30 m × 60 m while rapid gravity filters require 6 m × 8 m to 8 m × 10 m.
3. **Filtration Rate**: The main difference between slow and rapid gravity filters is the rate of filtration. Slow gravity filters typically have a filtration rate of 100 –180 litres/m^2/hour, while rapid gravity filters have a filtration rate of 4000 – 5000 litres/m^2/hour. As the names suggest, slow filters take longer to filter the water than rapid filters.
4. **Depth of Media/Quantity of Sand:** Slow filters have a deeper media bed than rapid filters. This is because slow filters rely on biological processes to remove impurities, and a deeper bed allows for more contact time between the water and microorganisms. Rapid filters rely more on physical filtration, so a shallower bed is sufficient. Thus, rapid filters require less quantity of sand.
5. **Maintenance**: Slow sand filters require less maintenance than rapid sand gravity filters, as the biological processes within the filter layer help to maintain the filter bed and prevent clogging. However, slow sand filters may require periodic cleaning or replacement of the filter layer. Rapid filters may need to be backwashed more frequently to remove accumulated sediment.
6. **Efficiency**: Slow sand filters are highly effective at removing suspended particles, organic matter, and some microorganisms, but they may not be as effective at removing dissolved contaminants or chemicals. While, rapid sand gravity filters are more effective at removing dissolved contaminants and chemicals, as well as suspended particles, but they may not be as effective at removing microorganisms.

7. **Cost:** Slow filters are generally more expensive to construct than rapid filters due to the deeper bed of media and requirement of large land.
8. **Supervision:** Slow sand filters does not require any skilled supervision while skilled supervision is needed in rapid sand gravity filter system.

5.4.5 Disinfection

Disinfection is an important step in water treatment to remove or kill potentially harmful microorganisms such as bacteria, viruses, and protozoa. In this process, water is treated with chemicals to make water safe to public use. It's important to note that disinfection alone is not enough to ensure the safety of drinking water. Proper water treatment, including filtration, is also necessary to remove sediment and other impurities from the water.

There are several methods for disinfecting water, including:

1. **Chlorination**: The most commonly used method for disinfecting water is chlorination. Chlorine gas, chlorine dioxide, or sodium hypochlorite can be added to water to kill microorganisms. Chlorine dioxide is an effective disinfectant that does not produce harmful by-products like chlorine. It is a commonly used method of water purification in many countries.
2. **Ultraviolet (UV) Disinfection**: UV light rays having wavelengths 100 – 400 nm can be used to kill microorganisms in water. The water is exposed to UV light, which damages the DNA of the microorganisms and prevents them from reproducing. This is very effective disinfectant and kill all disease producing bacteria.
3. **Ozonation**: Ozone gas can be used to disinfect water. Ozone is a powerful oxidant that destroys microorganisms by breaking down their cell walls. Ozone removes colour, odour, and taste from water.
4. **UV-ozone**: A combination of UV light and ozone gas can be used to disinfect water. This method is effective against a wide range of microorganisms and does not produce harmful by-products.
5. **Iodine and Bromine**: All pathogens can be killed if these are added to water.
6. **Boiling of Water**: The water can be disinfected by boiling it for 15 to 20 minutes. Boiling water can kill most types of bacteria and viruses. This method is simple and effective, but it can be time-consuming and may not remove all impurities.

5.4.6 Aeration

Aeration is a commonly used process in water treatment that involves the introduction of air into water to improve its quality. Aeration can be used to remove dissolved gases, such as carbon dioxide and hydrogen sulphide, and to increase the concentration of oxygen in the water. Aeration can be an effective way to improve water quality by removing dissolved gases, and thus reducing odour and taste problems, and promoting the growth of beneficial microorganisms. It can also be used in conjunction with other water treatment processes, such as filtration and disinfection, to provide a comprehensive approach to water treatment. The limitations of this process are that it is less economical, not very efficient method of removing or reduction of tastes and odours caused by non-volatile substances, and also there is possibility of air borne contamination of water.

Aeration can be achieved in a variety of ways, including:

1. **Bubble Aeration**: This involves the introduction of air bubbles into the water, typically through diffusers or mechanical means such as turbines or propellers.
2. **Spray Aeration**: This involves spraying water into the air to increase its surface area and promote gas exchange.
3. **Cascading Aeration**: This involves passing water over a series of steps or weirs to increase its exposure to air.
4. **Mechanical Aeration**: This involves agitating the water to increase its exposure to air.

5.5Design of Water Supply System

Designing a water supply system involves several steps, including assessing water demand, identifying water sources, and selecting appropriate treatment and distribution methods. During the design process, it is essential to consider factors such as water quality, cost-effectiveness, and sustainability. It is also important to comply with local regulations and standards to ensure that the system is safe and reliable. Here is a general overview of the steps involved in designing a water supply system:

1. **Assess Water Demand**: The first step in designing a water supply system is to determine the amount of water that will be needed. This involves estimating the population that the system will serve, as well as the water demand for different uses such as domestic, commercial, and industrial.
2. **Identify Water Sources**: Once the water demand has been estimated, the next step is to identify suitable water sources that can meet that demand. This may involve drilling wells, diverting surface water, or harvesting rainwater.
3. **Water Treatment**: Once the water has been sourced, it may need to be treated to remove impurities and make it safe for human consumption. Water treatment methods may include sedimentation, filtration, disinfection, and reverse osmosis.

4. **Water Storage**: Once the water has been treated, it will need to be stored in a reservoir or tank to ensure a constant supply of water. The size of the storage facility will depend on the water demand and the frequency of water delivery.
5. **Water Distribution**: The final step in designing a water supply system is to distribute the water to the consumers. This may involve laying pipes, constructing pump stations, and installing valves and meters.

5.6 Size of Overhead Storage Tank

Estimating the size of an overhead storage tank for a home depends on various factors such as the number of people residing in the house, the average daily water consumption per person, the availability of water supply, and the frequency of water supply. Whatever the source of water supply is, it must be capable of supplying the total quantity of water needed at the rate at which it will be used. Whenever, the rate of water required exceeds the supply, an overhead tank should be provided to store extra water for use at the peak demand period. The size of the storage tank should be such that the total yield of the water during the peak period, plus the water stored in the tank during periods of less demand, must be able to meet the total demand.

Example 5.1

Determine the size of an overhead water storage tank for a family of six members. Average daily water consumption per person is 150 litres per day, and water is supplied on every alternate day.

Solution

Average daily water consumption per person = 150 litres

Total number of persons = 6

Thus, total daily water consumption = 6 × 150 = 900 litres.

Water is supplied on alternate days.

Thus, number of days the tank should last = 2

The capacity of the tank should be such that the water should last in the tank before the next water supply.

Then, the required tank capacity = 900 × 2 = 1800 litres

It's recommended to add some extra capacity to the tank to ensure sufficient water storage in case of any emergency or unexpected water shortages. Let us assume that an extra 20 per cent capacity may be needed to meet the emergency demand.

Thus, final estimated tank capacity = 1800 + (1800 × 20 per cent) = 2160 litres

Therefore, in this case, the estimated capacity of the overhead tank for a home will be around 2160 litres.

Example 5.2

Determine the size of an overhead water storage tank for a farmstead demanding a maximum of about 20,000 litres of water per hour for 2 hours during noon and only about 10,000 litres per hour during rest of the period. The tube well is capable of supplying at the rate of only 14,000 litres per hour.

Solution

The supply of water only during the peak period is less than the demand. Thus, to meet the peak hours demand, an overhead water storage tank is needed.

The capacity of overhead water storage tank is calculated as follows:

Total quantity of water demand during peak period = 20,000 × 2 = 40,000 litres

Total quantity of water supply during peak period = 18,000 × 2 = 28,000 litres

Thus, storage capacity of the overhead tank = 40,000 – 28,000 = 12,000 litres

Example 5.3

Determine the size of an overhead storage tank for a farmstead, if 'A' is the average hourly demand of water calculated over 24 hours period, and the consumption pattern is as follows:

Time	*Consumption of 'A'*
6 a.m. to 10 a.m.	250 per cent of A
10 a.m. to 4 p.m.	Average hourly demand 'A'
4 p.m. to 8 p.m.	140 per cent of A
8 p.m. to 10 p.m.	60 per cent of A
10 p.m. to 6 a.m.	15 per cent of A

The pumping rate is 3A per hour and pumping timing is from 6 a.m. to 2 p.m.

Solution

Period	*Hours*	*Hourly Demand*	*Total Demand*	*Cumulative Demand*	*Pumping*	*Cumulative Pumping*	*Deficit or Surplus*
6 a.m. to 10 a.m.	4	2.50A	10A	10A	3A×4 = 12A	12A	+2A
10 a.m. to 2 p.m.	4	A	4A	14A	3A×4 = 12A	24A	+10A
2 p.m. to 4 p.m.	2	A	2A	16A			+8A

Period	Hours	Hourly Demand	Total Demand	Cumulative Demand	Pumping	Cumulative Pumping	Deficit or Surplus
4 p.m. to 8 p.m.	4	1.40A	5.6A	21.6A			+2.4A
8 p.m. to 10 p.m.	2	0.60A	1.2A	22.8A			+1.2A
10 p.m. to 6 a.m.	8	0.15A	1.2A	24A			-

Thus, the tank should have a capacity of 10A, and in this case the tank will be empty completely by 6.00 a.m.

WELLS AND PUMPS

5.7 Wells

A well is a hole, usually vertical, extending into the water-bearing layer in the ground. Wells have been a crucial part of human civilization for thousands of years. These structures, which allow us to access groundwater, have been used for a variety of purposes, including providing water for drinking, irrigation, and industrial processes. The earliest wells were likely dug by hand, using simple tools like shovels and picks. These wells were often shallow and relied on natural springs or rainfall to replenish the groundwater. Over time, people developed more sophisticated techniques for digging wells, including the use of animals like oxen to pull digging equipment. One of the most famous examples of early well construction can be found in the ancient city of Mohenjo-daro, which was part of the Indus Valley Civilization in what is now Pakistan. Archaeologists have uncovered a network of brick-lined wells in the city that date back over 4,500 years.

Wells continued to be an important part of human society throughout the middle ages and into the modern era. During this time, new technologies such as windmills and hand-operated pumps were developed to make it easier to extract water from deeper wells. In the 19th century, the invention of the steam engine and other industrial technologies led to even more advanced well-drilling techniques. These included the use of rotary drills and powered pumps, which made it possible to drill much deeper wells than had ever been possible before.

Today, wells are used around the world to provide access to clean, safe drinking water. They are especially important in areas where other sources of water are scarce or unreliable. Despite their importance, wells can also pose significant environmental and public health risks if they are not properly constructed and maintained. Poorly constructed wells can allow contaminants to seep into the groundwater, while abandoned wells can become breeding grounds for disease-carrying insects.

5.7.1 Types of Wells

There are several types of wells that can be used for drinking and irrigation purposes. The suitability of a particular type of well depends on various factors, such as the depth of the water table, the quality of the water, and the geology of the area. Wells are classified according to the

1. Type of aquifer supplying water,
2. Depth of well, and
3. Method of construction of the well.

5.7.2 Classification based on the Type of Aquifer Supplying Water

Wells can be classified based on the type of aquifer supplying water. Aquifers are geological formations that contain water and can supply it to wells. Based on the type of aquifer supplying water, the wells may be classified as gravity, artesian or a combination of artesian and gravity.

1. **Gravity Wells**: Gravity wells derives its water from an unconfined aquifer. The level of water in a gravity well is the water table. These wells are also known as percolation wells. These wells are typically shallow, and the water level in the well will fluctuate with changes in the water table.
2. **Artesian Wells**: These wells are drilled into a confined aquifer that is under pressure. When the well is drilled, the water in the aquifer is forced up to the surface, sometimes even above ground level, due to the pressure. These wells are also known as cavitation wells. The water rises to a height equal to the pressure level (piezometric level) of the water in the well.

5.7.3 Classification Based on the Depth of Well

Wells can be classified as shallow wells and deep wells based on depth of the pumping water table from the ground surface. Wells having a pumping water level at 6 meters or less from the ground surface are usually called as shallow well or surface well. While, deep wells are typically deeper than 6 meters.

5.7.4 Classification Based on the Method of Construction of the Well

Wells can be classified as dug, driven and drilled wells based on the method of construction or drilling used to create them. The method selected for constructing a well depends mainly on the depth to which it must be carried, ground water table, and nature of material through which hole is to be dug.

1. **Dug Wells**: Dug wells or open wells are shallow wells that are excavated by hand or with the help of a machine digging a hole into the ground until the water table is reached. These wells may have

walls lined with stones, bricks, or concrete to prevent collapse and contamination of the water source. The diameter of these wells vary from 1.5 to 4.5 metres. They are typically used in areas where the water table is close to the surface. They are generally less expensive to construct than drilled wells, but they may be more vulnerable to contamination from surface water runoff, sewage, and other pollutants. Dug wells have been used for thousands of years to provide a reliable source of drinking water in many parts of the world. However, their use has declined in many areas as more advanced technologies such as drilled wells, pumps, and water treatment systems have become available. Despite this, dug wells remain an important source of water for many rural communities and can be an effective solution for water access in certain circumstances.

2. **Driven Wells**: Driven wells are constructed by driving a pipe into the ground using a special driving tool. These wells are also called sand-point wells. The pipe has a screened or perforated section near the bottom that allows water to flow in. Diameters of driven wells vary from 3 to 10 cm. They can be constructed when it is known that the water table is comparatively shallow and where the overlying soil is not hard so that the pipe can be drawn in easily without injuring the well point.
3. **Drilled Wells (Tube wells)**: A tube well is a type of water well that uses a tube or casing to protect the well from collapse and to facilitate the extraction of water from underground aquifers. The tube well consists of a long, narrow casing made of materials such as PVC or concrete, which is driven into the ground until it reaches the water table. These wells are usually deep wells penetrating more than one aquifer. Diameters of tube wells usually range from 15 to 45 cm. These are created using a drilling rig, which can be mounted on a truck or trailer or set up as a permanent installation. There are several types of drilling methods, including cable tool, rotary, and percussion. The well casing in the aquifer is of perforated pipes or strainer pipes, whereas casing of plain pipes is used in non-water bearing layers. Water from the aquifer flows into the well through the perforations in the strainer pipe. Once the casing is in place, a pump is then installed and connected to a power source on the surface. The pump draws water up from the well and into a storage tank or distribution system, where it can be used for various purposes. The size and capacity of the pump are determined by the volume of water required and the depth of the well.

Tube wells are commonly used in areas where there is a lack of surface water or where groundwater is the only viable source of water. They are often found in rural areas where traditional hand-dug wells are not feasible due to the depth of the water table, and where drilling a deeper, more conventional well may be prohibitively expensive.

Location of Well Site

The location of a well site for water can depend on a variety of factors, including the geological conditions of the area, the availability and quality of groundwater, and local regulations. Here are some general considerations:

1. **Groundwater Availability**: The site should be located where there is a sufficient quantity of water available underground. This can be determined by studying the geology and hydrology of the area to estimate the potential yield of the well. It may be near a pond or reservoir or existing wells.
2. **Distance from Contamination Sources**: The site should be located at a safe distance from potential sources of contamination such as septic tanks, industrial sites, and waste disposal areas.
3. **Proximity to Water Users**: The well site should be located in proximity to the users of the water supply, to minimize the cost and complexity of laying pipes to distribute the water.

5.8 Pumps

Pumps are mechanical devices designed for uplifting and in the transportation of fluids or gases from one location to another. They are widely used in various industries, including water supply, chemical processing, oil and gas, and many others. Pumps come in various shapes and sizes, and each type has its unique advantages and disadvantages.

5.8.1 Types of Pumps

1. **Centrifugal Pumps**: Centrifugal pumps (Figure 5.7) are used in water supply systems, irrigation pumping, chemical processing plants, and wastewater treatment plants. They work on the principle of centrifugal force and are used for transferring liquids with a low viscosity. They are simple to operate, have a high flow rate, and are energy-efficient. These pumps are commonly used for lifting large volumes of water quickly and efficiently. Centrifugal pumps are often used in residential, commercial, and agricultural applications.

2. **Positive Displacement Pumps**: Positive displacement pumps are used for transferring high viscosity fluids, such as oils and syrups. They work by trapping a fixed amount of fluid and then discharging it through the outlet. Positive displacement pumps are further classified into two types: rotary and reciprocating.

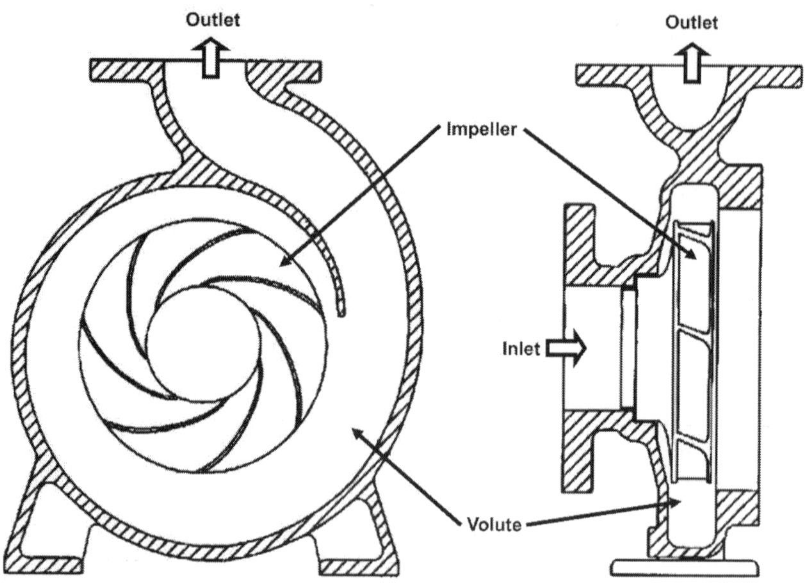

Figure 5.7. Centrifugal Pump.

(a) **Rotary Pumps**: Rotary pumps work by rotating a fixed volume of fluid between the pump and the discharge port. The most common type of rotary pump is the gear pump, which has two gears that rotate in opposite directions, trapping fluid between the teeth and the housing. Other types of rotary pumps include vane pumps, screw pumps, and lobe pumps.

(b) **Reciprocating Pumps**: Reciprocating pumps use a piston or diaphragm to displace the fluid. These pumps develop high heads, but their capacity is small compared to other types of pumps. These pumps are commonly used for home water supply.

3. **Submersible Pumps**: Submersible pumps (Figure 5.8) are designed to be immersed in the fluid that they are pumping. They are used for pumping water out of wells, boreholes, and other underground structures. Submersible pumps are usually electrically powered and can handle large volumes of water. They are often used in residential, commercial, and agricultural applications, as well as for sewage and wastewater treatment.

4. **Jet Pumps**: Jet pumps use a nozzle and venturi to create a vacuum that draws the fluid into the pump. They are used for pumping water from shallow wells and are commonly used in domestic water supply systems. Jet pumps are often used in residential and agricultural applications.

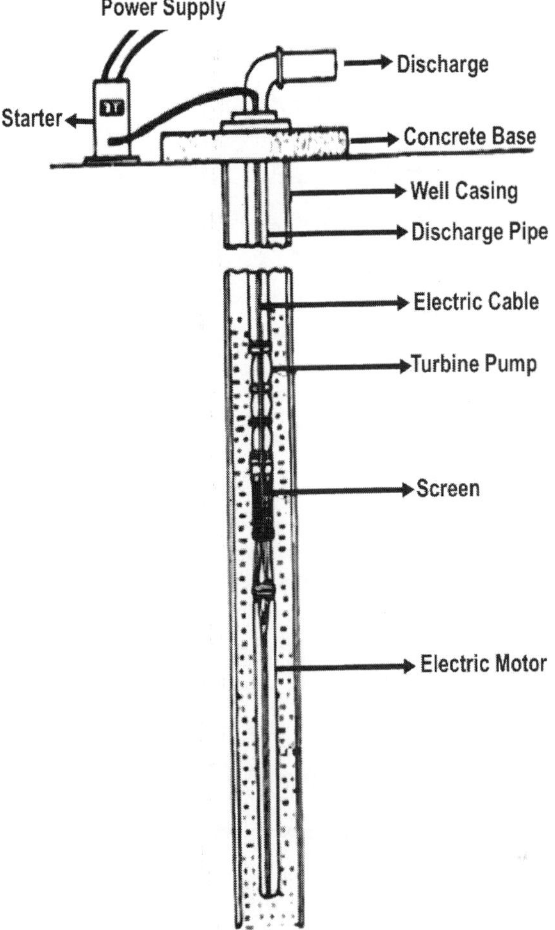

Figure 5.8. Submersible Pump.

5. **Solar Pumps**: Solar pumps are a type of submersible pump that is powered by solar panels. They are used in remote locations where electrical power is not available and are commonly used for irrigation. These pumps are commonly used for lifting water from wells, boreholes, and other sources in remote areas where there is no access to electricity. They are often used in agricultural and residential applications.

Overall, the type of pump that is best suited for a particular water lifting application will depend on a variety of factors, including the depth of the water source, the volume of water that needs to be lifted, and the available power sources.

5.8.2 Selection of Pumps

When selecting a pump, it is essential to consider the following factors:

1. The maximum rate of flow desired.
2. The total water required per day.
3. The maximum flow from the water source.
4. The vertical distance the water must be lifted to the pump.
5. The total head against which the pump must operate.
6. The properties of the fluid being pumped, such as viscosity, temperature, and chemical composition.
7. The efficiency of the pump to minimize energy consumption.

5.8.3 Pumping of Water

The systems in which pumping of water is done includes:

1. Direct pumping
2. Combined gravity and direct pumping

Direct pumping of water is done into water mains instead of storing it. Life of pump get reduced because they have to pump at varying rates. This system is less adopted as high power pumps are required. In combined gravity and pumping system, gravity forces affect the pressure of water to be distributed. Water is pumped firstly into a distribution reservoir (tank) which is located on natural high level ground. After that water is distributed under gravitational force.

5.8.4 Pump Characteristic Curves

A pump characteristic curve is a graph that shows the relationship between a pump's flow rate (Q) and its head (H), as well as its power consumption (P) and efficiency (η). The flow rate is usually measured in cubic meters per hour (m^3/h), while the head is typically measured in meters (m).

The pump characteristic curve is a fundamental tool for engineers and designers in the fluid handling industry. It provides valuable information on how a pump will perform under different operating conditions, such as changes in flow rate or head. By analysing the curve, engineers can select the best pump for a given application, optimize pump performance, and troubleshoot any problems that may arise.

The pump characteristic curve (Figure 5.9) typically consists of three main curves: the head-capacity curve, the power-consumption curve, and the efficiency curve. The head-capacity curve shows the relationship between the pump's head and its flow rate. The power-consumption curve shows the relationship between the pump's power consumption and its flow rate. The efficiency curve shows the relationship between the pump's efficiency and its flow rate.

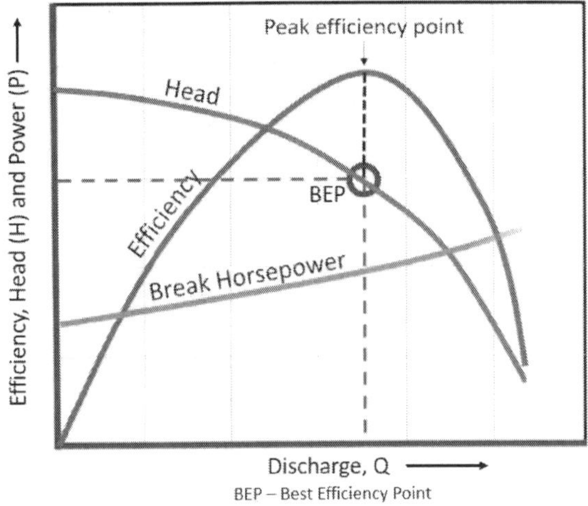

Figure 5.9. Typical Characteristics of a Centrifugal Pump.

5.8.5 Factors Affecting the Pump Performance

There are several factors that can affect the performance of a pump. Here are some of the most important ones:

1. **Type of Pump**: Different types of pumps have different performance characteristics. For example, a centrifugal pump will have a different performance curve than a positive displacement pump.
2. **Pump Speed**: The speed at which the pump is running can have a significant impact on its performance. Generally, higher speeds result in higher flow rates, but also higher pressure drops and higher power requirements.
3. **Pump Size**: The size of the pump can also affect its performance. Larger pumps may be able to handle higher flow rates and pressures, but may also require more power to operate.
4. **Fluid Properties**: The properties of the fluid being pumped can also impact the performance of the pump. Factors such as viscosity, density, and temperature can all affect the pump's efficiency and flow rate.
5. **System Resistance**: The resistance of the piping system and other components in the system can also affect the pump's performance. Higher resistance will require the pump to work harder to achieve the desired flow rate and pressure.

6. **Pump Impeller Design**: The design of the pump impeller can also have an impact on its performance. Factors such as the number of blades, blade angle, and shape can all affect the pump's efficiency and flow rate.
7. **Pump Condition**: Finally, the condition of the pump itself can affect its performance. Factors such as wear and tear, corrosion, and damage can all impact the pump's ability to operate efficiently. Regular maintenance and inspection can help ensure that the pump is operating at its optimal performance level.

5.8.6 Maintenance of Pumps

Regular maintenance of pumps is essential to ensure their longevity and efficiency. Regular maintenance helps to prevent breakdowns, reduce repair costs, and prolong the life of the equipment. Here are some essential tips for pump maintenance:

1. **Check the Pump Regularly**: Regular checks can help identify any problems early and prevent major issues. Inspect the pump for any signs of wear and tear, leaks, or unusual noises.
2. **Keep the Pump Clean**: Dirt and debris can cause damage to the pump and decrease its efficiency. Clean the pump regularly using appropriate methods, such as wiping with a cloth or using a pressure washer.
3. **Change the Oil**: Change the oil in the pump as per the manufacturer's recommendations. Dirty or low oil levels can cause the pump to overheat and fail.
4. **Replace Worn Out Parts**: Check for worn out parts such as seals, bearings, and impellers. Replace them immediately as they can cause the pump to fail.
5. **Lubricate the Pump**: Regular lubrication of bearings, gears, and other moving parts can help prevent wear and tear and keep the pump running smoothly.
6. **Check the Alignment**: Make sure the pump is aligned correctly with the motor. Poor alignment can cause vibration and lead to premature failure.
7. **Monitor the Motor**: Keep an eye on the motor and make sure it is running smoothly. Any unusual noises or vibrations may indicate a problem.

By following these tips, you can ensure the proper functioning of your pump and prolong its life. It is also important to refer to the manufacturer's instructions for specific maintenance requirements for your pump.

SANITATION

Sanitation refers to the practice of maintaining cleanliness and hygiene in order to prevent the spread of disease and promote good health. It involves the proper treatment, management, and disposal of waste, as well as the provision of clean water and safe food handling. Sanitation is essential for maintaining public health and preventing the spread of infectious diseases such as cholera, typhoid, and diarrhoea. Poor sanitation can lead to the contamination of water sources, food, and the environment, which can result in the spread of disease. Effective sanitation practices include proper hand washing, proper disposal of waste, and regular cleaning and disinfection of surfaces. Access to basic sanitation facilities such as toilets and hand washing stations is also critical for ensuring good hygiene and preventing disease.

In many parts of the world, inadequate sanitation remains a major public health concern, particularly in low-income and rural communities. Efforts to improve sanitation and hygiene practices through education and infrastructure development can have a significant impact on reducing the burden of infectious diseases and improving overall health outcomes.

5.9 Toilets

Toilets should take into consideration the specific needs and circumstances of the rural population. Here are some factors to consider:

1. **Affordability**: Many rural households in India have limited financial resources. Therefore, any toilet solution should be affordable and accessible to them.
2. **Water Availability**: In many rural areas, there may be limited access to clean water. Therefore, any toilet solution should minimize water usage or offer alternative methods of flushing.
3. **Climate**: Rural areas in India experience different weather conditions. Toilets should be designed to withstand harsh weather conditions, such as heavy rain, flooding, and extreme temperatures.
4. **Maintenance**: Maintenance is a crucial factor in ensuring the longevity and functionality of toilets. Any toilet solution should be easy to maintain, and its components should be readily available in local markets.
5. **Cultural Norms**: Toilets should also take into account the cultural norms of rural India. For example, some rural populations prefer squatting toilets, while others prefer sitting toilets.

One suitable toilet solution for rural India is the pour-flush toilet. This type of toilet uses a small amount of water to flush waste into a holding tank, which can then be emptied periodically. The pour-flush toilet can be built using locally available materials, such as bricks and cement, making it an affordable option for rural households. It also uses less water than conventional flush toilets, making it suitable for areas with limited water resources. Additionally, the pour-flush toilet can be designed to accommodate squatting or sitting positions, depending on local cultural norms.

5.9.1 Types of Toilets

Some toilet designs that can suit rural India:

5.9.1.1 Single-Pit Leach Pit Toilet

A single pit leach pit toilet is a simple and cost-effective sanitation system that is commonly used in areas with limited access to water and sewage infrastructure. It provides a safe and hygienic option for individuals and communities. It consists of a pit, usually between 1 and 3 meters deep, that is dug into the ground and lined with a non-permeable material such as concrete or plastic to prevent the seepage of faecal matter into the surrounding soil.

The toilet is built on top of the pit, and waste is deposited directly into the pit through a hole or chute. The pit is covered with an air-tight lid or a slab that has a vent pipe to allow for proper ventilation and prevent the build-up of toxic gases such as methane. The RCC slabs shall be used as pit cover. They are strong enough to be used for different household purposes. The pit covers ensure that the premises are free from foul smell or the nuisance of flies, cockroaches or mosquitoes. A toilet pan could be an Indian style squatting pan or a Western style commode.

Over time, the waste in the pit decomposes, and the liquid component leaches out into the surrounding soil. This process is aided by the addition of a layer of ash or soil after each use, which helps to absorb moisture and balance the pH levels in the pit. The remaining solid waste gradually reduces in volume, and the pit can be emptied and reused after a period of 2-5 years, depending on usage and the size of the pit.

Placement of Leach-Pit: In homogenous soil, bacteria are capable of moving up to about 3 metres horizontally and about 1 metre vertically. This varies with soil compaction. Therefore, leach pits can only be allowed in places which have water table 2 m below the pit even during rainy season. To prevent pollution of water sources, the pits need to be located at a minimum distance of 3 m from open wells and shallow hand pumps. If soils are coarse (effective size more than 0.2 mm), a 50 cm thick sand envelope of 0.2 mm sand has to be created all around the pit and the bottom of the pits need to be sealed by clay, polythene sheet, lean cement-concrete or cement stabilized soil to prevent contamination. Areas that are waterlogged, flood prone or having high subsoil water, the pits need to be raised so that the invert of pipe or drain is just above the likely water level. The raising of pits will necessitate raising toilet floor also. Earth need to be filled and well compacted all-round the pit.

The advantages of a single pit leach pit toilet are that it is simple to construct, low-cost, and requires minimal maintenance. It also helps to prevent the contamination of surface and groundwater, which can be a significant health hazard in areas without proper sanitation facilities. However, it is important to note that a single pit leach pit toilet may not be suitable in areas with high water tables or rocky terrain, where the leaching of waste may be impeded.

5.9.1.2 Twin-Pit Leach Pit Toilets

A twin pit leach pit toilet (Figure 5.10) is a type of sanitation system that is commonly used in areas where there is no access to traditional wastewater treatment facilities or sewage systems. It consists of two pits or chambers at a distance from one another or with an impermeable wall between them. Pits are typically constructed out of concrete or masonry that are used alternately for waste disposal. When one pit is full, it is left to decompose for a period of several months, during which time the waste is broken down by natural biological processes. The second pit is then used for waste disposal while the first pit is left to decompose. Once the first pit is fully decomposed, it can be

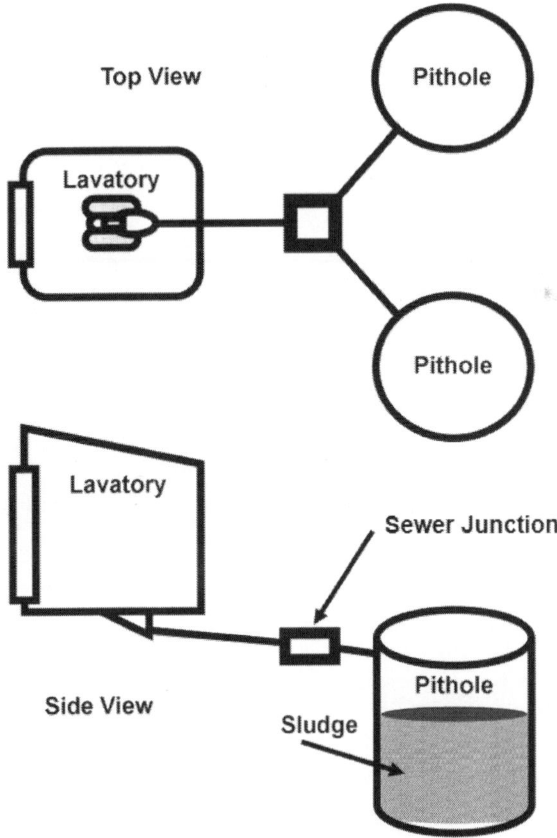

Figure 5.10. Twin-pit Leach Pit Toilets.

emptied and used again while the second pit is left to decompose. The long period of microbial digestion makes the manure pathogen free, odourless and safe for handling. This system requires less maintenance than single-pit toilets and can provide fertilizer for crops. This is the most popular model under the *Swachh Bharat Abhiyan*.

The leach pit part of the system refers to the fact that the liquid waste (or leachate) that seeps out of the pits is allowed to percolate into the soil, where it is further broken down by microorganisms and filtered by the soil. This helps to prevent groundwater contamination and the spread of disease.

Twin pit leach pit toilets are low-cost, low-maintenance, and environmentally friendly sanitation solutions that are particularly well-suited to rural areas and other regions where traditional wastewater treatment infrastructure is not available. However, they do require regular maintenance and attention to ensure that they are functioning properly and that the waste is being disposed of safely and effectively.

Shape of pits: As far as possible, separate circular pits need to be constructed as these are structurally more stable and the sludge is dry and safe to handle. If space is a constraint, pits of smaller diameter (not less than 75 cm) can be provided, but the depth need to be increased suitably to provide required storage volume and infiltration surface area. The minimum space between the two pits need to be 1 m or equal to the depth of pits below the level of incoming pipe or drain, whichever is more.

5.9.1.3 ECOSAN Toilets

ECOSAN (ecological sanitation) toilets are a type of dry toilet that is designed to treat human waste in a sustainable and environmentally friendly manner. These toilets are also known as composting toilets or urine-diverting toilets. This design is suited for areas where water logging and water scarcity is a concern. It is most effective in coastal and rocky areas, as it promotes soil fertility and crop production. Historically, such ecological methods of sanitation were utilized in Rome, China, Mexico, Peru and Yemen. Ecosan was implemented in Haiti as part of the emergency relief effort following the 2010 earthquake. In India, these designs are notably used in Tamil Nadu where the State Government provides subsidies.

The main idea behind Ecosan toilets is to separate the urine and faeces at the source. This is done through the use of two separate chambers or compartments in the toilet. The urine is collected in one compartment, while the faeces are collected in another. Once collected, the urine can be used as a fertilizer for plants, as it is rich in nitrogen and other nutrients. The faeces, on the other hand, are treated and processed to create a nutrient-rich compost that can also be used as a fertilizer.

Ecosan toilets have several advantages over traditional flush toilets. They require little to no water, which is a significant advantage in areas where water is scarce or expensive. They also do not require a sewer system or septic tank, which makes them a more viable option in areas where these systems are not available or too expensive to install. Ecosan toilets are also more environmentally friendly than traditional flush toilets, as they help to reduce water pollution and greenhouse gas emissions. Additionally, they can be used to create a closed-loop system, where the waste generated is used to fertilize plants, which can then be used for food production. Overall, Ecosan toilets offer a sustainable and environmentally friendly alternative to traditional flush toilets, particularly in areas where water and sanitation infrastructure is lacking or inadequate.

5.9.1.4 Biogas Linked Toilets

Biogas-linked toilets are a sustainable solution to two major challenges: sanitation and energy. These toilets use anaerobic digestion technology to convert human waste into biogas, which can be used for cooking, heating, and lighting. In a biogas-linked toilet system, human waste is collected in a tank where it undergoes anaerobic digestion, a process that breaks down organic matter in the absence of oxygen. The by-product of this process is biogas, a mixture of methane and carbon dioxide, which can be used as a source of fuel. The remaining material, called digestate, can be used as a nutrient-rich fertilizer.

Biogas-linked toilets have several advantages over traditional toilets. Firstly, they reduce the need for traditional sewage infrastructure, which can be expensive and difficult to install in remote areas. Secondly, they provide a source of renewable energy, which can be used to replace fossil fuels. Thirdly, they can help to address the problem of open defecation, which can lead to the spread of diseases.

Biogas-linked toilets are already being used in several countries, including India, Nepal, and Kenya. In India, for example, the government has launched a program to install biogas-linked toilets in rural areas.

5.9.1.5 Borehole Latrine

Borehole latrines are mainly provided in the acute response phase, when a large number of latrines are required quickly and the site conditions do not allow for the excavation of bigger pits. The borehole latrine is a modified version of the traditional pit latrine. A mechanical auger or a drilling machine is the main requirement for implementation. Borehole latrines are usually temporary solutions but depending on diameter, depths and number of users they can also be considered a longer-term solution with a potential life span of several years. The soil for the borehole needs to be stable and free of rock, gravel and boulders.

The size of pit is usually 25 to 75 cm in diameter, and 3 to 8 m deep. The pit is lined with a material that prevents the waste from contaminating the surrounding soil and groundwater. The superstructure can either be simple screens around the hole (as a temporary solution) or more solid cubicles. As it is not possible to easily ventilate the borehole, the superstructure should allow for air circulation to reduce potential odour problems. The hole should be covered with a slab or pedestal. A latrine 6 m or more in depth will last about a year if used by an average size family. When it fills to within 1.5 m, the pit may be closed with soil and then a second pit should be used. By the time second is filled, the first one can be emptied and reused. Thus, two pits can be alternatively used by a family for years.

The borehole latrine is relatively low-cost to construct and can be maintained easily. However, the effectiveness of the borehole latrine is dependent on proper installation and maintenance, as well as proper waste disposal practices by the users.

5.9.2 Maintenance of Toilets

Toilets are an essential part of our daily lives, and it's important to keep them clean and well-maintained to ensure proper hygiene and functionality. Regular cleaning is important to prevent the build-up of bacteria and unpleasant odours. Pour water in the pan after defecation and urination. The pan need to be cleaned once a day with a brush or a broom. Use a toilet cleaner or a mixture of baking soda and vinegar to clean the inside of the bowl, the tank, and the outside of the toilet. Kitchen, bathroom waste water or rain water need not be allowed to enter the pits. Non-flushable items like paper towels, feminine hygiene products, and wet wipes *etc.* also need not be thrown in the pan. This could block and damage the toilet.

5.10 Septic Tank

A septic tank is an underground closed water tight chamber that is designed to treat and dispose of household wastewater. It is a common on-site wastewater treatment system used in areas where centralized sewer systems are not available. Septic tanks are used in both rural and urban areas, and they are typically made of concrete, fiberglass, or plastic. With proper maintenance and care, a septic system can last for many years and provide reliable and cost-effective wastewater treatment for homeowners.

The basic idea behind a septic tank is to allow solids to settle to the bottom of the tank while allowing the liquid to flow out into the drain field for further treatment. When wastewater enters the septic tank, the solids sink to the bottom, while the liquid flows through the outlet pipe and into the drain field. The drain field is a series of perforated pipes buried in the ground, which allows the liquid to be distributed and treated by the soil.

The septic tank has two chambers, an inlet and an outlet, with a dividing wall separating them. As wastewater enters the tank, it flows through the inlet chamber, where the solids settle to the bottom of the tank. The liquid then flows into the outlet chamber, where it is released into the drain field for further treatment.

The tank is connected to the home's plumbing system, and as wastewater enters the tank, solids settle to the bottom, and oils and greases rise to the top. Bacteria in the tank help to break down the solid waste, but some of the solids still remain and accumulate over time. The liquid that remains in the middle layer is drained out of the tank and enters a drain field or leach field, where it is further treated by the surrounding soil.

Thus, these are the three important functions that take place within the septic tank.

1. Firstly, the removal of solids form the sewage as it enters the tank. The solids are retained in the tank and the remaining portion is discharged.
2. Secondly, the decomposition of the solid sewage under anaerobic condition, that is, in the absence of air.
3. Thirdly, the storage of the sludge and scum in the tank.

Sewage, sludge, and scum are all related to wastewater treatment and have distinct meanings.

Sewage is the liquid waste of a household or community including human excreta. Sewage contains a variety of contaminants, including organic and inorganic matter, bacteria, viruses, and other pathogens.

Sludge is the solid material that is left over after wastewater treatment processes have been completed, and settles at the bottom of the tank. It is in semi-solid condition. Sludge can be further processed into bio solids, which can be used as fertilizer or fuel.

Scum is the layer of floating material that can form on the surface of sewage or wastewater during treatment. It is typically composed of oils, fats, and other organic compounds that are less dense than water. Scum can be removed from the wastewater and treated separately from the sludge.

Septic tanks need regular maintenance to function properly, including regular pumping to remove accumulated solids and inspections to ensure that the tank and drain field are working correctly. Failure to properly maintain a septic system can lead to backups, sewage leaks, and groundwater contamination. It's important to follow proper guidelines for the use and maintenance of septic tanks to ensure they function properly and protect public health and the environment.

Benefits of a Septic Tank

1. Septic tanks are generally less expensive to install and maintain than a centralized sewer system. They are also more environmentally friendly, as the wastewater is treated and disposed of on-site instead of being sent to a treatment plant.
2. Septic systems can be used in areas where a centralized sewer system is not available, making them a popular choice in rural areas. They are

also typically more reliable, as they are not subject to the same types of disruptions that can occur in centralized sewer systems, such as power outages and equipment failures.

Location

The location of a septic tank can vary depending on the layout of the property, the type of soil, the water table level, and the distance from any water sources. But, generally, it is situated underground and outside of the house, open to sky, usually in the backyard. A septic tank should be located at least 30 m away from any wells, streams, or other bodies of water to prevent contamination. The location of the tank with respect to the exterior wall of the building should be as far as possible. It should not be closer than 1.5 m from any building foundation, and 3 m from any window opening. Surface and subsoil water should not find way into the septic tank. The septic tank should also be located downhill from the source of water supply to ensure proper gravity flow. In addition, the soil in the area should be permeable enough to allow for proper drainage and absorption of the effluent. However, the tank should not be located in areas liable to be flooded during rainy season. The tank should also be easily accessible for cleaning.

Construction of Septic Tank

The first step in the construction process is to excavate a hole in the ground where the septic tank will be located. Thereafter, floor and walls are constructed. The floor of the tank should be water tight and of adequate strength to resist earth movement, and to support the weight of the tank, walls and contents. The floor may be of cement concrete of minimum M15 grade (1:2:4) and a minimum slope of 1:10 may be provided towards the sludge outlet to facilitate de-sludging. The walls of the tank should be of such thickness as to provide adequate strength and water tightness. Walls built out of brick should not be less than 200 mm thick and should be plastered to a minimum thickness of 12 mm inside and outside with cement mortar not weaker than 1:3. Where the walls are built out of the stone masonry, they should have a minimum thickness of 370 mm. Water proofing agent shall be mixed in the cement mortar and concrete to make the walls and floor water tight. All inside corners of tank shall be rounded. The inlet pipe should be connected to the house's main sewer line, while the outlet pipe should be connected to the drain field.

Septic Tank Design

A septic tank is an underground chamber that is designed to treat wastewater from a house or building that is not connected to a public sewage system. The design of a septic tank is crucial to ensure proper treatment and disposal of the wastewater. A septic tank combines the function of sedimentation tank, a sludge digestion tank and a sludge storage tank. IS 2470 (Part 1) has provided the detailed guidelines for the design of septic tank. According to

Indian Standards, the sedimentation rate, sludge digestion rate, sludge storage period and inflow rate are the factors which decide the capacity of a septic tank.

Sewage Flow

The maximum flow to the tank is based on the number of plumbing fixtures discharging simultaneously. For this purpose various sanitation facilities are equated in terms of fixtures units. Fixture unit is a quantity in terms of which the load producing effect of different plumbing fixtures on the plumbing system are expressed on some arbitrarily chosen scale. In the design of septic tank, it is taken as 9 litres per minute.

Sedimentation

The surface area of the tank required will be 0.92 m^2 for every 10 litres per minute of peak flow rate at a temperature of 25°C. A minimum depth of sedimentation shall be 250 to 300 mm.

Sludge Digestion

Per capita suspended solids entering the tank may be taken as 70 g/day. The capacity required for sludge digestion is 0.033 m^3 per capita at 25°C. Volume of digested sludge is normally 0.00021 m^3 per capita per day or 0.07665 m^3 per capita per year.

Detention Time

A detention period of 24 to 48 hours based on an average daily flow of sewage is provided.

Dimensions of Septic Tank

Septic tank (Figure 5.11) must have minimum width of 750 mm, minimum depth of one metre below water level and a minimum liquid capacity of 1000 litres. For rectangular septic tanks, the length of the tank shall be 2 to 4 times the width. For circular tanks the minimum diameter shall not be less than 1.35 m and operating depth shall not be less than 1.0 m. Suitable sizes of septic tanks are given in Table 5.5.

Inlet and Outlet Pipes

The inlet pipe should be located near the top of the tank to prevent solids from entering the drain field. The design of septic tank inlets shall be such as to introduce the crude sewage with the least possible disturbance of the settled sludge or the surface scum. The outlet pipe should be located near the bottom of the tank to allow the clarified effluent to flow out of the tank.

Partitions

Where the capacity of a septic tank exceeds 2000 litres, the tank may be divided into two chambers by means of a fixed durable partition. The partition

Table 5.5. Recommended sizes of septic tank (IS 2470 – Part I)

Number of Users	*Length (m)*	*Breadth (m)*	*Liquid Depth (m)*		*Liquid Capacity (m^3)*		*Sludge to be Removed (m^3)*		*Depth of Sludge to be Withdrawn (m)*	
			For Cleaning Interval of Years							
			1	*2*	*1*	*2*	*1*	*2*	*1*	*2*
5	1.5	0.75	1.0	1.05	1.12	1.18	0.36	0.72	0.32	0.64
10	2.0	0.90	1.0	1.40	1.80	2.52	0.72	1.44	0.40	0.80
15	2.0	0.90	1.3	2.00	2.34	3.6	1.08	2.16	0.60	1.20
20	2.3	1.10	1.3	1.80	3.30	4.55	1.44	2.88	0.57	1.14
50	5.0	2.00	1.0	1.24	10.0	12.4	3.60	7.20	0.36	0.72
100	7.5	2.65	1.0	1.24	19.9	24.6	7.20	14.40	0.36	0.72
150	10.0	3.00	1.0	1.24	30.0	37.2	10.80	21.60	0.36	0.72
200	12.0	3.30	1.0	1.24	39.6	49.1	14.40	28.80	0.36	0.72
300	15.0	4.00	1.0	1.24	60.0	74.4	21.60	43.20	0.36	0.72

Note: A provision of minimum 300 mm should be made for free board.

$$\text{Depth of sludge to be withdrawn (m)} = \frac{\text{Sludge to be removed (m}^3\text{)}}{\text{Length (m)} \times \text{Breadth (m)}}$$

shall be located so that the capacity of the first chamber is twice that of the second chamber. Suitable openings rectangular or circular with minimum dia 100 mm and maximum 150 mm shall be provided in the partition at approximately 300 mm below top water level. For population of over 100, duplicating tanks, each providing half the total calculated capacity required, should be installed and operated in parallel. This arrangement permits the flow to be passed through one unit while the other is being desludged. To enable the top water to be decanted when desludging, a decanting valve should be provided in the wall dividing the two tanks; the invert of this valve should be 625 mm below top water level.

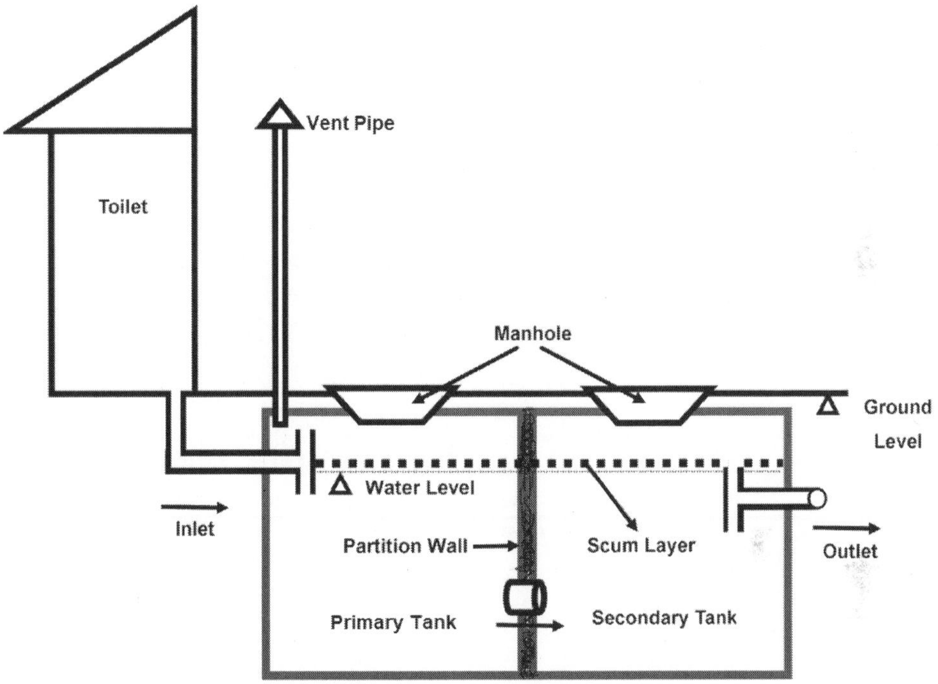

Figure 5.11. Septic Tank.

Free-Board

A minimum free board of 300 mm (0.3 m) should be provided above the liquid level in the tank, thus forming an air space.

Access Openings and Cover or Man Holes

Access ports should be provided for inspection and maintenance. These ports should be easily accessible and should be located at the ground surface. Each compartment of a septic tank shall be provided with a rectangular access opening measuring, not less 455 × 610 mm or a circular opening 500 mm diameter. Normally, these ports are located near the inlet and outlet pipes. The cover to access openings shall be of reinforced concrete or of cast iron. A cover shall incorporate a suitable lifting device and when in place after installation of the septic tank shall fit neatly and be sealed to prevent the ingress of water.

Ventilating Pipe

Ventilation is important to prevent the build-up of gases in the septic tank. A vent pipe should be installed to allow for the release of gases. Every septic tank shall be provided with ventilating pipe of at least 50 mm diameter. The top of the pipe shall be provided with a suitable cage of mosquito proof mesh. The ventilating pipe shall extend to a height which would cause no smell nuisance to any building in the area. Generally the ventilating pipe may extend to a height of about 2 m when the septic tank is at least 20 m away from the nearest building and to a height of 2 m above the top of the building when it is located closer than 20 metres.

Sludge Withdrawal/Cleaning

Half yearly or yearly desludging of septic tank is desirable. Small domestic tanks, for economic reasons, may be cleaned at least once in 2 years provided the tank is not overloaded due to use by more than the number for which it is designed. Frequent desludging inhibits the anaerobic action in the tank. Normally, the tanks are cleaned when the sum of the depth of the scum and the sludge is observed to exceed half the depth of the tank. A portion of sludge not less than 25 mm in depth should be left behind in the tank bottom which acts as the seeding material for the fresh deposits. Cleaning of the tank should be done through a sludge pump. Manual cleaning of the tank is to be avoided on account of likelihood of infection to the worker on account of toxic material, gases *etc.* When removal of the sludge is carried out the scum in the first tank should not be disturbed more than necessary, this scum is needed to ensure efficient operation. Sludge from septic tanks may be delivered into covered pits or into a suitable vehicle for removal from the site. Spreading of sludge on the ground in the vicinity should not be allowed.

Commissioning of Septic Tank

The sewerage system should be complete and ready for operation before connection is made to the building. The tank should be filled with water to its outlet level before the sewerage is let into the tank. It should, preferably, be seeded with small quantities of well digested sludge obtained from septic tanks or sludge digestion tanks. In absence of digested sludge a small quantity of decaying organic matter, such as digested cow dung may be introduced.

Disposal Field

A disposal field, also known as a leach field or drain field, is a crucial component of a septic system that is responsible for disposing of the effluent or wastewater that exits the septic tank. The disposal field typically consists of a network of perforated pipes laid in gravel trenches or beds, which are buried in the soil. As the effluent from the septic tank enters the disposal field, it is distributed evenly through the pipes and allowed to seep into the soil. The soil acts as a natural filter, removing harmful pathogens and contaminants from the effluent as it percolates through the layers of soil. The filtered water is then returned to the groundwater table, where it is further purified through natural processes.

It is important to properly maintain the disposal field to ensure its continued functionality. This includes avoiding the disposal of non-biodegradable materials, minimizing the use of water, and having the septic tank pumped regularly to prevent the build-up of solids that can clog the pipes and reduce the effectiveness of the disposal field.

Soakage Pit

A soakage pit, also known as a seepage pit or absorption pit, is a simple method of disposing of the effluent from a septic tank. The pit is essentially a hole in the ground filled with gravel or other porous materials, which allows the effluent to slowly seep into the surrounding soil.

To create a soakage pit for a septic tank, the following steps are typically taken:

1. Determine the appropriate size for the pit based on the number of people using the septic system, the soil type, and the amount of wastewater being generated.
2. Excavate a hole in the ground and fill it with a layer of large gravel or crushed rock.
3. Install an outlet pipe from the septic tank to the pit, ensuring that it is sloped appropriately to allow the effluent to flow.
4. Fill the remainder of the pit with smaller gravel or other porous materials, such as sand or crushed brick.
5. Cover the top of the pit with a layer of soil to prevent debris or other materials from entering the pit.

Over time, the effluent from the septic tank will slowly seep into the surrounding soil and be naturally filtered and treated by the soil and microorganisms. It is important to properly maintain the septic tank and limit the amount of non-biodegradable items that are flushed down the drain to ensure the soakage pit remains effective and does not become clogged.

Septic Tank Maintenance

Proper maintenance of a septic tank is critical to its longevity and effectiveness. Septic tanks need to be pumped out on a regular basis to remove the solid build-up that accumulates in the tank. The frequency of pumping depends on the size of the tank and the number of occupants in the house. In general, septic tanks should be pumped out every two to five years. In addition to regular pumping, there are a few things homeowners can do to maintain their septic tanks. These include:

1. **Conserve Water**: Septic tanks are designed to handle a certain amount of wastewater each day. Conserving water by fixing leaks, using low-flow toilets and showerheads, and not overusing the washing machine can help extend the life of the septic system.
2. **Avoid Flushing Certain Items**: Flushing items like baby wipes, feminine hygiene products, and other non-degradable materials can clog the septic system and cause it to fail.
3. **Avoid using Harsh Chemicals**: Harsh chemicals like bleach, drain cleaners, and other cleaning products can kill the beneficial bacteria in the septic system. These bacteria are responsible for breaking down the solids in the tank, so it's important to use septic-safe products.

Calculation of Capacity of Septic Tank as per British Standards

The size of the septic tank can be calculated using this formula,

$$C = A + P\,(rq + nS)$$

where,

C = capacity of the septic tank in litres,

A = constant,

P = number of users/persons,

r = minimum retention time of sewage (in days),

q = average sewage flow in litres per person per day,

n = number of years of sludge storage, and

S = amount of sludge accumulation in litres per person per year.

In the British code, A = 2000 litres, and rq + nS = 180 litres, so the rounding formula, we get

$$C = 2000 + 180P$$

A proper designing must be done, otherwise backflow of waste water might occur. Overall capacity of tank, dimensions and allowance for future consideration and emergency situations must be inculcated.

Example 5.4

Design a septic tank for 6 users with the help of the relationship as provided in British Standards.

Solution

Capacity of septic tank as per British Code is

$$C = 2000 + 180P$$

Then, the capacity of septic tank size for 6 users

$$C = 2000 + 180 \times 6 = 3080 \text{ litres} = 3.08 \text{ m}^3$$

Let the depth of the tank below water level = 1.0 m

So, the cross-sectional area of the tank = 3.08/1.0 = 3.08 m^2

Let the length of the tank is 2 times the width.

So, length (L) × width (W) = 3.08

or, 2W × W = 3.08

or, W ≈ 1.2 m

Then, length (L) = 2×1.2 = 2.4 m

The inside of the tank needs to be spaced 300 mm or 0.3 m (minimum) above the water level.

So, depth of tank from floor = 1.0 + 0.3 = 1.3 m

Size of the septic tank for 6 users = 2.4m × 1.2m × 1.3m with respect to their length, breadth and depth.

So, the tank capacity shall be 2.4m × 1.2m × 1.3m = 3744 litres (3.744 m^3).

Example 5.5

Calculate the septic tank volume for 20 persons as per IS standards. Consider the peak flow of total sewage as 30 L/min for 20 persons. The digestion rate per user at 25°C is 0.033 m^3 and the sludge clearance rate per user per year is 0.073 m^3.

Solution

Peak flow of total sewage = 30 L/min

As per IS standards, 0.92 m^2 area and 0.3 meters depth is needed for sedimentation when the sewage inflow rate is 10 litres per minute.

So, the area required for sedimentation for flow rate of 30 L/min = $0.92 \times \frac{30}{10}$

$$= 2.76 \text{ m}^2$$

and, volume needed for sedimentation = 2.76 m^2 × 0.3 m = 0.828 m^3

Digestion rate per user at 25°C = 0.033m^3

So, volume required for digestion for 20 persons = 0.033 m^3 × 20 = 0.66 m^3

Sludge clearance rate per person in one year = 0.073 m^3

So, sludge storage volume for 20 persons for one year = $0.073m^3 \times 20$
= $1.46\ m^3$

Space above liquid level (free-board) is 0.3m depth.

So, the volume of space above liquid level = $2.76\ m^2 \times 0.3 = 0.828\ m^3$

Hence, total volume of septic tank for 20 person with a sludge clearance period of one year is

= volume needed for (sedimentation + digestion + sludge + free-board)

= $0.828 + 0.66 + 1.46 + 0.828 = 3.776\ m^3$

Example 5.6

Considering detention time as 24 hours and cleaning period of 1 year, design a septic tank for 5 number of persons in a house. The water use per capita per day is 150 litres, and rate of sludge accumulation as 30 litres per person per year.

Solution

Number of persons = 5

Detention time = 24 hours

Total wastewater per capita per day = 150 litres

Rate of sludge accumulation = 30 litres per person per year

Wastewater

Total wastewater per day for 5 persons = $150 \times 5 = 750$ litres

For being on safer side, total waste water may be considered as 1000 litres.

Sludge Accumulation

Rate of sludge accumulation for 5 persons in a year = $30 \times 5 = 150$ litres

Total capacity required for detention and accumulation = $1000 + 150 = 1150$ litres = $1.15\ m^3$

Considering depth of septic tank to be 1.8 metres.

Plan area required = $1.15/1.8 = 0.64\ m^2$

Assuming Length (L) is twice of Width (W).

Since, Plan area = $L \times W = 2W \times W = 0.64$

Therefore, Width (W) = 0.565 metre. But as per code IS 2470, Minimum width must be 750 mm.

So, taking Width (W) = 750 mm = 0.75 m, and Length (L) = 1.5 m

Further, adding 300 mm to depth as free board.

Hence, Length = 1.5 m, Width = 0.75 m, Depth = $(1.8 + 0.3) = 2.1$ m

Total capacity of designed septic tank = Length × Width × Depth

$= 1.5 \times 0.75 \times 2.1$

$= 2.3625 \text{ m}^3 = 2362.5$ litres

Example 5.7

Estimate the septic tank size for a house of 10 persons considering detention time as 48 hours and cleaning period of 2 years. The water use per capita per day is 135 litres, and rate of sludge accumulation as 30 litres per person per year.

Solution

Number of persons = 10

Detention time = 48 hours = 2 days

Total wastewater per capita per day = 135 litres

Rate of sludge accumulation = 30 litres per person per year

Cleaning period after every 2 years.

Wastewater

Total wastewater per day for 10 persons = 135 × 10 = 1350 litres

Since detention period is 48 hours or 2 days, it means that the septic tank should have the capacity to retain the household water for at least 2 days.

So, total waste water in 2 days = 1350 × 2 = 2700 litres

Let us take 3000 litres capacity minimum for the house.

Sludge Accumulation

Rate of sludge accumulation for 10 persons in a year = 30 × 10 = 300 litres

Since, cleaning period is of 2 years, so total sludge = 300 × 2 = 600 litres

Total capacity required for detention and accumulation = 3000 + 600 = 3600 litres = 3.6 m^3

Considering depth of septic tank to be 1.8 metres.

Plan area required = 3.6/1.8 = 2 m^2

Assuming Length (L) is twice of Width (W).

Since, Plan area = L × W = 2W × W = 2

Therefore, Width (W) = 1 metre, which is more than minimum width of 750 mm.

So, taking Width (W) = 1 m, and Length (L) = 2 m

Further, adding 300 mm to depth as free board.

Hence, Length = 2 metres, Width = 1 metre, Depth = (1.8 + 0.3) = 2.1 metres

Total capacity of designed septic tank = Length × Width × Depth

$$= 2 \times 1 \times 2.1$$

$$= 4.2\ m^3 = 4200 \text{ litres}$$

Important Points to Remember

1. Where the water requirement is higher than 20,000 litres per hour, high yielding open wells, tube wells, are necessary to meet the need.
2. The water need of the farmstead is mainly met by the surface water and the ground water such as wells, ponds, streams and rivers.
3. As per IS: 1172 – 1993, water required for domestic purposes for average Indian conditions per head per day may be taken as 135 litres.
4. As a general rule of thumb, adult cattle require around 70-100 litres of water per day. Sheep and goats require around 4-14 litres per day, while pigs need around 10-20 litres per day.
5. Per capita demand is defined as total yearly consumption of water for a water supply scheme divided by the population of that area and the number of days in a year.
6. Water with a hardness below 60 mg/L is considered soft, while water with a hardness above 120 mg/L is considered hard.
7. The size of the storage tank should be such that the total yield of the water during the peak period, plus the water stored in the tank during periods of less demand, must be able to meet the total demand.
8. Rapid sand filters are widely used in municipal water treatment plants, as well as in many industrial applications.
9. The most commonly used method for disinfecting water is chlorination.
10. Aeration is a commonly used process in water treatment that involves the introduction of air into water to improve its quality.
11. Submersible pumps are designed to be immersed in the fluid that they are pumping.
12. Per capita water demand is calculated in litres per person per day.
13. Groundwater is usually free from suspended impurities.
14. Contaminated water contains pathogens.
15. For removal of temporary hardness of water, it is boiled.
16. Hardness of water is caused by the presence calcium and magnesium in water.
17. The temporary hardness of water is caused by bicarbonates of sodium and potassium.
18. BOD of safe drinking water should be nil.
19. Zeolite process can effectively remove permanent hardness of water.

20. For public water supply, hard water is not used because it consumes more soap.
21. The maximum permissible total solid content in water for domestic purposes should not exceed 1000 ppm.
22. Hardness limit for public supplies of water (potable water) ranges between 75 to 115 ppm.
23. In case of public water supplies, the permissible upper limit of chloride content is 250 ppm.
24. The tank which is generally used for plain sedimentation is of rectangular shape.
25. The best coagulant for removing of water is alum.
26. Alum increases acidity of water.
27. Chlorination of water is done for removal of bacteria.
28. When bleaching powder is added to water, its pH value increases.
29. Bleaching powder is hypo-chlorite of lime.
30. The process used for the removal of dissolved carbon dioxide from the supply water is aeration.
31. Potassium permanganate is used for improving colour.
32. The permissible pH value of potable water is 6.5 to 8.5.
33. The maximum permissible nitrate content for drinking water is 45 ppm.
34. In a 100 ml sample of drinking water, the coliform organism should not exceed 10.
35. Presence of calcium in drinking water is not considered dangerous.
36. A septic tank should be located at least 30 m away from any wells, streams, or other bodies of water to prevent contamination.
37. Septic tanks should be located downhill from the source of water supply
38. Septic tank must have minimum width of 750 mm, minimum depth of one metre below water level and a minimum liquid capacity of 1000 litres.
39. Half yearly or yearly desludging of septic tank is desirable. Small domestic tanks, for economic reasons, may be cleaned at least once in 2 years and cleaned before the sludge level reaches the bottom of outlet.
40. Connecting pipe for septic tank should not be less than 100 mm.
41. The minimum diameter of a manhole opening is 50 cm.
42. The detention period in a septic tank is usually 24 to 48 hours.
43. From septic tank the effluents are discharged into soak pit.

44. Three basic functions that take place in a septic tank are removal of solids from the sewage, decomposition of solid sewerage, and storage of sludge and scum.
45. The basic working principle of septic tank is that under suitable conditions bacteria will liquefy any organic material.
46. Baffle wall checks wave's formation in a septic tank.
47. Sewage is the liquid waste of a household or community including human excreta.
48. Sludge refers to that part of the sewage which after being acted on by the bacteria settles at the bottom of the tank.
49. Scum is a partially submerged mat of floating solids that may form at the surface of the fluid in the tank.
50. The location of the tanks with respect to the residence should not be closer than 1.5 m from any building foundation.

Exercises/Questions

1. What is hydrological cycle?
2. What are the various sources of water supply?
3. What is per capita demand?
4. Write a short note on drinking water standards?
5. What are the different impurities of water?
6. Write a short note on domestic water demand.
7. What are the main objectives of water treatment?
8. Explain the different water treatment systems suitable to rural community.
9. Discuss various types of filtration system used in water treatment.
10. What are the different types of filters being used for water treatment? Explain them.
11. Discuss different methods of disinfecting water.
12. Explain different types of wells.
13. Discuss different types of pumps.
14. What is the purpose of sanitation?
15. Explain different designs of toilets.
16. What is a septic tank? What are the functions of a septic tank? On which principle it works?
17. Define sewage, sludge and scum. Explain the construction and working of septic tank with the help of a neat diagram.
18. What factors should be considered while selecting a site for septic tank?

19. Write short notes on:
    (a) Pump characteristic curves
    (b) Ecosan toilet
    (c) Bore-hole latrine
    (d) Maintenance of septic tank

Chapter 6

Dairy Housing

Dairy farming has been a significant part of the agricultural scenario for thousands of years. India being a predominantly agrarian economy has about 65 per cent of its population living in villages in 2020, where livestock play a crucial role in the socio-economic life. A large number of rural families in India are engaged in dairy production, for whom this is an important source of secondary income. India is not only one of the top producers of milk in the world, but also the largest consumer of milk and milk products in the world. Dairy sector is instrumental in bringing socio-economic transformation in the country. It has created a lot of employment opportunities and also provides improved nutritional benefits. Animal husbandry is a major sub-sector of agriculture sector and contributes significantly to the agricultural value output. It has been noticed that during the last few years the growth rate in the livestock sector in India is much faster in comparison to agriculture and allied sector. Dairy farming has the potential to provide nutrition support, reduce rural poverty, inequity, ensuring food security for millions of rural households, and enhancing economic growth, particularly in rural areas.

Inclement weather conditions adversely affect animal's health, production and welfare. In order to prevent the adverse effects of the climate proper housing for dairy animals becomes necessary. Proper housing is very important aspect to increase the capability of animals to produce milk, meat, wool *etc.* The main objective of housing animals is to provide conditions for high productivity, health, comfort and convenience of management so that the animals can utilize their feed and genetic abilities for maximum output. Adequate animal housing improves the milk production, utilization of labour, prevention and control of disease, health of animals, care and management of the animals that enhance the quality of milk and helps in providing the better price from the market.

The structure of animal housing should be sound, adequate functionally and meet the thermal requirements. The structure shall have adequate floor space, sufficient air opening without draft, an ample supply of drinking water, good feed trough capacity and good drainage. The structure should be designed in such a way that the abrupt change in atmospheric temperature should not affect the animals as the cattle need environmental comfort for optimum milk production.

6.1 Terminology

Barn: An enclosed covered building for the keeping and care of livestock and/or storage of dry roughages and bedding.

Paddock or Yard: It is an open space with fences, wall or rails to keep the animals. It provides the open air exercise to the cattle. It has provision of the mangers and water troughs with or without ties to hold animals for various purposes.

Pen Barn: A structure for sheltering and restricting dairy animals. It may be divided into pens where the individual or groups of cows are bedded and fed.

Shed: A simple structure for shelter of livestock, storage or feed, or both, and open on one or more sides. It may be separate or attached to another structure.

Standing (Stall): It is an exclusive space inside the cattle shed for the individual animal to stand, sit or lie down.

6.2 Selection of Site

The following factors may be taken into consideration while selecting a site for a dairy shed:

1. **Topography and Drainage:** The location of a dairy facility should be dry, high, and well-drained to maintain a healthy environment for the animals. Avoid the use of fertile soil for the construction of shed.
2. **Protection from Wind:** The site shall be able to protect the animals from strong wind velocity whether hot or cold. The shed shall preferably be located at a place where there are considerable suitably placed trees to serve as wind-breaks and to provide shed. The row of trees should be planted against the wind direction at the boundary of farm to break the wind. The sheltered trees for animals need to be planted 4 m away from the cattle shed.
3. **Accessibility:** The farm should be located within 100 m of main road and should be accessible year-round.
4. **Availability of Water**: The site shall be such where ample and good quality water supply would be available.

5. **Availability of Electricity:** The site should be near to the existing electricity supply system as electricity is needed for lighting of shed and for operating various machines at the farm.
6. **Marketing:** The site should be in nearby areas of the market (or *mandi*) that facilitate the farm owner a regular sale of farm outlet with profit.
7. **Future Expansion:** There shall be provision for future expansion.
8. **Orientation**: (a) The site should have provision of east to west orientation to keep the structure cool during warm months in all areas of India except the temperate region like Himalayan and hilly regions.
   (b) In desert areas, the shed shall be oriented across the prevailing direction of the wind in order to protect the roof from being blown off by high wind.
   (c) In coastal areas, the shed shall be oriented across the prevailing wind direction in order to protect the roof from being blown off by high wind and at the same time to provide sufficient air movement in the shed.
   (d) In sub-mountainous region, the building should be so sited as to avail of the natural aeration and drying.

6.3 Types of Dairy Housing

The type of housing depends upon the geographical and local conditions as well as on the economic conditions of the farmer. Loose housing system is more suitable to Indian conditions.

The housing for dairy cattle can be classified as:

1. The stall/stanchion-barn or conventional barn system
2. The loose-housing
3. The open-air barn/free-range system

In the *stanchion barn*, each animal is kept tied up in separate stalls for resting, feeding, watering, and milking. The animals always stay inside their exclusive stall. This arrangement of the stall and animals is also known as *general purpose barn.*

The *loose housing barn* is comprised of covered or partially covered yard and a separate *milking parlour*. The animals are kept in a yard to fed, sit, stand or lie down whereas *milking parlour* is used to milk the animals. The *loose housing barn* is also known as milking house system. The adult animals are given access to a feeding area, a resting area, and an adjoining open lot. Other dairy animals may be in separate pens, lots or buildings.

The *open air barn* or *loafing barn system* is quite rare among the well-established or modern dairy farm. The animals stay inside the fenced area at the open field where the animals are fed and milked without any covered shelter.

6.3.1 Stanchion-Barn System

The stanchions, neck chains, or ropes used in the stanchions-barn system fastened the animals on a platform. The fodder and concentrate are fed to the fastened animals by using the mangers. The animals are also milked on the same stanchion. The stanchion barn is equipped with windows and ventilators at appropriate places inside a completely roofed space. This barn is suitable for the temperate region like Himalayan regions to keep animals safe from the prolonged winter.

The stanchion-barn system of housing offers several advantages over the loose housing barn. The extreme weather conditions do not affect the animals. It also takes less efforts to keep the animals clean. The animals are displayed to better advantage which may be of importance where milk is sold at retail at the farm. It is rather easy to detect and control the infections as the animals are under constant supervision. The other advantage is that the chances of fights among animals are lesser as they are tied and chained. There is no trouble from aggressive and bossy animals. Dehorning, though desirable, is not necessity under this system. The stanchion-barn is the most suitable barn on modern dairy farms having large herds, as a large number of animals can be handled in one barn unit. There is also no need to have a separate milking parlour. However, there are some disadvantages of this system of housing. The construction cost of stall barn system is more when compared to loose housing system. It is not possible to detect the animals in heat as they are unable to mount each other and thereby difficult to observe the heat symptoms. Moreover, the animals do not have freedom of movement. The stanchion-barn system is inappropriate for the hot and humid climates.

Depending upon the number of animals, the animals are grouped into the rows. Single row of cows inside a shed is generally preferred for the small number of cows. A two rows barn is preferred for housing of more than 8 cows. The rows can be so arranged that the cows either face out (tail to tail system – Figure 6.1) or face in (head to head system – Figure 6.2). In the case of enclosed barn, the face out barn is suitable for the milch animals. The stanchion-barn system is restricted to 80-100 cows inside a structure.

The width of the two row barn is 10-11m with sufficient length to keep all animals. Although, the length of the barn is limited to 40m that can house 72 cows (36 in one row) in one barn unit. In case the number of animals to be housed in the barn is large, a cross drive in the middle of barn shall be provided to improve the efficiency and convenience.

A small partition wall is provided to separate the stall. A metal bend pipe of same length as of stall is installed on the partition wall. The partition wall is made of concrete or brick to 12.5 cm height. Narrow and cramped stall cause discomfort or injury to animal by restricting their movement inside the stall. The size of stall should be decided on the basis of age and type of animal. The length of the stall may be variable so as to accommodate cows of various lengths. Figure 6.3 shows the section of face out type stanchion dairy barn.

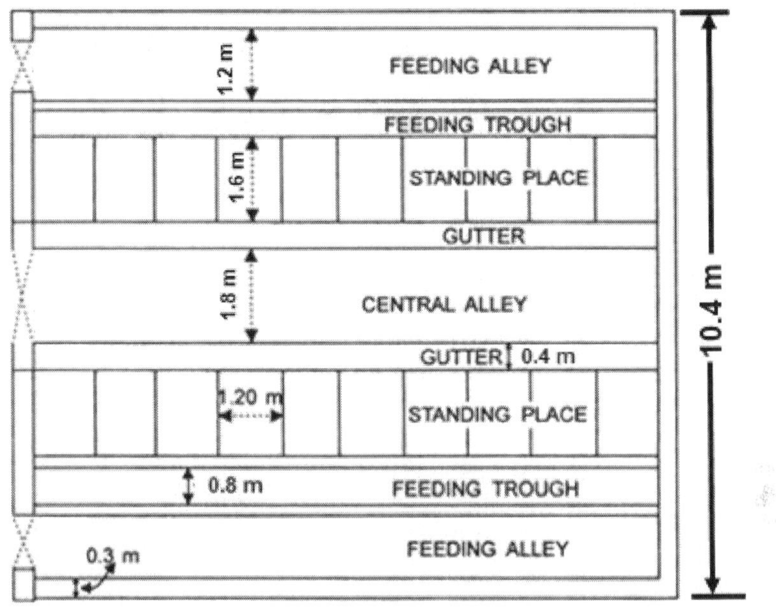

Figure 6.1. Face Out Arrangement (Tail to tail system).

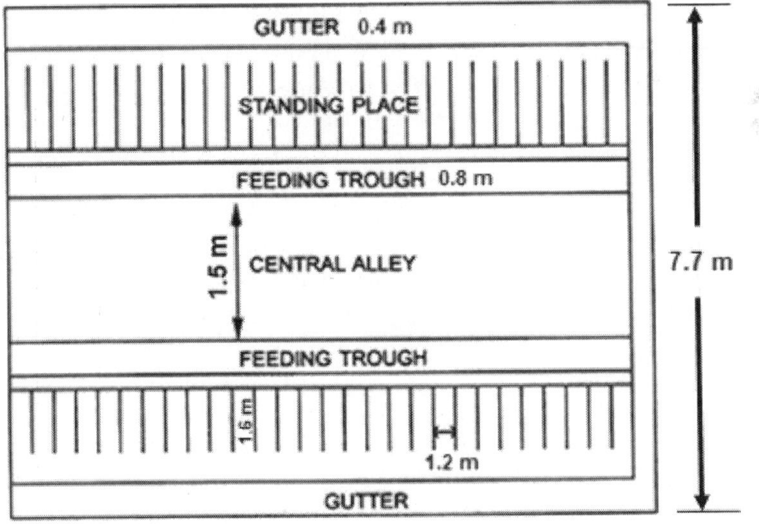

Figure 6.2. Face to Face Arrangement (Head to head system).

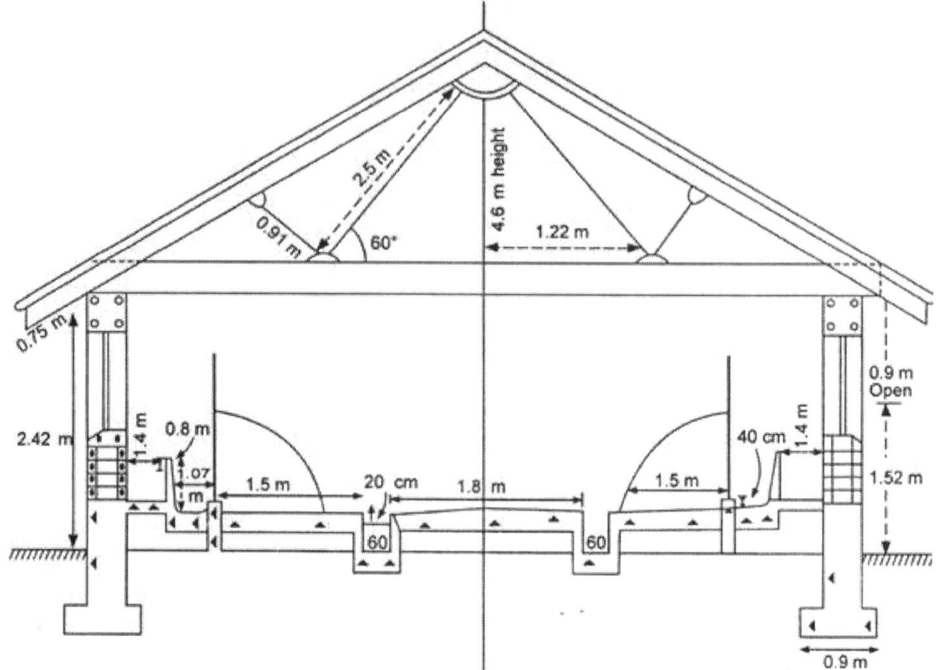

Figure 6.3. Stanchion Dairy Barn – Face Out Type.

Feed Alley

Feed alley is a path provided between the manger in the face in system or between the outer wall and manger in the face out system. It is used as a walkway for the cart full of animal feed like silage, straw, concentrate *etc.* In the case of face out system, the width of feed alley should be 1.2m. The central walk should have a width of 1.5 to 1.8 m exclusive of gutters when cows face out, and 1.2 to 1.5 m when they face in. The central walk should have a slope from the centre towards the two gutters running parallel to each other, thus forming a crown at the centre. The floors of the feed alley, central drive or manure alley and cow bed should be on the same level.

Advantages of Face Out System

1. This system is important since most of the time spent by the animal attendant is for the jobs to be performed from the rear of animals.
2. This system enables easy cleaning of barn.
3. This system provides better supervision of milkman while milking the animals.
4. Cows get more fresh air from outside.

5. The risk of disease transmission between animals is lower.
6. Minor disease can be detected quickly.

Advantages of Face in System

1. This system saves 20 per cent of the floor area and cost of construction as compared to face out type. The outside space of face in system is used as manure alley.
2. This system provides clearer space for cleaning of dung and floor and it is suitable for open barns. Face in type system is usually preferred for bullocks.
3. As one feed alley is used for both row, so face in system make feeding the cow easier.
4. The side by side head of animals on stanchion facilitate better display for visitors.
5. The stalls allow easy entrance for animals.
6. The gutters receive most of the Sun rays where they are needed the most.

6.3.2 Loose Housing System

Loose housing system (Figure 6.4) comprises of open and enclosed paddock or pasture that keep animal loose during day and night except at the milking parlour. The shed is also provided in the open paddock to shelter the animals during harsh weather like rainfall or excessive hot or cold. There is lesser dependency on labour in this system. The system is most economical for a large herd and moderate climate. Only a small area where feed mangers are built is paved. This type of animal housing system is not suitable for temperate Himalayan region and heavy rainfall areas.

A separate milking room or milking parlour shall be constructed in which cows are milked at milking time. In addition to milking parlour, the loose housing system should have feeding, resting and watering area, store room and maternity-cum-calf pens. The plantation of shady trees is preferred in open paddock to provide cool in summer and wind break in the winter. It is desirable to face north by the animal while feeding at manger. Fodder is made available in a common manger. The animals have continuous access to a common watering tank. The concentrates should be fed in the separate barn while milking the animal.

Half walls or fencing by plain wire of suitable height should be provided to protect the open paddock. The floor in the shed should be cemented or brick paved to easy cleaning of manure and drainage of urine. The floor should be rough to prevent the slip of animal. GI sheets or cement asbestos sheets are used as roofing material for construction of sheds.

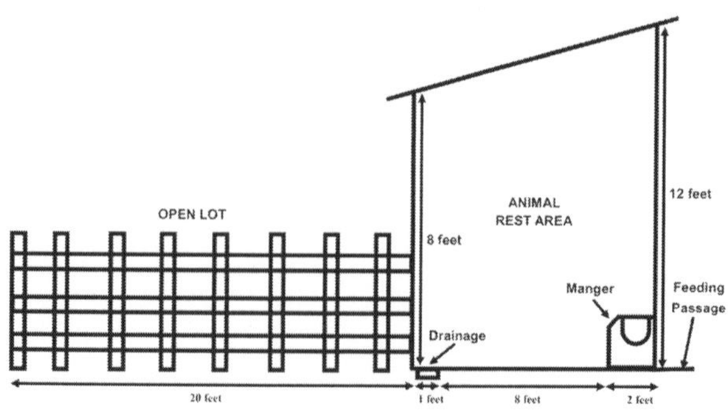

Figure 6.4. Loose Housing System.

Under the loose housing system, the animals need to be grouped on the basis of their age, sex and physiological condition. The animals can be classified into cows, bulls, heifers and the calves on the basis of age and sex. On the other hand, the milking cows, dry and pregnant cows, sick animal are the classification on the basis of physiological condition. The provision for the separate barn to these animals is desirable for their management, care, feeding and comfort. The space requirement per cow depends upon climate, size of cow, feeding arrangement (inside or outside) and the average production of the herd. The floor space requirement both for loafing as well as for feeding shall be 9 to 11 sq. m per cow. The width of feed manger shall be between 70 and 75 cm for each cow. The feeding platform at the side of the manger shall be paved. Water outlets are needed about every 20 to 30 meters to flush out the excreta from the yard.

Facilities for Resting

The sheltered area attached to open lot, is called resting area. This is a covered shed to protect the animals from inclement weather conditions. About 3.2 to 3.7 sq. m resting area for each cow is required except in very cold regions where provisions for about 4.6 sq. m per animal is needed. Open or paddock areas of sufficient space is provided to cattle and buffaloes in warm regions since these animals lie down during the night. Space used for mangers, passage *etc.* should not be counted as resting area. The resting area should be nearly as square as possible rather than long and narrow because the space near the open front do not provide good shelter. Pavement with mangers is preferable for maintaining a clean feeding area. Floor space requirement is given in Table 6.1.

It is preferable if the short side of the resting area faces the prevailing winds during the year. During severe cold weather, closing one side of the shelter is usually enough to keep it warm. The animals maintained in the loose housing system grow long coat during the winter months to protect themselves against cold. It is desirable to plant shady trees in the open lot for providing shade in summer. One tree (*Ficus* species) for every 100 sq. m shall be enough. The branches of the trees can be trimmed in winter for sun rays to pass.

Animals should have free access to the open lot. If it is not possible to brick-line the whole of the open lot, a strip of six-meter width in front of the open side of the resting area should be made *pucca*. As the animals remain loose all the time, there is no need for provision for tying the animals either in the resting area or in the open lot.

Comparison between Stanchion Barn and Loose Housing Systems of Housing

Basis	*Stanchion Barn*	*Loose Housing*
Floor area	Floor space requirement is less as compared to loose housing. Floor space of about 5.5 to 7.5 m^2 is required per animal.	Floor space requirement for loafing as well as feeding shall be between 9 to 11 m^2 per animal.
Cost of construction	It is more costly than loose housing system.	Cheaper as compared to stanchion barn.
Exposure to harsh weather	Animals are less exposed to harsh weather conditions as the animals are housed in a barn.	Animals are more exposed to harsh weather conditions.
Freedom of movement	The animals do not have freedom of movement	Animals are more comfortable as they move about freely.
Feeding and management of stock	Individual feeding and watering arrangement for each animal is required in this system of housing.	Feeding and management of stock is easier and more economical because of common feeding and watering arrangement.
Health and comfort	Cows do not feel as much comfortable as in loose housing barn. Sometimes cows get badly injured by slipping on floor.	Cows feel comfortable on bedded surfaces and do not develop troubles like injury due to slip.
Spread of disease	Easy to detect and control the infections.	The possibility of spreading disease is comparatively high.
Fights among animals	There is no chance of fight amongst animals in this system of housing.	There are chances of fights.
Detection of heat in cows	It is not possible to detect the animals in heat.	Detection of heat in cows is easier.
Feed consumption	Roughage consumption per cow is less in this system.	Fodder consumption may be slightly higher per cow because of larger appetite of animals due to freedom of movement.
Manure	The manure obtained from the stanchion barn is not of good quality.	The quality of manure produced in this barn is good.
Labour requirement	It requires more labour for cleaning the barn.	Comparatively less labour is required for cleaning.

Basis	Stanchion Barn	Loose Housing
Change in herd size	Additional facility need to be created for accommodating more animals.	More number of animals can be added without any inconvenience.
Display of herd	The herd display is good.	The herd display is not attractive as cows are always seen either moving or sitting in uncontrolled manner.
Milking parlour	There is no need of separate milking parlour as cows are milked where they are housed.	A separate milk parlour is needed for milking of cows.
Quality of milk	Daily cleaning of barn is required to produce good quality sanitary milk.	Separate milking parlour ensures production of high quality sanitary milk.

6.3.3 Open-Air Barn

In this type of housing, the animals move freely in a large estate. Open air barn is a natural and enclosed pasture with provision of water for animals. It is used for stock management of the animals that are not handled daily like horses, beef cattle *etc.* Although, the dairy animal can be kept in open air barn where they are fed and milked. This type of housing is not common for dairy cattle.

6.4 Floor Space Requirements

The floor space requirements of the animals depend upon their age and breed. *Murrah* buffalo requires more floor space as compared to other breeds due to difference in size. Bulls and calving animals are housed individually while other animals are usually housed in groups. The average floor required per animal is given in Table 6.1.

Table 6.1. Floor Space Requirements for different Types of Dairy Animals

Sl. No.	Type of Animal	Floor Space per Animal (m^2)		Manger Length per Animal (cm)	Water Trough Space per Animal (cm)
		Covered Area	Open Area		
1.	Calf/young calve (< 8 weeks)	1.0	2.0	40- 50	40- 50
2.	Adult calve/young stock (> 8 weeks)	2.0	4.0	40-50	40-50
3.	Cow	3.5	7.0	60-75	60-75
4.	Buffaloes	4.0	8.0	60-75	60-75
5.	Down calvers/calving	12.0	12.0	60-75	60-75
6.	Bull	12.0	20.0	60-75	60-75

Feeding and Watering Space Requirements

In the loose housing, unrestricted and adequate space for feeding and water should be provided. Animals should be comfortable and free of any competition while having feed, water or rest. The provision of feed trough should be for all animals at a time. Although, the water trough should accommodate the 10 per cent of the animals. Thus, the length of water tank is estimated by considering 10 per cent of the animals at a time. Thus the total length of water tank in a pen for 100 adult animals is calculated as:

Total number of animals = 100

Let us assume that the water trough length required per animal = 75 cm

Thus, total length of water tank required = 10 per cent of (100×75) = 750 cm

Similarly, water trough length for calves can be estimated.

6.5 Sheds

The dairy housing shall comprise of various types of sheds for housing different categories of animals. Various types of sheds that are required for organized milk producers and *Gaushala* are as follows:

1. Milch animals shed
2. Dry animals shed
3. Down-calver shed/Calving animals shed
4. Sick animals shed
5. Young stocks shed
6. Calf shed, and
7. Bull shed

The paddock or yard shall be attached to each of the sheds. Bulls, pregnant cows and calves under the age of six weeks require individual housing. All the other livestock can be managed conveniently in the common resting and feeding areas. There could be some variations of layouts with these units depending upon shape, size and topography of available land, size of enterprises, finances, *etc.* The constructional details of the sheds are discussed in the next section.

6.5.1 Milch Animals Shed

Tail to tail system with a central passage is preferred for the standings of milk animal. The size of the standing should be as per the size and age of animals. The standing should be 1.5-1.7 m in length and 1-1.2 m in width. The buffaloes with spreading horns commonly found in the coastal region need 1.2-1.3 m width. The width of central passage shall be 1.8 m with slope from centre to outward towards the drains. The either side of the central passage shall have drains for cleaning of urine. The continuous mangers need to be constructed in front of animal standings. For easy movement, the 1m cross passage shall be provided at the end of every eight standing. The floor of the standings on either side shall be sloped 1 in 60 towards the drains. The gable roof shall be used for the animal shed.

6.5.2 Dry Animals Shed

The head to head system with central manger shall be used in standings of the shed for the dry animals. The drain need to constructed behind the lines of animals. The standing shall face the wall in case of single line of animal.

The dimensions of the standings shall be same as specified for the milking animals. The bullock can be kept with the dry animals. Although, a separate bullock shed and cart shed can be constructed with the dry animal shed.

6.5.3 Calving Animals Shed

A calving pen is used to house pregnant animals 2-3 weeks prior to the anticipated date of calving. The cows should not be allowed to calve in the milking cowshed as it may lead to spread of disease and in-sanitary conditions for milk. Separate shed shall be provided for this purpose. The number of the calf room should accommodate the 5 per cent of adult female residing in the dairy farm. Two calving boxes shall be constructed with calf room at the end of the shed. The calf room need to be separated by partition wall from the animal standing. A 1.25-meter-tall partition wall must separate the two calving boxes. Each calving box must be 3×4 m in size. A 50 cm wide manger and a water trough shall be provided at two separate corners of calf box. The standings of the down-calver shed shall be built in such a way that animals are kept tethered facing the wall. The length and width of each standing shall be 2.0 m and 1.6 m respectively. Flooring throughout the resting area and the open lot should be of such material as to allow easy and through washing. A single-leaf door of 2.0×1.2 m size shall be provided for each of the calving boxes. With ample soft bedding, the shed should be provided with sufficient ventilation through windows and ridge vent. An attached attendant room for calving shall be provide with shed.

6.5.4 Sick Animals Shed

Sick animals due to contagious disease need to be separated from other animals of the dairy farm. The sick animals shed shall be constructed away from the shed of healthy animals. In addition, the sick animals shed should not be accessible to the heathy animals. The sick animals shed should have water trough and feed manager. The dimension of the sick animals shed is similar to the calving shed. A separate drainage disposal shall be provided to the sick animals shed.

6.5.5 Young Stock Shed

There may be a separate shed for young stock in order to economize the cost and space. The shed shall be nearby to the cow shed. The young stock shed shall have a central manager where young animals are fastened in two rows. The dimension of each animal standing inside the shed shall be 1.4×1m. The floor should be damp free and the water should not accumulate on the floor. The damp and ill drained floor shall be avoided to protect the calves from the respiratory diseases. The drains on either side of the standings shall be laid. Alternatively, a single line of the young stock against a wall can be maintained with drains. An open paddock shall be provided for exercise. There shall be provision of water trough inside each calf shed. The separate calf shed shall be provided to the group of calves as per their age. The calves of below one-year age shall be classified into three groups on the basis of their age *i.e.* <3 months, 3-6 months and >6 months.

6.5.6 Calf Shed

Calf shed is used for housing young calves under the age of six weeks separately. The calf shed may be annexed either at the end or on the side to the milking animal shed. This facilitates taking calves to their dams quickly. A suitable partition shall be provided between the calves and milking animals. A separate shed or unit can be constructed in the case of the large number of calves. A calf requires 1.2-1.5 m^2 space. The overall size of the calf shed depends upon the number of calves to be housed. The calves may be kept loose inside the shed. The manger shall be constructed along the walls of the shed. A central shallow saucer-shaped drain shall be constructed in calf shed. A water trough shall be provided at one corner of the shed. Young calves are very susceptible to inclement weather. During extreme cold days, one side of the shed shall be closed off with some temporary material. In addition to this, about 30 cm thick hay bedding shall be provided. On windless sunny days, the calves may be housed in portable calf pens in the open. During extreme hot days, protection from hot winds should be provided.

6.5.7 Bull Shed

Bull shed is meant for housing bulls separately on the farm because they are more active and sometimes vicious. The shed should be constructed towards one end of the farm. There shall be one shed for each bull. One bull is required for every 50 breedable females on the farm, if natural breeding is practiced. When artificial insemination service facilities are available, it is not necessary to keep the bulls on the farm. The bull shed shall have two boxes each measuring 4 × 3 m. Each box shall lead to separate paddock of 20 m^2 space. The height of the wall may be 1.5 m. A 50 cm wide raised manger with feed and water sections shall be provided in each box. Manger shall be constructed in the sheltered area. Strong material should be used for the construction. From the bull yard, the bull should be able to view the other animals of the herd so that it does not feel isolated. The bull should have a free access to an exercise yard/paddock which should be provided with a strong fence or a boundary wall.

6.6 Paddock or Yard

A paddock or yard shall be developed near to every shed for free movement/exercise of the animals. A paved brick-on-edge can be provided in the paddock. The minimum requirement of the space per animal is provided in following table:

Table 6.2. Minimum Space Requirement for Paddock for Different Categories of Animals

Category of Animal	*Minimum Space per Animal (m^2)*
Buffalo	8
Cow	7
Young stock	4
Calf	2
Calving	12
Bull	20

Trees shall be planted in the paddock to provide shade to the animals. Feeding and water troughs shall be provided within the paddock. A 2-meter wide paved platform shall be constructed away from trough to withstand the heavy treading of animals and allow easy washing and cleanliness. The paddock area shall be fenced either by brick wall or stone slab or a railing or wires. Fencing material should be cheap and locally available. The effective height of the fence for calf and cow may be 1.2 m, and for bull it may be 1.5 m. A gate of suitable type and size shall be provided. Figures 6.5 and 6.6 show the layout of dairy cattle sheds for rural milk producer.

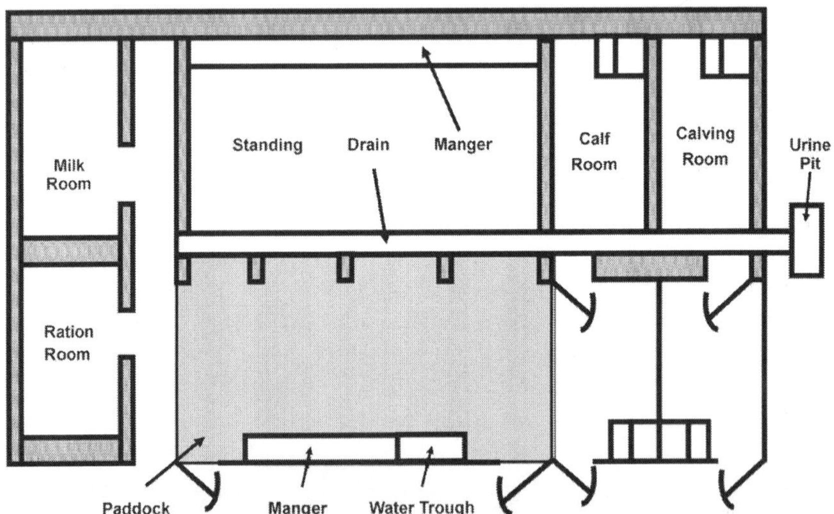

Figure 6.5. Layout of Dairy Cattle Shed for Rural Milk Producer (Lean to type shed).

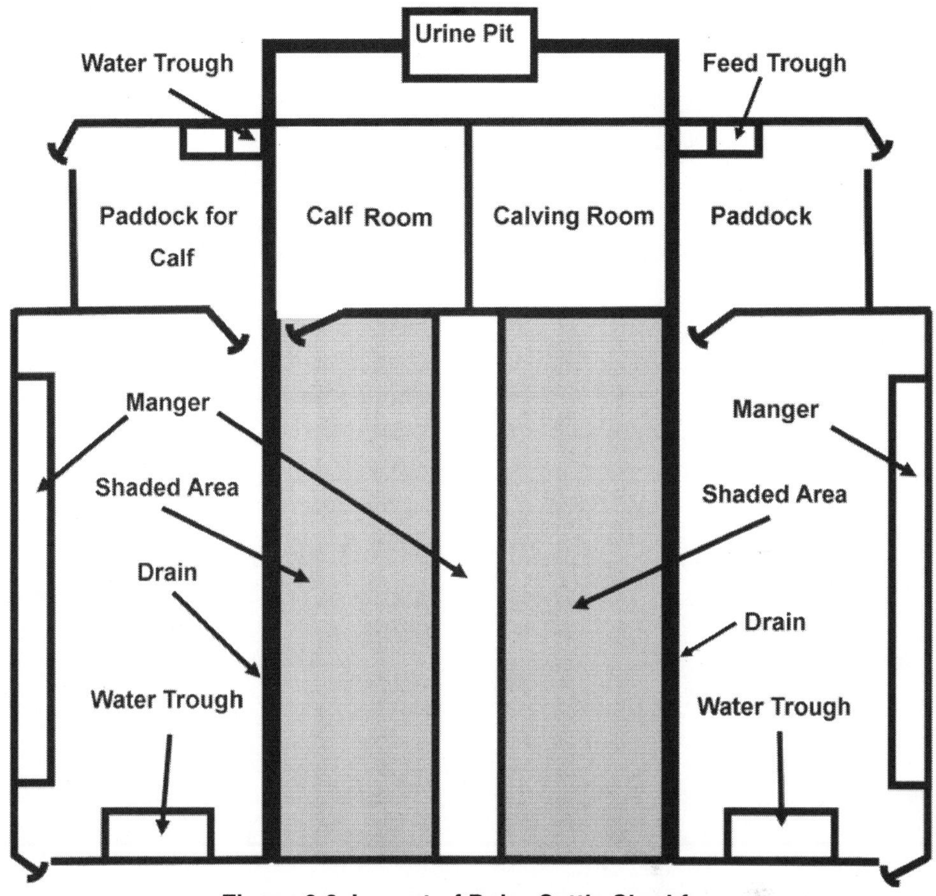

Figure 6.6. Layout of Dairy Cattle Shed for Rural Milk Producer (Gabled type roof).

6.7 Construction

Standing

The animals are generally kept in single row in a small farm. The standings of animals in a single row shall face the wall. The manger shall be constructed adjoining the wall. Whereas, the standing of animals in a large farm shall be arranged in rows. The size of standing depends upon the age of animals. Although, 1.5-1.7 m length and 1-1.2 m width per animals shall be preferred in small and large farms. In coastal areas, sometimes buffaloes with spreading horns are found. In that case, the width per animal may vary from 1.2 to 1.3 m.

Floor

The floor shall be either of *moorum, kankar*, cement concrete, bricks-on-edge, or stone slabs. The inside floor of the barn should be impervious which can be easily kept clean and dry, and is not slippery. The surface of the cement concrete or stone flooring shall be left rough. It must be ensured that there is no accumulation of water on it. While finishing the floor it is desirable to get a non-slip surface by pressing with an expanded metal piece. A slope of 1 in 60 shall be provided to allow drainage of water into the gutter. Steel rings may be provided on the floor near the manger for tying the animals.

Square shaped grooves shall be constructed in the cement concrete floor to avoid slipping of animals. A 15×15 cm size groove is for large animals and 10×10 cm is for the calves. The U shape groove shall be 12 mm of width and 10 mm deep (Figure 6.7).

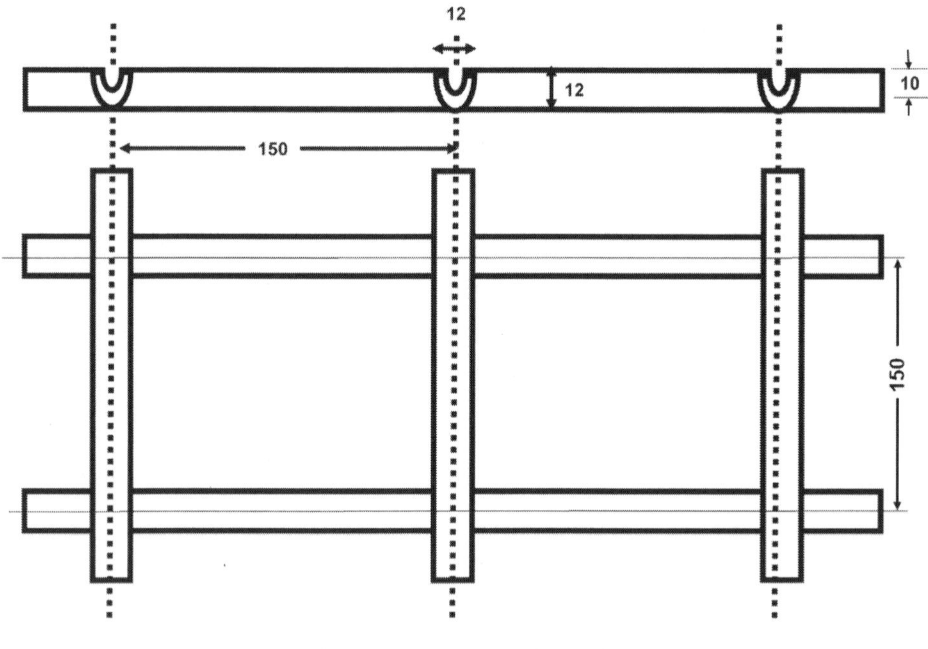

(All dimensions are in mm)

Figure 6.7. Cross-section of Standing for Larger Animals.

Manger

The manger shall be made either of cement concrete, stone slabs, wooden planks, or brick-in-lime or cement mortar. It shall be continuous type. Cement concrete mangers are preferred for reasons of strength and economy of space. The corners of the manger should be rounded off and finished smooth.

A common manger shall be provided for feeding roughages and concentrates. The manger shall be constructed in the open lot and/or in the sheltered area. The manger in shelter area can be used during the inclement weather. If cattle have access to roughage at all times, 30 to 45 cm of manger space per cow is enough. If the roughages are fed in batches so that all animals eat at the same time, each cow shall need 60 to 75 cm of manger space. The requirement for the calves shall be almost half of this. A paved strip of at least 3 m width shall be provided around the feeding area. Milking cows can be fed concentrates at the time of milking in the milking barn. Other animals can be fed concentrates in the same manger. To prevent the stronger animals from taking advantage, the concentrates shall be fed in batches.

The length of manger required of one cow shall be same to the width of the stall. The manger shall be demarcated using the wall of 12.5 cm height. For adult animals, 60 cm of inner width shall be provided inside the manger (Figure 6.8). The base of the manger shall be of 75 cm height above the floor to avoid throwing of fodder in feed alley by the cows. The angle iron shall be installed on the fore-curb of manger made of bricks. The size of the manger as per the age of animals and material of construction is provided in Table 6.3.

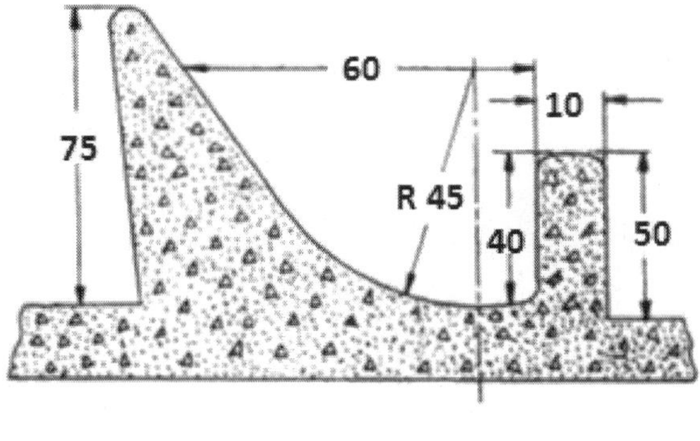

(All dimensions are in cm)

Figure 6.8. Cross-section of RCC/Bricks Manger for Adult Animal.

Table 6.3. Dimensions of manger (in cm)

Sl. No.	*Particulars*	*Construction Material*			
		RCC	*Bricks*	*Stone Slab*	*Wooden Plank*
1.	Height of fore curb (Maximum)				
	a. For adults	50	50	50	50
	b. For calves	30	30	30	30
2.	Thickness of fore curb (Minimum)	10	10	4	3
3.	Inner width of manger (Minimum)				
	a. For adults				
	i. One-way feeding	60	60	60	60
	ii. Two-way feeding	120	120	120	120
	b. For calves – one way feeding	40	40	40	40

Sl. No.	Particulars	Construction Material			
		RCC	Bricks	Stone Slab	Wooden Plank
4.	Depth of manger (Minimum)				
	a. For adults	40	40	40	40
	b. For calves	15	15	15	15

For large herds, self-feeding hay-bunks are suitable and labour saving. A wooden partition through which the animals can eat the fodder is removed to the edge of hay-bunk so that the animals eat only the portions accessible through the partition and do not spoil the fodder excessively.

Walls

The walls should be either of good quality bricks or stone slab laid in cement or lime mortar. The walls should be cement-plastered from inside. It shall be white-washed once in a year. The corners shall be rounded off. In the case of gable roof, the height of ceiling may be 5-5.5m for 10-11m wide dairy barn. The height of the wall shall be 2.4 m. Although, wall should be solid up to 1.25 m from the floor. The honey comb pattern of bricks shall be provided above 1.25 m of wall for air circulation and sufficient light in the shed. The thickness of the wall shall be at least 20 cm. In desert areas, the solid portion may extend up to 1.8 m in height. In case of lean-to type system, the end walls above the solid portion shall be left open or provided with a series of wire-netted windows.

Pillars

Pillars are made of either brick, stone masonry, iron pipe, mild steel I-section or timber. It is necessary to place pillars every two or three standings, depending on the width of each animal standing. It is necessary to paint iron pillars to prevent corrosion. Rectangular pillars must have all of their edges rounded off to a smooth finish. The minimum dimensions of the pillars shall be as follows:

Brick	-	45 cm × 35 cm or 40 cm × 30 cm
Stone masonry	-	30 cm × 30 cm
Iron pipe	-	10 cm dia
Mild steel I-section	-	10 cm × 10 cm
Timber	-	Square (10 cm × 10 cm), Round (15 cm dia)

Roof

The roof of the shed shall be either of lean-to type or gabled type. The roof material may either be galvanized steel sheets, asbestos cement sheets, asphalt roofing material, tiles, or locally available material. Extreme temperature fluctuations are experienced inside the shed due to usage of galvanized steel sheets. Galvanized steel or asbestos cement roofs may be overlaid with thatch to lessen the stress of extreme climate, particularly in hot climate. Alternatively, reflective paint may be applied on the top of the steel sheets to reflect sun rays.

Steel or wooden trusses or by series of pillars can be used to support the roof. An overhang or eave at least 50 cm shall be projected out from the pillars. The eave of roof shall be 2.2 m of height from the level of floor. The eave of roof can be extended to 75 cm to save the animals from the extreme weather (direct sun and rain). To make ventilation more effective a continuous ridge ventilation is considered most desirable.

Drains/Gutters

The drain of U shape (semi-circular) with 6 cm depth shall be constructed using the cement, mortar or stone. The drain shall have slope of 1 in 40 to 1 in 60 and width of 30-40 cm. It shall be constructed at the back of the animal standings. One drain for each row shall be provided in the larger dairy shed of two rows. The drain may be led to a common urine pit or to the field. The depth of urine pit should not exceed 40 cm, and it may be circular or rectangular.

Doors and Windows

In a double range stall barn, the doorway should be provided on both sides. The width of the door shall be between 2.4 and 2.7 m, while in a single range cowshed, the width shall be 1.5 metre. Usually, sliding or collapsible doors are used. All doors of the shed should lie flat against the external wall when fully open. In hot climates, doors are sometimes avoided. For adequate lighting in the barn, 0.37 sq. meter of window area per cow is desirable. However, this can also be accomplished either by eliminating the side walls completely or by making perforated walls all along.

Stanchion

It is used for tying the animals in the barn. Fastened animal using the stanchion can move its neck and body to reach the feed and water and to rest. Stanchion is a frame work made up of pipe or angle iron with an oval shaped adjustable ring kept one end hanging from the frame and the other hinged to the floor. It should be wide enough to pin it on easily to the neck to the animal and to lock it. A 25 cm wide and 1 m long frame loosely hanging from the top and fixed at the bottom is satisfactory. The bottom of the stanchion should not be more than 22.5 cm above the floor level. There may be various locking and unlocking devices used on the stanchion, but the simple hook technology is favoured. Galvanized iron pipes (3 cm dia) shall be used for the partitions of stalls. These pipes are connected to vertical posts supporting the water pipe lines and/or trusses. Instead of using oval shaped ring, chains are also used to fasten the animals to the stanchion. For this a strap is put around the neck of each animal.

Water Supply

Adequate clean water is essential for economic milk production. The amount of water intake depends upon the season, breed, types of feed, age/size of the animal and whether the animal is in milk or dry. Dry feed increases the demand for water. A milking cow requires 3 to 5 litres of drinking water for every litre of milk produced. The daily drinking water requirement is calculated at the rate of 50 litres per livestock unit per day. On the basis of number of animals housed, the daily requirement of water can be calculated. The requirements of a young-stock may be considered as half of that for a cow. The water requirement during the summer months is comparatively more.

A common watering tank (Figure 6.9) with a float valve near the feeding area and preferably outside the resting area is preferable. Watering tank shall be either rectangular or circular in shape. Figure shows the dimensions of rectangular watering tank facilities for 100 cows. The cows can drink on either side of the tank. The tank shall be constructed of reinforced cement concrete or brick-in-cement. About 3 m area around the watering tank shall be paved. A slope for proper drainage should be provided in the paved area. Animals may have to be tied at the time of washing. Thus, a suitable number of steel rings shall be permanently fixed in the cement concrete base at the time of construction. Watering tank should not be located under a tree as the leaves falling from the tree and also droppings of the birds on the tree are likely to make the water unfit for drinking purposes. For drinking purpose, it does not matter whether the water is hard or soft.

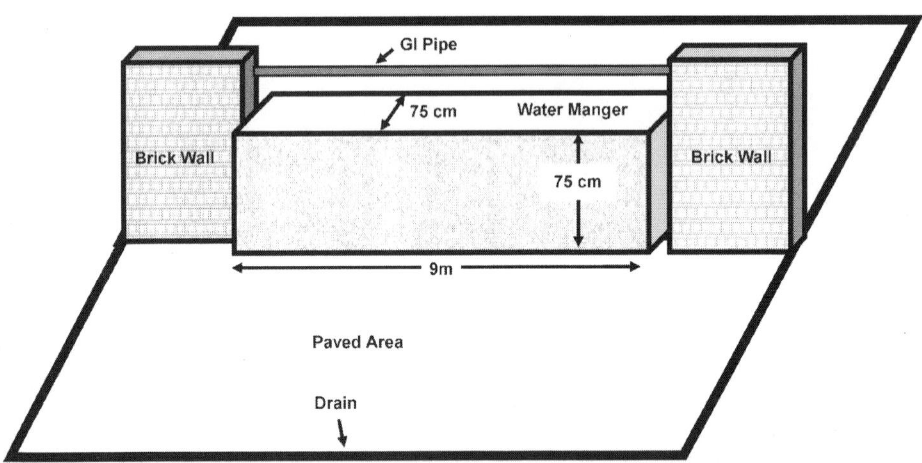

Figure 6.9. Watering Tank for 100 Animals (Cows can drink on either side).

For supplying water in the shed, water trough of 20 cm diameter shall be made on the manger wall. These troughs may be placed between the two standings. The railing on the sides of water troughs shall be provided so that animals may not try to step into it and contaminate the water.

Example 6.1

Calculate the dimensions of a watering tank for 100 animals.

Solution

Drinking water requirement per animal per day = 50 litres

Total water requirement per day = 50×100 = 5000 litres

Let the width and depth of watering tank as 0.75 m each

$$\text{Thus, the length of watering tank} = \frac{(5000/1000)}{0.75 \times 0.75} \approx 9.0 \text{ m}$$

6.8 Ancillary Structures

In addition to the animal shelters, a dairy farm must have a few ancillary buildings or structures. The layout of the structures shall be planned to maximize labour efficiency and animal productivity while minimizing the human and animal travel. To keep construction costs as low as possible, the farm buildings should be built with less expensive, locally sourced building materials.

The following ancillary structures shall be required for *gaushala* and organized milk producers.

1. Milk collection, recording and testing room,
2. Utensils room,
3. Ration room,
4. Feed store,
5. Office room,
6. Fodder processing and storage room,
7. Trevis or cattle crutch, and
8. Foot bath.

Milk Collection, Recording and Testing Room

A room of 4×3 m size shall be provided either inside or near the milking animal shed for collecting, recording and testing of milk. The floor of the room should be impervious and hard wearing. The doors and windows shall be made fly-proof. A suitable slab or platform should be provided for keeping testing apparatus.

Utensils Room

A room of 4×3 m size shall be provided for washing and storing milk cans. The floor of the room should be hard-wearing to withstand the frequent handling of heavy milk cans. The doors and windows shall be made fly-proof. The utensils room shall be located near to the milk recording and testing room.

Ration Room

A room of 4×3 m size shall be provided to store feed concentrates temporarily to meet the requirements of the animals for the day. This room shall be near to the milking animal shed. The room should be damp and rodent proof.

Feed Store

Feed store/room is a damp free and rodent proof facility used to store feed concentrates. Feed room shall have an attached room for the store keeper. The size of feed room depends on the assumption that is 0.2 m^2 storage space is required for each livestock unit. A concentrate-cum- feed mixing room shall be built away from the barn. A small feed ration room shall be built attached to the milking room.

Silos - Trench silos are practical in Indian circumstances. It is built by excavating the hard earth, and the sides and floor are lined with seepage-proof cement concrete, brick, or stone slab.

Hay/Straw storage – Hay and straw will be stored in a shed that has walls on three sides. The shed should be placed far from animal sheds due to the risk of fire.

Office Room

An office room of appropriate size shall be provided in the *gaushalas* and organized milk producers for keeping the record *etc.* It may also be used for keeping medicines and instruments for treating sick animals. A toilet may also be provided. A visitor's room may also be attached to office room.

Fodder Processing and Storage Room

The adequate provision for processing and storage of roughages shall be made.

Trevis or Cattle Crutch

A cattle trevis shall be provided to secure the animals for rendering first-aid and artificial insemination.

Foot Bath

A suitable foot bath at the entrance gate should be provided. It is also desirable to construct a shallow concrete floor at the main entrance of the gate. It will ensure that any vehicle entering the farm will pass through the shallow floor which contains medicated solution.

The other facilities required at farm are:

Segregation Room

A segregation room may be provided for keeping new animals introduced in the herd for some days before their mixing with the herd.

Lighting

Electric light is needed in all feeding areas and open lots. The lights should be controlled by a switch near the most use entrance. Desirable intensity for general illumination is 50 to 100 lux (lumens/m^2) at the point of work. A 25-Watt incandescent bulb/4-Watt LED bulb for each 10 sq. m floor space or 60 W incandescent bulb/9-Watt LED bulb for each 20 sq. m space or equivalent fluorescent tube light may be provided.

Waste Handling System

It is desirable to install a bio-gas plant of suitable size at the site. The animal waste such as dung and urine shall be effectively used for the production of cooking gas and manure.

Manure and Fodder Cart

The manure from the barn shall be removed by manure cart directly pulled through the barn by a tractor or bull. Wheel barrows or box type push carts shall be used for moving silage or freshly chopped fodder. The capacity of the cart shall be such that all the animals on one side of the two rows barn are served in one trip.

The removal of manure and the supply of chopped fodder and concentrate need special types of barn equipment. As far as possible the manure should be removed by manure car directly pulled through the barn either by a bull or by tractor. Silage or freshly chopped fodder can be easily moved by wheel barrows or box type push carts. The fodder, the concentrate and the manure should be handled only by forks and shovels and not with human hands.

6.9 Milking Parlour/Milking Room

The milking room is a necessary part of the loose housing barn. Although, it can be used in the stanchion barn. The milking room is a separate facility to milk few cows at a time. This facility needs to constructed with adjoining feed room to keep batch of awaiting cows for milking. The floor of stalls inside the milking room shall be equal or raised against the working floor of operator. The arrangement of the stalls of milking room, drain and feed alley shall be similar to the stall barns.

The floor level system is more popular for hand milking in India because of economy and tradition. However, floor level milking is inconvenient and tiresome as it involves bending and stooping on an average five times per milking. In elevated stalls, all the bending and stooping is eliminated. The room is covered on the sides and top to keep it clean. Sanitation can be maintained with little effort because the milking barn is small and cows are in it only for milking.

The number of stands is determined by the allowable milking time of the herd or time taken by the cows to eat the concentrate rations usually fed. One stall for every two or three milking cows shall be enough when the milking is done by hand. When milking is done by machine, one stall for five to seven milking cows shall be suitable.

The parlour should have a high-quality concrete floor and metal rails for durability and ease of cleaning. Walls should be plastered from inside. Milking parlour should be fly proof. To increase air circulation, upper parts of walls shall not be constructed, but this area should be screened against the nuisance of flies. The pit where the milker stands should have a floor level 90 cm below that of the cows for the most comfortable working position. Floors and gutters should be sloped to drains to assure a dry floor during the milking period. The milking barn shall be connected with the loose housing by means of a narrow passage, allows entry of cows in a single file. For hand milking, it is convenient if the cows are brought in batches, washed, cleaned, milked and then the next batch is taken in.

Adequate light is required for milking operations and proper cleaning of parlour. The intensity of illumination shall be 150 lux at the point of work. It is desirable to have good light at the flanks, udders and teats. Suitable number of 100 Watt bulbs spaced 3.6 to 4.6 meters above the operator alley and 6 to 9 meters apart above the centre of alley are sufficient. When there is only one row, light may be installed about 1.5 to 1.8 meters above the floor in wall opposite milking stalls to throw light on the important operations.

Several arrangements of stalls are in vogue. These are abreast parlour, tandem parlour, walk through parlour, Herringbone parlour, and rotary parlour. The most convenient and popular milking parlour is U-shaped arrangement of tandem stalls with either single string or two-string stall arrangement. In a single string arrangement, three elevated stalls are placed in a line with separate entrance and exit. Ramps are provided at entrance and exit. Similarly, for two strings of three cows each, there are two separate entrances and exit ramps provided. It is possible to milk 10 to 15 cows in each stall. The herringbone arrangement of stalls is more efficient and economical from construction point of view and is also widely used.

Abreast Parlour

The milker and the cows share the same floor space (Figure 6.10). Cows can enter and leave individually. In the case of hand milking using the bucket, the width of stands should be 1–1.1 m. On the other hand, the width should be 0.7-0.8 m for machine milking. This type of parlour is slow and tends to only be appropriate for smaller herds. The major disadvantages of this system are that the cow obstructs the milker and the milker need to walk longer between the milking point.

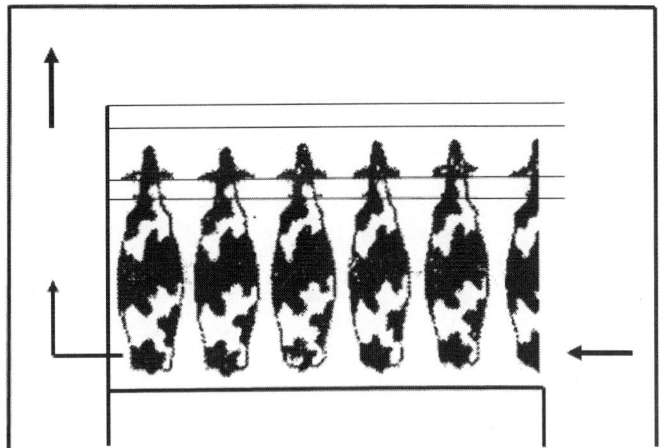

Figure 6.10. Abreast Parlour.

Tandem Parlour

In tandem parlours, cows are located parallel to the milking pit (Figure 6.11). Tandem parlours allow more individual attention to the cow, but are generally slow and only appropriate for smaller herds. However, where slow milkers are present, they do not delay the release of faster milking cows as is the case with herringbone parlours.

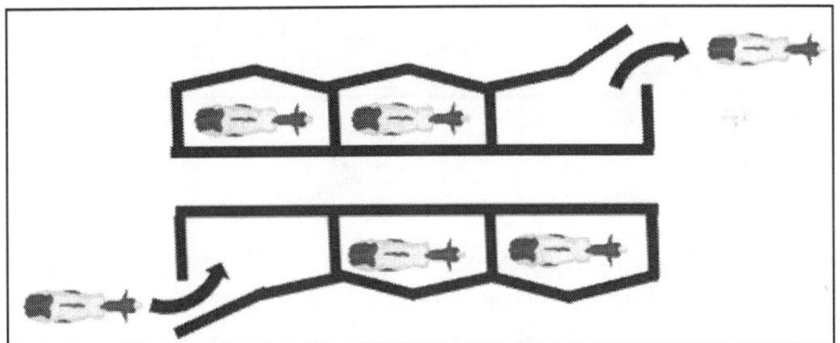

Figure 6.11. Tandem Parlour.

Walk Through Parlour

The narrower path in the exiting building can be constructed in which the batches of cow can enter and leave.

Herringbone Parlour

Herringbone parlours (Figure 6.12) are the most common design used on dairy farms with smaller herds. The cattle stand at a slight angle to the milker who is located in a pit between two lines of cows. This design offers the milker a different access point to the udder than the parallel or tandem designs, and also allows access for different types of equipment to be used. It is popular parlour due to its simple design and high capacity (cow milking per man hour). The compact working area keep movement of feeder near to the side wall. A total of 80 cows can be kept on each side of 3 or 4 stands. One disadvantage is that a cow that is slow to milk out delays the release of the rest of the cows in the line.

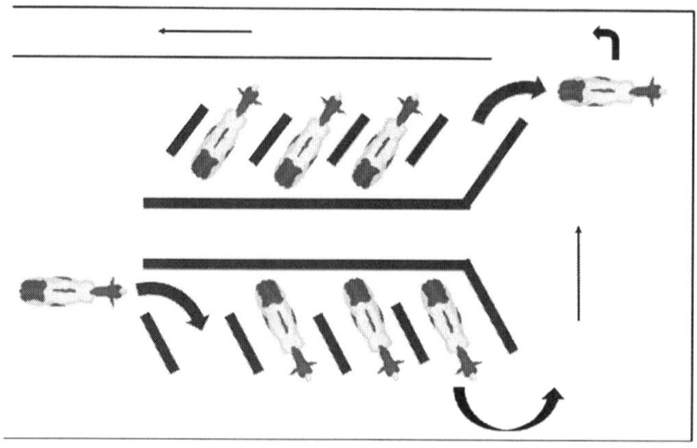

Figure 6.12. Herringbone Parlour.

6.10 Milk House

A milk house separated from the milking barn shall be provided to keep the milk. Milk house shall be located towards the main entrance of the farm as the milk is required to be hauled daily from the milk house. In order to save the labour, a loading platform for the loading of milk cans into the delivery vans shall be constructed. Equipment for automatic transfer of milk from the milking machine into the refrigerated milk tank through over-head pipe is convenient and sanitary though expensive.

It is essential to keep the milk and also to chill the milk in larger dairies having 450 to 700 litres production capacity that requires 3.7 m × 5 m size of room and an additional 0.37 m^2 for every 40 litres of milk production above 700 litres. For a smaller dairy unit below 100 litres a small room with a dimension of 3.7 m × 3 m can be adequate for storing milk and concentrate feed. The floor space as given in Table 6.4 can accommodate the milk recording equipment, milk cooling device in the form of bulk cooler, can-racks, milking pail racks, sinks *etc.* The space for pouring platform, boiler, compressor *etc.* shall be provided in addition.

The floor of the milk house should be *pucca* with a smooth finish to allow through cleaning. The floor should be sloped towards the drain. All the doors and windows should be protected against entry of flies and dust. Facilities for tap water, hot water, and steam under pressure should be provided. About 5 litres of water is needed for cleaning utensils and other operations in the milk-house for every litre of milk handled.

Table 6.4. Floor Space Requirement for Milk House.

Daily Milk Output (litres)	*Floor Space*
Up to 100	3.7 m × 3.0 m
100 to 200	3.7 m × 3.7 m
200 to 450	3.7 m × 4.25 m
450 to 700	3.7 m × 5.0 m

6.11 Design of Dairy House Shed

In case of commercial livestock production, animals may be kept in single row or double row styles with or without provision of feeding passages. The animals may be housed either as face to face or tail to tail in double row styles. The dimensions of the shed shall be calculated as follows:

1. Single row with feed alley

Width of the shed = wall + feed alley + manger + standing + gutter + wall

2. Double rows tail to tail (Figure 6.1)

Width of the shed = wall + feed alley + manger + standing + gutter + central passage + gutter + standing + manger + feed alley + wall

3. Double rows face to face (Figure 6.2)

Width of the shed = wall + gutter + standing + manger + central passage + manger + standing + gutter + wall

4. Length of the shed

Length = (Number of cows ÷ 2) × 1.2 + 0.3 × 2 m

or, Length = Area of the shed/Width of the shed

6. Length of the manger

Single row

Length of manger = Length of the shed

Double rows

Length of manger = Length of the shed × 2

7. Number of animals to be accommodated

Number of animals = (Length of the standing floor ÷ 1.2) × 2

Dimensions

Width of feed alley in case of face out system = 1.2 m

Width of central passage when cows face out =1.5 to 1.8 m exclusive of gutters

Width of central passage when cows face in = 1.2 to 1.5 m exclusive of gutters

Length of each standing = 1.5 to 1.7 m per animal

Width of each standing = 1.0 to 1.2 m per animal

Width of gutters in the stanchion barn = 30 to 40 cm

Minimum inner width of the manger = 60 cm

Thickness of wall = 30 cm

Example 6.2

Determine the dimensions of a dairy shed for housing 50 animals both under face out and face to face arrangements. Make reasonable assumptions wherever necessary, but state them clearly.

Solution

The animals shall be housed in double rows.

Assume

Width of feed alley in case of face out system = 1.2 m

Width of central passage when cows face out = 1.8 m exclusive of gutters

Width of central passage when cows face in = 1.5 m exclusive of gutters

Length of each standing = 1.6 m per animal

Width of each standing = 1.2 m per animal

Width of gutters in the stanchion barn = 0.4 m

Width of the manger = 0.8 m (including walls)

Thickness of wall = 0.3 m

Face out arrangement

The width of the shed shall be

Width of the shed = wall + feed alley + manger + standing + gutter + central passage + gutter + standing + manger + feed alley + wall

= 0.3+1.2+0.8+1.6+0.4+1.8+0.4+1.6+0.8+1.2+0.3 = 10.4 m

Face to face arrangement

Width of the shed = wall + gutter + standing + manger + central passage + manger + standing + gutter + wall

= 0.3+0.4+1.6+0.8+1.5+0.8+1.6+0.4+0.3 = 7.7 m

Length of the shed = (Number of cows ÷ 2) × 1.2 + 0.3 × 2

= (50/2) × 1.2 + 0.3 × 2 = 30.6 m

Thus the size of shed when animals are housed under face out arrangement = 10.4 m × 30.6 m

and, the size of shed when animals are housed under face to face arrangement = 7.7 m × 30.6 m

The layout of the shed shall be as per Figs. 6.1 and 6.2.

Example 6.3

Draw a layout plan of farm cattle shed for *gaushalas* and other organized milk producers.

Solution

A farm cattle shed consists of the sheds for different category of animals and ancillary structures. The sheds are for milking animals, dry animals, calving animals, sick animals, young stock, calf and bull shed. The ancillary structures shall be milk recording and testing room, utensils room, ration room, office room, fodder and chaffing shed, feed store, milk room, *etc.* Water trough and mangers shall be provided at suitable locations.

The floor space requirement for each shed is as follows:

Sl.No.	*Type of animal*	*Floor Space per Animal (m^2)*	
		Covered Area	*Open Area*
1.	Calf/young calve (< 8 weeks)	1.0	2.0
2.	Adult calve/young stock (> 8 weeks)	2.0	4.0
3.	Milking animals	4.0	8.0
4.	Dry animals	4.0	8.0
5.	Calving animals	12.0	12.0
6.	Sick animals	12.0	12.0
7.	Bull	12.0	20.0

Length of manger per adult animal = 60 to 75 cm

Length of manger per young animal = 40 to 50 cm

Water trough of sufficient size shall be provided depending upon the number of animals housed. On an average 50 litres of water per animal per day is required.

Ancillary structures

Milk recording and testing room - 4×3 m

Utensils room - 4×3 m

Ration room - 4×3 m

Office room - 4×3 m

Fodder and chaffing shed - Appropriate size for installation of machines

Feed store - 0.2 m^2/animal

Milk room - 4×3 m

A typical layout of the animal shed shall be as follows:

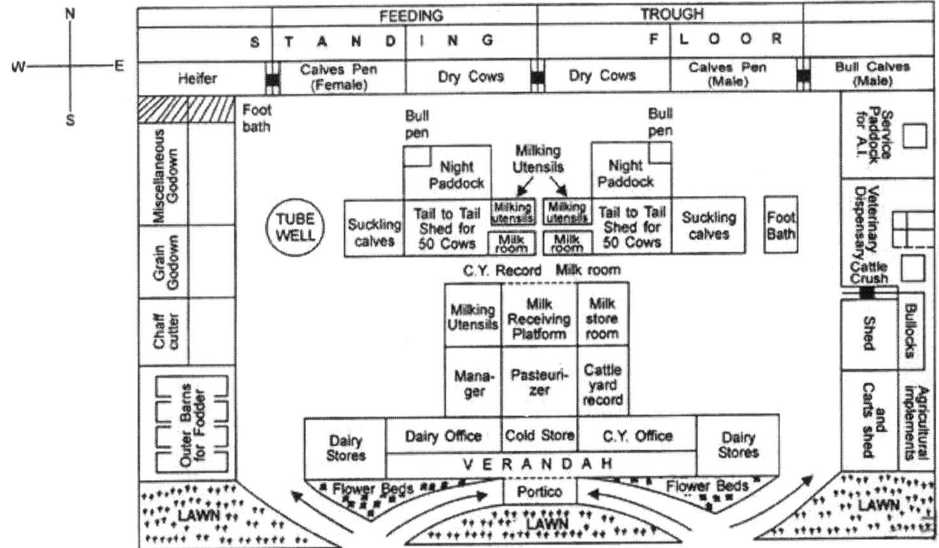

6.12 Environmental Requirements of Cattle

1. **Temperature:** Thermoneutral zone for dairy cows is typically 5°C to 25°C, while for beef cattle, it ranges from 0°C to 30°C.
2. **Relative Humidity:** Relative humidity should ideally be maintained below 70 per cent to prevent heat stress and reduce the risk of respiratory issues.
3. **Ventilation**: Proper airflow should be provided to remove moisture, gases, and airborne pathogens.
4. **Lighting**: Natural daylight or artificial lighting with sufficient intensity and duration should be available.
5. **Space:** Sufficient space for movement and resting, typically ranging from 9 to 11 square meter per cow is required.
6. **Shelter:** Shelter should be provided to protect from adverse weather conditions like rain, wind, and extreme temperatures.
7. **Bedding:** Clean and comfortable bedding material, such as straw or sawdust.
8. **Feed and Water**: Access to clean water and balanced nutrition according to dietary requirements.
9. **Security:** Secure fencing and gates to prevent unauthorized access.

Important Points to Remember

1. Paddock is an open area surrounded by walls, fences or rails for accommodating cattle.
2. Paddock is meant to provide open air exercise to the animals.
3. The site of dairy shed shall be oriented east to west in all areas except temperate Himalayan and hilly regions.
4. In desert and coastal areas, the dairy shed shall be oriented across the prevailing direction of the wind.
5. In the stanchion barn, the cows are housed and milked in the same building. It is also called as the general purpose barn system.
6. In the loose housing barn, the cows are housed in a covered or partially covered yard, but they are milked in a special milking house.
7. A barn in which cows run freely is known as loose housing.
8. Open air barn does not have any cover.
9. Open air barn is also termed as the loafing barn system.
10. The face in type barn gives a saving of 20 per cent in floor area of the barn as compared to the face out type barn.
11. The face out type barn is preferred for milch animals.
12. The maximum length of the barn should be restricted to 40 m so that nearly 72 cows can be housed in one barn unit.
13. Two rows face out dairy barns are commonly built 10 to 11 m wide.
14. The passage between the outer wall and the manger is called feed alley. A feed alley in case of face out system shall be 1.2 m wide.
15. The central passage should have a width of 1.5 to 1.8 m exclusive of gutters when cows face out, and 1.2 to 1.5 m when they face in.
16. In the loose housing barn, the floor space for loafing and feeding required is around 9 to 11 square meter per cow. The width of feed manger shall be between 70 and 75 cm for each cow.
17. Gutters in the stanchion barn are usually between 30 and 40 cm wide.
18. Stanchions are used for tying the animals in the barn.
19. For adequate lighting in the barn, 0.37 m^2 of window area per cow is desirable.
20. For designing manure pit, space requirement is calculated as 0.042 cum. per cow per day.
21. The length and width of each standing may vary 1.5 to 1.7 m and 1.0 to 1.2 m per animal, respectively.
22. The minimum inner width of the manger used in dairy barn shall have width of 60 cm for adult animals. The minimum back of the manger is 75 cm above the floor so that cows may not throw the fodder into the feed alley.

23. In stanchion barn a slope of 1 in 40 to 1 in 60 is provided in gutters to drain the water.
24. The floor space allotted to the animals in a barn depends on their age and breed.
25. The surface of the floor shall be left rough. In case of cement concrete flooring, the surface shall be properly grooved in order to avoid slipping of animals.
26. If gable roof is used, the maximum height of the ceiling may vary between 5 and 5.5 m for a barn 10 to 11 m wide.
27. All walls should be constructed solid up to 1.25 m height from the floor level. Above that height, the walls shall be constructed in honey comb pattern in order to provide sufficient air circulation and light in the shed.
28. Depending on age and breed, floor space required per cow is 5.6 to 7.5 m^2.
29. The width of the door in double range stall barn shall be between 2.4 and 2.7 m on both sides, while in a single range cowshed, the width shall be 1.5 m.
30. All doors of the dairy shed should lie flat against the external wall when fully open.
31. Pen barns are used for sheltering and restricting dairy animals.
32. Average water requirement of cows is 200 litre per capita per day, whereas the average daily drinking water requirement is 50 litre per animal.
33. In the loose housing barn, cows are housed in covered or partially covered yard and milked in milking parlour.
34. The labour requirement of stall barn is more than loose housing.
35. The size of the feed store is calculated based on the assumption that 0.2 m^2 storage space is required for each livestock unit.
36. A cattle trevis is used to secure the animals for rendering first-aid and artificial insemination.
37. Milk house should be built near milking barn.
38. Milking room is an essential part of loose housing but also be used in stanchion barn.
39. The bull shed shall have two boxes each measuring 4×3 m.
40. One bull is required for every 50 breed able females on the farm. Where artificial insemination service facilities are available, it is not necessary to keep bulls on the farm.
41. Herringbone design is used in the milking parlour.

Exercises/Questions

1. Compare in detail stall barn and loose housing barn.
2. List the various types of dairy barns and discuss one of them in detail.
3. Design a stanchion barn for 72 cows. Also draw its floor plan.
4. What points should be considered while selecting a location for dairy house?
5. What are the advantages of face out system of stanchion barn system in dairy house?
6. What are the advantages of face in system of stanchion barn system in dairy house?
7. List down the various types of sheds for housing different categories of animals.
8. Discuss the construction details of the floor, manger, walls, and gutters for the construction of dairy house.
9. What are the different arrangement of stalls in milking parlour? Discuss Herringbone parlour arrangement in brief.
10. Write short notes on:
    (a) Milk house
    (b) Milking parlour
    (c) Stanchion barn system
    (d) Loose housing system
    (e) Water supply at dairy farm

Chapter 7

Poultry Housing

Poultry is one of the fastest growing segments of the agricultural sector in India. The objective of poultry housing is to make the birds comfortable and safe from inclement weather. A well-constructed poultry house protects the birds from floods, rain, strong winds, extreme heat, and intense cold, and also protects them from predators like dogs, cats, snakes, birds of prey, rats, and thieves. In addition, good housing prevents mice, rats, and birds from eating their feed and transmitting disease. It also provides hens a safe room to lay their eggs. Birds raised in poultry farms are kept entirely inside buildings, which might be on the ground, the floor of wire netting cages, or slats. It is the most practical, cost-effective, and efficient approach for modern and large-scale poultry production. Modern poultry farms require a minimum amount of land, and farms might be situated close to a market.

7.1 Terminology

Broiler: Broiler is a young chicken usually 9 to 12 weeks of age. Broilers are reared to obtain meat.

Brooder: It is a device that provides artificial warmth to the chicks.

Chicken: Birds including chicks, broilers, layers, pullets, and cocks are called a chicken.

Chick: A young chicken is called a chick.

Cock or rooster: An adult male bird is called a cock or a rooster.

Cockerel: A young male chicken less than a year old is called a cockerel.

Hen: An adult female bird is called a hen.

Layer: Egg laying poultry is called a layer.

Pullet: A sexually immature chicken less than 22 weeks of age is called a pullet.

7.2 Typical Life Cycle of a Laying Chicken

A chicken lives for three to five years, except for some breeds which are known to live longer, and have four distinct life phases: egg, chick, pullet, and chicken stage. In each of these stages, a chicken will have differing needs to ensure they continue to develop healthily throughout their lifetime.

Phase 1: The embryo phase starts the moment the egg gets laid. It takes 21 days (3 weeks) for this phase of life.

Phase 2: The chick phase starts on day 21 when the chick hatches. The length of time it takes for this phase depends on the specific breed of chicken.

Phase 3: The pullet (or adolescent) stage is the teenager stage in a chicken's life cycle. A chicken is neither a chick nor an adult chicken. An adolescent female chicken might start laying eggs after 18 weeks.

Phase 4: In this stage, the male chickens become rooster and the female chickens become hen. The lifespan of an adult chicken depends on breed and also on husbandry.

A typical life cycle of a laying chicken is illustrated in Figure 7.1.

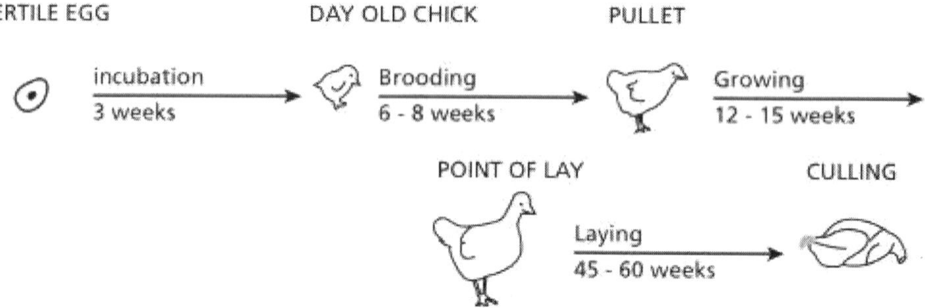

Figure 7.1. Typical Life Cycle of a Laying Chicken.

7.3 Poultry Housing Systems

There are three systems of poultry housing. These are the free - range or extensive system, the semi-intensive system and the intensive system (Figure 7.2).

Free–range or Extensive System: In this system, the birds are generally left free to roam during the day, and kept confined during the night. This system provides feed, water, and shelter to the birds. Shelter is generally provided by temporary roofing supported by ordinary poles. This system is common in rural parts of India. One can rear about 250 adult birds per hectare. The advantages of this system are that it requires less capital investment, less feed requirement as birds can consume fairly good amount of feed from grass land, and fertility of soil can be maintained. The disadvantages of free-range system are that the eggs are lost when laid inside the dense grasses, lack adoption of scientific management practices, more losses due to predatory animals, and possibility of diseases from wild birds.

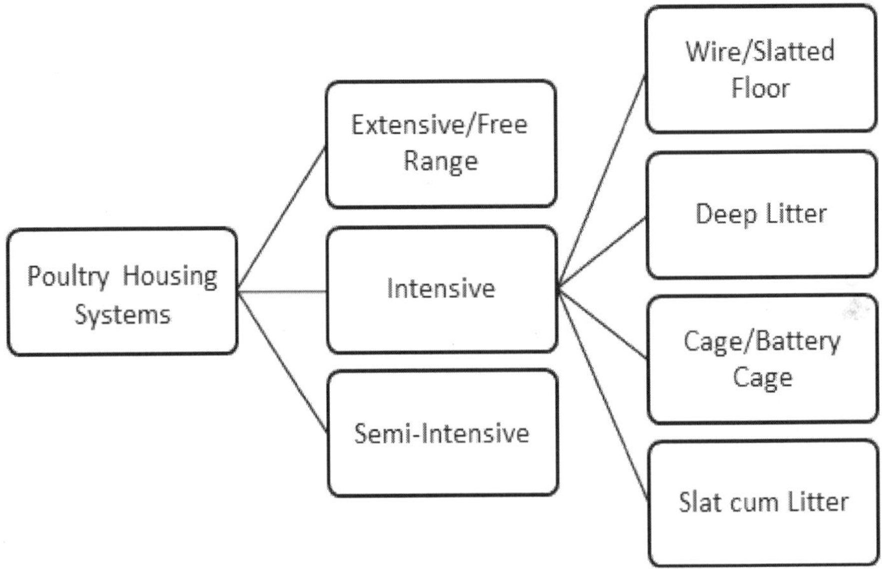

Figure 7.2. Types of Poultry Housing System.

Semi-intensive System: This system provides night shelters and fenced open spaces for birds, with a stocking density of 750 per hectare for adult birds. It is commonly used for duck rearing and offers more economical land use compared to free range systems. Feeding and watering facilities are provided in the pen, protecting birds from extreme climatic conditions. However, there are few disadvantages like high cost of fencing, and need for routine cleaning and removal of litter material from the pen.

Intensive System: In this system, the birds are totally confined to houses either on ground/floor or on wire-netting floor in cages or on slats. Arrangements for feeding, watering, nesting *etc.* are provided inside the house itself. It is the most efficient, convenient and economical system for modern poultry production with huge numbers. Intensive system is widely practiced in India. This system needs proper attention regarding nutrition and management. In this system land requirement is minimum, day-to-day management is easier, more output in terms of eggs and meat is achieved, farms can be located near the market, scientific management practices for poultry rearing can be applied, and the sick birds can be detected, isolated and treated easily.

The disadvantages of this system are that the birds cannot perform their natural behaviour like roosting, spreading wings, scratching the floor with legs *etc.*, and are not exposed to outside sunlight and feed sources. Also the chances for spreading of diseases are more.

7.4 Classification of Poultry Houses According to Age

The poultry houses are broadly classified into five categories namely the brooder house, grower house, layer house, broiler house and breeder house.

1. **The brooder house** accommodates chickens (layers or broilers) from day old to 8 weeks of age. It is used to brood and rear egg-type chicks. During this period birds are required to be reared under artificial heat.
2. **The grower house** is mostly for hens from 9 to 20 weeks of age which are reared to grow for egg purpose.
3. **The layer house** normally accommodates adult birds over 20 weeks of age, usually up to 72 weeks.
4. **The broiler house** is used to rear broilers up to 6 weeks of age.
5. **The breeder house** rears both male and female breeders in appropriate sex ratio.

The brooder house or layer house may also be used for housing growing stock after the brooding stage. In all-in all-out system, a layer house is used for brooding, rearing, and laying purpose.

7.5 Location

While selecting a location for the poultry house, the following points shall be considered.

1. Poultry house shall be located on an elevated and properly drained site, not liable to flooding or water-logging.
2. Orientation of the poultry house shall be such that avoids excessive solar heat and draughts in areas which experience high temperature for most part of the year. While in areas where the temperature is normally low, the house should be well exposed to sun but no draughts.
3. Poultry house should have proper ventilation and light facilities.
4. Poultry house should be located away from residential and industrial area.
5. Poultry house should have all the basic amenities like water and electricity, and shall have proper road facilities.

The poultry house orientation depends upon the climate where it is located. The orientation of the house should be located in such a way that long axis is in east-west direction in case of hot climates (Figure 7.3). This will prevent the direct sunshine over the birds. The long axis shall be in south-north direction in case of cold climates to allow for heating of the building.

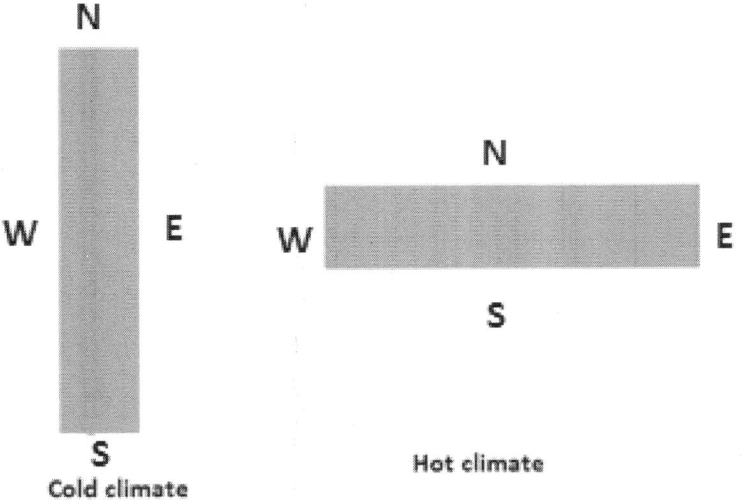

Figure 7.3. Housing Orientation for different Climates.

7.6 Environmental Requirements

The poultry birds should be provided with optimum environment to get the maximum output in terms of egg and meat production. Controlling the environment, particularly temperature, humidity, light, and litter quality, is crucial to chicken welfare.

Temperature

Extremes of hot or cold temperatures can cause chickens to stop laying eggs. Heat affects egg production in chickens more than cold. The most efficient temperatures for layers are between 20 and 24°C. Temperature above 24°C can cause an unhealthy amount of stress on the chickens, affect their organs, and potentially reduce their life expectancy. Because chickens do not have sweat glands, heat is trapped under their feathers. Their only cooling technique is to flap their wings and pant, so helping them maintain proper temperature is critical. The adverse effects of high environmental temperature (above 24°C) on poultry are decrease in growth rate, body weight, egg production, egg weight, egg quality, meat quality, semen quality, fertility and hatchability.

Humidity

The optimum range of relative humidity for chicken after brooding is between 50 and 70 per cent. However, during the brooding period the range is between 60 and 80 per cent. The higher humidity seems to favour better growth and feed conversion. If the relative humidity is too high, condensation can accumulate in the house thus resulting in the growth of micro-organisms. If humidity is too low, the environment will be dry. Dry environment produces dust which can become a major health concern for the birds.

Light

The light intensity at poultry house should be 15 to 30 lux in light-controlled houses and 30 to 40 lux in open-sided houses. The reproductive cycle of a chicken is stimulated by light, and chickens need 14-16 hours of light each day to maintain peak egg production. Any decreases in day length will adversely affect egg production. Brightness of light also has influence on egg production. The artificial light can be given either in the morning, evening or both morning and evening. The duration of artificial lighting depends upon the weather conditions and season. Incandescent or fluorescent bulbs shall be used for the purpose of artificial lighting.

Litter

Litter is used as the bed for the poultry birds in the deep litter system of housing. Paddy straw, rice husk, rice hulls, groundnut shells, wood shavings *etc.* are commonly used as litter material. The moisture content of good litter should lie between 15 to 25 per cent. High moisture of litter is harmful for poultry health which causes some problem like sores, and blister on poultry birds and produce ammonia gas which interrupts the respiratory system of poultry.

Sanitation

The design of the poultry house should be such that allows easy cleaning and spraying. Cleanliness and sanitation of the house are of extreme importance. The worst enemies of the birds like lice, ticks, fleas and mites are ample in poultry houses. Birds are subject to numerous diseases and insect infestations. Providing a suitable environment and cleaning the litter inside the house lead to sanitation. The construction of the house with smooth surfaces can reduce the accumulation of dirt and facilitate cleaning and disinfecting. There should not be any cracks and crevices in the walls. Wood should be properly treated with coal tar/insecticide before being fitted in the house. Equipment used in the poultry house, such as drinkers, nests and feeders, should be of a type that can easily be cleaned out and disinfected.

7.7 Space Requirements

Generally, the capacity of stationary layer house is in units of 100,200,500 and 1000 birds. The dimensions of a poultry house depend upon the number of birds and their floor requirement per bird.

Floor Space Requirement: The floor space requirement per bird in deep litter system shall be as given in Table 7.1.

Table 7.1. Floor Space Requirement per Bird in Deep Litter Intensive System

Age (Weeks)	*Minimum Floor Space per Bird (sq. m)*	
	Light Breeds	*Heavy Breeds*
0 to 8	0.07	0.09
9 to 12	0.095	0.18
13 to 20	0.15	0.225
21 and above	0.18	0.27

(IS: 2732 – 1985).

The space requirement as per Table 7.1 is enough for the winter whereas, extra floor space shall be provided to bird during summer. The floor requirement in the big poultry house is generally lower than the recommended floor space in Table 7.1. On the other hand, the bird requires extra space in small poultry house. The usable floor space in bigger poultry house or semi-intensive system is comparatively more than the deep litter intensive system. The space requirement is increased by 0.046 sq. m per bird in hot climate. For the design of poultry house, minimum floor space of about 0.09 square meter (one square foot) for each broiler, and 0.18 square meter (two square feet) for each layer is considered under deep-litter system of rearing. The floor space requirement in cages is about 50 per cent of the floor space in deep-litter.

Feeder Space Requirement: The feeder space requirement per bird depends on the age of the bird. Feeder space requirement increases with the increase in age. The minimum feeder space per bird varies from 2.5 to 12.5 linear cm depending upon the age of bird. The shape of the feeder may be rectangular or circular. The space requirement in circular feeders shall be half of the values of rectangular feeders. In circular feeders, length of the circumference is considered as space. A minimum of 5 cm length of feed hopper is adequate for each bird.

Watering Space Requirement: The quantity of water require depends upon the age of the bird and season. The consumption of water increases with increase in the age of bird. Also, the requirement of water will be more in summer as compared to winter season. Usually, 30 litres of water per day per 100 birds is required. Thus, the size of the trough shall be selected accordingly.

Brooding Space Requirement: The minimum brooding space requirement per bird shall be 45 sq. cm under a brooder. In case an electric brooder is used, an area of 65 sq. cm. shall be provided per bird.

7.8 Types of Poultry Houses (On the basis of material of construction)

7.8.1 Wire (or Slatted) Floor Poultry House

Wire or slatted floor system house (Figure 7.4) are comparatively cooler as compared to other types of housing systems but their cost is more. These type of houses are suitable for adult bird only. The operations of feeding, watering and egg collection are efficiently handled from the outside. No litter is needed in this system. Ventilation and manure removal are both facilitated as the house is built 0.8 to 1.0 meters above the ground. In this type of houses, bird density can be 6 to 8 birds per square meter.

Figure 7.4. Slatted Floor Housing System.

Galvanized mild steel wire is used for making the floor in wire floored houses. The steel wire of 2 mm or 12-14 gauge shall be used to make mesh of 12×12 mm. The wire floor shall be kept 45 cm above of the ground level. A slope of 15 per cent is recommended for the wire floor. The width of wire floored poultry house should be limited to about 2 meters to facilitate easy removal of manure, and adequate wall space for feed and nests.

While in a slatted floor, wooden pieces or iron rods are used as floor, usually 1.0 m above the floor of the house to facilitate fall of droppings through slats. Wooden pieces of 2.5 to 5.0 cm wide are placed 2.5 cm apart preferably run lengthwise to the house. Instead of wooden pieces, iron rods of 5.0 cm diameter can be used with interspaces of 2.5 cm between rods. The slatted floor must be strong enough for a person to walk on. Thus, a wider building is feasible.

Feeders can be placed completely inside where the birds have access to both sides of the trough. The floor is sectioned for easy removal during cleaning out of manure. The movable curtains or covers shall be provided in the outside wall for ventilation during winter or summer season. The roof of thatch or corrugated GI sheets shall be provided over the poultry house. The feed troughs should be equipped with hinged covers. Rat guards must be installed at the top of each pier.

The wire floored poultry house of 23×5.5 m dimension can house 400 birds with a service alley of 1.8 m in the centre. Feed, water trough and roosts shall be provided at suitable place of poultry house. The birds use the roosts to take rest. Along the whole length of the house, a number of nests are positioned in between each pillar. The wood or hard cloth shall be used for the floor of nests. The nest shall have slope of 12-15 per cent towards the central service alley. A 3 cm stopper shall be installed at the lower end of floor to collect the eggs.

There is increased sanitation and labour saving in this type of house. Also soil borne infection is controlled because no litter is used. The wire floored poultry house require high initial cost than the conventional solid floors. The higher fly problem and less flexibility in the use of building are the some of the disadvantages. In addition, any spilled feed is lost through slots.

7.8.2 Deep Litter Housing System

The deep litter technique originated in Ohio (USA) in the year 1946, and was an important development in poultry management because it significantly reduced poultry disease and the labour necessary to maintain chickens. In this method, poultry birds are reared inside a shed all the time with a reasonable investment. The trough for feed and water with nest shall be in the poultry house. The suitable litter material shall cover the floor to house birds (Figure 7.5). Litter usually consists of chopped paddy straw, paddy husk, saw dust, groundnut shells, wood shavings, coir pith, sugarcane bagasse or dry leaves. Straw of wheat or barley, dry grass or dry stalk of maize and *bajra* are also suitable as litter. Litter material is capable of absorbing moisture. When moisture is absorbed, there will be controlled microbial activity and odour will also be minimum. This method saves labour involved in frequent cleaning of the floor. It needs only periodic attention like stirring up the litter.

Figure 7.5 (a). Inside View of Deep Litter Poultry House.

Each broiler requires about 0.09 square meter (one square foot) of floor space while a layer requires 0.18 square meter (two square feet) of floor space under deep-litter system of rearing. The crowding of birds in the house should be

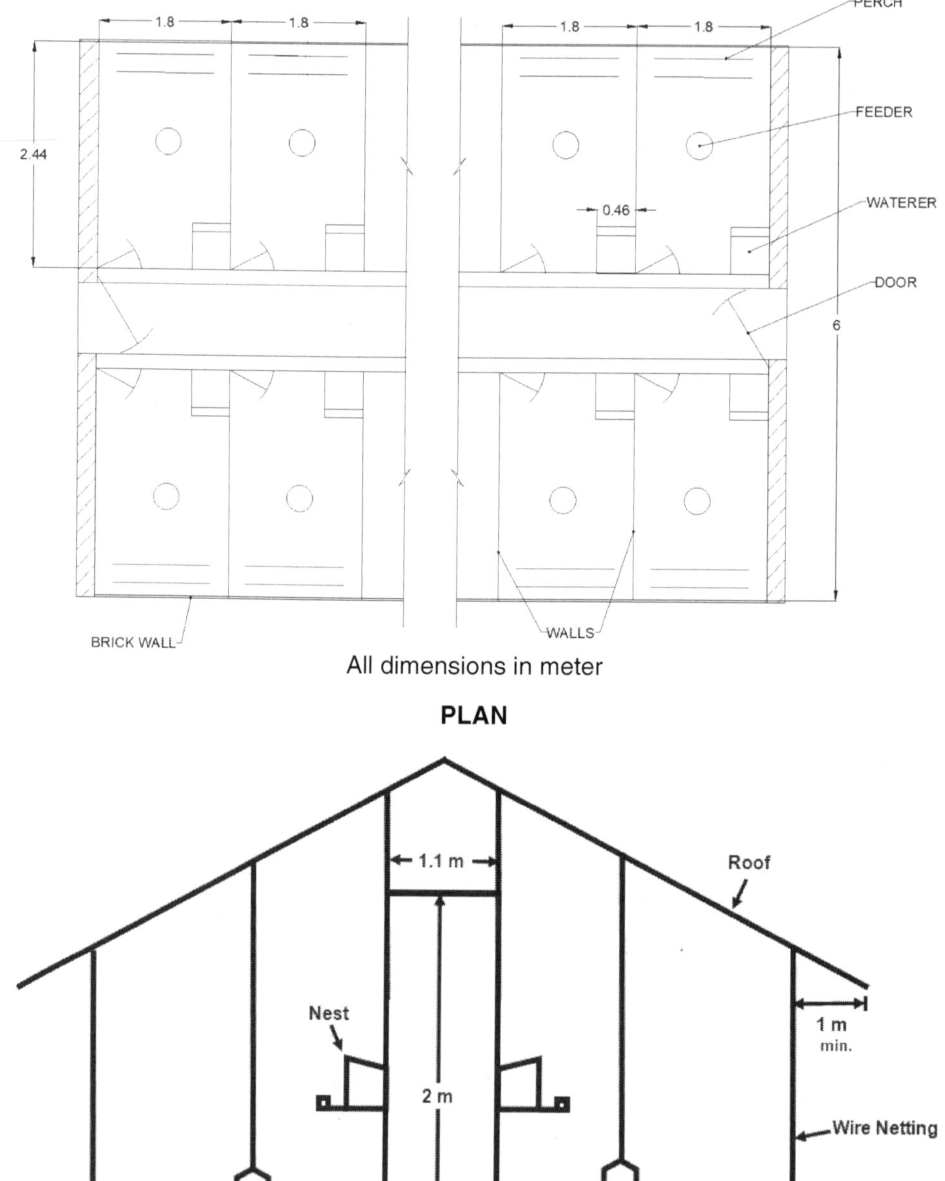

Figure 7.5 (b). Plan and Sectional End View of Deep Litter Poultry House.

avoided. The provision should be made for ample ventilation, and also provide an overhang on all the ventilators and openings to protect against sun and rains. Precautions should be taken to maintain the litter completely dry. The birds burrow themselves in deep litter to keep themselves cool in summer. During winter, the birds warm themselves as the deep litter provides extra insulation in colder temperatures, as well as extra heat from the decomposition of the litter. Birds often keep themselves busy in scratching and searching in the litter, and hence desist from feather picking.

Built-up Litter: An initial layer of litter is spread for the animals to use for bedding material and to defecate in, and as the litter is soiled, new layers of litter are continuously added. The usual procedure for built-up floor litter is to start with about 10 cm depth of fine litter material with additions of 2.5 to 5.0 cm every fortnightly without removal of the old. A depth of 20 to 30 cm is maintained by partial removals from time to time.

The litter and excreta is decomposed due to bacterial action. The decomposed material is of crumble form. Litter remains dry and warm due to heat of decomposition. The litter shall be replenishing if the quantity of litter shall be less than the droppings of birds. The litter shall be replaced after a year. The decomposed litter can be used as good quality manure. The deep litter pen should be started when weather is dry, and is likely to remain so for about 2 months for the operation of bacterial action.

Qualities of good litter material: Good litter material should readily absorb moisture, and should not cause any injury to birds. In addition, it should be non-toxic, easily spreadable, and decomposable; and should not cause dust pollution. Moisture level of litter should be less than 15 per cent.

The advantages and disadvantages of deep litter housing system are:

Advantages

1. Vit B2 and Vit B12 are made available to birds from the litter material by the bacterial action.
2. Maximises use of land.
3. The birds remain happy and busy as they get endless entertainment from scratching in litter, resulting in increases rate of growth of poultry birds.
4. Maximises use of labour, thus increases efficiency in poultry management.
5. Facilitates the management of very large flock.
6. Deep litter manure is a useful fertilizer. Also all of the chicken manure is harvested into the deep litter. It is estimated that 35 laying birds can produce in one year about 1 tonne of deep litter fertilizer.
7. Lesser nuisance from flies when compared to cage system.

8. There is less odour in the shed in comparison to concrete floor.
9. Reduces loss of eggs to vermin, snakes and thieves.
10. Protects birds from harsh weather conditions during hot and cold climates.

Disadvantages

1. Because of the direct contact between bird and litter, bacterial and parasitic disease may be a problem.
2. Respiratory problems in the birds may emerge due to dust from the litter.
3. Cost of constructing of deep-litter house is high and requires a large quantity of litter which adds to the cost of production.
4. Faults in ventilation can have more serious consequences than in the cage system.
5. The litter makes eggs dirty.
6. Cannibalism and pecking of eggs are common.
7. Difficult to determine the unproductive birds.

7.8.3 Cage or Battery System

Cage houses (Figure 7.6) are ideal for the area of moderate climate conditions where summer and winter temperatures respectively are neither too high nor too low. This system is generally adopted where land is limited and expensive. The cage system is very efficient for the rearing of the layers. The small compartment of cages is made by netting of the galvanized wires. The mesh of the floor is denser than sides and roof of cages. These cages may be hanged on the roof or kept on the stands. Cages are made in continuous rows. There may be one to four rows of these cages in a house depending upon its width. Cages are made of welded steel wires and are provided with a sloping floor from back to front so that the eggs as they are laid, rolled out of the cage to an egg retainer from where the eggs can be collected. Cages may be made either to house one or more birds at a time.

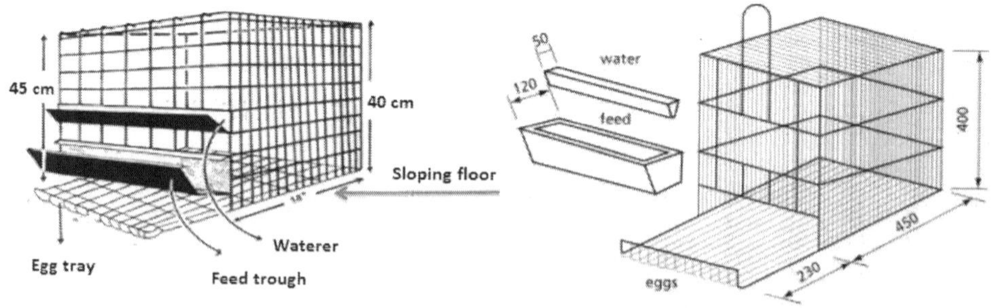

Figure 7.6. Cage Poultry House.

The size of cage is just large enough to permit very limited movement and allow bird to sit and stand comfortably. Cages of various sizes are available to house 3-5 birds in a cage. They are arranged in two or three such rows one above the other on either side. The suggested dimensions of the cages are:

45 × 30 cm	-	for 3 birds
45 × 40 cm	-	for 4 birds
50 × 35 cm	-	for 4 birds
55 × 45 cm	-	for 5 birds
60 × 37.5 cm	-	for 5 birds

A floor space of 0.042-0.045 m^2 per bird is required within the cage. The cages are of 50 cm and 40 cm height in front and back side. The troughs of water and feed are provided outside of the cage to facilitate easy refill to the workers.

This system utilizes auto-operated feeding trolleys and egg collection belts. The droppings are collected by trays, belts, floors, or deep pits under cages. The type of collection system used depends on the type of cage. The cages shall be hanged at height of 75-90 cm above the floor.

Cages are classified on the basis of the number of birds in a cage (single, multiple or colony cages); number of rows (single-deck, double-deck cages *etc.*); arrangement of cages (stair-step, M-type, L-type or battery cages); and on the type of bird reared (brooder, grower, layer or breeder cages).

Figure 7.7. Battery Cages.

Some of the advantages and disadvantages of cage housing system are:

Advantages

The advantages of cage housing system are:

1. Minimum floor space is required.
2. More and cleaner eggs are produced by the birds.

3. Small group size avoid injury from in-fighting within the flock.
4. There is less feed wastage and better feed efficiency.
5. There is ease of management and better working conditions.
6. There are fewer occurrences of soil borne illnesses, parasites and other exotic diseases.
7. Increased hygiene resulting in a much lower incidence of diseases in which the infectious agent is spread through the droppings.
8. Sick and unproductive birds can be easily identified and eliminated as they are housed in smaller groups.
9. Birds are protected from predators such as foxes, feral cats, and birds of prey.
10. There is no need of litter material.

Disadvantages

The disadvantages of cage housing system are:

1. Cage house has high initial investment cost.
2. Birds in cage system have reduced social interaction.
3. Birds are unable to roam and explore as they are confined to a relatively small place, which results in higher incidence of metabolic disorders.
4. Birds are unable to practice some natural behaviours like nesting and dust-bathing. It results in severe frustration for many birds each time an egg is laid.
5. Birds are usually unable to perch, which leads to lower bone strength which results in a higher incidence of foot lesions.
6. Handling of manure may be problem. Generally, flies become a greater nuisance.

7.8.4 Slatted-cum-deep Litter Floor Poultry House

This type of poultry house has about 60 per cent slat area and 40 per cent litter area. Drinkers and feeders are placed in both slat and litter area. The width of this type of house is limited to 3 to 4 meters. The area is raised above the concrete floor by 0.5 meters or more to accumulate manure below the slatted area, and thereby easy cleaning under the slatted portion from the outside.

This type of house offers some advantages over a simple deep litter house. The system saves on litter, reduces contact between birds and manure, and allows manure removal without disturbing the birds. The advantages of slat cum litter house are production of more eggs per unit floor, better fertility of birds than all slat house and improved ventilation. The main disadvantages of this system are the limited width, and higher housing investment.

7.9 Poultry House Construction

Foundation

Good foundation is necessary to prevent seepage of water into the poultry sheds. The foundation of the house should of concrete with 30 to 45 cm below the surface and 30 to 45 cm above the ground level.

Floor

The floor shall be made either of cement concrete or brick laid in cement mortar, or stone slabs set in cement mortar, or rammed earth. The floor level should be at least 25 cm above the ground level and free from dampness. The surface of the floor should be smooth without cracks, easy to clean, rodent proof and durable. The plinth protection on all sides of the walls of the house should be provided.

Walls

Brick cement or mud walls are built for poultry house. The one third lower portion of the wall from the floor on the longer side of house shall be brick masonry of 20 cm using the cement mortar. The pillars are made either of reinforced cement concrete or brick or angle iron or iron pipes up to roof level at an interval of 3-4 m. The wire netting of 12×12 mm size shall be made on the remaining portion of the walls. The wire shall be of galvanized mild steel of 1 mm diameter. The pillars or angle iron frame shall be used to support the wire netting. The height of brick masonry may be increased or decreased depending upon the climate of the region. The height may be increased when poultry house is constructed in cold regions in order to help conserve warmth. The wall of small height helps for better ventilation in warm region. More wire netted area of wall shall be provided in cage system for better ventilation. The partition wall of brick shall be 30 cm high and galvanized wire mesh shall be provided above that. A hessian or reed curtain should be provided outside the wire mesh to protect the birds from hot and cold winds, and also from direct sunshine.

Roof

The material for roof of the poultry house may be galvanized steel sheets, asbestos cement sheets, aluminium sheets, thatch, or corrugated tiles. For thatch roofs, the thickness of thatch should be at least 15 cm, and it should be made rain-proof and fire-proof. The roof may be lean-to or gabled type. Gabled type of roof with ridge ventilation is recommended for a poultry house having more than 10-meter width. Roof with metallic sheets may be covered with a layer of thatch. It will help in keeping the house cool in summer months, and warm in winter. The eaves of the roof should not be less than 1 meter in order to prevent the entry of rain water into the house. It shall also provide adequate protection from sun.

Doors

The door is usually made of wood or iron, and are provided in the end walls. The size of the door should be always large enough to allow a person to conveniently get through. The door must have a wire netting upper half and a solid lower half. The 30 cm space shall be between the bottom of door and floor. The foot bath filled with disinfectant shall be at entrance of poultry house. The door of the deep litter poultry house must open outside.

The usual size of doors are as follows:

1. Doors in side walls: 1.2×1.5 m
2. Doors in houses with central passage: 1.2×2.0 m
3. Doors in individual pens: 0.6×1.8 m

7.10 Brooder House

The brooder house is a poultry house used to keep and rear chicks during the first eight weeks of age. Here the chicks are raised until they have enough feathers to keep themselves warm. Brooder houses are usually of two types *viz.* stationary and portable. Brooding buildings should be isolated from the layer houses preferably by 35 meter, and be self-contained in terms of feed supplies and storage of equipment. The sheds should be so located that the fresh air first passes through the brooder shed, followed by grower and layer sheds. This prevents the spread of diseases from layer houses to brooder house. The constructional and other details are same as those of the layer house of the stationary type with addition of side screens made of suitable material like hessian cloth, tarpaulin, polyethylene *etc.* A concrete or mud flooring is preferred for a stationary house while a wire mesh or wooden floor is desirable for a portable house. The house should be well ventilated. Roof of the brooder house should be made waterproof. Care should be taken to shield the chickens from direct sun. The design of the brooder house should be such that it remains cool during summer. The feeders and drinkers are arranged on the outside and thus saved from being contaminated. Now-a-days nipple drinker system is followed from day-old itself. The minimum brooding floor space requirement per chick shall be 450 sq. cm under a brooder. Thus, total floor area required for 500 chicks shall be about 22.5 square meter. Feed and water should be available to the chicks as soon as they arrive. A provision of 2.5 cm of feeding space per chick at the hoppers is made initially and then increase to about 5.0 cm after chicks are 2 weeks old. After 8 weeks, provide 7.5 to 10 cm of feeding space for growing pullets. A hanging tube-type feeder of 37.5 cm in diameter will feed about 30 birds. Less feed is wasted by filling hoppers only half full and adjusting feeder height or size to bird size.

Heat sources for artificial brooding are incandescent bulbs, heat lamps, electric hovers, gas hovers, and hot water radiators. Each heat source works satisfactorily. The gas or oil is used as the fuel to supply heat in larger brooder house. The smaller brooder house prefers the electric heating element for small number of chicks. The infrared bulb can be used for heating of the brooder house. It is a convenient source of heat supply to the chicks. It shall be installed 60 cm away from everything. The lamp of 250 watts can keep 100 chicks on 35°C. The spare bulb/lamp shall be kept in house in the case of burn out of existing lamp. The temperature of air shall not be considered as indicator of convenience of chicks while using the infrared bulb.

Battery Brooder

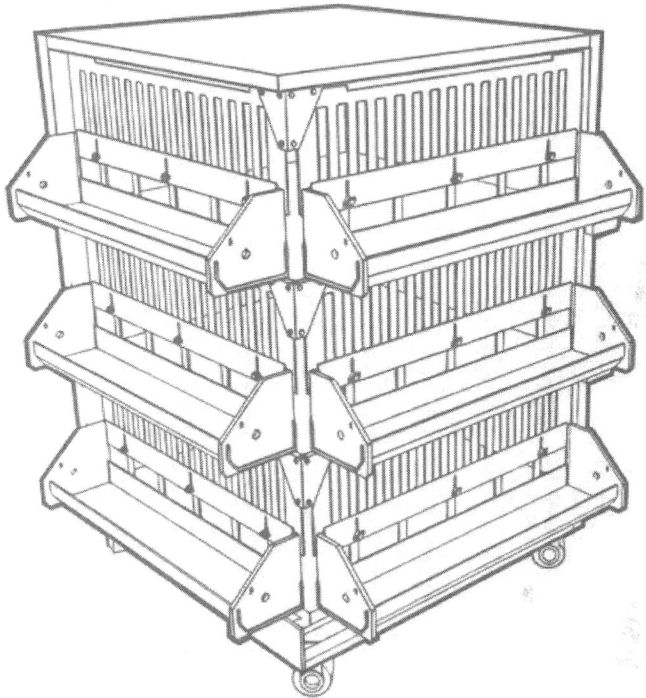

Figure 7.8. Battery Brooder.

It is one of the most popular form of chicken brooder. A battery brooder (Figure 7.8) looks like two or more cages stacked on each other which, in turn, are on wheels for mobility. It allows to keep lots of chicks in a small space. The battery brooders are provided with wire mesh floor. The advantage of wire mesh flooring is that the droppings fall through the wire and are caught by a tray provided under each compartment. Thus, the chicks are separated from their waste, and the waste is easily removed. The disadvantage is it reduces the opportunity for the chicks to develop immunities that floor-reared or litter-reared chicks have. Each level of the battery comes with feeders and drinkers along two outside edges that the chicks access by poking their head through feed guards. The floor of cage shall be covered with newspaper for first 7-10 days. During the first week, the feed shall be provided inside the cage. The floor space of at least 650 sq. cm. per chick be provided in a battery brooder. The chicks are usually kept for first three weeks in the battery brooders, thereafter it is not advisable to keep chicks in them. The chicks are put in other types of houses after their removal from the brooders.

7.11 Poultry Equipment

The salient features of the equipment needed in the poultry house is presented in this section.

7.11.1 Feed Troughs (Feeders)

Feed trough is fixed along the wall from the outside in such a way that the operator can easily fill the feed for chicken and clean it. Conventionally, long (linear) feeders are being used. The circular feeders are made of high impact plastic whereas wood or metal sheet is used for the linear feeder. Minimum 1 m distance shall be in between the circular feeders. One hen requires 100 g of feed mixture per day. One bird needs minimum 5 cm length of the feed hopper. The top edge of the feeder shall not be higher than the back of birds. The arrangement of the feeder shall not allow roost of the birds on it. The grill shall be provided to prevent entry of birds in the feeder. The dimensions of the linear feeder for broiler and chicks are provided below:

	Length	*Base Width*	*Top Width*	*Height*	*Spacing of Grill*
Linear broiler feeder	90-150 cm	15 cm	25 cm	7.5-10 cm	5 cm
Linear chick feeder	60-75 cm	7.5 cm	10 cm	5 cm	2.5 cm

The circular feeder is also known as hanging feeder. The circular feeder can be hanged on the roof of iron truss or separate pipe line shall be provided. These feeders are available in different capacities and shall not require refilling 4-7 days. The quantity of feed depends upon the age and numbers of birds. The circular feeder never kept on the litter floor.

7.11.2 Water Troughs (Drinkers)

The water requirement depends upon the season. Generally, 30 litres per day per 100 birds is required. Drinkers, similar to feeders can be either linear or circular. Linear water trough is usually made of metal sheet or cement concrete or PVC tube and placed along the entire length of the house in the service alley. One end of the trough is raised so that water does not spill out. They shall be kept at an appropriate height so that water is available at the level of back of the birds. Drinkers are fixed in equidistant rows alternating with the feeders' rows ensuring a minimum distance of 1 meter between two drinkers. No trough is provided near the doors.

Circular drinkers are hung from the truss or roof of the house and hence referred to as hanging drinkers. These are connected to an overhead tank. Thus, continuous supply of water is ensured. The drinker shall be installed at the level of back of the birds. Although, the drinker can be installed up to a height of crop of bird to minimize the spillage of water. The crop is the bottom of the neck of birds. The birds put their beak into the water and then heads up to move water from beak to mouth. The installation of circular drinker on litter shall be avoided.

7.11.3 Laying Nests

It is a box or cage where the chickens lay eggs in a clean place in relative peace and privacy. They are placed at convenient places in the house so that it is easy for layers to approach and the operator has no difficulty in collecting eggs. The layers have tendency to fly/jump/sit in the nest. Therefore, the nest of inclined roof shall be provided to allow layer to lay egg on the top of nests. The nest shall be arranged in such a way that eggs can be collected from central passage. The nests of inclined roof are also known as anti-perching nest. The laying nest are not required for the birds kept in cages.

Trap nest is a type of laying nest with specially design automatic front to separate the layer. It curbs the layer when they enter to lay eggs. Layer is confined until released by the attendant. Trap nests are useful in weeding out poor layers and increasing the average egg yield of a flock by selecting and breeding. They are not usually used on account of the high labour required to operate them. They are mostly used on breeding farms. Nest is made of wooden plank. They are placed at convenient places in the house so that operator has no difficulty in collecting eggs. Generally, trap nest is constructed in single or double tiers. Four nests are in one tier. One trap nest can keep 3 or 4 birds. The dimensions of the nest box are 37.5×30×37.5 cm. The wooden or wired bottom of trap nest is elevated to 45 cm above the floor. The sloping roof of the trap nests prevent the roosting of the birds.

7.11.4 Perches (Roosts)

A roost is the elevated wooden bar, or perch where a bird sleeps during night. The perch is either a circular or rectangular bar placed horizontally above the floor for the chicken to roost at night. They are either made of wooden pieces measuring 5×7.5 cm at their rectangular edges or round bamboos having a diameter of about 5 cm. The top corners of wooden pieces shall be rounded off. The length of a perch should not exceed 1.5 meter. They shall be placed at least 35 cm apart (centre to centre) and shall be fixed at least 25 cm away from the rear wall. They are fixed at a height between 55 and 75 cm from the floor. The perch space requirement is about 15 cm per chicken which may be raised to 25 cm per chicken for heavy breeds. Usually, portable perches are placed in the poultry shed for their easily cleaning and disinfecting. The perches shall be given a coating of anticorrosive or protective non-toxic paint. They should be rigid and firm, and all fixed at the same level.

Figure 7.9. Poultry Equipment.

7.12 Design of Poultry House

Size of the poultry house depends on number of birds to be reared. The assumptions for the design of poultry house are as follows:

(i) Floor Space Requirement

Cage type poultry house: On an average, a floor space of 0.045 m^2 per bird is required. Cages are built in continuous rows. Cages are made either to house one or more birds at a time. They are placed at a height between 75 and 90 cm from the floor.

Deep litter type poultry house: Each broiler requires about 0.09 square meter (one square foot) of floor space while a layer requires 0.18 square meter (two square feet) of floor space under deep-litter system of rearing. In other words, the stocking density per square meter for layers shall be 5, and for broilers it shall be 10.

Thus, by knowing the floor space requirement, the size of the poultry house can be estimated on the basis of number of birds to be reared.

(ii) Feeder Space Requirement

Feeder space requirement for 15 hens shall be one 1.2 m long feeder trough, *i.e.* 8 cm per bird.

(iii) Watering Space Requirement

30 litres of water per day per 100 birds is required. Watering of 10 litre capacity shall be used. Thus, for 100 birds, three number of drinkers shall be required.

(iv) Area for Nesting

Area for nesting shall be about one square meter for 50 layers.

(v) Thickness of Deep Litter

Thickness of deep litter shall be 15 to 20 cm.

Length: The poultry house can be of any length. The length depends upon the number of birds and the availability of land.

Width: The maximum width of open sided poultry house shall be 6-7.6 m in tropical countries. It provides a proper ventilation and aeration in middle of house. The ridge ventilation in the roof top with overhang is necessary in the case of shed width more than 7.6 m. The width of environment-controlled poultry house can be kept above 12 m. The exhaust fans in modern environment-controlled poultry house forced out the hot and obnoxious air. Whereas, in the normal poultry house, the aeration and ventilation depend upon the density difference and thermal buoyancy.

Height: The height of the sides from foundation to the roof line should be about 2.0 meter (eaves height) and at the centre 3.0 to 3.7 meter. In case of cage houses, the height is decided by the type of cage arrangements (3 tier or 4 tier). The height of side masonry wall shall be 0.80 meter.

Lighting: Light should be provided at 2.0 to 2.5 meter above the ground level and must be hanged from ceiling. If incandescent bulbs are used, the interval between two bulbs is 3 meter. In case of fluorescent lights (tube lights) the interval is 4.6 meter.

Equipment and devices required in a poultry house

1. Feeding troughs
2. Watering troughs
3. Nests
4. Roosts
5. Lighting traps

Example 7.1

Design a deep litter poultry house for 500 laying birds. Also list down the equipment needed for feeding, lighting and protecting the birds. Make reasonable assumptions wherever necessary.

Solution

Floor space requirement

Let us consider stocking density per square meter for birds as 5.

Floor area required for 500 layers =

$$\frac{\text{total number of layers to be accomodated}}{\text{stocking density of birds per sq meter}} = \frac{500}{5} = 100 \text{ m}^2$$

Feeder space requirement

Feeder space requirement for each bird is taken as 8 cm.

Total length of feed trough required = 500×8 = 4000 cm = 40 m

Watering space requirement

30 litres of water per day per 100 birds is required. Let the capacity of one water trough be 10 litre.

Quantity of water required for 500 birds = $\frac{30}{100} \times 500$ = 150 litres

Thus, 15 fountain type water trough each of 10 litre capacity may be provided.

Area for nesting

Area for nesting shall be about 1 square meter for 50 layers.

$$\text{Area for nesting required} = \frac{\text{total number of layers to be accomodated}}{\text{layers per 1 square meter}} = \frac{500}{50}$$

$$= 10 \text{ m}^2$$

Thickness of deep litter

Thickness of deep litter = 15 cm

Dimensions of poultry house

The maximum width of the poultry house should not exceed 7.6 meter.

Assume width of the poultry house as 7.5 meter.

Let the width of central passage be 1.2 meter.

The effective width of the poultry house = 7.5 – 1.2 = 6.3 m

$$\text{Thus, length of poultry house} = \frac{\text{total floor area required}}{\text{width of poultry house}} = \frac{100}{6.3} \approx 16 \text{ m}$$

The dimensions of poultry house shall be 16 m × 7.5 m.

Walls

Height of side masonry wall = 0.80 m

Brick walls on sides = 1.2 m

Wire-netting above brick wall should be supported with angle iron

Roof

The poultry house's roof must be 3.7 meters high at the centre and 1.8 to 2.4 meters high on the side walls, with a 1-meter extension to prevent rainwater entry. The roof can be made of corrugated galvanized iron (CGI), asbestos, or tile.

Floor

The poultry house's floor shall be strong and rodent-proof, with a slope towards the door. Cement concrete shall be used for the floor. The plinth shall be 0.6 m higher than the ground.

Doors

There shall be two doors 1.2 m wide and 2.0 m high placed opposite to each other.

Lighting

Lights shall be placed 2.0-2.5 meters above ground level and hanged from the ceiling. Incandescent bulbs have a 3-meter interval between bulbs, while fluorescent lights have a 4.6-meter interval.

Equipment and devices required in a poultry house are feeding troughs, watering troughs, nests, roosts, and lighting traps

The elevation and plan of the deep litter house for 500 birds shall be similar to Figure 7.5.

Important Points to Remember

1. Broiler is a young chicken usually 9 to 12 weeks of age.
2. An adult male bird is called a cock or a rooster.
3. Egg laying poultry is called a layer.
4. A sexually immature chicken less than 22 weeks of age is called a pullet.
5. In the free-range or extensive system, the birds are generally left free to roam during the day, and kept confined during the night.

6. In the semi-intensive system, the birds are provided with a night shelter surrounded by fenced open space.
7. In the intensive system, the birds are totally confined to houses either on ground/floor or on wire-netting floor in cages or on slats.
8. The most efficient temperatures for layers are between 20 and 24°C.
9. The optimum range of relative humidity for chicken after brooding is between 50 and 70 per cent.
10. Chickens need 14-16 hours of light each day to maintain peak egg production.
11. Wire floor type, deep litter type and cage house type are the three most popular poultry houses.
12. Deep litter poultry house is most economical based on initial investment as well as on operational cost basis.
13. The moisture content of good litter should lie between 15 to 25 per cent.
14. Thickness of deep litter shall be 15 to 20 cm.
15. Litter material used in deep litter poultry houses needs periodical stirring.
16. Deep litter poultry housing saves labour involved in frequent cleaning of the floor.
17. Each broiler requires about 0.09 square meter of floor space while a layer requires 0.18 square meter of floor space under deep-litter system of rearing.
18. The required floor area for hens is increased in hot climate.
19. Birds keep in good health and grow well in deep litter house.
20. Minimum floor area required for 100 adult hens in deep litter housing system is 27 m^2.
21. Stocking density per square meter for layers is 5, and for broilers it is 10.
22. About 30 litres water is required for 100 birds per day.
23. Wire or slatted floor system house are comparatively cooler as compared to other types of housing systems but their cost is more.
24. In the deep litter housing system, poultry birds are reared inside a shed all the time with a reasonable investment.
25. Cage houses are ideal for the area of moderate climate conditions where summer and winter temperatures respectively are neither too high nor too low.
26. Cage system is generally adopted where land is limited and expensive.
27. A floor space of 0.042-0.045 m^2 per bird is required within the cage.

28. The cages in poultry house are placed at a height of 75-90 cm from the floor.
29. The brooder house is a poultry house used to keep and rear chicks during the first eight weeks of age.
30. Generally, 100 grams of feed mixture per hen is required per day.
31. A roost is the elevated wooden bar, or perch where a bird sleeps during night.
32. Feeder space requirement is taken as 8 cm per bird for design purpose. Minimum length of a poultry feed hopper is 5 cm per bird.
33. Generally, single storied poultry houses are preferred in warm regions.
34. Area for nesting shall be about one square meter for 50 layers.
35. The width of open-sided poultry house in tropical countries should have a width not more than 6.0 to 7.6 meter in order to allow ample ventilation and aeration at the mid-portion.
36. The brooding house should provide temperature of 32.2°C to 37.7°C for three days old chicks.

Exercises/Questions

1. What are the different types of poultry houses being used in India? Which particular type can be popularized in rural areas and why?
2. Describe in brief the different poultry housing systems.
3. What points should be considered while selecting a location for poultry house?
4. What are the environmental requirements for poultry birds?
5. What is a litter type poultry house? What advantages has it got over cage house?
6. Discuss the essential features of poultry house construction.
7. Write short notes on:
   (a) Brooder house
   (b) Deep litter housing system
   (c) Cage house
   (d) Feeders
   (e) Water trough
   (f) Laying nest
   (g) Roost
8. Design a litter type poultry house for housing 1000 birds on a farm. Give details of dimensions, materials, facilities provided in shed. Draw neat sketches of its layout. Make necessary assumptions.

Chapter 8

Pig Housing

Pig farming is crucial for socio-economically weaker sections of society due to its high fecundity, feed conversion efficiency, early maturity, and short generation interval. Pig farming requires minimal investment on buildings and equipment, and has the potential to provide nutritional and economic security for these groups. Pig farming also offers faster economic returns to farmers. As a cheap source of healthy animal protein pigs can meet the requirement for the weaker section of the society. Pig population in India is estimated to be 10.29 million and it ranks 5th in the world. Pigs are also called swine.

Good housing and equipment for raising pigs are necessary to provide shelter against inclement weather, prevent diseases, control parasites and save labour. A good housing environment helps farmers rear 85 per cent or more live born piglets to market weight quickly. It provides adequate protection against sunlight, cold winds and continuous rain, as well as proper hygienic conditions for healthy growth. Pigs at different stages of their life need different environments. Piglets require special protection against very low temperatures for maximum potential growth and reproduction.

8.1 Terminology

Boar: An uncastrated male pig is called a boar.

Sow: A female pig that has farrowed at least once is called a sow.

Gilt: A female pig which has not farrowed is called a gilt.

Litter: A group of piglets born to a sow is called a litter.

Piglet: A young pig before weaning is called a piglet.

Weaners: The piglets that are separated from their mothers to be reared independently.

Hog: A domestic pig, especially a castrated male reared for meat is called a hog.

Sty: A dwelling place consisting of one or more of pens for pigs is called a sty.

Pen: A pen is a covered area with an open yard inside a sty. It shelters a single or group of pigs and has space for dunging, along with troughs for feed and water.

Dry sow and gilt sty: It is a dwelling place for dry sows and gilts.

Farrowing sty: Farrowing sty is a dwelling place for nursing sows and their litters.

Fattening sty: A dwelling place for young pigs kept for fattening purpose.

Sick sty: A dwelling place for sick pigs that separate the sick pigs from healthy stock.

Stud boar sty: A dwelling place for breeding boars.

8.2 Location

The site should be at an elevated and well-drained place that cannot be flooded by rain water. The site should not be clayey or slushy, and should be at least 15 m away from residences and factories. The pig housing shall be constructed 30 m away from other animal shelters and food grain storage structures. The site shall not be within 1 km of garbage dumping grounds, slaughter houses and tanneries. The site shall be well connected to reliable water and electricity sources. The orientation of the structures shall ensure ample of sun light and ventilation. The long axis of the structure shall run from east to west directions. Such structure shall be cooler in summer and warmer in winter. The site shall be well connected to the nearest roads. The farm should have enough space to have further expansion in future.

8.3 Space Requirements

The covered floor area, open yard area and the trough length for feeding and watering required for each animal shall be as given in Table 8.1. The suggested typical dimensions of the covered floor area for each animal is also given in the table. The open yard for the particular type of animal shall be adjoining to its dwelling. The width of the trough for feeding and watering shall not be less than 50 cm. The requirement of the water depends upon the weather conditions, type of feed used, and age/weight of the animals. On an average, the quantity of water required per animal per day is about 22 litres.

Table 8.1. Floor Space and Feeding and Watering Trough Length Requirement per Animal

Sl. No.	*Type of Animal*	*Floor Space Requirement (m²/animal)*		*Typical Dimensions (m)*	*Feeding and Watering Trough Length (cm/ animal)*	*Maximum Number of Animals per Pen*
		Covered Area	*Open Paddock*			
1.	Boar	6.25 to 7.50	8.8 to 12.0	2.5×2.5 to 2.5×3.0	60 to 75	Individual pen
2.	Farrowing sow	7.50 to 9.00	8.8 to 12.0	2.5×3.0 to 3.0×3.0	75 to 100	Individual pen
3.	Weaner/ fattening pig	0.96 to 1.80	8.8 to 12.0	0.8×1.2 to l.2×l.5	30 to 40	30
4.	Dry sow or gilt	1.80 to 2.70	1.4 to 1.8	1.2×1.5 to l.8×l.5	20 to 30	2 to 10

8.4 Effects of Environment on Production of Pigs

The thermal, social, and chemical environments significantly impact the production potential of the pigs. Although the impact of the thermal environment is comparatively higher than that of the social and chemical environment. The thermal environment includes providing a dry bed, protection from extreme temperatures, and sunlight. Other factors include adequate space, fresh air, hygiene, access to feed and water, and accommodation.

8.4.1 Influence of the Thermal Environment on Pig Production

The influence of the thermal environment plays a crucial role in pig production, affecting various aspects of the animals' health, well-being, and overall performance. Understanding and managing the thermal environment is essential for optimizing pig production and ensuring the welfare of the animals.

Temperature

Pigs are warm-blooded animals that can maintain a stable body temperature under fluctuating ambient conditions. New born piglets typically prefer temperatures between 27 and 35°C. However, their thin skin and body fat make them vulnerable to cold. The ability of piglets during their early life to withstand cold is limited, and life losses can occur quickly if the microclimate is below 16°C. Fatal chilling can occur within minutes without warmth below 20°C. Although, piglets may return to normal after 1-10 days.

Thermoneutral zone varies with age and weight, typically ranging from 18°C to 22°C for growing-finishing pigs. High temperatures can significantly impact pig production performance due to the lack of effective sweat glands. Increased air temperature can lead to rapid decreases in growth rate and feed efficiency, potentially causing weight loss and even death at 37°C and higher. Pigs at temperatures above 30°C are highly susceptible to heat stress, exhaustion, and sunstroke. However, the heat stress of pigs in dry climates can be reduced by drip or spray cooling with sufficient air movement. The excess heat of pigs consumed in evaporating.the water drops. Pigs also attempt to cool themselves through rapid breathing, spilling water from their troughs, and urinating on the floor.

Humidity

Humidity is crucial for microclimate conditions in piggeries, with an ideal range of 45-75 per cent. The pigs shall not be kept in humidity levels above 80 per cent and below 40 per cent. Humidity can be maintained using natural ventilation or controlled ventilation and proper management, ensuring optimal pig health. Excessive high humidity can lead to heat stress and reduced feed intake.

Ventilation

Fresh air is essential in a building to remove water vapour, carbon dioxide, ammonia, dust, bacteria, and odours from animals. It is important to ensure air circulation within the shed without direct flow on the pigs. The amount of air introduced depends on the number and type of animals.

8.4.2 Influence of the Social and Structural Environment

Pigs have a hierarchy of domination and submission within their group. Therefore, removing or adding pigs can lead to fighting and stress among piglets. Disturbances in this order negatively affect the growth and development of piglets.

Grouping

Pigs grow better in groups. The size of the group is influenced by factors like size, weight, environmental conditions, feeding, and pen-changing practices. The growth rate of pigs is negatively affected by keeping them in small spaces.

Changing pens

Changing pens involves transferring a litter or group of pigs from one type of pen to another, from birth to the weaning or slaughter stage. This process can be stressful for the pigs, as it could weaken growth rates and feed utilization. It may also promote disease spread. The main goal is to maximise the use of available housing, especially for growers and finishers. Pen changing is more important in larger or more intensive pig production units. Separate housing with sufficient space is needed for each life stage of the pigs. This ensures efficient use of available space and reduces the risk of disease spread.

8.4.3 Influence of the Chemical Environment

The intensified pig production system can lead to the production of harmful gases like carbon dioxide, methane gas, ammonia, and hydrogen sulphide. Polluted air can lower growth rates, feed utilisation, and health standards in pigs. Efficient ventilation is crucial to maintaining optimal temperature and climate conditions, ensuring an unpolluted air environment for pigs.

Hygiene

The health of pigs can be impacted by various diseases caused by microorganisms present in their gut. These microbes from their guts can infect them in unhygienic sheds through pig dung. Clean, dry conditions can reduce germ populations, improving the health and production of the pigs.

8.5 Sties

Stys are dwelling places for pigs. It consists of one or more pens for single or group animals based on age, sex, and purpose. Stys has a covered area and an open yard. The number and dimensions of pens depend on the number of animals and the floor area recommended per animal. In extensive pig-keeping systems, sties and pens are constructed separately. For smaller farms, animals can be housed under the same roof with separate pens for different animals. For better management on large-scale commercial pig farms, different types of sties may be used for different animal housing, as indicated below.

1. Boar sty
2. Dry sow and gilt sty
3. Farrowing sty
4. Fattening sty
5. Weaner piglets sty
6. Sick sty

The requirements of different sties shall be as follows:

Boar Sty

Maximum 24 pens are in the boar sty under one roof. One pen shall accommodate only one animal.

Dry Sow and Gilt Sty

A maximum of 40 pens are arranged in two rows in the dry sow and gilt sty under one roof. Each pen shall accommodate two, five, or ten animals based on the pregnancy stage.

Farrowing Sty

A maximum of 40 pens are in farrowing sty under one roof. Each pen shall accommodate only one animal with the litter.

Fattening Sty

A maximum of 20 pens are in fattening sty under one roof and each pen shall accommodate 16-32 animals.

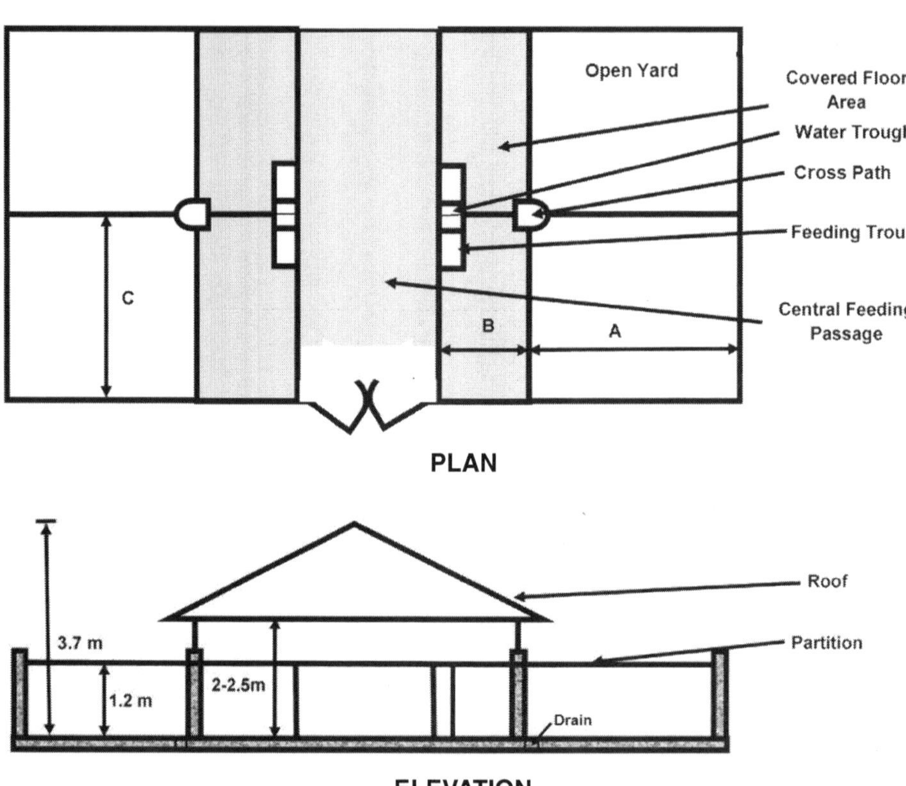

Type of Animal	Number of Animals per Pen	Covered Area per Animal (m^2)	Open Yard per Animal (m^2)	A (m)	B (m)	C (m)
Boar	1	7.5	12.0	4	2.5	3
Dry sows	2	2.7	1.8	2	3	1.8
Fatteners	16	1.8	12.0	32	4.8	6
Weaners	10	1.8	12.0	20	3	6

Figure 8.1. Layout Plan of Boar, Dry Sow, Fattening and Weaner Sty.

Sty for Weaner Piglets

A weaner piglet sty typically has no more than 30 pens. Each pen shall accommodate 10-20 animals aged between 2-6 months.

Sty for Sick Animals

At least 5 per cent of farm stock must be provided with accommodation facilities for sick animals. The sick sty shall be situated at a reasonable distance to ensure effective isolation from healthy animals. Feeding and watering facilities to the sick animals should be provided in the sty itself.

8.6 Constructional Details

Floor

The floor of the pig house should be smooth but not slippery. It should be so designed and constructed that it does not cause injury or suffering to pigs standing or bedding on it. Well-designed flooring can help prevent injury from abrasions, disease and even help keep a piglet clean and warm. The floor of a house should be rigid, rough, impervious to water, sloping toward the drain, and easy to clean. It should be made of cement concrete or paved with tiles, stone slabs, or bricks, set in cement mortar or lime mortar. Joints made of stone slabs or bricks should be filled with binding material. Plastic slat floors are now available. The floor should be raised about 60 cm above the ground. In intensive pig production systems, all pigs should be raised on concrete floors for a clean and sanitary environment. In semi-intensive systems, concrete floors are used in pens for finishing pigs and farrowing pens. Earth floors or deep litter bedding can be used in other pens and yards. The use of litter on concrete floors is desirable in the farrowing pens.

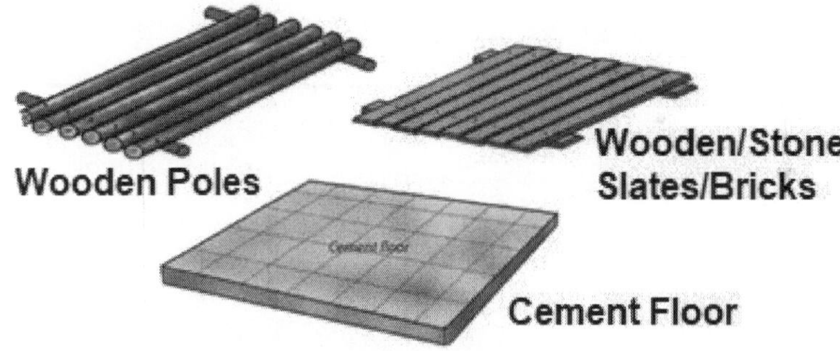

Figure 8.2. Different Types of Floors.

Walls

The floor and walls should be strong to withstand the rooting habits of pigs. The walls may be built of bricks with cement mortar or lime mortar and finished smoothly. The height of walls of the pig house shall be 2 to 2.5 m from the floor level. The brick walls are constructed up to two-thirds of the height. The brick walls should have a minimum height of 1.5 m. In the remaining portion, wooden rafters with adequate clearance are provided. The open space may be covered with wire netting.

Roof

The roof should be light, strong, durable, weather-proof and a bad conductor of heat. Roofs can be made of asbestos cement sheets, aluminium or galvanized iron sheets, clay tiles, thatch, wood or bamboo. The roof must ensure maximum shade inside the house. It helps to reduce penetration of oblique sun rays and entry of rain when blown by the wind.

Doors

The number of doors in each sty depends on its dimensions. The door must be strong and fitted close to floor level. Wooden planks or iron shall be used in doors. The width of individual pen doors ranges from 0.75–1 m, while sty doors range from 1.2–1.5 m and height from 1.2-2 m.

Ventilators

Pig house needs good ventilation. Without good ventilation, stale air can makes pigs ill and trigger respiratory diseases. Ventilation provides oxygen, controls temperature, and remove moisture, excess heat, gases and dust from the enclosure. Proper ventilation keep pigs healthier and happier. Ventilators are required for closed-side spaces. The size of ventilators depends on the dimensions of the sty. Each ventilator should be 0.6 m high and 1m long, with a top 0.5 m from the highest point of the wall. Ventilators shall be covered by wire netting from the outside.

Drainage

For proper drainage, each sty should have at least one drain on one side, with a minimum width of 25 cm and a slope of 1 in 40.

8.7 Ancillary Facilities

Large size pig farms may require additional auxiliary buildings/facilities like weighing yard, store building, loading and unloading ramp, skim milk tanks, manure pit, garbage boiling vat, boiler room, wallowing tank *etc.* The details of each building are as follows:

Weighing Yard

A shed for the installation of the weighing scale shall be provided. A weighing scale is required to record the weights of the animals regularly.

Store Building

The purpose of the store building is to keep the feed for the animals, and other miscellaneous items such as farm equipment and tools. The store building may comprise of a feed store, a mixing room to mix the feed for the animals, a store-keeper's office, and a room for miscellaneous items. The size of feed store depends on the number of animals to be housed.

Loading and Unloading Ramp

A permanent or portable ramp at the farm shall be provided for loading and unloading of animals. The slope of the ramp should be below 20 degrees to load and unload pigs.

Skim Milk Tanks

Skim milk can be used as a source of feed for pigs. The farm shall have a permanent or portable tank to store the skim milk. The capacity of the tank shall match the necessary amount of skim milk for two days, depending upon the number of animals.

Manure Pits

The manure pits shall be constructed away from the sties. The minimum distance from the farthest end of the sties shall be 15 meter. The manure pit shall be so located at the farm that the collected manure is easily available for cultivation purpose. Similarly, suitable provision for the disposal of the liquid manure shall be made for irrigating the fields.

Garbage Boiling Vat and Boiler Room

The food waste or garbage fed to swine must undergo proper cooking and handling to prevent possible disease transmission. In order to kill any harmful viruses or bacteria, the food waste must be cooked to 100°C for 30 minutes before feeding it to the pigs. A shed is required for boiling the garbage either in big vessels or in jacketed vats. The steam needed for boiling is supplied by the boiler. Direct fire method may also be used for boiling purpose instead of boiler.

Wallowing Tank

The purpose of wallowing tank is to keep the animals cool during summer. It is usually provided in the farms situated in hot zones. The tank shall be made of cement concrete with minimum dimensions of $2.5 \times 1.2 \times 0.15$ m. Provision of overhead sprinklers or showers can also be made to keep the animals cool.

Light

Artificial lighting is not usually required in piggeries. In intensive farming system, translucent flaps or shutters usually allow enough daylight to enter the building. Openings in the pig house allow enough daylight into the building.

Example 8.1

Design a floor plan and layout of a hog house. The herd size is as follows:

1. Male pig (Boar) = 2
2. Female pigs (farrowing sows) = 10
3. Fatteners = 32
4. Weaners = 20
5. Dry sow or gilt = 4

Make reasonable assumptions wherever necessary, but state them clearly.

Solution

Location: Site should be on fairly raised and well drained site, not liable to flooding. The site should not be clayey or slushy, and should be at least 15 m away from residence, 30 m away from the dairy animal or poultry house and food grain storage. It should have plenty of sunlight and circulation of fresh air. The house shall be well connected to the roads.

Space Requirement

The covered space requirement per animal for boar, farrowing sows, fatteners, weaners, and dry sow are respectively 7.5 m^2, 9.0 m^2, 1.8 m^2, 1.8 m^2, and 2.7 m^2.

The open space requirement per animal for boar, farrowing sows, fatteners, weaners, and dry sow are respectively 12.0 m^2, 12.0 m^2, 12.0 m^2, 12.0 m^2, and 1.8 m^2.

Number of animals usually housed in one pen

Boar – One

Farrowing sows – One animal with litter

Fatteners – 16

Weaners – 10

Dry sow - 2

Type of Animal	Number of Animals	Covered Area per Animal (m^2)	Open Yard per Animal (m^2)	Number of Animals per Pen	Number of Pens	Total Covered Area (m^2)	Total Open Area (m^2)
Boar	2	7.5	12.0	1	2	15.0	24.0
Female pigs	10	9.0	12.0	1	10	90.0	120.0
Fatteners	32	1.8	12.0	16	2	57.6	384.0
Weaners	20	1.8	12.0	10	2	36.0	240.0
Dry sows	4	2.7	1.8	2	2	10.8	7.2

Type of Animal	Covered Area per Pen (m^2)	Open Area per Pen (m^2)	Size of One Pen (m)	
			Covered	Open
Boar	7.5	12.0	2.5×3	4×3
Farrowing sows	9.0	12.0	3×3	4×3
Fatteners	28.8	192.0	9.6×3	64×3
Weaners	18.0	120.0	6×3	40×3
Dry sows	5.4	3.6	3×1.8	2×1.8

Thus, draw a plan similar to Figure 8.1.

Important Points to Remember

1. An uncastrated male pig is called a boar.
2. A female pig that has farrowed at least once is called a sow.
3. A group of piglets born to a sow is called a litter.
4. A young pig before weaning is called a piglet.
5. Piglets which have been separated from the mother for the purpose of independent rearing are called weaners.
6. Farrowing sty is a dwelling place for nursing sows and their litters.
7. Fattening sty is a dwelling place for young pigs kept for fattening purpose.
8. The width of the trough for feeding and watering shall not be less than 50 cm.
9. The most favourable temperature for new born piglets is between 27 and 35°C.
10. At temperatures higher than 30°C, a pig is extremely vulnerable to heat stress, bodily exhaustion and sunstroke.
11. The relative humidity range of between 45 and 75 per cent is ideal to keep pigs.
12. Sty is a dwelling place consisting of one or more of pens for pigs.
13. Each pen may house a single animal or group of animals depending upon age, sex and purpose. Each sty has a covered place and an adjoining open yard.
14. The covered space requirement per animal for boar, farrowing sows, fatteners, weaners, and dry sow are respectively 7.5 m^2, 9.0 m^2, 1.8 m^2, 1.8 m^2, and 2.7 m^2.
15. The open space requirement per animal for boar, farrowing sows, fatteners, weaners, and dry sow are respectively 12.0 m^2, 12.0 m^2, 12.0 m^2, 12.0 m^2, and 1.8 m^2.
16. The floor of the pig house should be smooth but not slippery.

17. The floor and walls should be strong to withstand the rooting habits of pigs.
18. The roof should be light, strong, durable, weather-proof and a bad conductor of heat.
19. Proper ventilation keeps pigs healthier and happier.
20. Wallowing tank is used to keep the animals cool during summer.

Exercises/Questions

1. What points should be considered for selecting a site for pig housing?
2. What are the effects of thermal environment on production of pigs?
3. Discuss the influence of the social and structural environment on production of pigs.
4. What ancillary facilities are required for large size pig farm?
5. Draw a layout plan for a pig house. Make necessary assumptions.

Chapter 9

Sheep and Goat Housing

Sheep and goats are important species of livestock for India. They contribute greatly to the agrarian economy, especially in areas where crop and dairy farming are not economical, and play an important role in the livelihood and financial security of a large proportion of small and marginal farmers and landless labourers. Sheep and goats provide wool, meat, milk, skin, hair and manure. Their rearing ensures self-employment and acts as a cushion in distress situations like drought and famine.

In India, goats are often referred to as the cows of poor men. The goat is domesticated for meat, milk, and wool. The meat of the goat, or Chevon, has high domestic demand. Goat milk is also known for its medicinal properties and health benefits for humans. It is very similar to human milk due to its easy digestibility and health-promoting traits. The economic viability has promoted the rearing of goats under intensive and semi-intensive conditions in recent years.

Proper shelter is crucial for sheep and goats to protect them from inclement weather and maintain their health. This ensures animal comfort, better care and supervision, increased herd efficiency, and higher labour efficiency. Minimum provisions in shelters for goats and sheep can increase animal productivity, especially in protection against weather conditions and predation. During fair weather, flocks are often penned in the open, while temporary shelters are used in monsoons and winters. Building requirements for sheep and goats are similar, with additional buildings needed for milk-raised goats.

9.1 Terminology

Buck: A male goat is called a buck.

Doe or nanny: A female goat is called a doe or nanny.

Ewe: An adult female sheep is called a ewe.

Goatling: A young goat from 10 to 15 months of age is called a goatling.

Kid: A young goat up to four months of age is called a kid.

Lamb: A young sheep up to four months of age is called a lamb.

Ram: An adult male sheep is called a ram.

Weaner: A young sheep or goat from 5 to 9 months of age is called a weaner.

Yearling: A young sheep from 10 to 15 months of age is called a yearling.

9.2 Location

The housing for sheep and goats shall be built-up on a dry, elevated and well-drained sites. The open part of the shelter should face away from the prevailing winds in order to protect the animals from severe winds mainly during summer and winter months. A place which provides a longer period of shade in a day shall be preferred. Longer walls should face the north-south directions in plains as it ensures cooler environment. The sheds should be easily accessible for deliveries.

9.3 Space Requirements

The rearing units are generally constructed in a yard. The type of unit can be constructed by considering the age and purpose of the animals. The total number of units depends on the number of animals to be kept under shelter. The floor area of the unit shall be calculated using the space requirement per animal and the number of animals. The minimum space requirement per animal is provided in Table 9.1.

Table 9.1. Floor Space Requirements per Animal

Type of animals	*Minimum Floor Space per Animal (sq. m)*
A male sheep or goat in groups	1.8
A male sheep or goat - individual	3.2
Sheep or goat up to 4 months of age in groups	0.4
Sheep or goat from 5 to 9 months of age in groups	0.8
Sheep or goat from 5 to 9 months of age - individual	0.9
Sheep or goat from 10 to 15 months of age	0.9
Sheep or goat from 10 to 15 months of age - individual	1.0
A female sheep or goat in groups	1.0
A female sheep or goat - individual	1.2
Ewe with lamb	1.5

Nearly 1.0 square meter floor area is needed per animal, if they are housed together. An ewe (adult female sheep) requires 1.2 square meter of living space. Pregnant sheep (lambing pens) require quite a bit of space ranging from 1.5 to 2.25 square meter. Feeder lambs need 0.8 square meter. The open space requirement per animal for different types of animals shall be about two-times of the covered space. Lambing by sheep requires a dry and wind-free area. Sheep shall have access to a large grass pasture for exercise and feeding. Sheep who stay outside for more time are less likely to suffer from respiratory problems. On an average, the water requirement per animal is about 10 litres per day which depends mainly upon the age of the animal and the season. The feeding space requirement shall be 40 to 50 cm for adult animals and 30 to 35 cm for kids. Water trough shall preferably be kept on a *pucca* floor otherwise the area surrounding the trough may become muddy.

9.4 Sheds and Rooms (for housing the sheep kept for wool and meat production, and goats kept for meat production)

The sheds and rooms of the following types are required for housing the sheep kept for wool and meat production, and goats kept for meat production.

1. General flock shed (Ewe/Doe shed)
2. Ram or buck shed
3. Lambing or kidding shed
4. Lamb or kid shed
5. Sick animal shed
6. Shearing shed and storeroom, and
7. Attendant's room

General Flock Shed (Ewe/Doe Shed)

The general flock sheds are meant for housing ewes or doe kept for breeding purpose. The dimensions of the shed shall be 15 m × 4 m and should accommodate not more than 60 animals. The height of the shed shall be 3 m.

Ram or Buck Shed

This shed is used for housing rams or bucks kept for breeding purpose. The dimensions of the shed shall be 4 m × 2.5 m and should accommodate not more than three animals. The height of the shed shall be 3 m. The animals are housed individually in these sheds. The shed shall be partitioned lengthwise with the help of wooden planks or *ballis* to form three equal compartments. The partitions should not exceed more than one-metre high from the floor. There should be a gap of 10 to 15 cm between each plank or *balli* to enable the animals to see one another.

Lambing or Kidding Shed

A lambing shed is a maternity room for pregnant ewes or doe. It is also known as the kidding shed. The dimensions shall be 1.5 m × 1.2 m and a height of 3 m. Each kidding shed has a manger for feed and hay, a bucket for water, and a room heater to protect the new born from cold in winter.

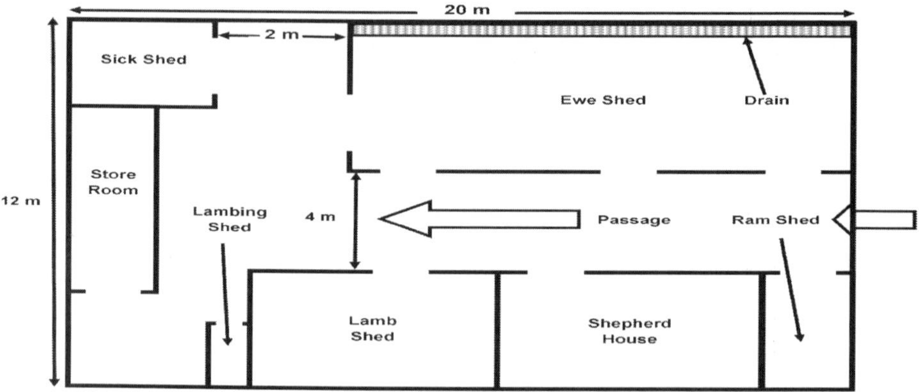

Figure 9.1. Layout Plan of Sheds for Sheep (kept for wool and meat production) and Goats (kept for meat production).

Lamb or Kid Shed

This shed is used for housing lambs or kids until they attain maturity and are weaned. It is also used for those animals which are weaned but have not attained maturity. The dimensions of the shed shall be 7.5 m × 4 m and 3 m high to accommodate not more than 75 animals. The shed shall be partitioned breadthwise dividing into two compartments. The compartment having the dimensions of 5 m × 4 m shall be used to keep the unweaned animals, and other compartment with dimensions of 2.5 m × 4 m shall be used for keeping the weaned animals.

Sick Animal Shed

This shed is used for segregating ailing and disabled animals. The dimensions of the shed shall be 3 m × 2 m and 3 m high. The sick animal shed should be away from the other sheds. A door of 1-meter wide and 2 meters high should be provided on one of the broad sides of the shed. The lower and upper half of the doors in the shed shall be made of wooden planks and wire nets, respectively. A window of 0.7 m by 1.2 m with a wire net cover shall be in each shed for ventilation. The upper half of the door can be used for supervision of the shed.

Shearing Shed and Storeroom

The shed is divided into two compartments by a partition wall. One compartment is used for storing wool and shearing equipment, and the other is for keeping feed and medicines. The room measures 6 m × 2.5 m × 3 m, with a 1 m × 2 m door on the front side. The wooden planks shall be used for door leaf. The room may have two windows, one on each of the long sides. The dimensions of each window shall be 0.7 m × 1.2 m, covered with wire netting.

Attendant's Room

The attendant's house or room must be located in the yard and have dimensions of 6m × 4m × 3m. It has a door of 1 m × 2 m dimensions. The wooden planks shall be used in fabricating the door leaf. The room can have four windows, one facing the passage of the yard and three facing the outside. Each window shall be 0.7 m × 1.2 m and covered with wire netting. The size of the different sheds and maximum number of animals which can be housed in a single shed is summarized in Table 9.2.

Table 9.2. Size of different Sheds

Sl. No.	*Type of Shed*	*Size of the Shed (m)*	*Height of the Shed (m)*	*Maximum Animals in a Shed*
1.	Ewe/doe shed	15×4	3.0	60
2.	Ram/buck shed	4×2.5	3.0	3
3.	Lambing/kidding shed	1.5×1.2	3.0	3
4.	Lamb/kid shed	7.5×4	3.0	75
5.	Weaner shed	7.5×4	3.0	75
6.	Yearling shed	10×5	3.0	50
7.	Sick animal shed	3×2	3.0	1
8.	Shearing shed and store room	6×2.5	3.0	
9.	Attendant/Shepherd's room	6×4	3.0	

9.5 Sheds and Rooms (for housing goats exclusively kept for milk production)

The following types of sheds and rooms are required for housing goats exclusively kept for milk production.

1. Dairy goat shed
2. Buck shed
3. Kid shed
4. Kidding shed
5. Milk room
6. Store room

Dairy Goat Shed

A dairy goat shed shall have separate stalls for the milking doe. These stalls shall be arranged in two rows with a passage between them. The dimensions of the stall shall be 1.2 m × 1.4 m. The overall length of stalls depends upon the number of milking doe. The shed should have partitions made of wooden planks, galvanized steel sheets, or asbestos sheets, at least 15 cm above floor level. Racks shall be provided in each shed for hay and green for milking doe.

Buck Shed

This shed is used to keep bucks. A partition made of either wooden planks or *ballis* shall be provided to divide the shed into two equal compartments for housing two bucks. The dimensions of the buck shed shall be 2.5×3.0 m.

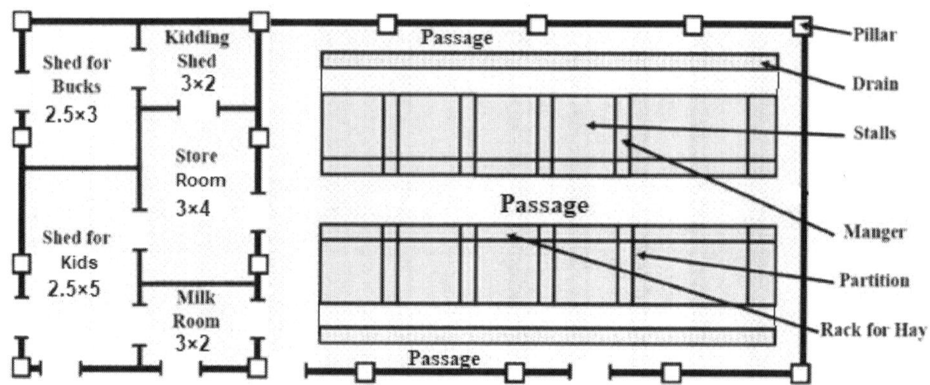

Figure 9.2. Layout Plan of Sheds (kept for milk production).

Kid Shed

Kid shed is used to accommodate young goats. The dimensions of the kid shed shall be 2.5×5.0 m.

Kidding Shed

Kidding shed is used as maternity room for doe. The dimensions of the kidding shed shall be 3.0×2.0 m.

Milk Room

It is the room where the excess of milk is kept for disposal. It is sometimes also used for milking doe. The dimensions of the milk room shall be 3.0×2.0 m.

Store Room

Store room is meant for keeping feed and other material. The dimensions of the store room shall be 3.0×4.0 m.

9.6 Constructional Details

Floor

The flooring for sheep and goat housing may be either of strong wooden battens or of *moorum*. Moorum flooring is considered satisfactory for covered area as well as for open yard. Concrete flooring is costlier but keep the shed cleaner. The salient features of these floors are as follows:

Wooden batten flooring: Wooden batten flooring is preferred in the regions where rainfall is quite heavy. The planks shall be 7.5-10.0 cm wide and 2.5-4.0 cm thick, with well-rounded sides. The clearance between the two planks shall be 1.0-1.5 cm to facilitate dung and urine disposal. The flooring should be constructed at least one meter above ground level. An appropriate ramp or steps of wooden planks shall be provided for easy access to the floor.

***Moorum* flooring**: The *moorum* is the cheapest flooring material and can be only adopted for ground floor. It has the same merits as that of mud flooring. It maintains and adjusts the temperature in both summer and winter. For its construction, a plinth walls between 15 cm and 30 cm in height shall be provided. The *moorum* is a form of disintegrated rock with a binding material. First of all, the subgrade is prepared up to 15 cm thick and then it is laid in layers of 8 to 15 cm thickness over a prepaid subgrade. It is well consolidated and well water every time.

The floor of the rooms for shearing, store, and attendant shall be made of moorum or brick in cement mortar. The floor shall be levelled properly and have a drain at the end of the floor. The drain shall have a 30–40 cm width and a 7.5 cm depth. The kachha, or sandy floor, shall be made in hot and dry conditions in arid and semi-arid areas. The floor for goats kept for milk production should be made of cement concrete tiles, or of movable or immovable slatted wooden platform. The floor is raised 45 cm above the ground. The floor shall have a slope of 1 in 40. For each row of stalls, a suitable drain made of brick masonry should be provided at the end. The size of the drain shall be 7.5 cm deep and 30 to 40 cm wide.

Walls

The walls can be made of brick in lime or cement mortar up to 2/3rd of their height from the floor level. In the remaining upper 1/3rd portion of the walls, the bird-proof netting is provided. The chain-link netting shall be supported by angle irons and brick masonry columns placed alternatively at 1.5 m centre to centre. The upper 1/3rd part of lamb or kid shed may be of wire netting supported by angle irons at the four ends. The back and side walls of the shearing and store room shall be made of brick in cement mortar. The front wall of these rooms shall be constructed up to half using brick in cement mortar. The upper half shall be of wooden batten or wire netting. For goats for milk production, the walls of bricks in cement mortar should be up to 1.2 m high. The remaining portion shall have wire-netting supported by angle iron or wooden posts. Stalls adjacent to the passage should have walls of 1.5 m height.

Roof

The lean-to or gabled type of roof can be made of plain or corrugated galvanized steel sheets or asbestos cement sheets, corrugated asphalt, and thatch. The thatch roof must be rain and fire-proof, with a minimum thickness of 15 cm. In heavy rainfall areas, gunny sacks treated with a cement-lime mixture can be used. The eaves should project out to prevent rainwater from entering the shed.

Gate

The number of gates to be provided in the shed depends upon its dimensions. Gates are provided either on the long or broad sides of the sheds. The size of each gate shall be 0.8 m wide and 1 m high. The gate shall be made of wooden battens or any other suitable material, and shall fit the entrance closely.

Manger

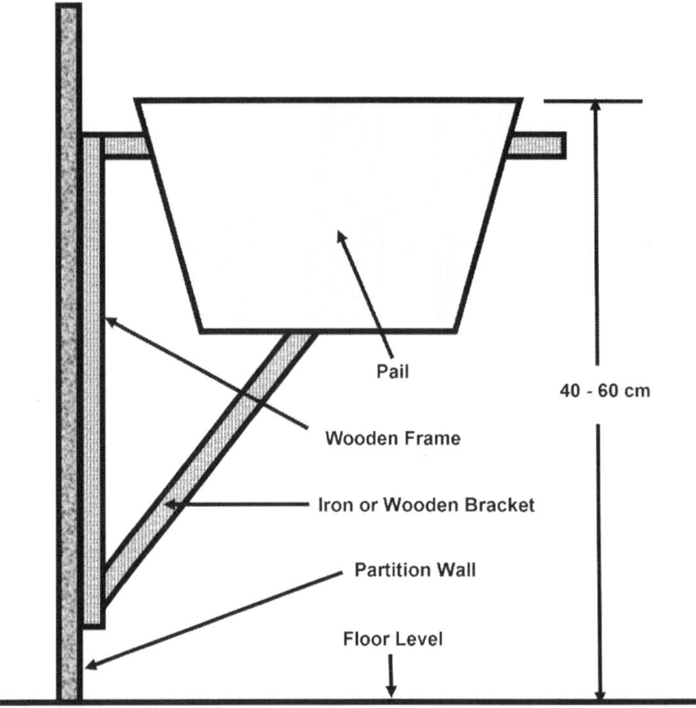

Figure 9.3. Manger for Goats.

The manger should be constructed from cement concrete or wood. It shall have two compartments for feed and hay. It may be fixed or portable type. The height of the manger shall be adjustable. With the help of clamps, the manger may be raised within the height ranging between 40 and 60 cm from the ground. The separate racks for hay and greens can be installed at or below the heads of animals for easy approach. Water troughs can be made of cement concrete or galvanized steel. The water trough can be hung from a hook fixed to the walls. The total number of mangers and water troughs depends upon the number of animals.

The design of the manger for goats kept for milk production is slightly different. The manger is a raised wooden frame with openings for galvanized steel pails for goat feed and water. These mangers are supported using iron or wooden brackets. It can be built on partition walls. The manger shall be raised at least 45-60 cm from the ground. A rack shall be fixed in the front wall to keep hay and other greens to enable the goats to pull out the greens as and when require.

Dipping Tank

The dipping tank is provided to save animals from infection. It can be made of galvanized steel sheets or stone or brick in cement mortar. Steel sheets should be fixed and protected in the soil to prevent the bulging of the sides of a full tank. If soil is unstable, cement concrete can be used. The tank can be placed on one side of the yard.

Foot Bath

A footbath at the yard entrance is required to protect animals from foot-rot disease. It is made of galvanized steel sheets or brick in cement mortar, and embedded in suitable soil.

9.7 Environmental Requirements of Sheep and Goats

Temperature: Thermoneutral zone ranges from 0°C to 30°C.

Relative Humidity: Relative humidity should be kept below 70 per cent to avoid heat stress and respiratory problems.

Ventilation: Sheep shelters should be naturally well ventilated. Sheep are susceptible to respiratory problems such as pneumonia and bronchitis often result from poor ventilation. Sheep prefer colder surroundings as opposed to a heated environment. Barns should not be heated or closed up. If there is presence of ammonia odours in the shed, it may be due to inadequate ventilation. Sheep require a dry and draft-free shelter. Cold housing with natural ventilation is preferred over mechanical. Over-ventilation is better than under-ventilation, as sheep including healthy baby lambs are cold tolerant.

Lighting: Natural daylight or artificial lighting for 12 to 16 hours per day.

Space: Sufficient space for movement and resting, typically ranging from 1 to 2.25 square meter per animal.

Bedding: The purpose of bedding is to provide warmth, insulation, and comfort to housed animals. Materials like straw, hay, dried corn stalks, corn cobs, peanut hulls, cottonseed hulls, oat hulls, sawdust, wood shavings, wood chips, pine shavings, sand, paper products, peat, hemp, and leaves are used for bedding for sheep. There are certain advantages and disadvantages of each bedding material. One of the requirement of the bedding material is that it should be clean and dry.

Feed and Water: Access to clean water and balanced nutrition suitable for the specific needs of sheep and goats.

Shelter: Protection from adverse weather conditions, especially during lambing or kidding seasons.

Security: Fencing and enclosures to protect against predators and theft.

Example 9.1

Draw a layout plan for a goat house for the following requirement.

1. Number of goats = 100
2. Number of bucks = 4
3. Number of kids = 50

Make reasonable assumptions wherever necessary, but state them clearly.

Solution

Location: It should be located on dry, elevated and well drained place with natural shed.

Space Requirement

A goat house comprises of general flock shed, buck shed, lambing or kidding shed, lamb shed, sick animal shed, shearing shed and store room, and attendant room.

Nearly 1.0 square meter floor area is needed per animal, if they are housed together. A buck needs 3.2 m^2 floor area when housed individually. Kids need 0.4 m^2. Lambing animals require quite a bit of space ranging from 1.5 to 2.25 square meter. The height of the shed shall be 3 meter. The open space requirement per animal for different types of animals shall be about two-times of the covered space. Provision of water troughs and feeding racks shall also be provided within the shed as well as in open yard.

Sl. No.	*Type of Shed*	*Number of Animals*	*Area Required per Animal (m^2)*	*Total Area Required (m^2)*	*Suggested Size of the Shed (m)*
1.	General flock shed	100	1.0	100	20×5
2.	Buck shed	4	3.2	12.8	4×2.5
3.	Lambing/kidding shed	1	1.8	1.8	1.5×1.2
4.	Lamb/kid shed	50	0.4	20.0	5×4

Sl. No.	Type of Shed	Number of Animals	Area Required per Animal (m^2)	Total Area Required (m^2)	Suggested Size of the Shed (m)
5.	Sick animal shed			6.0	3×2
6.	Shearing shed and store room			15.0	6×2.5
7.	Attendant/Shepherd's room			24.0	6×4
	Total area required			**179.6**	

Total covered area required = 179.6 $m^2 \approx 180$ m^2

Open space area required = 2×180 = 360 m^2

The different sheds shall be arranged as per Figure 9.1.

Example 9.2

Draw a layout plan for 50 number of sheep.

Solution

Nearly 1.0 square meter floor area is needed per animal, if they are housed together. The height of the shed shall be 3 meter. The open space requirement per animal for different types of animals shall be about two-times of the covered space. Provision of water troughs and feeding racks shall also be provided within the shed as well as in open yard.

Covered area required by 50 animals = 50×1 = 50 m^2

Open space area required = 2×50 = 100 m^2

The size of covered shed shall be 5 m × 10 m and that of open yard shall be 10 m × 10 m.

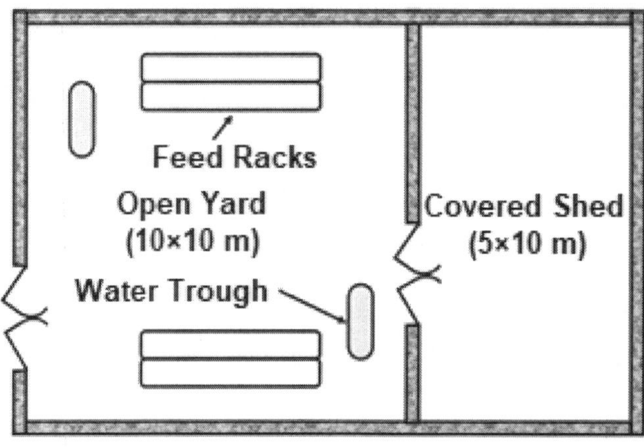

Plan for Sheep Shelter for 50 Sheep.

Important Points to Remember

1. An adult female sheep is called an ewe.
2. An adult male sheep is called a ram.
3. Goat is often termed in India as poor man's cow.
4. A female goat is called a doe or nanny.
5. A male goat is called a buck.
6. The location for housing sheep and goats shall be dry, elevated and well-drained sites.
7. Sheep barns and goat houses should be located not very far from the farm house.
8. Adequate fencing should be provided to protect the sheep and goats from the wild animals such as jackals, wolves, *etc.*
9. If sheep are housed together, nearly 1.0 square meter of floor area is needed per sheep.
10. An ewe (adult female sheep) requires 1.2 square meter of living space. Pregnant sheep (lambing pens) require quite a bit of space ranging from 1.5 to 2.25 square meter. Feeder lambs need 0.8 square meter.
11. The open space requirement per animal for different types of animals shall be about two-times of the covered space.
12. The feeding space requirement for sheep and goat shall be 40 to 50 cm for adult animals and 30 to 35 cm for kids.
13. The sick animal shed should be away from the other sheds.
14. The purpose of dipping tank is to protect the animals from infection.
15. Foot bath is provided at the entrance of animal house to protect them from foot-rot disease.
16. Sheep shelters should be naturally well ventilated.

Exercises/Questions

1. What should be an ideal location for sheep and goats housing?
2. Discuss in brief about the different sheds and rooms for housing the sheep kept for wool and meat production, and goats kept for meat production.
3. Discuss in brief about the different sheds and rooms for housing the goats exclusively kept for milk production.
4. What is the purpose of dipping tank?
5. What is the purpose of foot bath?
6. What are the ventilation requirement for sheep barns?

Chapter 10

Farm Fencing and Gates

Farm fencing and gates are essential components of agricultural operations, serving as first line of defence for farms. These essential structures play a crucial role in safeguarding livestock, crops, and property, ensuring both the security of farm assets and well-being of animals. Farm fencing meet the diverse needs of farmers, offering protection against predators, boundary delineation, and efficient livestock management. Similarly, farm gates provide access points that are integral to the daily functioning of a farm, facilitating the movement of animals, machinery, and personnel. Various types of fencing and gates being used at farm are discussed in this chapter.

FARM FENCING

For a farm, fences may be used to confine animals in it, or to prevent trespass of stray or wild animals. It is also used to protect and distinguish the boundaries, and to improve its appearance. Various types of fences such as hedges, masonry walls, ditches, wire *etc.* are used on the farms. Fences may be permanent or temporary depending on their intended use. Wire fences are commonly used as they are easy to install, most effective and cheaper as compared to masonry walls. Based on the construction design, wire fencing may be classified as:

1. Barbed wire fencing
2. Plain wire fencing
3. Welded wire fencing
4. Woven wire fencing
5. Interlocking fencing/Chain-link fencing

6. Razor wire fencing
7. Electric fencing/Solar electric fencing

10.1 Barbed Wire Fencing

Barbed wire, also known as barb wire, is a type of steel fencing wire constructed with sharp edges or points arranged at intervals along the strands. The barbs shall be so finished that the four points are set and located or locked as far as possible at right angles to each other. The barbs should be well formed, tightly wrapped and shall have a length of not less than 13 mm and not more than 18 mm. The point shall be sharp and cut at an angle not greater than 35° to the axis of the wire forming the barbs. The spacing between two barbs varies between 75 and 150 mm.

Two types of wires are used to manufacture barbed wire. Wire of specified diameter of which barbed wire is made is known as line wire, while wire of specified diameter of which barbs are made is known as point wire. The diameter of line wire shall be 2.50 mm (12 gauge) or 2.24 mm while that of point wire shall be 2.50 mm or 2.00 mm (14 gauge). The barbed wire shall be formed by twisting together two-line wires, one or both containing the barbs.

Barbed wires are classified as:

Type A (Iowa type): It is formed by twisting two-point wires, each two turns, tightly around both line wires making altogether four complete turns. The barbs shall have four points. (Figure 10.1)

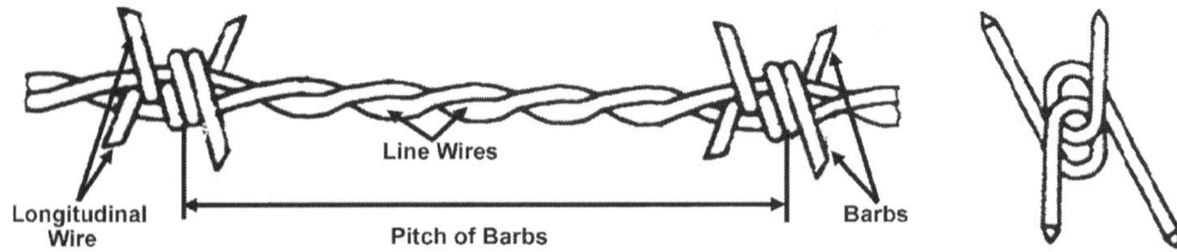

Figure 10.1. Type A (Iowa type) Barbed Wire.

Type B (Glidden type): It is formed by twisting two-point wires, each two turns, tightly around one-line wire making altogether four complete turns. The barbs shall have four points. (Figure 10.2)

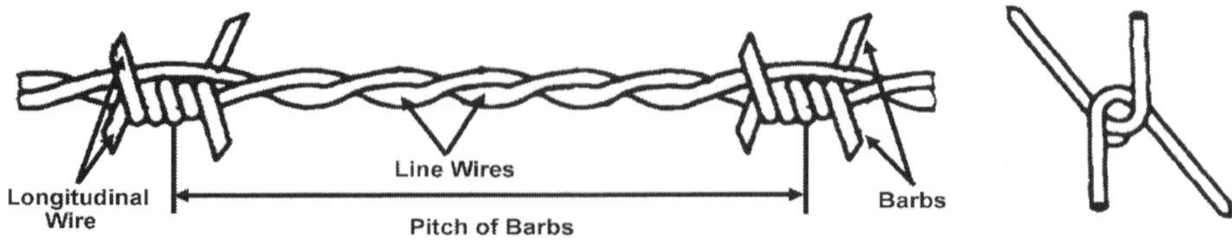

Figure 10.2. Type B (Glidden type) Barbed Wire.

Barbed wires are available in different sizes and weight. The details as provided in IS 278:2009 are given in Table 10.1.

Table 10.1. Size and Mass of Completed Barbed Wires

Size Desig-nation	Diameter of Wire (mm)		Distance between Two Barbs (mm)	Mass of Completed Barbed Wire (kg/100 m)	
	Line Wire	Point Wire		Maximum	Minimum
1	2.50±0.08	2.50±0.08	75±12	15.5	13.6
2	2.50±0.08	2.50±0.08	150±12	12.0	10.8
3	2.50±0.08	2.00±0.08	75±12	12.5	10.8
4	2.50±0.08	2.00±0.08	150±12	10.3	8.9
5	2.24±0.08	2.00±0.08	75±12	10.6	9.7
6	2.24±0.08	2.00±0.08	150±12	8.5	7.8

The mass of 100 meters completed barbed wire for Size 1 made of 2.50 mm (12 gauge) diameter line and point wires having 75 mm distance between two barbs varies between 13.6 to 15.5 kg. Thus, one meter of barbed wire of 12-gauge size has about 0.145 kg (145 g) weight. Barbed wires are available in a roll of about 30 kg. Total length of one roll made of 12 gauge and 14-gauge diameter wire are about 200 m and 320 m long, respectively. Standard barbed wire fences usually have posts spaced 3 to 4 meter apart and use three to six strands of wire. The spacing between different barbed wire strands is shown in Figure 10.3.

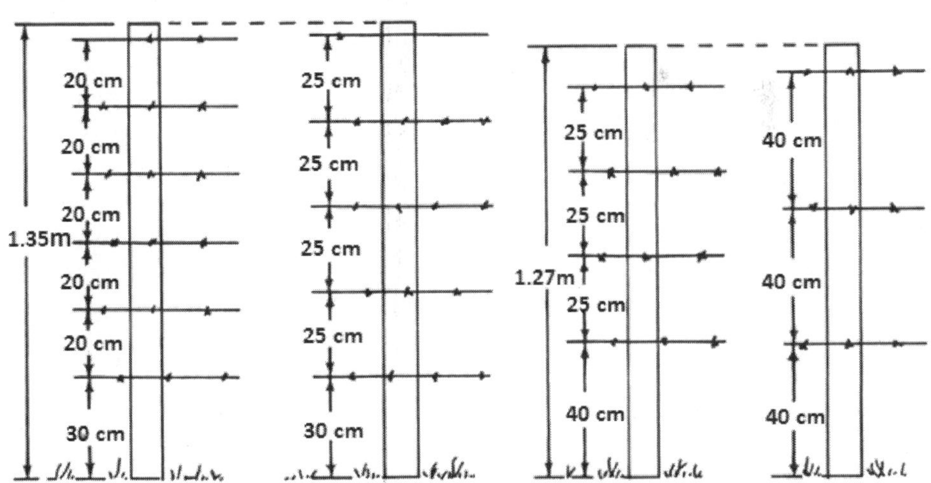

Figure 10.3. Spacing between Barbed Wire Strands.

Barbed wire is particularly effective for confining the large cattle. It cannot effectively contain goats, pigs, sheep or rabbits and cannot be used for poultry.

10.2 Plain Wire Fencing

Plain wire fencing is an affordable fencing option that is safe and strong. It can be used alone or with barbed wire or wire mesh. It is most suitable for large cattle. It will keep stock contained, deter trespassers and is an effective way of defining the boundaries. The plain wire fencing is available in different options - high tensile, medium tensile, galvanized, zinc aluminium coated, nylon wire. The diameter of standard galvanized plain wire varies from 2.50 mm to 4.00 mm diameter. The spacing between the horizontal wires is same as with barbed wire fencing. PVC coated wire has a 3 mm wire coated with 1 mm of PVC which is ideal for horses.

10.3 Welded Wire Fencing

Welded wire fencing is similar to woven wire. The only difference is that fence is permanently welded at all or most intersecting joints. This welding process makes the material more stable and rigid, and allows it to hold a more uniform shape. This type of fencing is suitable for demarcating the boundary of a farm. These are used for security, screens, cages, machinery guards, *etc.* These are available in different sizes having heights up to 2.4 meter having wire diameter varying from 2 mm to 4.9 mm.

10.4 Woven Wire Fencing

A woven wire fence is made up of several smooth wire horizontal lines spaced apart by vertical wires. Woven fencing is produced by weaving pieces of wire into various configurations. They are large mesh type, close mesh type and hexagonal wire type. It features a loose structure that stretches and bends with great flexibility. The main advantage of woven wire fencing over other fence types is its security, particularly for animals. These are specially designed to keep animals in or out. The only way animals get through is if the woven wire fence is actually broken. The woven wire fencing is available with a number of configuration of different size of wires and height of the fence. The size and jumping ability of animal helps in deciding the height of the fence.

Large-mesh type fence: Spacing between horizontal line wires may vary from as close as 7.5 cm at the bottom for small animals to 22.5 cm at the top for large animals (Figure 10.4). Spacing of the wires generally gets wider as the fence gets higher. Vertical wires are spaced 15 cm apart for small animals and 30 cm for large animals. The height of fencing materials ranges from 125 to 150 cm. Number of horizontal wires varies from 9 to 11. These fences are particularly used for large animals, cattle, buffaloes, horses, *etc.*

Close-mesh type fence: At the bottom, the spacing between the horizontal wires varies between 2.5 cm and 7.5 cm while at the top, it varies from 10 cm to 12.5 cm (Figure 10.5). The spacing between vertical wires is about 10 to 12.5 cm. In this type of fencing the horizontal wires are made heavier than the vertical wires and the vertical wires are wrapped round the horizontal ones. These fences are considered as an ideal fence for small animals and birds like poultry, rabbits and goats.

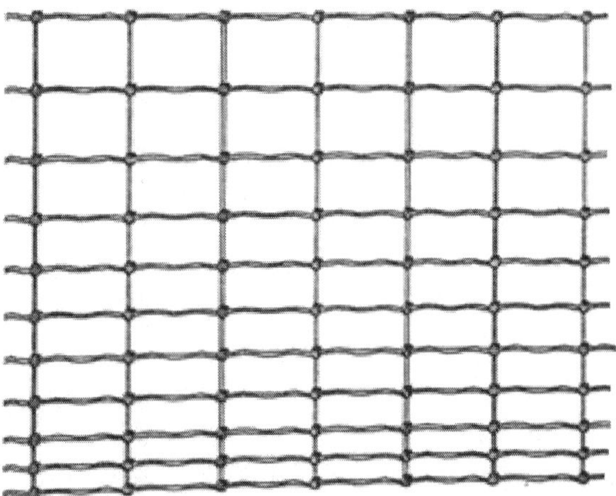

Figure 10.4. Woven Wire Large-Mesh Fence.

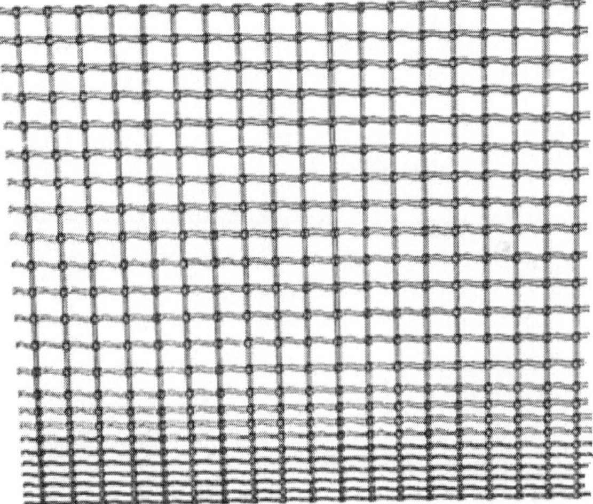

Figure 10.5. Woven Wire Close-Mesh Fence.

Hexagonal-wire type fence: Hexagonal woven type wires are uniform all through their height. The shape of woven wires is hexagonal in shape (Figure 10.6). These are also called chicken wire mesh or *Murga Jali,* because of its usage in poultry farms. These are made from stainless steel and galvanized iron drawn wire. The height of these type of fences is 2 meter and above. These are available either in 2.5 or 5.0 cm hexagonal mesh made of 1 mm (20 gauge) wire.

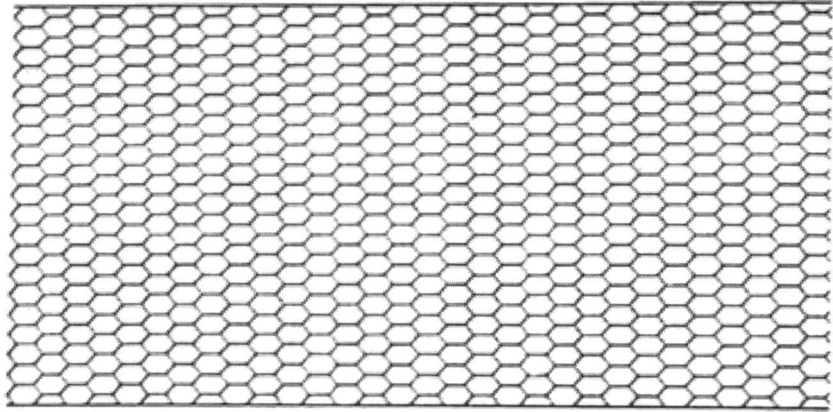

Figure 10.6. Woven Wire Hexagonal Chicken Fence.

10.5 Interlocking Fencing/Chain-Link Fencing

Chain-link fencing is an interlock fence used for providing barricades at different perimeters such as barricades of parks, green belts, *etc.* (Figure 10.7). Galvanized as well as PVC coated chain link fence are perfect for protecting boundaries from trespassing and intrusion. The closely spaced diamond mesh construction gives fence linear strength and springy texture to protect animals from injury. These are available in different sizes. The diameter of wire varies from 2.00 to 5.00 mm while the mesh size ranges from 25 mm to 150 mm. The width of these fences may vary from 0.90 m to 3.00 m and are available in the lengths of 5.0 m to 25.0 m. (IS 2721:2003)

Figure 10.7. Interlock Fencing/Chain-Lock Fencing.

10.6 Razor Wire Fencing

Razor wire (Figure 10.8) is manufactured with galvanized or stainless steel wire materials. It comes in variety forms, such as flat razor wire, straight type and concertina wire. Razor barbed wire as peripheral security fencing plays a safe deterrent effect. Razor wire is more effective than conventional fences. Razor wire fences are widely used for safety and security in civil, industrial and military uses, and find an extensive application in tea estates and high security sites. It prevents cattle, goat and human trespass.

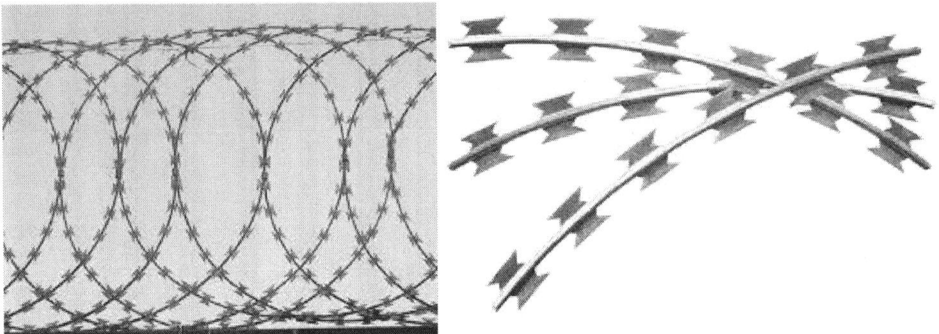

Figure 10.8. Razor Wire Fencing.

10.7 Electric Fencing/Solar Electric Fencing

There are three types of electric power fences – electric grid powered, battery powered and solar powered. Each relies on a component known as an electric fence charger to convert the incoming charge into a safer form of electricity. The difference among them is how the charger is powered. Electric powered fences are used to contain livestock or to keep unwanted animals out of field. These fences are psychological barriers and not physical barriers for all classes of cattle. The electric fence has multiple strands (one or more) of plain wires and metal/cement/wooden posts to hold the strands in position. The current of high voltage is provided in the fence wires. In the case of any infringement inside the protected area, the fencing system activates the alarm for information to the owner.

Animal or human being receives a sharp, short, painful but safe electric shock when they come into contact with electric fence. The pulsating current shall be in the live wire to avoid electrocution of animal/human. The current pass for only 1 milli-second at every 1-1.2 seconds to provide enough time to stay away from fence. The intensity of electric shock depends upon the voltage and ampere generated by the energizer, duration of wire and animal contact and sensitivity of animals. The contact of fence and animal for milli-second is completely safe and non-lethal due to pulsating current.

The components of an electric fence are controller, battery, earthing rod, bare wire, insulators and posts. In case of solar fencing, in addition to above, solar panel is needed to charge a regular 12-volt battery. The controller is also known as charger or energizer. The controller is the heart of the electric fence and it is responsible for the electric shock. The charger steps up the low voltage of battery up to 10,000 volts. Generally, 6-7 kV voltage is enough for most of the animals. However, voltage can be transform as per the size of the animals. The selection of the energizer depends upon the intensity of electric shock, single or multi wire fence, length of the fence, area under the fence. As a thumb rule, 10 km of electric fence can be powered by the 1 joule of the output energy. The current flowing through the wires must be as low as possible but should not exceed 0.008 ampere.

The current produced by the energizer stays in the live wire. In the case of the animal contact to the live wire, the current return back to the energizer through the animal. The contact of live and ground complete the circuit that give electric shock to animal. The intensity of the electric shock depends upon the efficacy of the earth system. The earth shall be of low resistance for adequate electric shock. The earth system is made of a number of earthing rods. The number of earthing rods depend upon soil types, mineral content, ground moisture and fence load. There is two type of the earth system being used in electric fence.

Fence Wire Return System. The fence is constructed using both live and earth wires. The animal touches the live wire the current flows through the animal and the ground back to the earth rods. This system is also practiced to keep away monkeys with earth wire under the live wire. The touch of earth or ground along with the live wire shock the monkey (Figure 10.9).

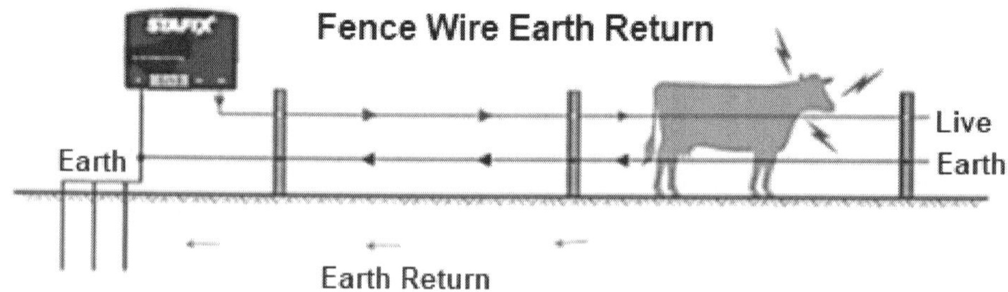

Figure 10.9. Fence Wire Earth Return.

Ground Earth Return System: Practiced where soil is conductive. This system is suitable for most moist soils. Current flows through the animal and the ground back to the earth rods (Figure 10.10).

Installation of electric fence begins with the corner posts, which can be made of concrete, wood, or metal. The intermediate posts are positioned roughly 10 m apart by temporarily drawing a wire between the end posts. In order to prevent large gaps that would make it easy for animals to enter the fence, insulators are placed in close proximity to the posts. The voltage of electric fence shall be between 5-10kV. The fence shall be checked for short circuit or improper connections in the case of voltage drop less than 5kV. A metal rod about 60 cm below the ground level is placed for earthing. Copper wire of 14 SWG should be used for connection between the earth terminal on the controller and the earth.

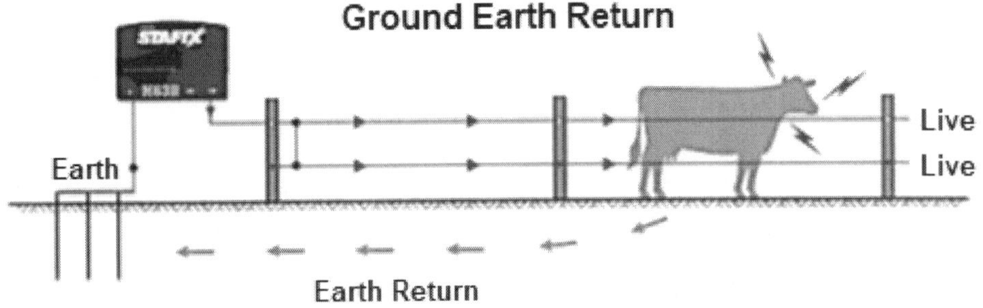

Figure 10.10. Fence Wire Ground Earth Return.

Heavy vegetation in contact with charged (live) wires may cause short-circuit thus making the system ineffective. The contact drops the voltage of energizer by completing the loop between earth and fence wire. The intensity of the shock to animal will also reduce due to the voltage drop. It is advised to keep control over the growth of vegetation near to the electric fence.

Advantages and Disadvantages of Electric Fencing

The low initial cost, low operating cost, low maintenance cost, easy to install or dismantle, long life and portability are the advantages of the electric fence. The barbed or woven wire fences are costly option than the electric fence. These fences are highly reliable and do not cause any physical harm to animals or human beings. The existing wire fence can be renovated to the electric fence. The wire spacing and design of electric fence can be modified as per the size of animals. The added advantage of solar powered fencing is that it can be installed in remote areas where there is no grid supply.

Disadvantages of electric fencing are only approved fence chargers of standard make should be used. In the case of cattle and sheep, the fence charger shall be active all the time. The ground wires shall be used to keep fence effective in the dry weather. The installation of electric fence in swampy and wooded shall be avoided due to heavy vegetation. The contact of vegetation and fence wire can short circuit and that deactivate the fence. In case of solar powered, fencing system is dependent on weather.

10.8 Fencing Posts

The posts support the entire fence and thus should have enough strength to resist lateral or bending pressure. Most commonly used posts are made of wood, angle iron, circular pipes, reinforced concrete or stone pillar. Wood or stone pillars are best suited for temporary fencing. For permanent fencing, reinforced concrete or concrete set angle iron posts are widely used. Stone pillars are used in the areas where they are locally available in abundance. Used circular pipes can also be used as posts for fencing. These posts are comparatively cheaper.

Wood posts: The advantages of wood posts are strength and resistance to bending. The disadvantages are that they should be treated regularly so as not to be affected by termites, and does not last long. The diameter of wood posts varies from 6 cm to 20 cm or more. The larger the top diameter, the stronger the post. Corner and gate posts should have a top diameter of at least 20 cm. Brace posts should be 12.5 cm or more in diameter.

Angle iron posts: The advantages of angle iron posts are less weight, fire proof and maintenance free. They also help ground the fence against lightning when the soil is moist. The disadvantage of steel post is that they do not have as much strength against bending as wood posts. The size of angle iron posts may be 50×50×6 mm or 40×40×6 mm while the end posts should have more strength and may have the size of 100×100×6 mm. The weight per meter of 50×50×6 mm, 40×40×6 mm and 100×100×6 mm size angle iron are respectively 4.5 kg, 3.5 kg and 9.2 kg.

Concrete posts: Concrete posts are simple to build and can be made on the farm. The concrete may be of two shapes *i.e.* square and tapered square post. The size of square post shall be 10×10 cm whereas, the tapered square post shall be 10×10 cm and 7.5×7.5 cm at bottom and top, respectively. The length of the posts shall be 1.8-2.1 m. The concrete post last longer and of low cost than the iron and wooden post. Iron posts are susceptible to corrosion whereas, wooden post can be damaged by termite.

Erecting Fence Posts and Wire

The post of fence shall be fixed underground to 52.5-60 cm using the concrete mix of 1:2:4. Extra 15 cm shall be added to the height of the post along with the underground depth of post and the height of the top wore of fence. The fastened wires on the post shall be well stretched. The load of the fence wires required to be supported at the corner post using the braces to the ground (Figure 10.11).

FARM GATES

Farm gates are used to control the movement of people, animals, and equipment in and out of a farm or ranch property. They come in a wide range of shapes, sizes, and materials, and are often tailored to fit the unique needs of the farm where they are used. The ideal farm gate should be durable, simple in construction, secure, easy to use and maintain, visible, weather resistant, aesthetic, cost-effective, and of sufficient heights.

One of the primary functions of a farm gate is to provide a barrier to keep livestock within a defined area. This is especially important for larger animals like cattle and horses, which can cause damage to crops or neighbouring properties if they are allowed to roam free. A well-designed farm gate can also help to prevent predators from entering the property and preying on the livestock. Another important function of farm gates is to control the movement of people and vehicles in and out of the property. This can be important for security reasons, as well as for preventing unwanted visitors from disturbing the livestock. It can also be important for regulating the flow of traffic on the farm, especially if there are multiple access points or if the farm is located near a busy road.

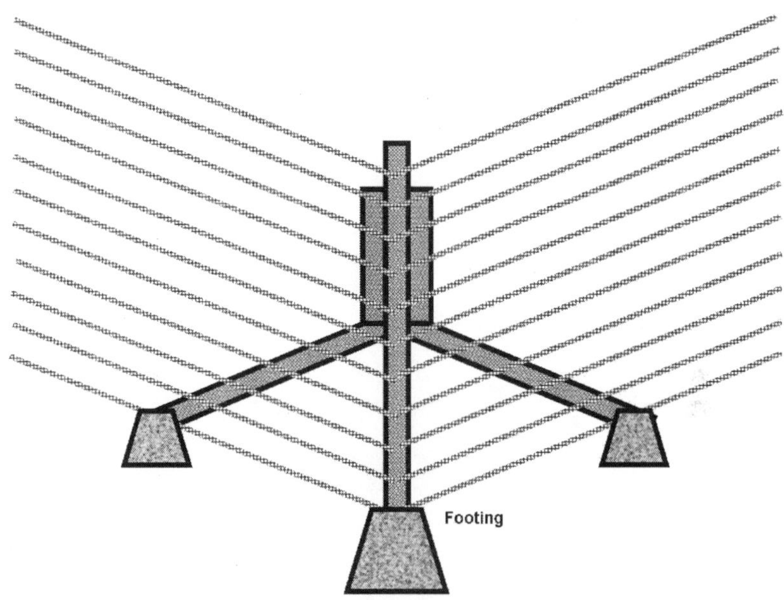

Figure 10.11. Fixing of Corner Post.

Gates should be located in a place that is easily accessible for vehicles and people. The proper location will result in saving the farmer's time, unnecessary work, and inconvenience. The location should also provide adequate security for the farm.

When selecting a farm gate, there are several factors to consider, such as the size of the gate, the type of opening mechanism, and the level of security needed. Some gates are designed to swing open on hinges, while others slide open on tracks. Some gates are equipped with locks or other security features to prevent unauthorized access. In addition to traditional farm gates, there are also a variety of specialized gates that are designed for specific purposes.

The width and height of farm gates can vary depending on the specific needs of the farm, the type of livestock being raised, and the size of the entrance or enclosure. Generally, the width of a farm gate should be wide enough to allow for easy passage of vehicles and equipment, while the height should be tall enough to contain the animals and prevent them from jumping over the gate. A common width for farm gates is around 3-5 metre, but they can be wider or narrower depending on the size of the entrance or the type of vehicle or equipment that needs to pass through. The height of the gate will depend on the type of livestock being raised. For cattle, a typical height is around 1.2-1.5 metre, while for horses or sheep, a height of around 0.9-1.2 metre may be sufficient (Figure 10.12). Sometimes, a small side gate of 1-metre wide is provided to facilitate the movement of people (Figure 10.13).

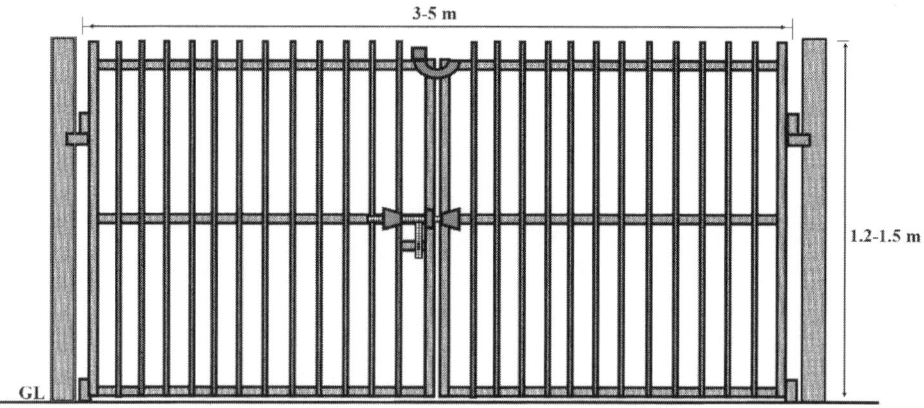

Figure 10.12. Farm Fencing Main Gate.

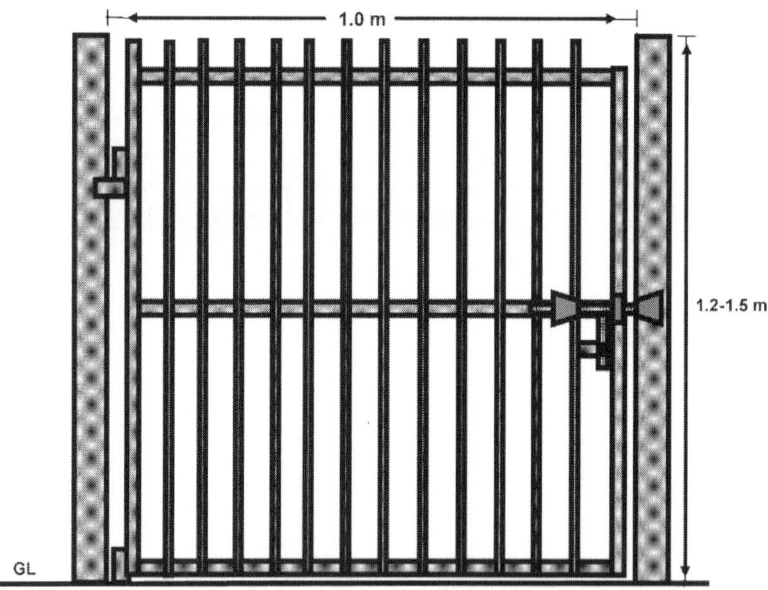

Figure 10.13. Farm Fencing Side Gate.

10.9 Construction Details

Farm gates come in a variety of sizes and materials, but most are constructed with similar basic elements. Here are some common construction details of farm gates:

1. **Frame:** The frame of the gate is typically made from either wood or metal, with metal being more common for larger gates or those that will be exposed to heavy use or harsh weather. The frame provides the structural support for the gate and is often designed with diagonal bracing to add strength. Wooden gates are popular for their rustic appearance, but they require regular maintenance to prevent rot and decay. Metal gates are more durable and can be designed to withstand harsh weather conditions, but they may require periodic painting or rust treatment.
2. **Hinges:** The hinges on a farm gate are typically heavy-duty and designed to support the weight of the gate over time. Common types of hinges include strap hinges, T-hinges, and post hinges. The hinges are usually attached to the gate frame with bolts or screws, and to the post or other supporting structure with lag bolts or through-bolts.
3. **Latch:** The latch is the mechanism used to secure the gate in the closed position. There are a variety of latch designs, including sliding bolts, gate hooks, and gravity latches. Some gates may also include a lock for added security.
4. **Materials:** The materials used to construct a farm gate can vary depending on the application and budget. Common materials include wood, metal, vinyl, and composite materials. The choice of material will depend on factors such as durability, weather resistance, and aesthetic preferences. Generally, woven or plain welded wire filler is securely tied around the frame work. However, the filler material should be selected so that even small animals may not have access to the farm from outside.

Example 10.1

Calculate the cost of barbed wire fencing of the 40-hectare farm having length and breadth in 4:3 ratios. Make reasonable assumptions wherever necessary, but state them clearly.

Solution

Assumptions

Number of strands = 6

Weight of 1 meter barbed wire = 0.145 kg

Cost of 1 kg of barbed wire = Rs 80/-

Distance between two angle iron posts = 3 meter ($\approx$ 10 feet)

Length of one angle iron post = 2.5 meter

Weight of one meter of angle iron = 4.5 kg

Cost of 1 kg of angle iron = Rs 50/-

Cost of digging pit = Rs 200 per cubic meter

Cost of 1 m^3 of cement concrete = Rs 1500

Number of gates = 1 main gate and 3 small gates

Cost of main gate including fixing = Rs 20,000

Cost of one small gate including fixing = Rs 5,000

Labour charges for filling cement concrete = Rs 500 per cubic meter

Labour charges for fixing and stretching of barbed wire = Rs 5 per meter

Labour charges for installation of posts is included in the labour charges for filling cement concrete.

(Prices/rates prevailing in your area may be considered while estimating the cost of fencing)

Step 1: Calculate the perimeter of the farm

Total area of the farm = 40 ha = 4,00,000 sq. meters

Let the length of the farm = 4x

Thus, width of the farm = 3x

Area of the farm = $4x \times 3x$ = 4,00,000

or, $12x^2$ = 4,00,000

or, $x \approx 182.6$ m

Thus, length of farm = 4x = 4×182.6 = 730.4 m

and, width of farm = 3x = 3×182.6 = 547.8 m

Perimeter of the farm = $2\times$ (length +breadth) = $2\times(730.4+547.8)$=2556.4 m $\approx$ 2557 m

Step 2: Calculate the quantity of barbed wire

Length of wire needed = perimeter of the farm $\times$ number of strands

= 2557×6 = 15342 m

Weight of wire needed = 0.145×15342 = 2224.6 kg

Add 5 per cent for tying and wastage *etc.* = 2224.6 + $0.05\times2224.6 \approx$ 2336 kg

Step 3: Calculate the quantity of angle iron posts

Number of posts

$$= \frac{\text{perimeter of the farm}}{\text{distance beteen two posts}} + 8 \text{ number extra corner posts} = \frac{2557}{3} + 8 = 861$$

(2 extra posts at all four corners of the farm to support the corner poles from bending due to wire tension)

Weight of 1 angle iron posts = 1×2.5×4.5 = 11.25 kg

Weight of 861 angle iron posts = 861×11.25 = 9686.25 kg

Step 4: Calculate the quantity of excavation for digging of pits

Size of pit = 0.3×0.3×0.45 ≈ 0.04 m^3

Total volume of pits to be dug = 861×0.04 = 34.44 m^3

Step 5: Quantity of cement concrete

Quantity of cement concrete = Total volume of pits = 34.44 m^3

Step 6: Number of iron gates

There is a provision of 1 main gate and 3 small (side) gates.

Material Cost

Sl.No.	*Description*	*Quantity*	*Rate (Rs)*	*Unit*	*Amount (Rs)*
1.	Barbed wire	2336 kg	80	kg	1,86,880
2.	Angle iron posts	9686.25 kg	50	kg	4,84,313
3.	Cement concrete	34.44 m^3	1500	cubic meter	51,660
4.	Iron gate (main) including fixing	1 no.	20,000	number	20,000
5.	Iron gate (small) including fixing	3 nos.	5,000	number	15,000
	Total				**7,57,853**

Labour Cost

Sl.No.	*Description*	*Quantity*	*Rate (Rs)*	*Unit*	*Amount (Rs)*
1.	Earthwork excavation	34.44 m^3	200	cubic meter	6,888
2.	Filling of cement concrete	34.44 m^3	500	cubic meter	17,220
3.	Stretching and fixing of barbed wire	15342 m	5	running meter	76,710
	Total				**1,00,818**

The total cost of fencing the land

= total material cost + labour cost = 7,57,853 + 1,00,818 = Rs 8,58,671

$$\text{Cost of fencing per running meter} = \frac{\text{total cost}}{\text{perimeter of farm}} = \frac{8,58,671}{2557}$$

$$\approx \text{Rs } 336 \text{ per meter}$$

Important Points to Remember

1. Types of wire-fencing for the farm are plain wire, barbed wire, woven wire and electric.
2. The length of barbs in barbed wire fencing should not be less than 13 mm and not more than 18 mm.
3. The spacing between two barbs varies between 75 and 150 mm.
4. The weight of one meter barbed wire of 12-gauge size is about 0.145 kg or 145 grams.
5. Barbed wires are available in a roll of about 30 kg.
6. Total length of one roll (30 kg) made of 12 gauge and 14-gauge diameter wire are about 200 m and 320 m long, respectively.
7. Barbed wire fencing is preferred for containing large cattle. It is suitable on dairy farms.
8. Barbed wire fencing cannot be used for small animals like poultry, goats and rabbits.
9. Plain wire fencing is most suitable for confining the large cattle.
10. In the large mesh woven wire fence, number of horizontal wires varies from 9 to 11.
11. Woven wire close-mesh type fences are considered as an ideal fence for small animals and birds like poultry, rabbits and goats.
12. Woven wire hexagonal type fence is also known as chicken wire mesh or *Murga Jali*.
13. Woven wire hexagonal type wire fencing is used for poultry.
14. The current in electrical fencing is pulsating and passing at every 1 to 1.2 second and only for a mille-second of time.
15. Most of the animals can be kept away with voltage of 6000 – 7000 volts.
16. In electric fencing, the current flowing through the wires must be as low as possible but should not exceed 0.008 ampere.
17. Fence wire return system of electric fencing is practiced in dry or sandy soil.
18. Ground earth return system of electric fencing is suitable for most moist soils.
19. Cost of electric fence is much less than that of comparable barbed or woven wire fences.

20. Corner posts are specially braced to take up the loads from the sides exerted by the fencing wire.
21. Farm gates should be located in place that is easily accessible for vehicles and people.
22. The width of farm gates should be about 3-5 meter.
23. The height of farm gates for cattle is around 1.2-1.5 meter, while for horses or sheep, a height of around 0.9-1.2 meter may be sufficient.
24. The total length of 6 lines of barbed wire for fencing 4 hectare square farm would be 4800 m.

Exercises/Questions

1. List various types of fencing used on farm. Explain in detail the construction and working of electric fencing. Compare it with barbed wire fencing.
2. Discuss various types of fencing and gates used for cows and buffalo confining.
3. Calculate the cost of fencing a square farm of 25 hectares fenced by barbed wire using angle iron posts. Make necessary assumptions.
4. What are the advantages and disadvantages of electric fencing?
5. What should be the characteristics of the ideal farm gate? Discuss about its location and size. Give the construction details of the farm gates.
6. Write short notes on:
   (a) Barbed wire fencing
   (b) Woven wire fencing
   (c) Interlock fencing
   (d) Electric fencing
   (e) Razor wire fencing
   (f) Fencing posts
   (g) Farm gates

Chapter 11

Storage Structures

Storage structures on a farm are essential components for preserving the quality of products, reducing waste, and ensuring the farm's long-term sustainability. These structures play a crucial role in ensuring the efficiency, sustainability, and profitability of the farm. In this chapter, different storage structures and their design features for storing of grains, fodder, and perishables and semi-perishables are discussed.

STORAGE OF GRAINS

Food grain production in India since independence has witnessed remarkable growth and transformation, with a shift from food scarcity to self-sufficiency. The country has achieved self-reliant in the production of food grains. Since Independence, India's food grain production has registered an over a six-fold increase, from 50 million tonnes in 1950-51 to about 315 million tonnes in 2020-2021. Insufficient and outdated storage facilities, including warehouses and silos, often lead to spoilage and damage of food grains due to pests, moisture, and temperature fluctuations. Post-harvest losses of food grains in India have been a long-standing issue and continue to pose significant challenges to the country's food security and agricultural sustainability. Proper and scientific storage of grains can bring down the losses to a considerable extent.

India has a massive population, and ensuring a stable and adequate food supply is crucial for national food security. Grain storage helps mitigate the impact of crop failures, natural disasters, and price fluctuations. It ensures a buffer stock of food grains that can be distributed in times of need, such as during droughts, floods, or other emergencies. Adequate storage facilities enable farmers to store their produce and sell it at a more opportune time when prices are favourable, reducing the pressure to sell immediately after harvest when prices may be low. Storing food grains plays a vital role in both meeting the nutritional needs of the population and supporting the agricultural sector and overall economy.

11.1 Factors/Parameters Influencing the Shelf Life of the Stored Product

The shelf life of stored products is the cumulative result of the overall management of the factors that are responsible for grain spoilage. The influence of biotic and abiotic factors starts right from the field and continued till consumption. The shelf life is comprised of physical, chemical, and biological properties that change with the violation of limits and ranges of the biotic and abiotic factors. Physical changes include changes in colour, size, shape, surface roughness *etc.* whereas chemical changes include the development of rancidity, gluten content, sedimentation value, protein content *etc.* Germination is the main change that is classified under biological changes.

Biotic Factors

All living organisms that affect the quality and quantity of the stored products are categorized as biotic factors. Insects, microbes, rodents, birds and pests are the example of these factors.

Abiotic Factors

The non-living parameters like moisture content, pH, fat content, colour, kernel damage, temperature, relative humidity, gas composition, storage time, and type of storage structure influence the quality of stored parameters.

The biotic and abiotic factors can further be classified as per their origin with respect to food grains.

Intrinsic Factors

The moisture content, pH, fat content, colour, and kernel damage are the intrinsic parameters of the food grains. The grains are susceptible to deterioration due to uncontrolled changes in these parameters.

Extrinsic Factors

Temperature, relative humidity, gas composition, storage time, type of storage structure, insects, microbes, and rodents are the factors that affect the quality and quantity from outside of the food grains.

11.2 Quality Deterioration Mechanisms and their Control

Quality is a package of various physical, chemical and biological parameters. These parameters help to assign grades to the food grains for further use/sale in the food industry/market. The independent parameters *i.e.* temperature, relative humidity, gas composition, type of storage structures *etc.* deteriorate the food over time. Respiration and senescence are the two major deterioration mechanisms for food.

Respiration

Biological respiration refers to the viability of the grains. It is the main criteria that differ the grains and seeds. The grains/seed consume carbohydrates with oxygen and produce carbon dioxide and water along with heat.

$$C_6H_{12}O_6 + 6O_2 \rightarrow 6CO_2 + 6H_2O + \text{Energy } (2835\ kJ/mol)$$

The rate of respiration of food grains depends upon the temperature, relative humidity, damage to the kernel by insects or mechanical handling, moisture content, presence of foreign material and insects, gas composition under a modified atmosphere, type of storage system *etc.* The foreign matter in stored grains should be below 1 per cent. In silos, a 1.2-1.5m distance is recommended between the grain mass and lid of the silo for better air exchange to avoid condensation of moisture. The grain should be dried to 12 per cent for cereals, 10 per cent for legumes/pulses and 8 per cent for oilseeds. The higher rate of respiration decreases the germination of the seeds.

Senescence

The ageing of the grains is known as senescence which is accompanied by biochemical changes, and chemical, physical and biological changes that occur during processing and storage. All factors associated with deterioration interact in a very complex way. Hence, concentrating on one factor can cause the commercial failure of the product in the market. The rate of the deteriorative mechanisms is controlled by independent parameters. Here are the reactions that take place under the senescence process of the food grains.

- **Enzymatic Reactions:** The enzymes are present in the aleurone layer of food grains that help in the synthesis or degradation of the starch present in the endosperm.
- **Chemical Reactions:** Sensory quality, lipid oxidation, non-enzymatic browning, colour changes, flavour changes, nutritional quality, degradation of vitamin, proteins and lipids
- **Physical Changes:** Mechanical damage to the product during transportation, handling and packaging
- **Microbiological Changes:** Food grains are generally stored below 12 per cent of the moisture content, therefore the microbes *i.e.* bacteria, mould *etc.* are unable to degrade the quality of stored grains. Although, in the case of poor aeration or inadequate drying, the respiration of the food grains generates heat and water vapour that develop a conducive environment for the mould. The aeration using dry air and drying before the storage helps to control the microbiological incidence in

the stored food grains. The high moisture foods are susceptible to microbiological changes.

- **Macro Biological Changes:** The insects, rodents, birds *etc.* are the macro biological agent that is responsible for the direct and indirect losses to the food.

11.3 Storage Losses

Biotic factors like insects, mites, fungi, rodents, and birds damage the food grains during storage of food grains. The insects are of three types *i.e.* primary, secondary and tertiary. Primary insects damage the outer layer of the grains and secondary insects eat the endosperm of pre-damaged grains. Whereas, tertiary can feed on the damaged, undamaged whole grains and flour. The mites are smaller organisms than the insects and feed on the surface of the grains. Fungi grow in a moist environment in which the food grains have water activity above 0.7. Rodents damage the storage structure and feed on the grains. Birds can enter the godowns and eat the whole grain. In addition, abiotic factors like environmental changes affect the quality by the increase of storage temperature or moisture content of the food grains. These biotic and abiotic factors are responsible for losses in stored food grains that can be classified into two categories:

113.1 Quantitative Losses

Insects, mites, fungi, rodents, birds *etc.* eat the grains and it reduces the weight of the stored material. The temperature hike increases the respiration rate that consumes the dry matter of the food grains and reduces the thousand kernel weight. The temperature and relative humidity changes also affect the moisture content of the stored material. Although, the change in the moisture content is reversible and follows the hysteresis effect of equilibrium moisture content (EMC). Higher moisture content is also linked with a higher rate of respiration that result in loss of dry matter. High moisture content also harbours insects and fungi that develop the hotspots inside grain mass. Hotspots are the particular areas of condensed moisture inside the grain mass. The insects eat the endosperm and cause the weight loss of the stored commodity. The respiration of food grains causes dry matter losses of 1 per cent over the storage period.

11.3.2 Qualitative Losses

The biotic and abiotic factors affect the quality of the stored commodities. The quality of the stored commodities is defined by their intrinsic and extrinsic changes. The intrinsic changes are more related to the chemical changes inside the stored commodities. Whereas, the colour or visible damage due to fungus or insects is categorized under the extrinsic changes. The germination percentage of seeds, gluten quality of grains, development of aflatoxins in groundnut, oil content of oilseeds, the nutritive value of stored food *etc.* are the changes that occur during the storage period. Although, changes depend upon the type of storage structures used. The hermetic bags keep 85 per cent or more germination in stored wheat after nine months of storage. The insect infestation decreases the carbohydrate content of the stored commodity as insects eat the endosperm. The protein content of the stored food grains generally remains the same for a period of one-year storage. The quality is also associated with the use of chemical fumigants for insect control. The use of phosphine in stored commodities is responsible for the carcinogenic chemical residues. The use of physical methods of insect control like the use of hermetic bags, the leaves of the *neem*, use of a modified atmosphere kept the quality intact and generally fetch more prices in the market. The use of hermetic bags also keeps grains brighter over the metal silos and godowns.

The losses can be classified into direct and indirect losses. The direct losses are mainly monetary losses to the stakeholders in the supply chain of the food grains due to qualitative and quantitative losses. Whereas, indirect losses are happened due to the use of low-quality food grains as human and animal food that cause health issues.

11.4 Safe Storage Moisture Content

Moisture content of the grain is basic physio-chemical variable that limits the development of bacteria, actinomycetes, yeasts, fungi, mites and insects which are the main factors responsible for spoilage of stored grain. For practical storage purposes, the maximum permissible moisture content for safe storage of food stuff should not exceed in equilibrium with relative humidity of 70 per cent. Chemical and physical changes take place more slowly in low equilibrium relative humidity. Most fungi do not develop below 70 per cent relative humidity. Insects are more tolerant to low relative humidity but most species are adversely affected to some extent.

Moisture contents of grain below 13 per cent arrest the growth of most micro-organisms and mites and moisture contents below 10 per cent limit development of most stored grain insect pests. Therefore, to preserve grain we must prevent the initial infection by drying the grain to less than 13 per cent moisture content, storing at low temperature and storing the grain in an air tight environment.

Table 11.1. Maximum Permissible Moisture Content for Safe Storage of Various Commodities

(Moisture content in equilibrium with 70 per cent RH at about 27°C)

Commodity	*Maximum Permissible Moisture Content (per cent)*
Wheat	13.5
Maize	13.5
Sorghum	16.0

Commodity	*Maximum Permissible Moisture Content (per cent)*
Millet	16.0
Paddy	14.0
Rice	13.0
Beans	15.0
Cowpeas	15.0
Groundnut (shelled)	7.0
Cottonseed	10.0
Cocoa beans	7.0
Copra	7.0
Palm kernel	5.0

11.5 Climatograph and Deterioriation Index

The graphical representation of the ambient condition of a particular location is known as a climatograph. It is used to compare the climate of different places and to predict the arriving climate by past data. The data under the climatograph include average temperature and average precipitation (rainfall and snowfall) against the time. The climatograph is the combined word made of climate and graph. The climatograph had data of more than 30 years.

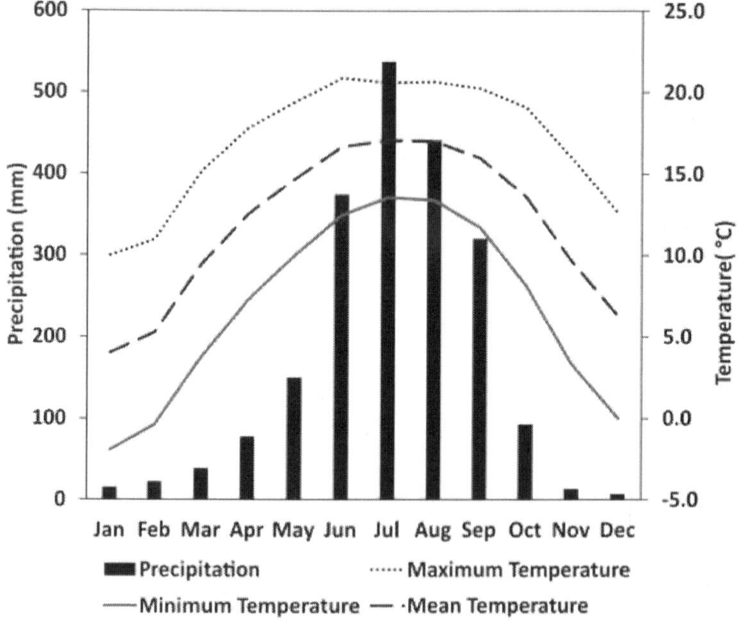

Figure 11.1. The Climatograph of Sikkim, India.

In a climatograph, temperature is usually shown as a line graph, while precipitation is shown as a bar graph. The temperature scale is typically shown on the right-hand side of the graph, while the precipitation scale is shown on the left-hand side (Figure 11.1).

Figure 11.2 suggests the temperature range for the safe storage without effect on germination, growth of microbes and insects. The moisture content of the food grains has a direct relationship with the relative humidity of the ambient. The relative humidity increases the moisture content of the food grains. Although this relationship is more valid for the non–hermetic storage structures. The equilibrium moisture content (EMC) describes the absorption and desorption of water content from/to the food grains due to changes in the ambient relative humidity which is known as the hysteresis effect (Figure 11.3).

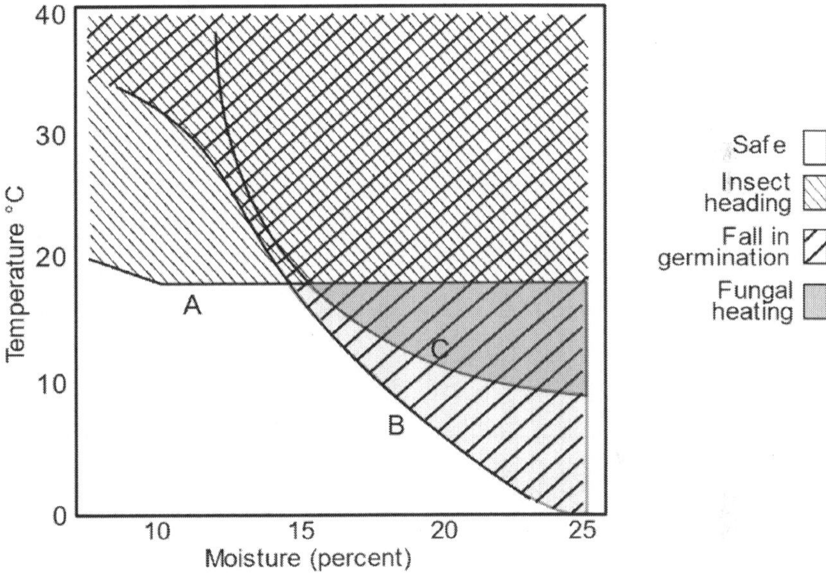

Figure 11.2. Interaction of Temperature and Moisture Content on Grain Quality Parameters (Flor *et al.*, 2022).

11.6 Moisture Migration in Bins

The changes in the ambient atmosphere affect the storage temperature and develop the convection current inside the storage bins. The moisture present in convection current condense at the top and bottom of the silo and creates damp heating or hotspots. The damp heating is the result of the higher rate of biological respiration and moisture condensation. Biological respiration produces heat, carbon dioxide and water vapour. The moist environment with heat and carbon dioxide creates a favourable environment for fungal growth and also for anaerobic degradation. The insect also attracts these hotspots for their growth. The damp heating develops the off odour in the stored grains and also changes the qualitative parameters like germination percentage, thousand kernel weight, gluten content *etc.*

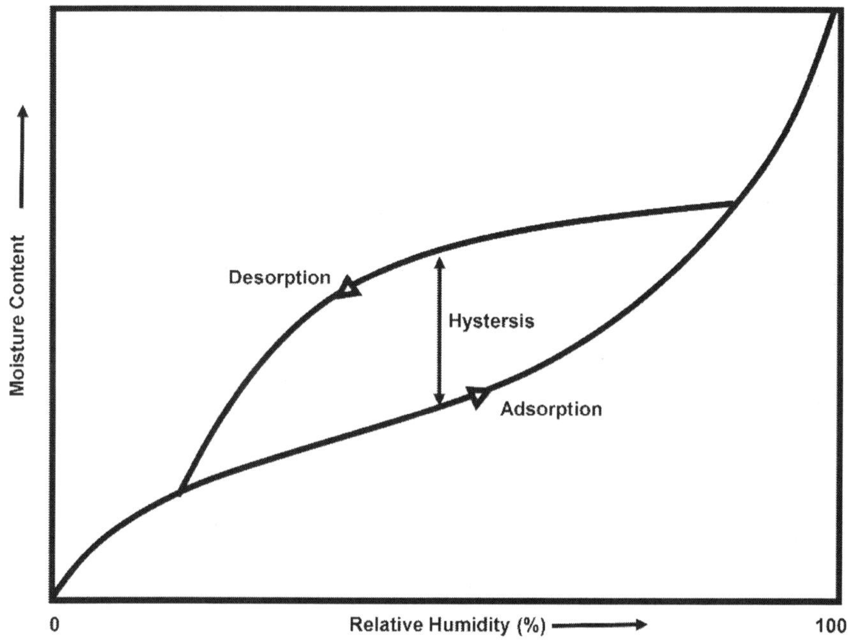

Figure 11.3. Hysteresis Effect Due to Moisture Changes in Food Grains.

There are two cropping seasons in the country *i.e. rabi* and *kharif*. *Rabi* crops are grown in winter season and harvested before the onset of summer (from March to April) whereas *kharif* crops are grown in monsoon season and harvested before the onset of winter (from September to October). Thus, the grains are fed in the bins either in warm or cold conditions.

Moisture Migration in Storage Bins when Grains are Harvested and Stored in Summer

When grains are harvested in hot, dry conditions in the summer and stored, grains at that time are warm. During winter season, the atmospheric temperature is lowered down. After passage of time the air at the surface of storage bin cools, because the outside atmosphere temperature is lower than the temperature in the center of bin. The cool air at inside surface of bin moves down along the edge of the bin, across the bottom and then upward near the center of the bin where the air and grain are warm (Figure 11.4). The cold air is denser and hence it moves downward. The air moving through the center of the bin picks up moisture from the grains and moves across the top of the bin.

At this location (top of the bin) the surface of the grain is cold and the moisture condenses on the grain. The moisture content of the grain at top surface is consequently raised. Because of increased grain moisture, spoilage takes place at top of the bin during the winter seasons.

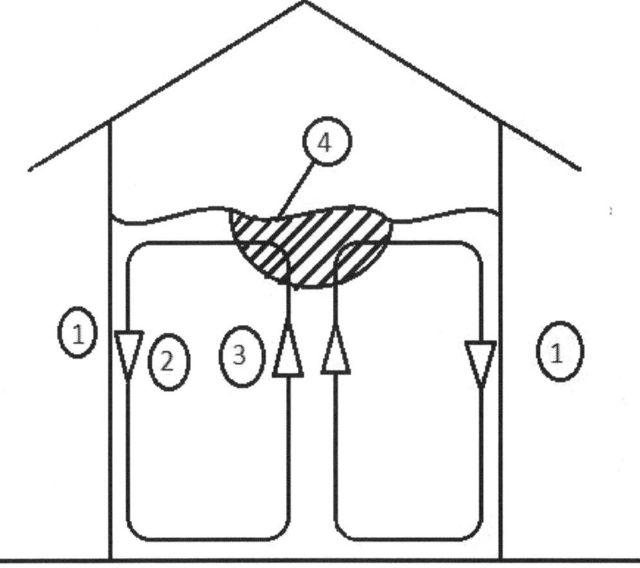

1. Ambient cool temperature; 2. Cool grain; 3. Warm grain; 4. Spot of moisture accumulation.

Figure 11.4. Moisture Migration in Winter Season.

Moisture Migration in Storage Bins when Grains are Harvested and Stored in Winter

The grains stored in winter season are cooler. In spring and later on, the atmospheric temperature starts to warm up. Thus, the bin wall also tends to warm, which in turns warms the adjacent grains. This results in the air adjacent to the bin wall warming up as well. As a result, the warm air moves upward along the circumference of the bin (Figure 11.5) The warm air is lighter and hence it rises upward. This air cools down and starts to move down through the center of the bin to its bottom. During the movement, the warm air absorbs the moisture of the food grains. The bottom of the silo is generally cooler than the top and sides of the silos. Therefore, the convection current of higher humidity gets condensed in the bottom of the bin. Due to moisture accumulation at the bottom of the bin, spoilage of grain takes place at bottom of the bin during the summer seasons.

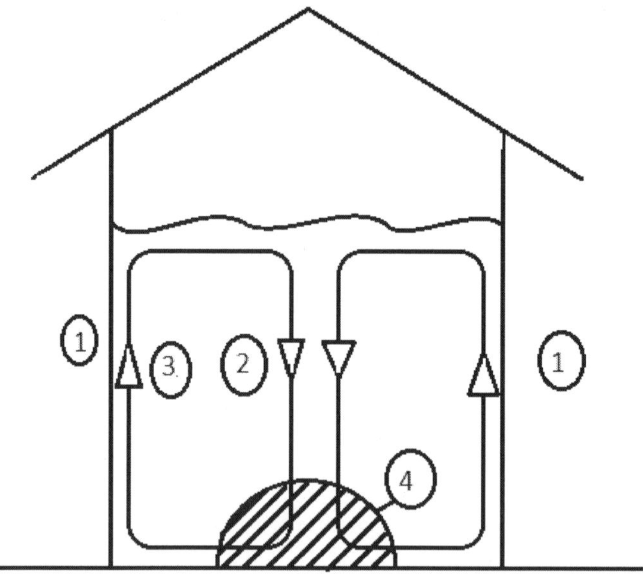

1. Ambient warm temperature; 2. Cool grain; 3. Warm grain; 4. Spot of moisture accumulation.

Figure 11.5. Moisture Migration in Summer Season.

11.7 Functional and Structural Design Considerations

Functional and structural design considerations for food grain storage structures are essential to ensure the safe and efficient storage of food grains for extended periods. The following are some critical considerations for designing such structures:

Structural Design Considerations

- ☆ The structure must be able to bear the weight of the stored food grains.
- ☆ The design must be able to withstand natural forces such as wind, rain, and seismic activity.
- ☆ The structure must be designed to protect against pest infestations and theft.
- ☆ The ventilation system must be adequate to prevent mould and moisture build-up.
- ☆ The design must be able to withstand temperature fluctuations.

Functional Design Considerations

- ☆ Access for loading and unloading food grains should be designed to minimize damage and loss.

- ☆ The design must provide adequate lighting for loading and unloading.
- ☆ The structure must be easy to clean and maintain.
- ☆ The design must incorporate safety measures to prevent accidents during loading and unloading.

11.8 Terminology

The parameters which are useful in the design of storage structures are explained below.

Bulk Density

Bulk density is defined as the weight of the grains divided by the total volume they occupy. It depends on moisture content, the amount of impurities, and degree of filling. It is one of the most important properties in the design of storage bin.

$$\text{Bulk density} = \frac{\text{weight of the grains}}{\text{total volume they occupy}} = \frac{W}{V_g + V_v}$$

where, V_g = grain volume, and V_v = void volume

True Density

True density represents the actual density of particle. True density is the weight of a kernel dividing by its volume, excluding pores. It is determined by, liquid displacement method. True density is dependent on moisture content. It is an important factor for heat and mass transfer analysis through grains.

$$\text{True density} = \frac{\text{weight of the grain}}{\text{grain volume}} = \frac{W}{V_g}$$

where, W = weight of grain, and V_g = grain volume

Apparent Density

The bulk density is increased when the grains are compressed. During this process reorientation and settling of particles take place. As a result, air percolates out of the bulk mass and the volume decreases which results in increase of the bulk density. The density of compressed mass is called the 'apparent density'. The value of apparent density is higher by 20 per cent than the normal bulk density. The apparent density of grain is taken into account in the design of a silo. Because of compaction, the bottom layer or base layer of silo has maximum density whereas the top layer has minimum density.

External Friction/Wall Friction

External friction is friction between grain mass of kernels and the wall of bin or a surface. External friction arises when two bodies in contact with each other try to move or there is an actual relative motion between the two. It is also called contact friction.

Internal Friction

Internal friction is friction of the grains against each other. It is an internal force that resists the movement between the particles of a material.

Angle of Internal Friction (ϕ)

The sliding stress between the grain to grain gives the frictional resistance per unit of surface area of the grain against the grain. This is called the angle of internal friction. It is the ability of a unit of solid material to withstand a shear stress. It is an important property which helps to estimate the lateral pressure in storage silos.

Coefficient of Friction (μ)

If 'F' is the horizontal force or pull and 'N' is the vertical force caused by the total weight of grain, the coefficient of friction 'μ' can be given by

$$\mu = \frac{F}{N}$$

Coefficient of friction between granular materials is equal to tangent of the 'angle of internal friction' for that material. Coefficient of friction depends on grain shape, surface characteristics, and moisture content. Higher values of angle of internal friction indicate that the material is cohesive. The lower values of angle of internal friction are indicative of easy flowing products.

Figure 11.6. Angle of Repose.

Angle of Repose: It is the angle between the base and the slope of cone formed on a free vertical fall of the granular material to a horizontal plane (Figure 11.6). Angle of repose depends on size, shape, moisture content and orientation of the particles. The angle of repose is low when the grains are smooth and rounded; it is high with very fine, angular or sticky particles. For homogenous mass, angle of repose is equal to angle of internal friction. In practice, the angle of repose is smaller than the angle of internal friction because the grains at the exposed surface are most loosely packed than those inside the mass.

11.9 Deep and Shallow Bins

Depending upon the relative dimensions of the container, silo/bins are classified as deep bins and shallow bins. The structure behaves as a deep or a shallow bin is decided from the concept of plane of rupture. The plane of rupture is that surface down which a wedge of material bounded by one wall face, the free surface, and the plane of rupture would start sliding if bounding wall were to move. A bin whose relative dimensions are such that the plane of rupture meets the grain surface before it strikes the opposite side wall is called 'shallow bin' (Figure 11.7). A bin in which plane of rupture meets the opposite side wall before it emerges from the grain is called 'deep bin' (Figure 11.8).

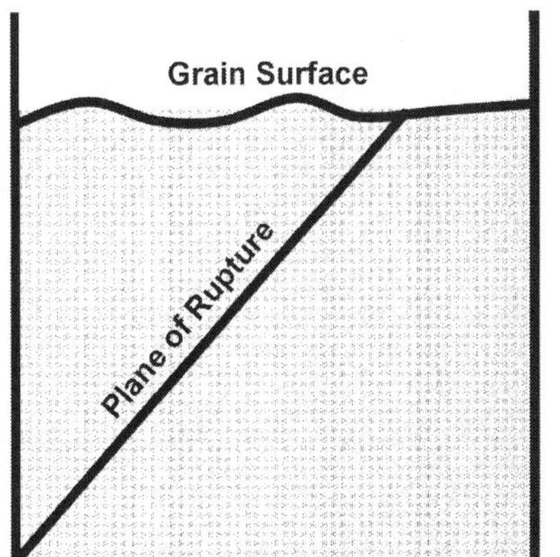

Figure 11.7. Shallow Bin.

Figure 11.8. Deep Bin.

The difference between the deep and shallow bin is tabulated as under:

Sl.No.	*Principle*	*Shallow Bin*	*Deep Bin*
1.	Plane of rupture	When the plane of rupture strikes on the grain surface in the case of removing one of the sides of the bin	When the plane of rupture strikes the wall in case of removing one of the sides of the bin
2.	Hydraulic radius	$h < 4R$	$h \geq 4R$
3.	Rankine theory	$h < D \tan (\frac{90+\phi}{2})$	$h \geq D \tan (\frac{90+\phi}{2})$
4.	Issacson and Boyd	$\frac{h}{D} < 0.75 (\frac{1}{\mu k})$	$\frac{h}{D} \geq 0.75 (\frac{1}{\mu k})$

Example 11.1

A cylindrical grain bin of height 3 m and diameter 2 m is to be designed to store grains which have angle of repose of 36°. Find out whether the grain bin is a shallow or a deep bin?

Solution

Angle of repose (ϕ) = 36°

$$\tan\left(\frac{90+\phi}{2}\right) = \tan\left(\frac{90+36}{2}\right) = 1.963$$

$$\frac{h}{D} = \frac{3}{2} = 1.5$$

Since, $\frac{h}{D} < \tan\left(\frac{90+\phi}{2}\right)$, the bin is shallow.

11.10 Food Grain Storage Structures and its Requirements

The requirements of good storage structures are:

1. Storage structures should have adequate strength and capacity.
2. These should provide adequate protection from rodents, birds, insects, mites and other harmful microorganisms.
3. These should facilitate proper cleaning and should be self-cleaning, if it is a silo.
4. These should permit aeration and fumigation when required.
5. These should prevent losses due to moisture and temperature.
6. These should permit easy inspection.
7. These should have efficient loading and unloading systems.
8. These should be economical on unit storage cost basis.

Sl.No.	*Type of Storage System*	*Name of Storage Structures*
1.	Farm level	CAP, Silo Bags, Bunker storage
2.	Bag storage	Warehouse, CAP, Controlled atmosphere storage (CAS), modified atmosphere storage (MAS)
3.	Bulk storage	Bunker, Tower silo, silo bags (hermetic bags),
4.	Domestic level	Metal silos, traditional storage structures
5.	Temporary	CAP, bunker storage, silo bags (hermetic bags)
6.	Permanent	Warehouse, Controlled atmosphere storage (CAS), modified atmosphere storage (MAS), Tower silo
7.	Outdoor	CAP, Silo Bags, Bunker storage, Tower silo
8.	Horticultural	Controlled atmosphere storage (CAS), modified atmosphere storage (MAS)

Food grain structures are classified as traditional storage structures, improved storage structures and modern storage structures. The details of each type of structure are presented below:

11.10.1 Traditional Storage Structures

The traditional structures used for storage of food grains are of different type, shape, material, and capacity. These are small buildings or structures used to store grain in rural and remote parts of India. These structures are made of the locally available materials like wood, bamboos, clay, straw, jute bags, bricks *etc.* Some of them are raised above the ground to protect the grain from moisture and pests. In some areas, pits are dug in the ground and used to store grain. The pits are lined with straw or other materials to prevent moisture due to seepage and spoiling the grain. This method helps keep the grain cool and dry and protects it from pests and other environmental factors. The traditional structures need renovation every year or before use. These structures are also not able to protect grains from moisture, insects and rodents. The shade is required for these traditional structures to save them from the direct impact of rain. It can store grains only for a short period.

The region-wise list of traditional storage structures used in India is as follows:

- **North India**: *Kaccha kothi/Mud bin, Kuthle, Bharola, Peru, Theka, Kotha, Petti, Kothar, Bukhari, Khatti*
- **East India**: *Dhana, Ollya, Amar, Dholi, Ghama, Khari, Basta, Kachha kothi, Bohuri/Kunda, Mora, Muda, Kuthle, Gota, Pool, Jabbar*
- **West India**: *Obri/Kuthal/Kuthla, Bukhari, Khani, Thakka, Gare ki kothi, Dholi/Kanagi, Kothi/Kangi/Ranjan, Pev, Kanaja*
- **South India**: *Puri, Kotlu, Gade/Gum/Borem, Jadi/Kunda/Golem, Pathara, Basta/Sanchi, Seru, Kottal, Kalanjiam, Pathayam, Kudir, Thombe, Jadi, Koni, Kanaja, Panatha, Ambarkani, Hagevu, Goni Cheela, Ara, Vallam, Mankalam*
- **Central India**: *Bindi, Mud bin, Kuthla, Banski Bindi, Banda, Dholi*
- **North East India**: *Bhorai, Duli, Topa/Tom, Mer, Gam, Bakash, Bharal, Dekuti, Dola*

The details of some of traditional structures are provided as below.

Morai Type Storage Structure

Morai type of structures is used in the rural areas of eastern and southern regions of India for storing of food grains. The structure is locally called as 'Puri' in Andhra Pradesh, 'Morai' in West Bengal, 'Tom' in Assam, and 'Mora' in Bihar. The capacity of this storage structure ranges from 3.5 to 18 tonnes. The shape of the structure is similar to inverted truncated cone.

Construction details: The improved Morai storage structure consists of a circular wooden plank floor supported on pillars by means of timber joints. Pillars made of wood or masonry is used to support the structure. A 22 gauge corrugated metal cylinder of 90 cm height is placed all around the wooden floor and fixed with the help of nails. Ropes (7.5 cm dia) made of paddy straw or similar material is placed inside the cylinder up to a height of 90 cm from the floor level. Thereafter, the bamboo splits (equal to the height of the structure) are placed vertically along the inner surface. No gap is left between the ropes and bamboo splits. Then, grain is poured inside the structure up to the height of metal cylinder which helps in keeping the bamboo splits in erect position. Thereafter, the winding of the rope as well as the pouring in of grain are done simultaneously till the grains are filled up to the required height. About 1 cm thick layer of mud plaster is applied over the rope. The structure is then covered with conical roof with ample overhang all around. Rat proofing cones are provided on all the posts to avoid entry of rats in the structure. The Figure 11.9 depicts the cross-section of a Morai type grain storage structure.

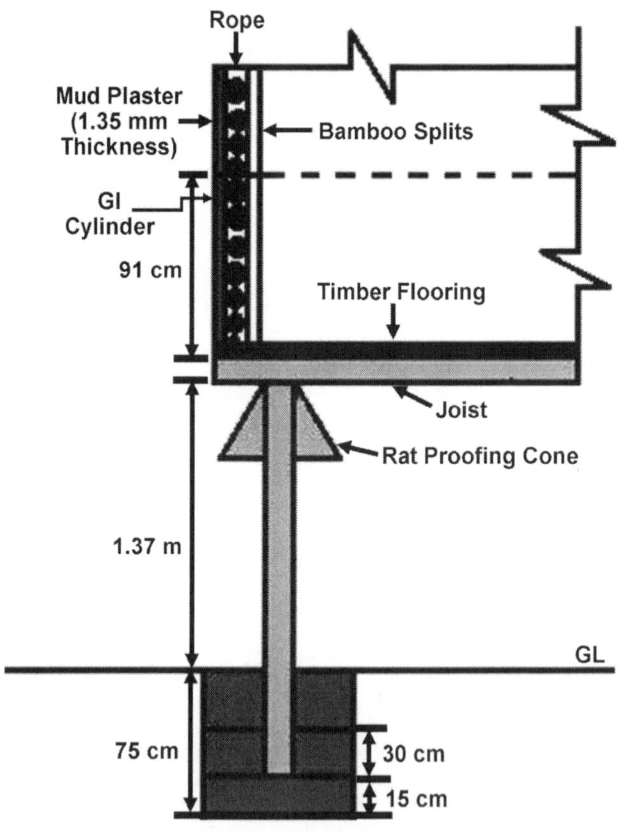

Figure 11.9. ***Morai*** **Type Storage Structure.**

Kothar Type Storage Structure

Kothar type storage structures are used in the rural areas of Jammu and Kashmir, Himachal Pradesh and Uttar Pradesh for the storage of wheat, paddy, maize, sorghum, and barley. The capacity of these storage structures ranges from 9 to 35 tonnes.

Construction details: *Kothar* type of storage structures are made of wooden planks. The roofs are made of thatch or tiles to protect the food grains from the sun and rains. In the improved storage structure, 5 cm thick wooden plank and wooden beams are used. The walls and floor are so made that no gap exists between the planks. The gabled roof on the top may either be made of planks or 22 gauge corrugated metal sheet having sufficient overhang on all sides. Roof of metal sheet is preferred to planks. The structure should be moisture resistant as well as free from the danger of moisture rising from the ground surface. Rat proofing cones are provided on all the posts to avoid entry of rats in the structure. The Figure 11.10 depicts the cross-section of a *Kothar* type grain storage structure.

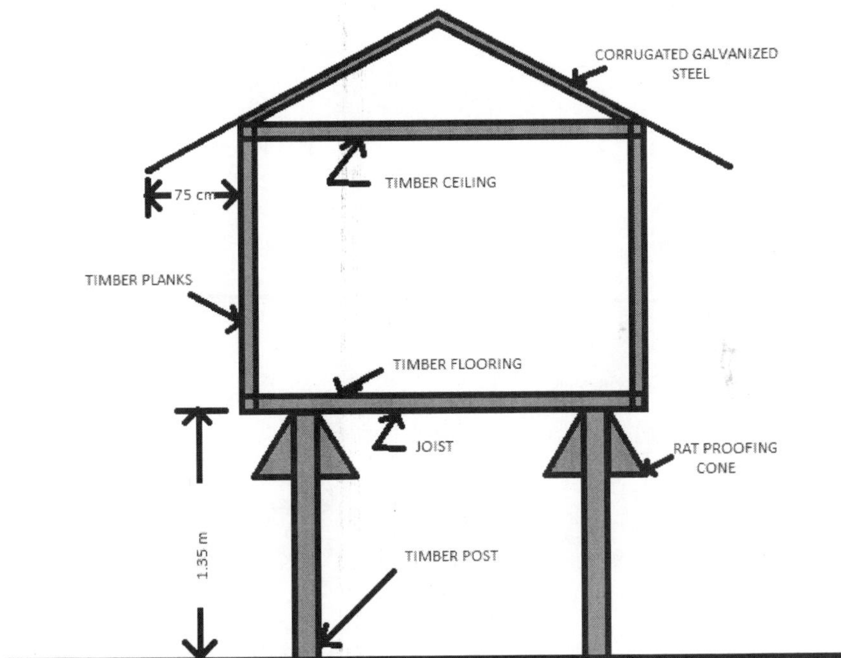

Figure 11.10. *Kothar* Type Storage Structure.

Bukhari Type Storage Structure

Bukhari type storage structures are generally used for storing of wheat, paddy, gram, maize and sorghum *etc.* Their capacity generally ranges from 3.5 to18 tonnes. For quantity less than 3.5 tonnes, the small size structures are also used.

Construction details: *Bukhari* type storage structure is cylindrical in shape and is made of mud alone or combination of mud and split bamboo. The whole structure is place on a wooden or masonry platform to avoid direct contact with the ground. The floor of the storage structure is constructed of either timber planks or bamboo splits, plastered over with mud mixed with cow-dung and paddy straw. The walls are made of timber or wooden frame work and bamboo matting, over which mud straw plaster is applied on both inside and outside surfaces. The walls are further strengthened by timber posts which are fixed in the ground, extending up to full height of walls. A cone type roof made of bamboo frame work and paddy or similar straw is provided over the structure. Sufficient overhang is provided all around to protect it from rains.

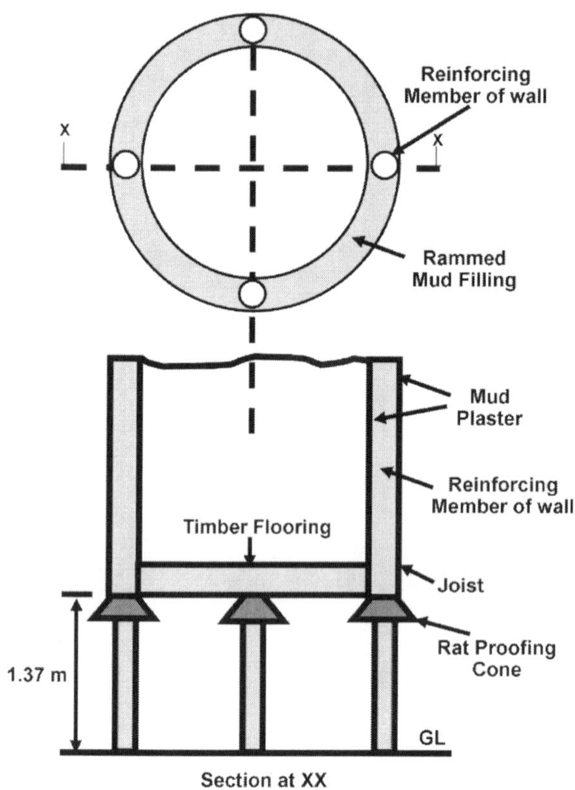

Figure 11.11. *Bukhari* Type Storage Structure.

An improved storage structure is similar in design except method and material used in construction to enable the structure more durable and safe. The floor of the structure is either made of wooden planks joined by lap joints or by a double layer of bamboo splits closely set at right angles to each other. About 5 cm thick mud plastering is done over the floor. The walls are made of two sets of strong bamboo frame work and inter-space is filled with mud. Mud plaster is applied on both sides of the walls. A conical roof made of bamboo frame-work is covered with thatching material like paddy straw. The top of conical roof is covered with 4 to 5 cm thick mud layer to provide protection from rains.

The whole storage structure is raised on timber or masonry pillars. Rat proofing cones are placed on the pillars to avoid rats entering the storage structure. The Figure 11.11 depicts the cross-section of a *Bukhari* type grain storage structure.

Mud *Kothi* or *Katcha Kothi*

These are indoor structures of 2 to 5 quintals capacity. The structures are usually circular in shape. The *katcha kothi* is popular in the states of Bihar, Uttar Pradesh, Punjab and Madhya Pradesh.

Construction details: The mud *kothi* is made of clay mixed with wheat straw, paddy straw or husk. The floor and the bottom portion of the wall section is constructed in the form of rings which can be put one over the other so as to make the structure of desired capacity. Inside and outside surfaces of the structure are plastered with mud and cow-dung (Figure 11.12).

Figure 11.12. Mud *Kothi*.

11.10.2 Improved Storage Structures

Improved grain storage structures of storing grain have been developed to overcome the limitations of traditional storage methods. Some examples of improved grain storage structures include Pusa bin, brick and cement bin, metal bin, CAP, bunker storage *etc.*

Bins are made from steel or aluminium/cement with bricks and are designed to be more durable and resistant to pests and moisture. Metal bins are designed to prevent air and moisture from getting inside up to a certain extent, which can help keep grain fresher for longer periods.

Overall, improved grain storage structures are designed to be more efficient, durable, and effective at preserving grain for long periods. Pusa bins, metal bins and cement bins are often used in rural and urban areas for domestic and on-farm storage at the farmer level. The Cover and plinth (CAP) is being used by the FCI, CWC and private industries for temporary storage of food grains. A bunker with a plastic liner is also a semi-permanent and temporary storage structure and is used for bulk storage of grains in farms.

Pusa Bin

Pusa bin (Figure 11.13) is one of the most suitable structures for storing food grains in rural areas. The bin has been developed at the ICAR-Indian Agricultural Research Institute (IARI). It is a low cost improved storage structure. The materials requirement for its construction of bin is unburnt bricks (*Katcha* bricks), burnt bricks (*Pucca* bricks), polyethylene film, and a small pipe (9 cm diameter) made out of galvanized iron or plastic for outlet of grains.

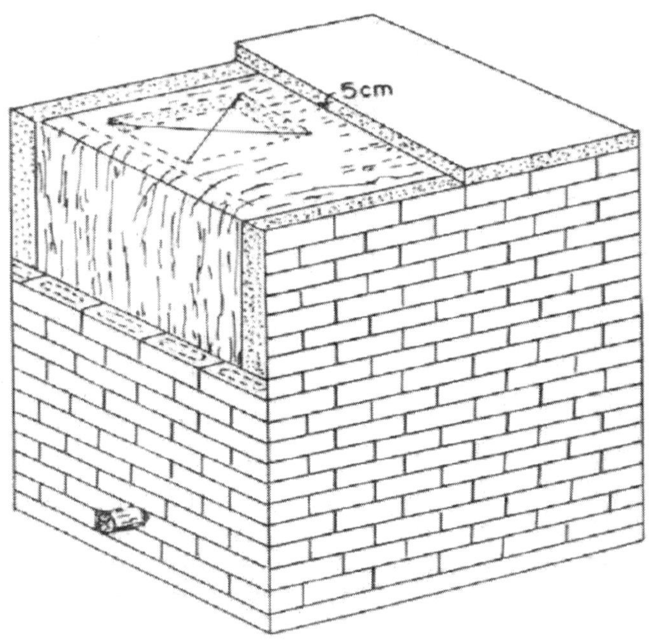

Figure 11.13. Isometric View of Pusa Bin.

Construction details: The Pusa bin is constructed on a hard surface to prevent rodent attack. First step is to make a platform of burnt bricks. The LDPE film (700 gauges) is spread over the platform, extending at least 6 cm from all four sides. Another platform of unburnt bricks is constructed over the LDPE layer. Pusa bin is comprised of two walls of bricks in which 700 gauge (175 micron) LDPE (low density polyethylene) film is sandwiched. LDPE film acts as moisture barrier and does not permit transmission of moisture from outside the atmosphere to grains. Unburnt bricks are used for erecting the inner walls to the required height depending on the capacity of the structure. Unburnt bricks/mud bricks has good insulation property and thus keep the grains cool. To protect the structure against rats, outer walls up to 45 cm is constructed with the help of burnt bricks. Rest of the portion of outer walls is made of unburnt bricks. A wooden frame is placed on the top of the inner walls to support the roof. A mud slab (5 cm thick) is placed over the inner walls to serve as roof. An open space of 50 cm×50 cm is left for loading of grains while an inclined plastic or steel pipe is fixed at the bottom of bin for delivery of grains. The surfaces of the walls and roof are then plastered with mud. The bins can be made of various capacities as per the requirement. The quantities of raw materials required for construction of Pusa bins of different capacity are given in Table 11.2.

Table 11.2. Raw Material Requirements for Construction of Pusa Bins

Sl. No.	*Capacity*	*Internal Dimension (cm)*			*Quantities of Material Required*		
		Length	*Breadth*	*Height*	*LDPE Film (m^2)*	*Unburnt Bricks (Number)*	*Burnt Bricks (Number)*
1.	500 kg	61	61	136	8.1	600	80
2.	1000 kg	84	84	140	11.7	900	95
3.	2000 kg	140	100	160	15.3	1150	100
4.	4000 kg	172	152	160	20.9	1650	210

Brick and Cement Bin

The walls are made of brick and plastered with cement mortar. The combination of these materials provides an excellent environment for preserving the quality and freshness of grains, ensuring they remain safe for consumption over extended periods. The roof is made of RCC, and the slope is provided for ease of unloading. The thick walls of these bins offer insulation against temperature fluctuations, helping to maintain a relatively constant internal climate that discourages mould growth and insect infestations. When constructing a brick and cement bin for grain storage, it's essential to consider factors such as location, size, and design. Proper placement away from direct sunlight and water sources is crucial. The size should be determined based on the quantity of grains you intend to store, and the design should incorporate elements that facilitate easy loading, unloading, and inspection. Regular maintenance is key to ensuring the longevity of these storage structures. Periodic checks for cracks, leaks, and signs of degradation should be carried out. The capacity of such bin varies from 1.5 to 60 tonnes.

Bunker Storage

Bunker storage involves creating large, sunken storage areas in the ground or utilizing existing depressions in the landscape for storing grains. This method offers several advantages for preserving the quality and quantity of grains over time. One of the primary benefits of bunker storage is its capacity to handle substantial quantities of grains (40 to 1600 tonnes), making it suitable for bulk storage needs. The natural insulation provided by the earth helps maintain a relatively stable temperature within the bunker, reducing the risk of temperature-related spoilage. Additionally, bunker storage minimizes exposure to sunlight, moisture, and external contaminants. This reduced exposure significantly contributes to preserving the grains' freshness and nutritional value. Bunkers are covered with materials like tarpaulins (plastic sheet) to provide further protection against the elements. Plastic sheets are covered with soil at their edges. Bunkers are made moisture proofing from the bottom side.

Cover and Plinth (CAP) Storage

The word 'CAP' is used for cover and plinth, plinth from the bottom and cover from the top. This type of open storage is considered as transit storage and serves the purpose of storage of food grains in bags for short periods. It is the most economical storage structure and is being widely used by the Food Corporation of India (FCI) and other procurement agencies for bagged grains. This type of storage is cheaper as compared to conventional bag storage godowns.

CAP is an improvised arrangement for storing food grains in the open, generally on a plinth (constructed on a high elevated ground) which is damp and rat proof. The grain bags are stacked in a standard size on wooden sleepers with one or two layers of bamboo mat cover on the top as dunnage. The stacks are covered with 250 microns (1000 gauge) LDPE sheets from the top and all four sides to protect grains against rain and sun. The cover is tied to the stack with the help of plastic ropes to prevent from blowing off with high velocity wind. The gutters are provided all around the area to drain off the rain water easily. Food grains such as wheat, maize, gram, paddy, and sorghum are generally stored in this type of storage for 6-12 month periods.

Normally the stack is built over a space of 9.11 m × 6.1 m with a height of 18 bags which gives the storage capacity of about 150 tones. The LDPE cover having a dimension of 9.4 m × 6.4 m × 5.5 m normally weighs around 52 kg.

Indoor Steel Bins/Metal Bins

Metal bins/steel bins (Figure 11.14) are made of galvanized steel. These bins are suitable for storage of food grains. The bins with capacity up to 2 m^3 can be kept indoors.

Construction details: The thickness of metal sheet used for wall, top and bottom plates, inlet and outlet shall be 0.63 mm (24 gauge). For the construction of main body of the bins, the sheets of the size 2.50×0.90 m shall be used. The thickness of sheet shall be 1.0 mm (20 gauge), 1.60 mm (16 gauge) respectively for clasp rings and clasps. A circular inlet of 225 mm diameter or 450 mm diameter is fitted to the top plate. A horizontal or incline outlet of 150 mm diameter is provided in wall panel at the base. The volume and dimensions for steel bins shall be as given in Table 11.3. The dimensions of the bins have been worked out taking into consideration the available sizes of sheets in the market and minimum of wastage.

Table 11.3. Volume and Dimensions of Bins

Sl.No.	Volume of the Bin (m^3)	Diameter (mm)		Quantity (approx. in kg)		
		Height	Diameter	Wheat	Paddy	Maize
1.	0.42	900	769	315	230	298
2.	0.68	990	938	510	374	483
3.	082	1785	769	615	451	582
4.	1.35	1950	938	1013	743	959

Source: IS: 7147 (Part – I) – 1973.

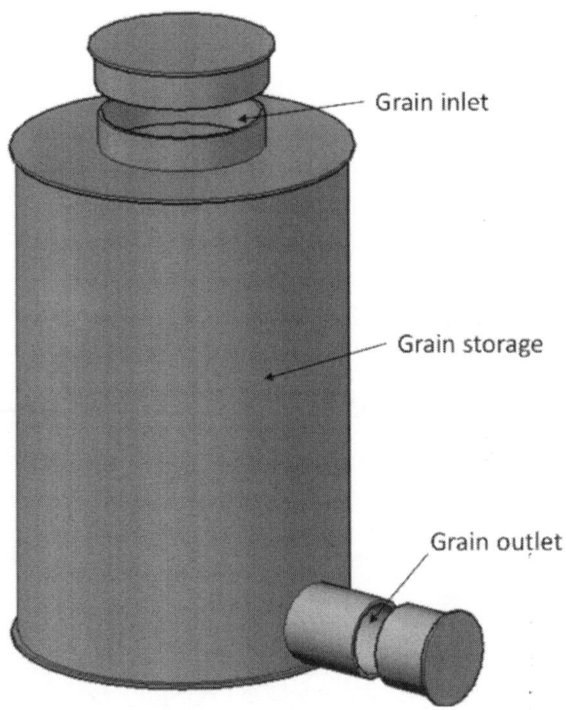

Figure 11.14. Metal Bin.

Ferro-cement Bins

The Structural Engineering Research Centre, Roorkee has developed ferro-cement bins of 1, 2 and 3 tonnes capacity for storage of food grains. Ferro-cement is a highly versatile form of reinforced mortar in which closely spaced layers of wire mesh reinforcement are impregnated with rich cement mortar mix. These bins are strong, light weight, durable, and are damp-proof, rodent-proof, and fire-proof. These structures need no maintenance and can be easily cast. The ferro-cement bins are cylindrical in shape. There is no corrosion problem in this type of bins as posed by steel structures, and also the crack widths are much less as compared to reinforced concrete. The ferro-cement bins are cheaper compared to steel bins, reinforced bins, and aluminium bins.

Construction details: The ferro-cement bins are assembled out of prefabricated units *viz.* reinforced concrete base slab, ferro-cement wall unit, and dome-shaped ferro-cement roof unit. A wall unit has a diameter of 1.2 m and a height of 1.0 m. Depending on the reinforcement, 1, 2, 3 tonne bin may be assembled by erecting one, two or three wall units one over the other and filling up the joints with cement mortar. A manhole is provided in the roof unit for loading and an outlet is provided in the bottom most wall unit, for unloading the grains. Rubber gaskets are provided both for inlet and outlet openings to make the bins air-tight. Provision of locking is also provided. The external surface of bins is painted with two coats of bituminous aluminium paint to make them damp proof.

11.10.3 Modern Storage Structures

Modern grain storage structures refer to the latest advancements in technology and design for the storage of grains, often used in large-scale commercial operations. Some examples of modern grain storage structures include silos (tower, bunker and bags), Controlled atmosphere storage (CAS), modified atmosphere storage (MAS), warehouses (or godowns), hermetic storage *etc.*

Silo Types, Location, Filling and Packing, Size and Capacity

Silos are the modern storage structures for food grains. Tower silo, bunker silo and silo bags are three types of silos used based on their way of construction and manufacturing material.

Tower silos are elevated structures over the ground and are generally made of concrete, metals and sometimes wood. Now a day, metal silos are preferred over concrete due to investment and the dead weight of the concrete. This silo takes less space but it is a permanent structure. As per the Department of Food and Public distribution, modern silos of 50,000 tonnes capacity are available in India.

Bunker silos are horizontal trenches made on the ground. The grains are filled inside the trench and covered using a plastic cover. The plastic cover is restrained over the trench using the soil. The initial investment is very low in comparison to the tower silos. The trench of the bunker is a permanent underground structure made of concrete, although frames of stainless steel, wood and plastic are available in the market to develop temporary bunkers. A bunker silo can be used for the storage of the silage also. The bunker is generally

6-18m in width and 12-60m in length with a capacity between 40-1600 tons of food grains.

Silo bags are flexible and big PVC bags used for grain storage in the field like a bunker (half on the ground and half will be inside the soil) or on the ground. The initial investment for the silos bags is comparatively very low than the tower and bunker silos. It is a temporary structure erected in the catchment of food grains as per the requirement for longer and shorter periods of storage. Although, these silo bags are susceptible to damage by pests and rodents. Suitable arrangements for avoiding rodent attacks need to be done for successful storage. Silo bags of big capacity is also known as cocoons and are of 1000 tonnes capacity.

The silos are generally constructed or erected in the catchment area of the food grains. It saves the transportation cost of the raw material from the field to the silos. The silos are also treated as collection and primary processing centres for agro commodities. The collected grains are cleaned, dried and stored for the further use of government, private industries, export *etc.* The site of the silo needs to be cleaned, well-drained and connected to the main roads. The site should have proper arrangement of utilities like electricity, water, security, sewer *etc.* for the staff of the silo operation and maintenance. Electricity is also required in managing the automated tower silos for fumigation and aeration, loading and unloading, and operation of sensors and actuators.

Tower silos generally have automated loading and unloading facilities. The unloading system of the tower silos uses gravity for unloading whereas conveyors are used for loading of silo. In the case of the bunker silo and silo bags, trucks or trolleys with special unloading chutes or grain stackers are used for filling the bunker. Grain scoops or grain sweepers are generally used for unloading bunkers and silo bags. The grain scoop is a screw conveyor with an unloading arm operated by the electric motor.

Hermetic Storage

The airtight storage of the food grains develops a carbon dioxide enriched environment due to the respiration of food grains. The modified atmosphere controls the growth of insects without the use of any fumigant. It is a physical and organic way of insect control. The hermetic storage is suitable for low moisture durable commodities like food grains. The food grains should have moisture below 12 per cent for longer storage inside hermetic bags. The multilayer laminates of HDPE, flexible PVC, and EVOH laminates are the plastic materials used for the development of flexible bags for food grains. Hermetic bags of 1000 tonnes capacity is also available in the market. The hermetic bags are susceptible to puncture by rodents, during handling and transportation, and pest attacks therefore regular monitoring of the bags is critical for the maintenance of the modified atmosphere. Although, instrumentation can be used for the detection of puncture by regular monitoring of the interstitial atmosphere and generation of alerts for corrective actions. The equilibrium moisture content of grains kept inside the hermetic bags is comparatively stable in comparison to other non-hermetic storage systems. There are three variants of hermetic storage being used in the world *i.e.* organic hermetic storage, vacuum hermetic fumigation and gas hermetic fumigation.

Organic hermetic storage is about keeping the food grains inside the hermetic bags and respiration consumes the available oxygen and developed the modified carbon dioxide enriched environment. In vacuum hermetic fumigation, the vacuum is used as a fumigant to control the insects. The vacuum of 100 mm Hg was found sufficient to kill all forms of insects. Although, the exposure time for mortality differs with the forms of insects. In gas hermetic fumigation, the gasses like carbon dioxide, biogas *etc.* are utilized to kill the insects. The interstitial atmosphere is modified by the desired gas(es) with specific concentrations or gasses combinations that kill the insects. The exposure time for the complete mortality depends upon the gasses used and their combination. The effectiveness of the gas hermetic fumigation can be increased with the manipulation of the temperature of the interstitial atmosphere. The vacuum and gas hermetic fumigation, the mortality of the insects is comparatively fast than the organic hermetic storage. Due to the cost of creating a vacuum and using gasses, vacuum hermetic fumigation and gas hermetic fumigation are costly and reduce the life of the hermetic bags.

Overall, modern grain storage structures are designed to be more efficient, automated, and flexible than traditional or improved storage structures. They are often used in large-scale agricultural operations and require significant investment in technology and infrastructure.

11.11 Grain Pressure Theories

The pressure exerted by stored grains can impact the structural integrity of these storage systems, so understanding grain pressure is crucial for their safe design. There are three primary theories related to food grain pressure.

11.11.1 Rankine Theory

Rankine theory, also known as Rankine's earth pressure theory, is a concept that relates to the lateral pressure exerted by granular materials against retaining walls. Although originally developed for soil mechanics and geotechnical engineering applications, it can also be adapted to the design of storage structures like silos and bins, which deal with lateral pressures from stored grains. Rankine's theory provides formulas to calculate the lateral pressure exerted by the granular material at different depths. The pressure distribution is not uniform and depends on factors like the value of coefficient of friction for the grains, the coefficient of friction between the solids and the silo wall, and the way solids are placed in the silo.

The grain pressure in bins were first calculated as being a semi liquid of same density as the grains. The lateral pressure was calculated using the hydrostatic formula.

$P_l = \omega h$

where,

P_l = lateral pressure exerted by the grain on the bin wall, kg/m^2,

ω = density of grain, kg/m^3, and

h = depth of grain from the top of the bin surface, m.

The above formula has serious deficiencies as many structures buckled under the vertical load arising from the friction of the grain on the walls. The formula was modified to incorporate a factor 'K ' known as Rankine's earth pressure coefficient, and this is known as Rankine's formula.

$P_l = K\omega h$

where, $K = \dfrac{1-\sin\phi}{1+\sin\phi}$

and, ϕ = angle of internal friction of the grain

Rankine's formula reduced the magnitude of the lateral pressure, but did not take into account the vertical load on the wall resulting from friction effect. The whole weight of the grain was assumed to be transferred to bottom of the beam.

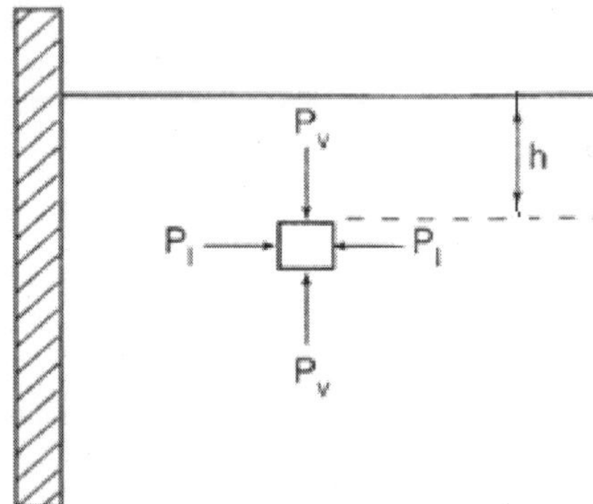

Figure 11.15. Pressures on an Element of Soil behind a Vertical Wall.

The Rankine theory with horizontal soil surface behind a vertical wall is shown in Figure 11.15. Consider an element of soil at depth 'h'. The vertical pressure acting on this element 'P_v' and the lateral pressure 'P_l' is active pressure at failure. By constructing Mohr circle, we get

$$P_l = \left(\frac{1-\sin\phi}{1+\sin\phi}\right)P_v = KP_v$$

$$K = \frac{P_l}{P_v} = \frac{1-\sin\phi}{1+\sin\phi} = \frac{\text{lateral pressure}}{\text{vertical pressure}} = \tan^2\left(45-\frac{\phi}{2}\right)$$

The above Rankine formula is used for determination of pressure induced by granular materials against retaining wall in shallow bins. The Rankine coefficient is the ratio of lateral and vertical pressure experienced by the silo filled with the grains.

Example 11.2

For a mass of grain stored in a bin, the angle of internal friction of the grain is 30°. What will be the ratio of normal pressure to the applied pressure within the bin? (GATE 2011)

Solution

Ratio of normal pressure to applied pressure, $K = \frac{P_l}{P_v} = \frac{1-\sin\phi}{1+\sin\phi} = \frac{1-\sin 30}{1+\sin 30}$

$$= \frac{1-0.5}{1+0.5} = 0.33$$

Example 11.3

A steel bin 3 m deep and 3.6 m in diameter is filled with wheat. Calculate the lateral pressure at bottom of the bin. The angle of internal friction of the grain is 24°. (ARS, 1985)

Solution

Diameter of silo (D) = 3.6 m

Height of silo (h) = 3 m

Angle of internal friction of the grain (ϕ) = 24°

Let the bulk density of wheat (ω) = 800 kg/m^3

The bin is shallow.

According to Rankine's formula

Lateral pressure exerted by the grain on the bin wall (P_l) = $K\omega h$

$$K = \frac{1-\sin\phi}{1+\sin\phi} = \frac{1-\sin 24^o}{1+\sin 24^o} = \frac{1-0.4067}{1+0.4067} = 0.4217$$

$P_l = 0.4217 \times 800 \times 3 = 1012\ kg_f/m^2$

11.11.2 Airy Theory

The scientist Airy developed a theory used to estimate the pressure exerted by granular materials, such as food grains, on a storage container. The Airy theory assumes that the grains in a storage container are in a state of equilibrium, where the forces acting on each grain are balanced. According to the theory, the pressure exerted by the grains on the container walls is proportional to the square of the depth of the grain column. The Airy theory is a simplified model that assumes that the grains are homogeneous, mono-disperse, and incompressible. The lateral pressure exerted by grains in a shallow bin can be given by following Airy's equation:

$$P_l = \omega h\left[\frac{1}{\sqrt{\mu(\mu+\mu')}+\sqrt{(1+\mu^2)}}\right]^2$$

μ = coefficient of friction between the grains = tan ϕ,

μ' = coefficient of friction between the grains and bin wall= tan ϕ',

ϕ = angle of internal friction,

ϕ' = angle of wall friction,

ω = bulk density of grain, kg/m^3, and

h = depth of grain, m

Example 11.4

Wheat weighing 900 kg/m^3 is loaded in circular concrete silo of 3 m internal diameter and a clear height of 8 m. The angle of internal friction for wheat is 25° and that for wheat and concrete is 24°. Applying Airy's theory, calculate the maximum lateral pressure at the bottom of bin section.

(ARS 1984 and 2006)

Solution

Unit weight of wheat (ω) = 900 kg/m^3

Height of bin (h) =8 m

Coefficient of internal friction for wheat (μ) = tan25° = 0.466

Coefficient of internal friction for wheat and concrete (μ') = tan24° = 0.445

Lateral pressure at the bottom of bin (P_l) using Airy's formula,

$$P_l = \omega h\left[\frac{1}{\sqrt{\mu(\mu+\mu')}+\sqrt{(1+\mu^2)}}\right]^2$$

$$= 900 \times 8 \times \left[\frac{1}{\sqrt{0.466(0.466+0.445)} + \sqrt{1+(0.466)^2}} \right]^2$$

$= 2338.2 \text{ kg}_f/\text{m}^2 = 2338.2 \times 9.8 = 22914 \text{ N/m}^2$ (Pa)

11.11.3 Janssen Theory

The Janssen (1895) formula is widely used for the design of deep bins to calculate the pressure on the walls at different depths. The key idea behind Janssen's theory is that when granular material is stored in a container, the pressure exerted by the material on the walls of the container increases with depth until it reaches a point of equilibrium. This pressure increase is due to the weight of the material above pressing down on the lower layers.

Bins designed by this formula is safe because of higher factor of safety (3 to 4 times high). Janssen took into account the friction between the grain and bin wall. The main assumptions of Janssen's theory are that the material is uniform in texture, has a definite angle of repose, and that the coefficient of friction between material and side wall is constant. It also assumes that the mass and size of the particles in the material are constant. The theory also assumes that the grains are not in motion and that there is no friction between the grains. This theory cannot predict the emptying pressure while unloading the grains.

The vertical pressure on the bin bottom increases with the level of filling up to certain value. Beyond this level of filling, additional solids have no effect on the pressure at the base. The total mass, of course, increases if more solids are added, but the additional mass is carried by the walls and foundation, not by the floor of the silo. At this level of filling, the vertical pressure attains a maximum value and the pressure becomes constant.

Derivation of Janssen Equation

Let

dh = thickness of elementary layer considered,

h = depth of grain,

P_v = vertical intensity of pressure acting at the top of the layer,

$(P_v + dP_v)$ = vertical intensity of pressure acting on the bottom of the layer,

P_l = lateral (or horizontal) pressure,

ω = unit weight of grain,

R= hydraulics mean radius = $\dfrac{\text{cross-sectional area}}{\text{perimeter}} = \dfrac{A}{P} = \dfrac{D}{4} = \dfrac{r}{2}$,

D = diameter of the bin and, r is the radius of the bin,

A = cross-sectional area of material stored,

P = interior perimeter of bin,

$K = \frac{\text{lateral pressure}}{\text{vertical pressure}} = \frac{P_l}{P_v}$,

μ = coefficient of friction between the grains and bin wall = tan ϕ, and

ϕ = angle of friction on the walls of bin.

During storage, the lateral pressure increases constantly while the bin is being filled. The lateral pressure has to be taken by the bin walls. For a particular depth 'h', the total lateral wall load 'P' per unit of perimeter can be given by

$$P = \int_0^h P_l \, dh$$

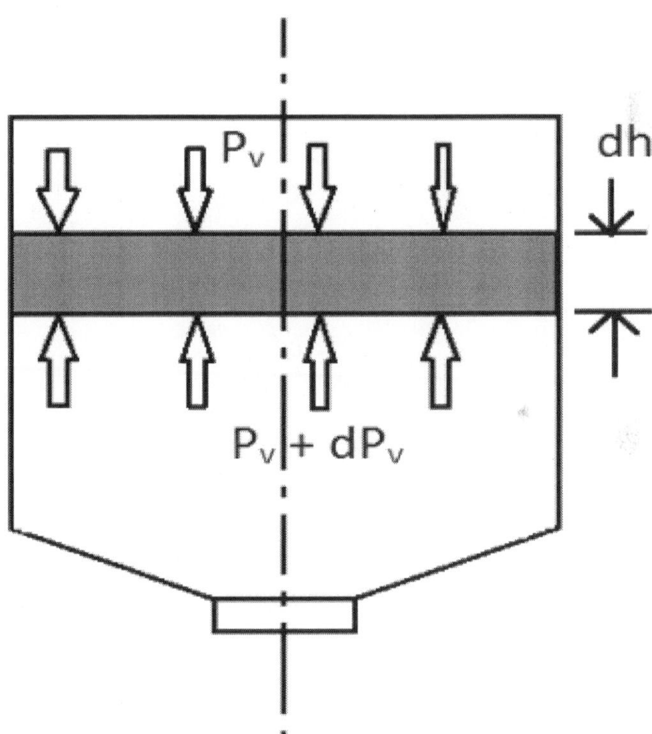

Figure 11.16. Vertical Forces on Elementary Disc.

Also, the vertical wall load 'P_v' per unit perimeter can be given by

$$P_v = \mu P = \mu \int_0^h P_l \, dh$$

Consider an element disc of grain in a deep bin as shown in the Figure 11.16.

The weight of elementary disc = density × volume of the disc = $\omega\pi r^2 dh$

$= \omega A dh$

Consider an equilibrium of the vertical forces acting as shown in Figure 11.16.

Thus,

$P_v.A + \omega\pi r^2 dh = (P_v + dP_v).\ A + \mu P_l\, 2\pi r\, dh$

Substituting $P_l = KP_v$ in the above equation, we get

$\omega A\, dh = AdP_v + \mu\, KP_v\, 2\pi r\, dh$

$$\text{or, } A\,dh\left[\omega - \frac{\mu KP_v}{R}\right] = AdP_v$$

$$\text{or, } dP_v = \left[\omega - \frac{\mu KP_v}{R}\right] dh$$

$$\text{or, } \frac{dP_v}{\left(\omega - \dfrac{\mu KP_v}{R}\right)} = dh$$

On integrating the above equation both sides, we get

$$\frac{\log\left(\omega - \dfrac{\mu KP_v}{R}\right)}{-\dfrac{\mu K}{R}} = h + \text{constant (C)}$$

$$\text{or, } \log\left(\omega - \frac{\mu KP_v}{R}\right) = -\frac{\mu K}{R}h + C$$

When h = 0, $P_v = 0$. By putting these values in the above equation, we can find out the value of C.

Therefore, $C = \log \omega$

Thus, the equation becomes

$$\log\left(\omega - \frac{\mu KP_v}{R}\right) = -\frac{\mu K}{R}h + \log\omega$$

$$\text{or, } \log\left(\frac{\omega - \dfrac{\mu KP_v}{R}}{\omega}\right) = -\frac{\mu K}{R}h$$

$$\text{or, } 1-\frac{\mu}{\omega}\frac{KP_v}{R}=e^{-\frac{\mu Kh}{R}}$$

$$\text{or, } P_v=\frac{\omega R}{\mu K}\left(1-e^{-\frac{\mu Kh}{R}}\right)$$

$$\text{or, } P_l=\frac{\omega R}{\mu}\left[1-\exp\left(\frac{-\mu Kh}{R}\right)\right]=\frac{\omega R}{\mu}\left[1-e^{-\frac{\mu Kh}{R}}\right], \text{ because } P_l = K.P_v$$

$$\text{Maximum lateral pressure, } P_{lmax}=\frac{\omega R}{\mu}$$

Example 11.5

A cylindrical silo, 3 m in diameter and 20 m high, is filled with barley having bulk density of 625 kg/m³. Coefficient of friction between grain and bin wall is 0.45 and the ratio of lateral pressure to vertical pressure is 0.4. Calculate the lateral pressure at the base of the bin in kPa.

(GATE 2017)

Solution

Diameter of silo (D) = 3 m

Height of silo (h) = 20 m

Bulk density of barley (ω) = 625 kg/m³

Coefficient of friction between grain and bin wall (μ') = 0.45

Ratio of lateral pressure to vertical pressure (K) = 0.4

$$R = \text{Hydraulic radius} = \frac{D}{4}=\frac{3}{4}=0.75\,m$$

h>4R, hence deep bin.

$$P_l=\frac{\omega R}{\mu'}\left[1-e^{-\frac{\mu' Kh}{R}}\right]=\frac{625\times 0.75}{0.45}\left[1-e^{-\frac{0.45\times 0.4\times 20}{0.75}}\right]$$

= 1033.097 kg_f/m^2 = 10.13 kPa (As 1 kgf/m² = 9.81 N/m² = 9.81 Pa)

11.12 Bagged Storage Structures (Godowns/Warehouses)

A warehouse or godown is a very popular bagged storage structure in India. The salient design features of the bagged storage structures are given below. IS 607:2009 provides detailed design features of the bagged storage structure.

Location

The structure shall be located on a land where there are no chances of any flood and shall be on a raised well-drained site. The area should have proper hygienic and sanitary conditions. The structure shall be at least 0.5 km away from kilns, bone crushing mill, garbage dumping grounds, slaughter houses, tanneries, dairy and poultry farms; and at least 150 meter away from factories and other possible source of fire. There shall be no tree near the structure to avoid the roots to enter the foundation. Branches of trees, telephone/electricity poles shall be at least 3 meter away from the structure to stop the entry of rats and squirrels. Anti-termite treatment shall be carried out to provide the structure with a chemical barrier against the termites before and after construction. The structure may preferably be situated near main road or near to motor able road. There shall be sufficient space for parking near the structure. If the structure is situated near railway station, airport *etc.*, sufficient loading and unloading facilities shall be made available.

Capacities and Dimensions

The recommended capacities and dimensions of bag storage structures (godown) are given in Table 11.4. The godowns up to 2500 tonnes capacity is considered as small godown. The large godowns above 2500 tonnes may be divided in suitable compartments depending upon the availability of land. The capacities have been estimated by considering 16-bag high stacks.

Table 11.4. Capacities and Dimensions of Bagged Storage Structures

Number of Bags (approx.)	*Capacity (tones)*	*Internal Dimensions (meter)*		*Remarks*
		Length	*Breadth*	
1,120	100	12.0	7.5	—
2,700	250	20.0	9.0	—
5,400	500	34.0	12.0	—
10,500	1000	35.5	18.0	—
28,510	2500	97.19	14.48	—
57,020	5000	129.74	21.34	For 4 compartment design
57,020	5000	125.01	21.34	For 3 compartment design

Foundation

Depth and width of foundation depends on soil conditions of each site. The minimum depth of foundation shall be one meter unless the hard rock is met at a depth less than one meter. Lean cement concrete (1:5:10) for the walls and reinforced cement concrete (1:2:4) footings for reinforced cement concrete columns shall be used.

Plinth

The plinth has to be 0.80 meter above the ground level both for road- and rail-fed structures. The width of platform shall be at least 3.05 meter (10 feet) for rail siding and 2.45 meter (8 feet) for road-fed godowns. The platform along the length of the godown shall be provided with a slope of at least 1 in 40 from the wall to its outer edge. It shall prevent rain water from getting inside the godowns. It is always better to cover the platform. The plinth shall be filled with good quality earth in layers. Each layer should be well rammed and consolidated. Avoid use of black cotton soil and other unsuitable soils for refilling. The plinth shall be constructed either brick or stone masonry in cement mortar (1:6) or lime mortar (1:2). It shall be provided with damp proof course (DPC) of well graded concrete with waterproofing compound. Plinth protection of at least 90 cm (~ 3 feet) shall be provided around the structure of godown. It shall have minimum outward slope of 1 in 48 for satisfactory drainage of rain water.

Floor

Floor of the godown should be damp proof, rigid, durable and free from any cracks or crevices.

Walls

The longitudinal walls shall be of brick or stone masonry in cement mortar (1:6). These walls shall be at least 5.6-meter high for road-fed and 6.35 meter for rail-fed godowns from the floor level. The thickness of brick walls shall be at least 34.0 cm with pilasters (rectangular column) provided at truss support. RCC columns at the spacing of 4.65 meter (from centre to centre) shall be provided under trusses instead of pilasters for span of 10 meter and above. The gable walls shall be of brick or stone masonry in cement mortar (1:4), and shall be at least 46.0 cm in thickness. Pilasters or RCC columns shall be provided at intervals for gable walls and partition walls. Smooth plastered shall be provided both on the outer and inner surface of the walls with cement mortar (1:6). The inside edges of the walls where they meet the floor; all corners shall be rounded off to a radius of at least 5.0 cm.

Doors and Ventilators

Rolling shutters not less than 1.83 m × 2.45 m (6′×8′) is most suitable types of doors for godowns. Doors shall preferably be provided opposite each alleyway. Tow steel ventilators of opening size not less than 1.494 m×0.594 m (~ 5′×2′) for both longitudinal walls are satisfactory. These shall be provided in each bay between RCC columns spaced at 4.5 meter from centre to centre. The ventilators shall be fixed 15 cm below the top edge of the wall, and shall be provided with glazed centre hung with fixed wire-mesh shutters. Air inlets of steel ventilators of the size 62 cm×62 cm (~ 2′×2′) shall be provided at the bottom of the wall at 60 cm from the floor level of the godown. These are fixed in between the top ventilators, and glazed open able outside double-hung. These ventilators shall be protected by expanded metal from outside. On gable walls, suitable number of steel ventilators glazed fixed with fixed wire-mesh shall be provided. All ventilators should be airtight when closed. For this purpose, suitable beading should be provided on the frames of the ventilators. Sunshade of at least 46.0 cm (~ 1.5′) shall be provided on these ventilators to protect the structure from direct sunlight and rain water. Timber in place of steel can be used where it is easily available. In coastal regions, where steel may be subjected to salt action, timber ventilators may be provided.

Generally, a godown has a capacity of 5000 tons and consists of 3 compartments, each having a span of not less than 21·7 m with a clear height of 5·4 m.

Roof

Due to durability, corrugated galvanized iron sheets or corrugated aluminium sheets or galvalume sheet not thinner than 0.56 mm may be used as roofing material. Asbestos sheets are recommended for coastal areas. The sheets shall project at least 46 cm from the outer face of the longitudinal walls. In areas liable to excessive heat, a heat reflecting paint may be applied on the roof. Fibre glass (translucent) sheets are provided for about 2 per cent of the roofing area for natural light.

Drainage

On the road side platforms, generally rain-water pipes are not provided to drain out the water. Rain water falls from the roof directly on the road surface. To protect the scouring of the road, a strip of 90 cm brick paving may be provided along the platform walls. A suitable saucer drain of 30 cm diameter may be provided by the side of platform wall to drain away the rain water from the storage structure. On the rail-side platforms, rain-water pipes shall be provided at each bay for drainage of rain water from the roof. The rain-water gutter shall be provided to receive the rain water, and suitably connected to the down-take rain water pipes. The rain water shall be drained off fairly away from the storage structure.

Finishing

The internal faces of the walls shall be cement plastered, and external faces shall be smooth plastered. The internal surfaces may be white washed and external faces provided with colour washed. All steelwork and woodwork should be provided with two coats of paints over a coat of primer. The galvanized iron sheets shall also be painted with two coats of paints over a coat of primer. The paint to be used inside the godown shall resist the adverse effects of fumigants.

Lighting

Sufficient lighting shall be provided inside and the outside the godown to facilitate loading and unloading operations.

Stack Plan

Floor space is marked into equal rectangles to ensure uniformity in stacking and in accounting. Bagged food grains are arranged in stacks with a base of 6 m × 9 m, leaving 27 per cent free space of the floor area for alleyways generally. Stack heights shall be decided on the basis of grains to be stored (Table 11.5). It shall leave a free space of 2 m between the stacks and 1.0 m between walls to stack for easy movement of person(s) for loading and unloading as well as for observations. Bags are placed in blocks containing one layer lengthwise and other layer width wise.

Dunnage

The stacks of the gunny bags are laid on the wooden planks or polythene sheets. Wooden planks of 1.5 m × 0.9 m size are convenient to place below and are easy to carry. The height from the ground shall be 0.2 m. These should be strong enough to bear a load of about 0.07 tonne per square meter. Polythene sheet of 0.03 mm thickness shall be used.

Table 11.5. Stack Heights for different Food Grains

Grain	*Number of Layers*	*Stack Height (m)*
Paddy	24	5.64
Wheat, barley, maize, gram, jowar, wheat bran and other millets	20	5.04
Rice and milled pulses	16	3.96
Milled cereals and pulses	12	2.13
Seed grains	10	2.00

Example 11.6

Design and draw the plan and elevation in section of a grain storage structure to store 4000 bags of grain. Each bag is 85 cm long, 60cm wide and 30 cm high.

(ARS 1986)

Solution

Assuming capacity of bag of 85×60×30 cm be 75 kg

Quantity of grain stored in 4000 bags = 75×4000 = 300000 kg

Assuming that there are 10 layers in a stack, 10 bags in length and 10 bags in width, total number of bags in a stack = 10×10×10 = 1000

Number of stacks required = $\frac{4000}{1000}$ = 4 stacks

Space required by each stack

Length = (10×0.85) = 8.5 m

Width = (10×0.6) = 6.0 m

Height = (10×0.3) = 3.0 m

Clear distance between walls and end of stack = 1.0 m

Clear distance between stacks = 2.0 m

Hence,

Length of floor = (4×6.0) + (3×2.0) + (2×1.0) = 32 m

Width of floor = (8.5×1) + (2×1) = 10.5 ≈ 11 m

Therefore, the overall dimensions of the storage structure may be taken as 32×11 m.

The height of the walls may be kept as 4.0 m above floor level.

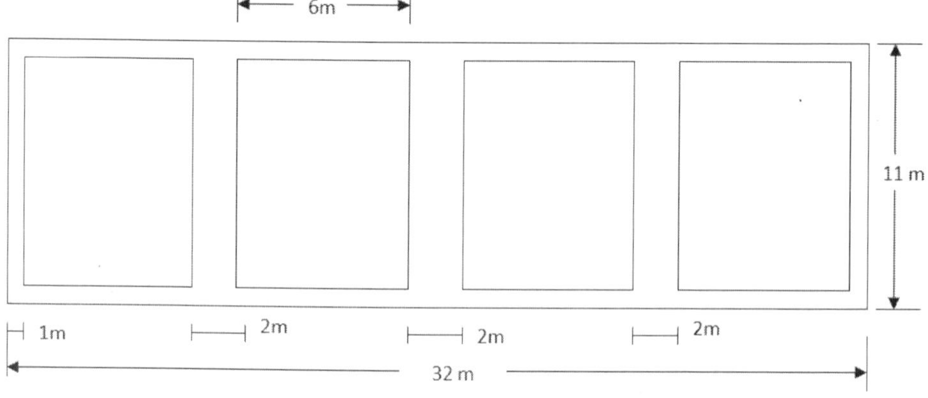

Advantages of using Silos in Place of a Traditional Warehouses

1. **Improved Grain Quality**: Silos provide a sealed environment that helps protect stored grains from external factors such as moisture, pests, and temperature fluctuations. This controlled environment reduces the risk of spoilage, mold growth, and insect infestations, leading to better grain quality retention.
2. **Reduced Spoilage and Losses**: Traditional warehouses may suffer from improper ventilation, inconsistent temperature control, and exposure to pests, leading to higher rates of spoilage and grain losses. Silos' airtight design minimizes these issues, resulting in fewer losses due to deterioration.

3. **Better Pest Control**: Silos can be equipped with various pest control measures, such as aeration systems, fumigation, and insect-repellent coatings. These methods are more effective in silos compared to open warehouses, where pests have easier access.
4. **Less Land Requirement**: Silos are designed to maximize vertical space utilization. This is especially beneficial when the available land area is limited. Traditional warehouses often require more horizontal space due to the need for aisles and stacking limitations. The land area requirement for silos is almost half of the warehouses for the storage of same quantity of grains.
5. **High Level of Mechanization**: Silos often have integrated loading and unloading systems, such as augers or conveyors. This streamlines the process of moving grains in and out, reducing labour requirements and operational time.
6. **Lower Labour Cost**: The mechanized loading and unloading processes in silos reduce the need for extensive manual labour compared to traditional warehouses, leading to cost savings in the long run.
7. **Longer Storage Durations**: Due to the controlled environment and reduced exposure to external factors, silos can enable longer storage durations without significant deterioration of grain quality.
8. **Customized Storage Conditions**: Some silos are equipped with temperature and moisture monitoring systems, allowing for precise control of storage conditions. This is crucial for maintaining grain quality over extended periods.
9. **Flexibility in Storage Capacity**: Silos come in various sizes, allowing storage capacity to be tailored to specific needs. New silos can be added as storage demands increase, providing scalability without disrupting ongoing operations. Large quantity of grains can be stored.
10. **Reduced Handling:** Silos allow for "first in, first out" (FIFO) storage, which reduces the need for excessive handling of stored grains. This minimizes the risk of damage and maintains grain quality.
11. **Easy to Shift**: Metallic structures of silos can be easily shifted from one place to another.
12. **Less Environmental Impact**: Silos' airtight design can help minimize the need for chemical pesticides and preservatives, which can have environmental implications. Additionally, the reduced spoilage and losses mean less grain wastage, which is beneficial from both economic and environmental perspectives.

Disadvantages of Using Silos in Place of a Traditional Warehouses

While using silos for storing grains offers numerous advantages, there are also some disadvantages and challenges associated with this storage method. Here are some of the potential drawbacks:

1. **Higher Initial Cost**: Constructing and equipping silos can be more expensive upfront compared to traditional warehouses.
2. **Limited Storage Flexibility**: Silos are designed for specific types of grains and storage capacities. Switching between different types of grains may require cleaning and potentially modifying the silo's interior, which can be time-consuming and costly.
3. **Maintenance**: Silos require regular maintenance to ensure their functionality and to prevent issues such as corrosion, mechanical failures, and airtight seal breaches. Maintenance can involve specialized knowledge and resources.
4. **Risk of Contamination**: If not properly cleaned and maintained, silos can become breeding grounds for mould, bacteria, and pests. Contaminated silos can lead to spoilage and reduced grain quality.
5. **Limited Accessibility**: Silos often have limited accessibility, especially for manual inspection and maintenance. This can make it challenging to address issues promptly and perform routine checks.
6. **High Operational Skill Requirement**: Operating and maintaining silos effectively requires a certain level of technical expertise. Personnel must be trained to use monitoring systems, aeration equipment, and other specialized machinery.
7. **Cleaning of Silos**: Emptying and cleaning a silo can be a labour-intensive and time-consuming process. Ensuring that all residues are removed is essential for preventing contamination and maintaining grain quality.
8. **Inflexibility for Certain Grains**: Some grains may not be well-suited for silo storage due to their unique properties. For example, very fine grains might have difficulty flowing through loading and unloading equipment or might form bridges within the silo.

11.13 Insects, their Control and Fumigation

The production of the crops is generally time specific in a year and to ensure their availability in the off seasons or throughout the year, it is important to store the food products in suitable storage structures. These storage structures need to be managed for insect control to curb the quantitative and qualitative losses in stored products. As per Post-Harvest study at ICAR-CIPHET Ludhiana, 4-6 per cent food loss is reported in India during flow of cereals in various unit operations. An estimate of 1-2 per cent of cereals is lost due to lack of storage structures and unscientific storage techniques. Storage losses are comparatively higher in the case of legumes (6-9 per cent) and oilseeds (3-10 per cent). Overall about 10-15 per cent of the total production is lost during the various unit operations prior and during the storage. During the storage, insects, pest, mites and rodents are the major culprits for food grain losses. Mould may damage stored food grains but its infection depends upon the moisture content of stored products. Stored product should be maintained on or below the safe moisture content for longer storage to control mould. Moreover, the safe moisture content prevents the condensation problem due to seasonal temperature difference at top and bottom of storage structure. The safe moisture content for storage is 8 per cent, 10 per cent and 12 per cent for oilseeds, legumes and cereals, respectively. Storage of safe moisture content keeps grain brighter, odour free and of high quality.

Insects

Insects are free-living small living organisms and can be visualized by the naked eyes. The life stages of the insects are *i.e.* egg, larvae, pupa and adult. The egg and pupa are the non-mobile life stage of the insects whereas the adult and larva are the mobile life stages. The moisture content 12-14 per cent and temperature of 26-34°C are the optimum conditions for insect growth in stored food grains. The mobile life stages of the insects are the most dangerous stage for infestation of the stored food grains. The insect is a global problem and causes qualitative and quantitative losses in stored food grains. The reduction in weight of stored food due to eating by the insects is the quantitative losses whereas the nutrient losses and addition of their dropping and dead bodies are considered as qualitative losses. The losses may be categorized as direct losses and indirect losses. The direct losses are the presence of open or concealed live and dead insects, their droppings and fragments whereas allergens, microorganisms, low market value are the indirect losses.

There are five sources of the insects in the stored food grains. Grains grow at farms and insects enter in the different parts during drying, threshing and other unit operations. The harvested and threshed crops in the form of food grains filled in the gunny bags or similar type of bags made of different material and then loaded in the transportation cart/tractor trolley. If the gunny bags are old or not disinfected may be responsible for the food grains infestation. The crevices and small apertures or holes in the transportation cart/tractor trolley may have insects and infest the loaded material. Beside it, the storage structure should be disinfected to keep insects away from the food grains. Sometimes, old stocks are stored with the new stock. The old stock may have insects and these insects infest the new stock. Therefore, old stock should not be stored with the new stock of food grains. The first in and First out (FIFO) principle of godown management should be followed to control insects inside the storage facility.

Type of Insects

The classification of different types of the insects depends upon their eating behaviour the stored food grains.

1. *Primary Insects*

These insects are also known as external feeders. It damages more to the stored food grains than secondary insects in a short period of time. The primary insects damage the outer hard layer of the food grains and allow the secondary insects to eat the endosperm part of grains. Rice weevil (*Sitophilus oryzae*), Lesser grain borer (*Rhyzopertha dominica*) *etc.* are the examples of the primary insects.

2. *Secondary Insects*

Primary insects are followed by the secondary insects. The endosperm part of damaged grains is eaten by the secondary insects. The food grains may be damaged by the primary insects' infestation or during the threshing operation. Red flour beetle (*Tribolium castaneum*), saw toothed grain beetle (*Oryzaephilus surinamensis* L.) are the examples of secondary insects.

3. *Tertiary Insects*

This category of the insect mostly infests or feeds on the semi processed or processed food products. Red flour beetle (*Tribolium confusum*), Khapra beetle (*Trogoderma granarium*) *etc.* are the main tertiary insects that infest wheat flour, noodles *etc.*

Factors that Affect the Growth of Insects

The factor that affect the growth of the insect in stored commodities may be divided into the biotic and abiotic factors. Biotic factors are of grain itself. The quality of grains, presence of insects, mould or fungus, are one of the biotic factor whereas type of storage structures, presence of foreign material, environmental factors such as temperature and relative humidity, gas composition *etc.* covered under the abiotic factors. Some of these factors are discussed under.

1. *Temperature*

Stored-grain moulds may grow in grain with a_w more than 0.65, that is, when the moisture content of cereal grain is more than 14.5 per cent (w.b.). Some strains of the grain mould species can produce mycotoxins, which are secondary metabolites of fungi. Mycotoxins are an important issue worldwide in terms of public health for the agro food industry with huge economic implications. The optimum temperature for growth of stored-grain moulds is around 25–30°C, but some grow well at 35–37 °C or above, such as *Aspergillus* spp. Grains are contaminated by spores of storage fungi during harvest, transport, and handling operations. In storage, provided that temperature and moisture content (a_w) meet their minimum requirements, fungi germinate and fungal growth occurs. The rate of growth is correlated with the levels of water activity and temperature.

2. *Relative Humidity*

Food grains absorb or desorb the moisture from/to environment to achieve the equilibrium moisture content (EMC). Relative humidity is the parameter that shows the water content of the environment. Although relative humidity also depends on the temperature. Absorption of water from high humidity environment promote the growth of insect, fungus and degrade the quality of stored commodity.

3. *Quality of Grain*

Presence of chaff, broken grains *etc.* attract the insects and promote their growth. Secondary insects generally eat the grains which are damaged by the primary insects. Broken grains provide the food to the secondary insects. Although the proper cleaning and drying reduced the insect growth in stored commodities. Sometimes old grains may be added with the new grains during the storage. The old grains may be infested with insects and of low quality that results the growth of insects. The transportation vehicle, gunny bags and godown may be a source of insect infestation. These are also disinfected or fumigate before using for newly harvested grains to avoid re infestation. Quality of grains also cover the moisture content of the store commodity. The food grains should be properly dried before the storage in a proper storage structure.

4. *Storage Structure*

The storage structure is the most important factor on which the storability of stored commodity relies. A properly dried and cleaned grain may be infested or deteriorated in case of faulty storage structure. Grain absorb or desorb the water from environment and storage structure should prevent this attainment of equilibrium between environment and grains. Airtightness of storage structure increase the carbon dioxide gas inside the storage structure due to biological respiration. Depletion of oxygen results the downfall of the insects, mould and mites. Now a day, sensor may be installed inside the storage structure that provide the real-time data for monitoring the quality of stored commodity. Temperature, relative humidity and gas are the major parameters that were recorded with the time. Airtightness of storage structure also helps in effective fumigation. Phosphine gas generally takes up to five days for complete kill of insects. Provision of aeration in the storage structure also helps in maintaining the quality of stored products and helps in prevention of moisture condensation due to seasonal temperature difference.

11.14 Control of Micro-organisms

Micro-organisms affecting stored grain are fungi, bacteria, yeast, actinomycetes, and protozoa. Fungi out of all the microorganisms are responsible for maximum damage/spoilage of food grains in storage. Presence of fungi causes heating, off odour, kernel discoloration, loss of nutritive value, change of palatability and loss in viability. In some cases, fungi produce poisonous substances known as mycotoxins which are harmful and may be fatal to human beings and animals.

The reproduction of fungi is through spores of either sexual or asexual nature. The factors which constitute a favourable environment for the development of spores are as follows:

1. Moisture content of grains (above 12 per cent)
2. Temperature (30 to 32°C)
3. The degree to which the grain has been invaded by fungi/insects before or after storage
4. Amount of foreign matter (dust, dirt, chaff, straw *etc.*) present in the grains

Moisture is one of the important factors which is essential for growth of fungi. The moisture content of the grains must be reduced to safe level to minimize the chances of fungal growth. Reduction of temperature to a range of 23°C to 30°C reduces growth rate of a number of species of fungi.

To avoid damage to food grains from micro-organism during storage, grains should be properly cleaned, dried and cooled before storage. If hot grain is put in storage, it will start respiring at a faster rate resulting in generation of more heat. Thus, it is advisable to put the grains in the storage structures during morning time. Storage structures should be made airtight so that there is no moisture increase during rainy season.

11.15 Control of Rats

One rat eats about 10 kg grains in a year (about 24 grams per day), and also spoils almost the same amount of grains. Methods used to minimize the damage are proofing, repelling, trapping, and poisoning.

1. **Proofing:** Storage structures should be made rodent-proof so that rats may not be able to enter into it. Silos made of either steel or concrete are sufficiently rodent-proof. The storage structure should be on high plinth. While constructing the godowns, a plinth of about 1 meter without steps, with a U-shaped drain at the ground level and 12″ inversed L-shaped projection at the top shall be provided. All windows, ventilators, gutters, drains should be fitted with 24 gauge expanded metal mesh.
2. **Repelling:** Use of sound waves emitted by ultrasonic devices are unbearable by rats and mice and repel the rats without any harm to human beings as these waves are unheard by humans.
3. **Trapping:** Rat traps may be used to control rat menace inside the storage structures. They are effective only when the rat population is low.

4. **Poisoning:** Chemicals may be used to kill rats. Compounds which kill the rats by their chemical action are known as rodenticides. Zinc phosphide, barium carbonate, red squill and ANTU are being used as rat poisons. Poison bait is prepared by mixing 1 part of zinc phosphide with 49 parts of popped corn/rice/fresh meat/peanut butter *etc.*

11.16 Control of Birds

Birds consume about 8 to 25 grams of food grains per day. Besides eating grains in the fields and godowns, birds are also responsible for spoilage, contamination and creating nuisance and unhygienic conditions in the warehouses. Some of them are also responsible for spreading diseases. The methods for the control of birds are physical, biological, chemical and scaring.

1. **Physical Methods:** All the windows, ventilators and other possible entries should be made bird proof by putting meshes. The other methods like use of air guns, bird traps and nets can be used for the control of birds.
2. **Biological Methods:** Other predator animals/birds (hawks, owls *etc.*) could be utilized for bird control. Such methods give limited results.
3. **Chemical Methods:** Chemicals mixed with food grains or any other eatables could be used to kill the birds. Since the birds feed in open, there is possibility of health hazards to other birds or pets also.
4. **Scaring Methods:** Birds can be scared by using different methods. Dummy similar to man are pitched in the fields to scare the birds. One can control the birds manually also by scaring the birds by throwing small stones or other materials. This process is very tedious. Other methods are mechanical bird scarers and electronic devices. In mechanical method, a mixture of oxygen and acetylene gas is burnt which creates a bumping noise at irregular intervals. The device is kept at height in godown or on the top of trees for getting better results. Electronic device such as tape recorder can be used for scaring birds. The different sounds are played at high pitch in the infested areas.

11.17 Grain Protection from Insects

During storage, grain needs protection from biotic and abiotic factors to prevent quantitative and qualitative changes. There are physical, chemical and biological ways to protect the grains from deterioration.

1. Physical Methods

Separation of insects from grains using sieving, use of inert dust like silica dust, temperature manipulation by aeration, changes in gas composition under modified atmosphere storage and controlled atmosphere storage, the killing of insects using ionizing and infrared radiation, storage of grains under low and high atmospheric pressure, attract and catch insects using UV traps (Figure 11.17) are some of the physical ways of controlling the insects inside the grains storage structures. In comparison to the chemical methods, the physical methods of grain protection are costly, need monitoring, laborious and time-consuming. This is the main reason behind the low popularity of the physical methods of insect control.

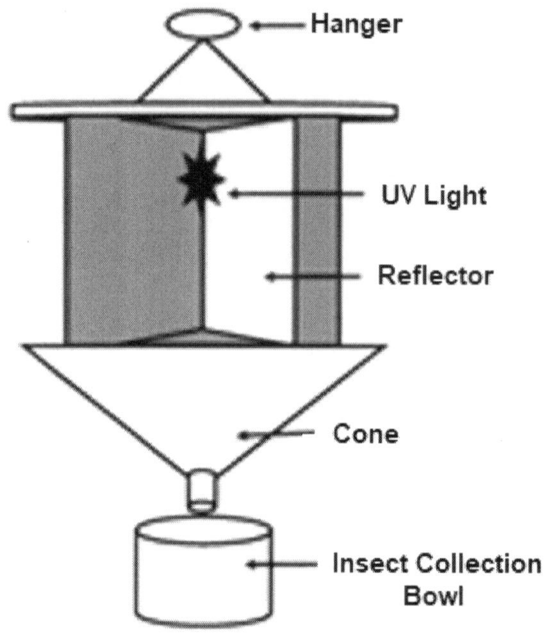

Figure 11.17. UV Based Insect Traps.

2. Chemical Methods

Insecticides made of aluminium phosphide are being used in India widely for the control of insects. This insecticide is available in the market under the brand name Celphos, Cellofumes, Phostoxin *etc.* These insecticides react with the moisture available in the interstitial atmosphere of the storage structure and produce phosphine gas. The phosphine gas can kill all forms of insects. Initially, aluminium phosphide is available in tablet form. Later on, it is available in powder form in sachets due to increasing human suicide using this insecticide. It is slow releasing insecticide and may develop insect resistance to the phosphine gas if not applied in proper dosage and exposure time. The storage structure needs to be airtight for its successful application. The 10g sachet per ton should be applied for 7-10 days for complete mortality of the insects. In the case of silo bins, the flour paste needs to be applied in all ways of air infiltration or exchange. The sand snakes are useful in the Cover and Plinth storage and for the fumigation of stacks in the godown. The sand snakes kept the polythene cover in place and air-tight for successful exposure to the insecticide. The phosphine is 1.2 times heavier than the air therefore, the sachet should not be placed in the bottom/ centre of the stacks. The chemical residues in grains, the development of insect resistance, and the inorganic nature of fumigants are the problem associated with the use of chemical fumigants.

3. Biological Methods

Pheromones or semiochemical, biological derivatives like *neem* oil and light-based insect traps are some of the biological methods that can be used to control the insect activity inside the stored food grains. The Pheromones are the chemicals whose vapour pressure is higher than the ambient and therefore diffuse in the interstitial atmosphere of storage. This chemical is responsible for the communication between the insect for reproduction, warning about predators, finding food *etc.* These chemicals either attract or repel the insects and make them suitable for monitoring, mass trapping and detection. The light of a specific wavelength attracts the insects and the mass trap system made of a cone or sticky gel caught the insects. *Neem* oil repels the insects from the stored grains due to its strong smell.

Methods for Disinfestations of Vacant Food Grain Storage Structure

Any one of the following method may be used for disinfestations of vacant food grain storage structure:

1. Dusting
2. Smoking
3. Fumigation

Dusting can be done of any food grain storage structure, smoking only of those structures which can be made reasonably air-tight, and fumigation of only those which can be made perfectly air-tight.

1. **Dusting**: Benzene hexachloride (BHC) dust containing 0.65 per cent gamma isomer is used for dusting. It is applied at the rate of 25 g per 100 cm^2 surface area of the floor, walls and the ceiling. A hand below duster, a rotary dust blower or a power duster shall be used for dusting the structure. The type of duster is selected on the basis of area to be dusted. Keep the nozzle of the duster at a reasonable distance from the surface to be dusted. Ensure that the deposit of the material on the surface is uniform. During the process of dusting, all the openings such as doors, windows, ventilators *etc.* of storage structure shall be kept closed.
2. **Smoking**: A suitable formulation of BHC which on ignition gives out smoke charged with BHC fumes is used for smoking. All the openings of the food grain storage structure should be properly closed (with the help of mud plaster or strips of gummed paper) except one opening

for the operator to come out. Ensure that the whole structure has been made reasonably airtight. Put the required quantity of material on a piece of galvanized sheet and place it in the center of the structure. Ignite the material. The operator shall wait till the material is well-ignited, and the smoke has started to evolve in large quantities. After ascertaining these points, the operator should immediately leave the structure through the opening left for the purpose. Close and put lock on this opening and make this opening also air-tight. Ensure that smoke may not escape from any point in the structure. Keep the structure closed till the following morning.

3. **Fumigation**: All the openings of the food grain storage structure should be properly closed (with the help of mud plaster or strips of gummed paper) except one opening for the operators to come out. Ensure that the whole structure has been made reasonably airtight. Ethylene dichloride carbon tetrachloride (EDCT) mixture in the proportion of 3:1 is used for fumigation. EDCT is required at the rate of 325 g per 1 cubic meter. Depending on the size of the structure, calculate the quantity of fumigant required. Take out this calculated quantity in suitable number of buckets. Give one bucket to each operator. All the operators must be asked to quickly throw the fumigant on the floor of the structure simultaneously, and immediately leave through the opening left open for this purpose. Close and put lock on this opening. Before locking the opening, ensure that no operator has been left inside the structure. Put mud plaster on all the possible points of leakage of fumigant vapours. Ensure that no fumigant vapour leaks from any of the points in the structure. Keep the structure closed for at least 48 hours.

After the expiry of this period, open the lock and then push or pull the door wide open. Run away to a distance not less than 15 meter from the opening. Keep it open for at least 4 hours. During this period, nobody should be allowed the come within a radius of 15 meter from the opening. After the expiry of this period, two operators wearing gas masks may be allowed to enter the structure to open all the other openings. After this, the operators should leave the food grain storage structure. Keep the structure open for a period of another 6 hours, and during this period do not allow any person to enter the structure.

Fumigation shall always be done under the supervision of an experienced person. The operators shall be well trained, and shall wear gas masks while handling the fumigant.

11.18 Aeration

Aeration involves moving a relatively low volume of air through the grain. It is an indispensable practice in the post-harvest management of food grains. It helps control moisture and temperature, reduces the risk of spoilage, and prolongs the shelf life of grains. Proper aeration techniques, combined with monitoring and control systems, play a critical role in maintaining the quality and nutritional value of food grains, ensuring they reach consumers in a safe and wholesome condition.

One of the primary objectives of aeration is to control moisture levels within the stored grains. Aeration systems help maintain an optimal moisture level, typically between 12-14 per cent for most grains, by removing excess moisture or adding controlled amounts of dry air when necessary. Proper aeration also involves regulating the temperature of the stored grains. Temperature fluctuations can impact the rate of biological and chemical reactions within the grains. Aeration helps to maintain a consistent temperature throughout the storage facility, reducing the risk of hotspots and temperature-related spoilage. Aeration systems assist in controlling insect infestations by creating unfavourable conditions for pests. Aeration systems distribute air uniformly throughout the grain mass, ensuring that all grains receive adequate ventilation. This prevents the formation of condensation, which can lead to localized spoilage. Uniform aeration also helps maintain consistent grain quality.

11.19 Quality Monitoring and Control of Stored Product

The quality is a compilation of the different attributes of food grains and the selection of these attributes depends on the orientation of the quality. There are two types of quality orientation. The consumer-oriented quality target the quality as per their eating habit, location, age *etc.* whereas product-oriented quality is related to the longer shelf life with minimum changes in its quality attributes. Quality monitoring is process-oriented and prevents quality hazards whereas quality control focus on the product that detects the quality hazards. Monitoring of process parameters like the temperature of the dryer, air velocity of the blower/aspirator *etc.* assure the development of a quality product. These product parameters are optimized in lab and pilot trials before being used for mass production in the food industry. Instead of quality assurance by quality monitoring, there are chances of quality failure due to errors in machinery, sensors, actuators, human error *etc.*, therefore quality control over the developed product is also important. The quality of the stored material can be monitored by the two methods:

1. Destructive Methods

These methods involve the destruction of the samples used for the analysis. These methods are labour-intensive, time-consuming, costly and need skilled personnel for the laboratory. For example, the determination of the moisture content of food grains can be performed using the air oven method. Although, the vacuum oven method and infrared method are also available which is comparatively fast than the air oven method. In addition, the oven methods consume samples during the determination of the moisture content. The air oven method is often more accurate and is used for the validation of non-invasive or non-destructive methods of moisture determination.

2. Non-Destructive Methods

In these methods, the electronic equipment equipped with sensors is used for the determination of the quality parameters. These methods are fast, easy and non-invasive. The moisture meter that works on the electrical resistance/capacitance is the non-invasive equipment for moisture determination. This method enables moisture determination at the field, godown, grain market *etc.* and provides real-time data for spot decisions for harvesting, price determination, requirement of aeration *etc.* In addition, personnel with nominal training can use a moisture meter. The non-destructive/non-invasive methods also provide the possibility of automation in the godown through sensors like temperature, relative humidity, gas composition *etc.* These sensors provide data that helps in taking early/preventative action for aeration and fumigation to prevent moisture accumulation or insect growth inside the grains silos.

11.20 Modelling of Metabolic Activities and Prediction of Storage Life

Metabolic activities are the result of changes in the independent parameters involved in a particular biological phenomenon. For example, changes in the germination percentage of the stored food grains depend upon the storage temperature, relative humidity, storage structures *etc.* Germination is the combination of various metabolic activities that change the chemical composition of the grains. Here is the list of independent and dependent parameters involved in grain storage.

1. **Independent parameters**: Storage temperature, relative humidity, type of storage structures, initial moisture content, foreign material, storage time *etc.*
2. **Dependent/response parameters**: Germination percentage, moisture content, thousand kernel weight, insect activity, milling yield, sedimentation value, gluten content, protein content *etc.*

The combination of all dependent parameters defines the storage life/shelf life/post-harvest life of stored food grains. The monitoring of the dependent parameters is a costly, time-consuming and destructive process. The variations in the dependent parameters are the result of independent parameters. The independent parameters can be controlled and automated in comparison to the dependent parameters. Therefore, it is required to monitor these independent parameters during the storage to certain the quality of the food grains.

The monitoring of independent parameters needs specific ranges, combinations of independent parameters, alert points, protocols *etc.* This combination and ranges of parameters are determined by lab experimentation as per their effect on dependent parameters. The data from the lab experiment is fitted in the different types of models to establish their relation with each other. The coefficient of determination (R^2) is the main indicator that is used to judge the efficacy of the model for the data. There are three types of models available for modelling the storage data.

1. Linear Models

As per the linear models, the dependent parameter is the function of one or more independent parameters. It relates two parameters using a straight line. Regression is a statistical procedure used to develop linear models.

2. Non-Linear Models

The real phenomenon is generally non-linear. Non-linear models are useful when linear models are unable to capture or describe the relationship of parameters. These models are comparatively more complex and provide more accurate results. The log model, power model *etc.* are examples of non-linear models.

3 Artificial Neural Network

It is a type of non-linear model used for modelling the data that is not fitted properly in linear and non-linear models. ANN can identify any type of pattern inherited in the data and develop a model to relate the parameters. Software like SPSS, MatLab *etc.* can be used for the development of the ANN models.

STORAGE OF FODDER (FARM SILOS)

Fodder serves as a crucial resource for livestock farmers, ensuring the well-being and productivity of their animals. Proper storage of fodder is paramount to maintaining its nutritional value and preventing spoilage. Fodder is usually stored in a specialized storage structure known as farm/fodder silo. These silos play a crucial role in agricultural and livestock management, providing a means to store large quantities of feed in a controlled environment. Fodder silos offer several benefits that contribute to the overall efficiency and health of livestock operations. Fodder silo enable farmers to store and manage animal feed efficiently, leading to improved feed quality, reduced wastage, and overall enhanced productivity in livestock operations.

11.21 Silage

Silage is a highly valuable and nutritious feed option for livestock farmers, especially during periods of scarcity. Silage is a type of preserved forage that is commonly used as feed for livestock, particularly ruminants like cattle, sheep, and goats. It is produced through a process called ensiling, which involves fermenting and storing high-moisture crops under anaerobic (low-oxygen) conditions. The primary goal of making silage is to preserve the nutritional content of the harvested crop for feeding animals during periods when fresh forage is scarce, such as winter or drought. Silage is more nutritive than dries stalk.

Silage can be made from various crops, including grasses, legumes, and cereal crops like maize and sorghum (*jowar*). The chosen crop is harvested at the optimal stage of maturity, when its nutrient content is high and the moisture content is suitable for ensiling. The harvested crop is chopped into small pieces to increase its surface area, which aids in compaction and fermentation. The chopped crop is packed tightly into a storage structure, such as a bunker silo, tower silo, or silage bag. Proper packing is crucial to eliminate as much oxygen as possible, as the fermentation process that follows is anaerobic. Once packed, the lack of oxygen encourages beneficial bacteria (lactic acid bacteria) to thrive. These bacteria convert the plant sugars into organic acids, primarily lactic acid, which lowers the pH of the silage. The pH of around 4.0 to 4.5 is best to ensure effective preservation.

This acidic environment preserves the silage and prevents the growth of spoilage microorganisms. The top of the silage is tightly sealed with plastic, tarp, or other materials to create an airtight environment, further preventing the growth of spoilage organisms. The silage is allowed to ferment for several weeks or months, during which the pH continues to drop and the desirable acids accumulate, contributing to its preservation. When it's time to feed the silage to the livestock, it's removed from the storage structure. Proper feed out practices are important to minimize exposure to oxygen and maintain the quality of the remaining silage. The density of silage depends upon its dry matter contents, depth, fineness of chop and degree of compaction. Its value ranges from 450 to 800 kg/m^3.

11.22 Classification of Silos

Silos, used for storing and preserving silage, come in various types and designs, each with its own advantages and considerations. Silos can be classified based on their construction, shape, and location. Some of the common classifications are:

On the basis of Construction Material

- ☆ **Concrete Silos:** These are large, permanent structures made of reinforced concrete. They are durable and can last for many years. Concrete silos offer good insulation and protection against weather and pests.
- ☆ **Steel Silos:** These are typically constructed using steel panels bolted together. They are relatively easy to install and can be used for both temporary and permanent storage. Steel silos are well-suited for areas prone to seismic activity.

- ☆ **Wooden Silos:** While less common today, wooden silos were historically used. They are often made of staves or planks and have limited durability compared to concrete or steel.

On the Basis of Purpose

- ☆ **Temporary silos:** Some silos are constructed for temporary use during specific seasons or for specific crops. They may be disassembled and moved to different locations as needed.
- ☆ **Permanent silos:** Permanent silos are built to last for many years and are intended for ongoing silage storage.

On the Basis of Shape

- ☆ **Tower Silos:** These tall, cylindrical structures are usually made of brick, concrete or steel. Tower silos are loaded from the top and can have unloader systems to remove silage from the bottom. Loading of tower silo is difficult as it needs a mechanical loader or a large capacity blower for elevating the cut fodder. Tower silos are preferred in the areas where water table is very close to the ground level. The cost of construction of these silos is comparatively much higher than that of horizontal types.
- ☆ **Bunker Silos:** Also known as trench silos, bunker silos are low-profile, elongated structures built into the ground. Silage is placed in long trenches and compacted before covering with plastic or other materials to create an airtight seal.
- ☆ **Horizontal Silos:** These silos are constructed both surface and below ground. Pit, trench and surface silos are the examples of horizontal silos. They may have limited capacity compared to tower or bunker silos. These silos are cheaper, easy to construct, and can be easily filled and unloaded without any special equipment. Horizontal silos are successful in areas having deep water table. Surface silos are rarely used for storing silage.
- ☆ **Silo Bags:** Silage can also be stored in long plastic bags, often referred to as silage bags or silo bags. The chopped silage is packed tightly into the bag and then sealed.

11.23 Location

Silos for the storage of fodder, also known as silage, are typically located on or near livestock farms where the fodder is produced or used to feed animals. The location of these silos can vary based on factors such as farm size, available land, and the types of livestock being raised. Factors to be considered for the location of silos are:

- ☆ **Proximity to Livestock**: Fodder silos are usually located near the areas where livestock are housed or grazed. This reduces the distance and effort required to transport the stored feed to the animals.
- ☆ **Accessibility**: Silos should be easily accessible for loading and unloading. They may be situated near roads, driveways, or other transportation routes to facilitate the movement of feed to and from the silos.
- ☆ **Drainage**: Proper drainage is essential to prevent water from accumulating around the silos, which could lead to spoilage of the stored fodder. Silos should be placed on well-drained ground.
- ☆ **Near Grazing Areas**: In some cases, silos may be situated near pastures or grazing areas where animals are kept. This placement allows for convenient access to the stored fodder during feeding times, and reduces the time and travel required for feeding.
- ☆ **Close to Feed Preparation Areas**: If the fodder needs to be processed or mixed with other ingredients to create a balanced diet for the animals, silos may be located near the feed preparation areas or feed mills.

11.24 Loading and Packing

Filling and packing silos for the storage of fodder (animal feed) is an important process to ensure the quality and longevity of the stored feed. Proper filling and packing techniques help prevent spoilage, maintain nutritional value, and reduce the risk of contamination.

Chopped silage is transported to the pit or trench using loaders or other equipment. The silage is evenly spread and compacted in layers as it is filled. Proper compaction is crucial to eliminate air pockets and create an airtight environment. This prevents the growth of spoilage microorganisms, and ensures proper fermentation. Tractors, specialized compaction equipment, or even the weight of the silage itself can be used for compaction. Silage is typically loaded in thin layers, and each layer is compacted before adding the next. This helps in achieving uniform density and exclusion of oxygen.

Once the pit or trench is filled and compacted above the ground level, it is covered with airtight plastic sheeting. The plastic helps create an anaerobic environment, which is crucial for proper fermentation and preservation. To keep the plastic in place and aid in creating a tight seal, the edges of the plastic are often weighted down with tires or other heavy materials.

Feeding

Typically, a cow is fed about 3 kg of silage per 100 kg of her body weight per day. Thus, the normal silage allocation ranges from 14 to 18 kg per cow per day depending upon the size of cows.

11.25 Horizontal Silos

Horizontal silos are surface as well as below ground (underground) types of storage structures used on most of dairy farms as temporary or permanent storage structures for silage. They are classified as pit, trench and surface silos. The spoilage of silage and dry matter losses of these silos ranges between 20 to 30 per cent.

Pit silos and trench silos are often more cost-effective to construct compared to above-ground silos. They require minimal equipment for loading and compaction. The earth surrounding the pit/trench provides natural insulation, helping to maintain a more stable temperature inside the silo. Pit or trench silos can be adapted to the available space on a farm and can be constructed in areas where above-ground structures may not be feasible.

11.25.1 Pit Silo

A pit silo is a type of silo used for storing and preserving silage. It is a below-ground storage structure that offers certain advantages in terms of cost-effectiveness, ease of loading, and insulation. They are constructed where the water table is very low. Pit silos are a practical option for farmers seeking efficient and cost-effective silage storage solutions, particularly when considering the factors of space, resources, and farm management practices. Proper construction, loading techniques, and sealing are essential for successful pit silo operation.

A pit silo is created by excavating a depression in the ground. Pit silos are typically circular in shape. The walls of a pit silo can be made of compacted soil, bricks, stones, concrete, or concrete panels. The floor can be made of soil or concrete. The entire surface coming in contact with the silage shall be plastered to make it smooth, air and water tight. For this purpose, concrete walls and floors offer better durability and protection against moisture. A simple roof over the silo shall be provided to protect the silage from sun and rains. To facilitate the loading of silage into the pit, ramps are often built at one or both ends. These ramps allow tractors or loading equipment to drive onto the silage surface for filling.

Size and Capacity

The size and capacity of pit silos can vary widely based on factors such as the type of crop being stored, the number of animals being fed, and the farm's feeding needs. To prevent spoilage, silage should be removed at the rate of 10 cm per day. Diameter of the silo is determined by the quantity of silage to be fed daily. It is generally limited to 6 m, and its depth is kept 2 to 3 times that of the diameter. The dry matter losses ranges from 20-30 per cent.

Example 11.7

Find out the economical diameter and depth of a pit silo to store sufficient quantity of silage for a herd of 300 dairy cows, 40 *murrah* buffaloes and 40 heifers having average body weights of 450 kg, 680 kg and 180 kg respectively. Each cow and heifer will be fed 3 kg silage/100 kg body weight, whereas each buffalo will be fed 4 kg silage/100 kg body weight. All the above animals will be fed silage for 150 days in a year. The thickness of silage fed each day will be 10 cm only. One cubic meter silage weighs 650 kg. *(GATE 1991)*

Solution

Number of dairy cows = 300

Number of *murrah* buffaloes = 40

Number of heifers = 40

Average body weight of dairy cow = 450 kg

Average body weight of *murrah* buffaloes = 680 kg

Average body weight of heifers = 180 kg

Rate of feeding for cow and heifer = 3 kg/100 kg body weight

Rate of feeding for buffaloes = 4 kg/100 kg body weight

Per day requirement of feed

For 300 dairy cows = $\frac{450 \times 3}{100} \times 300$ = 4050 kg

For 40 *murrah* buffaloes = $\frac{680 \times 4}{100} \times 40$ = 1088 kg

For 40 heifers = $\frac{180 \times 3}{100} \times 40$ = 216 kg

Thus, total feed requirement per day = 4050+1088+216 = 5354 kg

Density of silage = 650 kg/m^3

Volume of silage fed per day = $\frac{5354}{650}$ = 8.24 m^3

There are 15 per cent losses, the actual volume required = $\frac{100}{85} \times 8.24$

= 9.69 m^3

Assume the daily consumption is 10 cm from silage

Therefore, volume of silage removed per day = $\frac{\pi}{4} \times d^2 \times 0.1 = 0.0785d^2$

Thus, $0.0785d^2 = 9.69$

d= 11.1 m

Since, diameter of silo is more than 6.0 m, consider three pits of equal size.

Thus, $0.0785d^2 = (9.69/3)$

d= 6.4 m

Depth of silo = 0.1×150 = 15 m.

Hence, three pits of 6.4 m diameter and 15 m depth shall be required.

11.25.2 Trench Silo

A trench silo (Figure 11.18) is an in-ground storage structure used for preserving and storing silage. It is created by excavating a trench or depression in the ground, and it offers several benefits such as cost-effectiveness, easy loading, and natural insulation. Trench silos are a practical option for farmers seeking efficient and cost-effective silage storage solutions, especially considering factors such as available space, resources, and farm management practices. Proper construction, loading techniques, and sealing are essential for successful trench silo operation.

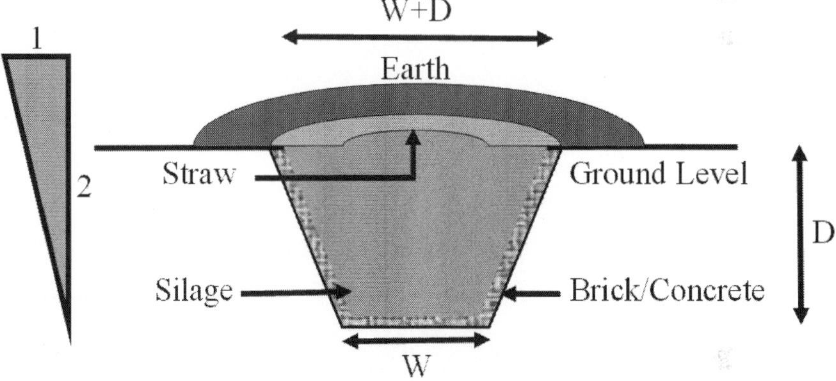

Figure 11.18. Cross-section of Trench Silo.

Trench silos are typically elongated in shape, resembling a trench. Trench silos can either be lined or unlined. The walls of a trench silo can be made of compacted soil, bricks, concrete, or concrete panels. The floor can be the natural soil or a layer of compacted gravel. Concrete walls and floors provide better durability and protection against moisture. Trench silos often have ramps at one or both ends to facilitate the loading of silage. These ramps allow tractors or loading equipment to access the silage surface for filling. An unlined trench silo although cheap but results in more spoilage of silage, and also likely to cave in due to heavy rains. The drainage water should not be allowed to enter or collect near the silo.

Size and Capacity

The size and capacity of a trench silo can vary based on factors such as the type of crop being stored, the number of animals being fed, and the farm's feeding needs. The cross-section of the silo depends upon the number of animals being fed, and the length is determined by the number of days the silage is fed in a year. The following dimensions are generally considered for the design of trench silo:

Length of silage fed per day	= 15 cm
Depth of silo	= 2.5 m
Side slope	= 1:2

The dry matter losses ranges from 20-30 per cent.

Example 11.8

Design a trench silo for preparing silage at a government farm. The daily requirement of silage is 2 tonnes for 160 days in a year. The average density of silage may be taken as 800 kg/m³.

Solution

Daily requirement of silage = 2 tonnes = 2000 kg

Density of silage = 800 kg/m³

$$\therefore \text{ Volume of silage fed per day} = \frac{2000}{800} = 2.5 \text{ m}^3$$

Let 'd' is the depth of the trench, 'W' be the bottom width and 1:2 side slope

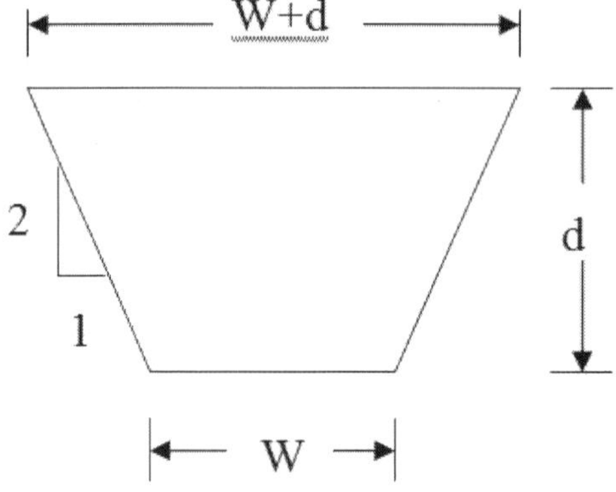

$$\text{Thus cross-sectional area of trench (trapezium)} = \frac{1}{2} \times [W+W+d] \times d$$

Assume depth of trench as 2.5 m and length of silage fed per day as 15 cm

$$\text{Thus, } \frac{1}{2} \times [2W + 2.5] \times 2.5 \times 0.15 = 2.5$$

or, W = 5.4 m (width at bottom)

Width at top = 5.4+2.5 = 7.9 m

Length of trench = 160×0.15 = 24 m

Assuming 20 per cent losses, the required length of trench = $24 \times \frac{100}{80} = 30$ m

Thus, dimensions of a trench silo shall be

Bottom width of trench	= 5.4 m
Top width of trench	= 7.9 m
Depth of trench	= 2.5 m
Length of trench	= 30 m

STORAGE OF PERISHABLES AND SEMI-PERISHABLES

The storage of perishables and semi-perishables stands as a critical practice that bridges the gap between abundance and sustainability. Perishable and semi-perishable items, ranging from vibrant produce to delicate dairy products, require meticulous attention to maintain their freshness and nutritional value while minimizing waste. Proper storage practices can significantly minimize waste and ensure optimal consumer satisfaction. The losses in postharvest sector are estimated to be from 10 to 25 per cent in durables, semi-perishables and products like milk, meat, fish and eggs. The estimated losses in fruits and vegetables are higher and reached from 30 to 40 per cent.

Perishable foods can be kept at room temperature for only few hours or 1 or 2 days before spoiling. The examples are milk and milk products, meat, fish, poultry, fruits, leafy vegetables and cooked food. **Semi -perishable** foods can be stored for a couple of weeks or even a month or two without any detectable signs of spoilage. Temperature and humidity of the environment affects the shelf stability of such foods. Proper handling and storage can result in fairly long storage without spoilage. Examples are all cereal and pulse products like wheat flour, semolina, vermicelli, broken wheat, Bengal gram flour, and some fruits and vegetables like citrus fruits, aonla, apples, pumpkin, roots and tubers, yams, potatoes, onions, garlic *etc.*

11.26 Environmental Conditions for Fruits and Vegetables

Most fruits and vegetables have an optimal temperature and humidity range for storage. Generally, a cool temperature, but above freezing, is ideal. Humidity levels play a crucial role in preventing moisture loss or excess moisture, both of which can lead to spoilage. High humidity is typically beneficial for vegetables like lettuce, cucumbers, and broccoli, as it helps retain moisture. However, low humidity is preferred for fruits like berries and grapes to prevent mould growth. Generally, a humidity range of 85-95 per cent is recommended for most produce. The desirable storage conditions for selected fruits and vegetables are given in Table 11.6.

Table 11.6. Desirable Storage Conditions for Fruits and Vegetables

Product	*Temp. (°C)*	*RH (per cent)*	*Storage Life (Days)*	*Product*	*Temp. (°C)*	*RH (per cent)*	*Storage Life (Days)*
Apple	-1 to 4	90-95	30-180	Cabbage	0	90-95	21-42
Banana	13 to 15	90-95	7-28	Carrot	0	90-95	60-150
Grape	-1 to 1	90-95	14-56	Cauli-flower	0	90-95	14-28
Guava	5 to 10	85-90	14-21	Garlic	0	65-70	30-240
Lemon	10 to 13	85-90	30-180	Onion	0	65-70	30-240
Mango	13	90-95	14-21	Potato	4 to 10	85-90	120-180
Orange	0 to 9	85-90	56-84	Tomato (ripe)	7 to 10	85-90	4-7

11.27 Maturity Indices and Ripening Stages for Sale/Storage

The maturity indices are a bunch of indicators that show the readiness of the horticultural crops for harvesting. There are two types of maturity indices *i.e.* physiological and commercial maturity and that depends upon the type of fruit/vegetable. The fruit/vegetables are classified into two groups as per their ripening methodology. Some of the fruit/vegetables ripe only on the plant/tree are known as non-climacteric whereas climacteric fruits/vegetables can be ripened on plant/tree and even in a ripening chamber using an artificial hormone named ethylene.

	Fruits	*Vegetable*
Climacteric	Mango, banana, Papaya, apple, peach, guava	Tomato, Potato
Non-Climacteric	Orange, grape, pineapple, strawberry, litchi, kinnow	Lemon, cucumber

In the case of non-climacteric fruits/vegetables, both physiological and commercial maturity are the same. Physiological maturity is about attaining the final size, shape and colour whereas commercial maturity is related to the demand of consumers in the market. The physiological and commercial maturity is different in the case of climacteric fruits/vegetables. The fruit/vegetable becomes softer, changes its colour, increases in soluble solids, and decreases in titrable acidity after the ripening.

Lemon attains its physiological maturity on tree/plant but green-coloured lemon is not required in the market. Therefore, it is necessary to attain both physiological and commercial maturity in the case of non-climacteric fruits/vegetables. The lemon cannot be ripe after harvesting hence, lemon is used to harvest after their ripening on the tree. On the other hand, tomatoes are green/

golden yellow in the stage of physiological maturity and turn red in the stage of commercial maturity. Tomatoes are climacteric vegetables and can be ripened in ripening chambers. Therefore, tomatoes are harvested in their physiological maturity stage and then ripen as per the demand in the market. The harvesting of tomatoes in their physiological maturity stage also increases their shelf life. Controlled ripening of the tomatoes using the ripening chamber provides ripen tomatoes of comparatively uniform colour than natural ripening. The ripened tomatoes using artificial hormones fetch more prices in the market due to their uniform colour, fresh state and controlled supply and demand.

Fruit/ Vegetable	*Ripening Temperature (°C)*	*Exposure Time (h)*	*Ethylene Dosage (ppm)*
Mango	20-22	24-28	100
Papaya	20-25	24-28	100
Banana	15-18	24-28	100
Tomato	18-20	24-28	100

11.28 Cooling Load Calculations

The cooling load is the amount of heat that needs to be removed from the building to maintain it at desired temperature. Normally air conditioning cooling load estimates are done on an hourly basis and cold storage load calculations are done on a 24-hour basis. The plant sizing is based on 16/18/20-hour operation and depends on the temperature/humidity to be maintained and the defrosting methods employed. All the heat which comes to the cold room must be removed to maintain the temperature. The capacity of refrigeration plant depends on the total heat load to be removed. Each load shall be calculated as kWh/day or kJ/day (1 kWh = 3600 kJ).

The heat load is usually divided into following four components:

1. **Transmission Load** – sensible heat gain through the floor, walls and roof from the temperature difference across these surfaces.
2. **Infiltration Load** – sensible and latent heat that carried into the refrigerated space from the air exchange through the door openings.
3. **Product Load** – heat removal required to lower the incoming product to the space temperature in a given amount of time.
4. **Miscellaneous Load** – load from lights, people and equipment in the space.

11.28.1 Transmission Load

This is the thermal energy transferred through the roof, walls and floor into the cold room. Heat always flows from hot to cold and the interior of the cold room is obviously a colder than its surroundings, so heat is always trying to enter the space because of that difference in temperature. The heat flow rate (Q_t) by conduction through walls, ceiling and floor is given by

$$Q_t = UA(T_o - T_i)$$

where, Q_t = heat flow rate (W),

U = overall heat transfer coefficient (W/m²-°C),

T_o and T_i = outer and inner room temperatures, respectively (°C), and

A = surface area of walls, roof and floor (m²)

The overall heat transfer coefficient is given by

$$\frac{1}{U} = \frac{x_1}{k_1} + \frac{x_2}{k_2} + \frac{x_3}{k_3} + ...$$

where, x_1, x_2, x_3 = thickness of different layers of wall, and

k_1, k_2, k_3 = thermal conductivity of the respective materials

The conversion factor to convert the units in Watts to kWh/day is 24/1000 (Students are advised to verify this conversion factor). Thus, multiply the above term by 24 and divide by 1000 to get the transmission load in kWh/day.

Example 11.9

Find the refrigeration load expressed in tons of refrigeration which is caused by heat loss from the four side walls of a small cold room 2.4 m×3.0 m×2.4 m. The walls are made of 20 cm brick, 20 cm cork board and 1.25 cm cement. Inside wall temperature is 21°C and outside is – 30°C. Add a suitable safety factor for losses through joints. The thermal conductivity of brick, cork and cement plaster are 0.6 W/m °C, 0.04 W/m °C and 0.8 W/m °C respectively.

(ARS 1991 and 2003)

Solution

$$\frac{1}{U} = \frac{x_1}{k_1} + \frac{x_2}{k_2} + \frac{x_3}{k_3} = \frac{0.2}{0.6} + \frac{0.2}{0.04} + \frac{0.0125}{0.8} = 5.3486$$

or, Overall heat transfer coefficient (U) = 1/5.3486 = 0.187 W/m²-°C

Surface area = 2×[2.4×2.4+3×2.4] = 25.92 m²

Rate of heat loss (q) = UAΔT = 0.187×25.92×[21 – (- 30)] = 247.2 W

Consider safety factor for losses through joints as 2.

Thus, total heat loss = 247.2×2 = 494.4 W = 0.4944 kW

$$\text{Refrigeration load} = \frac{0.4944}{3.5} = 0.14 \text{ ton (as 1 ton of refrigeration} = 3.5 \text{ kW)}$$

11.28.2 Infiltration Heat Load

Infiltration heat load is the heat gain due to air leakage into the refrigerated space. This occurs due to the infiltration of warm air through openings like doors, windows, and gaps. The other consideration is ventilation. Fruit and vegetables give off carbon dioxide so some stores will require a ventilation fan, this air needs to be cooled down so one must account for this if it's used. The infiltration heat load (Q_i) in kWh per day is calculated as

$$Q_i = \frac{ACPD \times V \times E \times (T_o - T_i)}{3600}$$

where, ACPD = number of air changes per day,

V = volume of the cold store, and

E = energy per cubic meter per degree Celsius (kJ/m^3-°C).

11.28.3 Product Load

The product load refers to the amount of heat that needs to be removed from the refrigerated space to lower the temperature of the product from initial temperature to storage temperature. It is the energy required to cool, freeze and further cool after freezing. If there is just cooling of the product, then consider only the sensible heat load. For freezing the product then one need to account for the latent heat also as a phase change occurs. There is additional energy required to further chill this food down below the freezing point which is again sensible heat. The product load can be calculated as

Total amount of heat removed (Q_p) = sensible heat removal from initial temperature to freezing temperature + latent heat to convert water into ice + sensible heat removal from freezing temperature to final storage temperature

$$Q_p = \frac{mc_{pb}(T_1 - T_f) + m\,\lambda_f + mc_{pa}(T_f - T_2)}{3600}$$

where,

Q_p = total amount of heat removed in kWh/day (divided by 3600 to convert kJ to kWh),

T_1 = initial temperature (°C),

T_2 = final temperature (°C),

T_f = freezing temperature (°C),

c_{pb} and c_{pa} = specific heat of the product before and after freezing respectively (kJ/kg K),

λ_f = latent heat of crystallization (kJ/kg), and

m = mass of the product (kg).

Product respiration also need to be accounted for if fruits and vegetables are to be stored. Fruit and vegetables are alive and they will generate some heat due to respiration. It is calculated as

$$Q_r = \frac{m \times R}{3600}$$

where,

Q_r = heat load due to respiration in kWh/day (divided by 3600 to convert kJ to kWh),

m = mass of product in storage (kg), and

R = respiration rate of the product (kJ/kg-day).

Total product load (Q_P) = $Q_p + Q_r$

Example 11.10

One thousand units of mixed fruit bar, each weighing 100 g with a surface area of 0.01 m^2, are frozen from 70°C molten mass condition to -20°C frozen storage condition within 3 hours. The specific heat capacity values of the bar are 3.6 kJ kg^{-1} K^{-1} and 1.97 kJ kg^{-1} K^{-1} before and after freezing point (0°C) respectively. If the latent heat of crystallization is 250 kJ kg^{-1}, what will be the cooling capacity of the refrigeration unit required in tons of refrigeration? *(GATE 2012)*

Solution

Weight of one fruit bar = 100 g = 0.1 kg

Total mass of fruit bars (m) = 1000×0.1 = 100 kg

Initial temperature (T_1) = 70°C

Final temperature (T_2) = - 20°C

Freezing temperature (T_f) = 0°C

Specific heat capacity before freezing (c_{pb}) = 3.6 kJ kg^{-1} K^{-1}

Specific heat capacity after freezing (c_{pa}) = 1.97 kJ kg^{-1} K^{-1}

Latent heat of crystallization (λ_f)= 250 kJ kg^{-1}

Total heat extracted = sensible heat removed for cooling upto 0°C + latent heat to convert water into ice + sensible heat removed for freezing from 0°C to – 20°C

$$\text{Cooling load (Q)} = mc_{pb}(T_1 - T_f) + m\,\lambda_f + mc_{pa}(T_f - T_2)$$

$$= 100 \times 3.6 \times (70 - 0) + 100 \times 250 + 100 \times 1.97 \times [0 - (-20)]$$

$$= 54140 \text{ kJ}$$

$$\text{Rate of cooling} = \frac{54140}{3} = 18046.7 \text{ kJ/h} \approx 5 \text{ kJ/s} = 5 \text{ kW} = (5/3.5)$$

$$= 1.42 \text{ ton of refrigeration}$$

(since 1 ton of refrigeration = 3.5 kW)

11.28.4 Miscellaneous Loads

(i) Internal Heat Load - People

This load is the heat given off by people working in the cold room. It is calculated as

$$Q_{m1} = (N \times t \times H)/1000$$

where,

Q_{m1} = heat load due to occupants (kWh/day)

N = number of people working in the cold store,

t = length of time spent in the cold store per person in a day (hours), and

H = heat losses per person per hour (Watts)

(ii) Internal Heat Load – Lighting

Lighting fixtures generate heat within the cold storage area. The heat load (kWh/day) due to lighting can be calculated as

$$Q_{m2} = \frac{\text{number of lamps} \times \text{daily usage hour} \times \text{power rating of the lightning in watts}}{1000}$$

(iii) Equipment Load

The refrigeration equipment also add heat in the room. The rating of the fan motors and how long they run for each day must be known to calculate heat generation of the fan motors. The heat generation (kWh/day) of the fan motors in the evaporator is calculated by the following equation:

$$Q_{m3} = \frac{\text{number of fans} \times \text{fan running time per day (hours)} \times \text{fan motors nominal power (Watts)}}{1000}$$

Total miscellaneous load $(Q_m) = Q_{m1} + Q_{m2} + Q_{m3}$

11.28.5 Total Cooling Load

Sum up the individual heat load components to obtain the total cooling load that the refrigeration system needs to remove from the cold storage.

$$Q_T = Q_t + Q_i + Q_P + Q_m$$

Safety factor is added to account for errors and variations from design. Typically, it is taken as 10 to 30 per cent onto the calculation to cover this.

11.28.6 Refrigeration Cooling Capacity

The capacity of refrigeration unit in kW is estimated by dividing the total cooling load per day (kWh/day) by the number of running hours of the refrigeration unit. The capacity in terms of ton of refrigeration can be calculated by dividing the capacity in kW by 3.5 (as 1 tonne of refrigeration = 3.5 kW).

Example 11.11

Calculate total cooling load of a cold store of size 8×5×4 m, where 6000 kg of products are placed daily and 30000 kg of apples are stored permanently inside. The temperature of ambient air is 30°C with 50 per cent RH and the store is maintained at 1°C and 95 per cent RH. The walls, roof, and floors are insulated with 80 mm polyurethane with a U = 0.28 W/m^2–K. The floor temperature is 10°C. Assuming that the door will create 5 volume air changes per day due to the product entering and leaving the warehouse. Each cubic meter of new air adds 2kJ/°C heat. Every day, 6,000 kg of apples arrive at the cold store at a temperature of 10°C and a heat capacity of 3.65 kJ/kg. The average respiratory heat of apple is 1.9 kJ/kg per day. In the cold store, 2 people work for 4 hours daily. The heat losses per person per hour is 271 W. There are 3 lamps at 120 W each in the cold store, running 4 hours a day. This cold room evaporator uses 3 fans of 300 W each and run 16 hours a day. Take safety factor as 1.2.

Solution

1. *Transmission Load*

Overall heat transfer coefficient (U) = 0.28 W/m^2–K

Outside temperature (T_o) = 30°C

Inside temperature (T_i) = 1°C

Floor temperature (T_f) = 10°C

Surface area of walls and roof = 2×(8×4 + 5×4)+ 8×5 = 144 m^2

Surface area of floor = 8×5 = 40 m^2

$$Q_t = UA(T_o - T_i)$$

Heat gain due to walls and roof = 0.28×144×(30 – 1) = 1169.3 W

Heat gain due to floor = 0.28×40×(10 – 1) = 100.8 W

Total transmission load = 1169.3 + 100.8 = 1270.1 W

= 1270.1×(24/1000) ≈ 30.5 kWh/day

2. *Infiltration Load*

Number of air changes per day = 5

Volume of the cold store = 8×5×4 = 160 m^3

Energy per cubic meter per degree Celsius = 2 kJ

$$Q_i = \frac{ACPD \times V \times E \times (T_o - T_i)}{3600}$$

$$= \frac{5 \times 160 \times 2 \times (30 - 1)}{3600} \approx 12.9 \text{ kWh/day}$$

3. *Product Load*

Mass of the product (m) = 6000 kg per day

Specific heat of the product (c_p) = 3.65 kJ/kg

Initial temperature of apple (T_1) = 10°C

Final temperature of apple (T_2) = 1°C

Total amount of heat removed

$$Q_p = \frac{mc_p(T_1 - T_2)}{3600} = \frac{6000 \times 3.65 \times (10 - 1)}{3600} \approx 54.8 \text{ kWh/day}$$

Product respiration

Mass of product in storage (m) = 30,000 kg

Respiration rate of the product (R) = 1.9 kJ/kg per day

Product respiration, $Q_r = \frac{m \times R}{3600} = \frac{30{,}000 \times 1.9}{3600} \approx 15.8$ kWh/day

Total product load (Q_P) = $Q_p + Q_r$ = 54.8+15.8 = 70.6 kWh/day

4. *Miscellaneous Loads*

(i) Internal Heat Load - People

Number of people working in the cold store (N) = 2

Length of time spent in the cold store per person in a day (t) = 4 hours

Heat losses per person per hour (H) = 271 W

Heat load due to people working in the cold room

Q_{m1} = (N × t × H)/1000 = (2 × 4 × 271)/1000 ≈ 2.2 kWh/day

(ii) Internal Heat Load – Lighting

Number of lamps = 3

Daily usage = 4 hours

Power rating of lamp = 120 W

Heat load due to lighting

$$Q_{m2} = \frac{\text{number of lamps} \times \text{daily usage hour} \times \text{power rating of the lightning in watts}}{1000}$$

$$= \frac{3 \times 4 \times 120}{1000} = 1.44 \text{ kWh/day}$$

(iii) Equipment Load

Number of fans = 3

Fan running time = 16 hours

Fan motor power = 300 W

Heat generation of the fan motors in the evaporator

$$Q_{m3} = \frac{\text{number of fans} \times \text{fan running time per day (hours)} \times \text{fan motors nominal power (Watts)}}{1000} = \frac{3 \times 16 \times 300}{1000}$$

$$= 14.4 \text{ kWh/day}$$

Total miscellaneous load (Q_m) = $Q_{m1} + Q_{m2} + Q_{m3}$

= 2.2 + 1.44 + 14.4 = 18.04 kWh/day

Total Cooling Load

Total cooling load is sum of transmission load, infiltration load, product load and miscellaneous loads.

$Q_T = Q_t + Q_i + Q_P + Q_m$ = 30.5+12.9+70.6+18.04 = 132.04 kWh/day

Safety factor = 1.2

Thus, total cooling load = 132.04×1.2 ≈ 158.5 kWh/day

Refrigeration Cooling Capacity

$$\text{Capacity of refrigeration unit} = \frac{\text{Total cooling load per day}}{\text{Number of running hours}}$$

$$= \frac{158.5}{16} = 9.9 \text{ kW} = 2.82 \text{ ton of refrigeration}$$

(as 1 tonne of refrigeration = 3.5 kW).

11.29 Controlled Atmosphere Storage

It is a type of cold store that has provision to control the temperature, relative humidity, and gas composition as per the requirement of the stored commodity. It is a costly technology due to the use of refrigeration systems, heating systems, humidifiers and gas cylinders for a closed chamber. The food grains are of low moisture content (about 12 per cent) and are categorized under durable commodities. Therefore, food grains are less susceptible to deterioration than fruit and vegetables. Generally, CAS are being used for the storage of fruit and vegetables, although food grains can be stored in the CAS and it will prolong their shelf life.

The food commodities are stored inside the CAS in their safe storage condition comprised of storage temperature, relative humidity and gas composition. The fruit and vegetables are of high moisture content therefore, about 90 per cent of the RH is required to be maintained. Fresh commodities consume oxygen during respiration therefore minimum of 1 per cent of the oxygen is very important to maintain inside CAS. The presence of oxygen inside CAS will prevent fermentation.

Here are some of the design considerations for the controlled atmosphere storage:

1. The gross volume is about 50 per cent greater than the net volume, and the gross area (same concept as volume) is about 25 per cent greater than the net area.
2. Heat load include field heat, heat of respiration, service load and heat loss by store.
3. A cubical shape is recommended for minimum surface area.
4. The multi-storey building is recommended to save space and lower the heat load.
5. An anteroom should be provided in front of the storage space.
6. Pre-cool the fresh produce before cold storage to reduce the heat load and increase the shelf life.
7. 95-98 per cent relative humidity is recommended for fresh produce like mango, grapes *etc.*
8. 2-6 air exchange per hour ventilation rate is recommended inside the cold store.
9. The carbon dioxide should be below 5000ppm during the ripening process otherwise the higher level of carbon dioxide will slow down the ripening process.
10. The fruits should not occupy above 75 per cent of the ripening chamber.
11. The crates of fruits should be 4-6 inches away from the wall of the ripening chamber.

11.30 Modified Atmosphere Storage

The atmosphere is comprised of 78 per cent nitrogen, 21 per cent oxygen, 0.03 per cent carbon dioxide and other gasses. Oxygen is required for the respiration of fresh fruits and vegetables and they produce carbon dioxide. The shelf life depends upon the rate of biological respiration. The process of respiration consumes the carbohydrate of fruits/vegetables and produce water along with heat. It is also a process of senescence that ultimately spoil the agro-produce. The ripening process is also dependent upon respiration that increases the concentration of ethylene that ripen the agro produce. The respiration depends upon many factors like temperature, age of fresh produce, moisture content, damage level, atmospheric gas composition *etc.* Therefore, the rate of respiration can be controlled by using the desired composition of the gasses and this modified atmosphere increases the shelf life of the produce. The composition of gasses depends upon the moisture content and type of the agro-produce.

Modification of Atmosphere Composition

The modification of the gas composition can be achieved in two ways:

1. Active modified atmosphere
2. Passive modified atmosphere

An active modified atmosphere is utilizing the active compounds for absorption or emission of desired gasses. The active atmosphere can also be achieved by using machinery that replaces the normal atmosphere gasses with specially designed gas composition as per the stored/packed commodities. The sensors monitor the gas composition and maintain it by using sachets/pouches of active ingredients. The use of silica gel sachets for absorption of extra moisture is an example of the active modified atmosphere.

A passive modified atmosphere is usually used for the respiring agro produce. In the case of food grains, lower moisture content (12 per cent) took a comparatively longer time to attain the desired gas composition. The killing or control of insects is the main target in the case of food grains. The hermetic bags are an example of the passive modified atmosphere for durable or low moisture agro produce. Whereas, the higher concentration of carbon dioxide suppresses the rate of respiration and limits the production of ethylene. Modified atmosphere storage and packaging are already being practised for fresh fruit and vegetables. The modified atmosphere storage and packaging accompanied by low temperature add more days in the post-harvest life of agro produces.

11.31 Zero Energy Cool Chamber

Traditional methods of storing fruits and vegetables often involve refrigeration, which consumes significant energy and contributes to environmental concerns. Refrigerated cool storage is not only energy intensive and expensive, but also involves large initial capital investment, and cannot be afforded by small and marginal farmers. However, the zero energy cool chamber offers a sustainable and eco-friendly solution to this challenge. The concept behind the zero energy cool chamber is ingeniously simple yet effective. It harnesses the principles of passive cooling and thermal insulation to create an environment that naturally maintains low temperatures, thus extending the shelf life of perishable produce without the need for electricity. Once constructed, the maintenance costs are relatively low compared to traditional cooling systems.

Zero energy cool chamber (or evaporative cool chamber) was first developed at ICAR–Indian Agricultural Research Institute, Pusa, New Delhi which works on the principle of evaporative cooling. Evaporative cooling is the reduction in temperature resulting from the evaporation of a liquid, which removes latent heat from the surface from which evaporation takes place. It is named as zero energy cool chamber because there is no need for electrical and mechanical energy for the functioning of this storage structure. This chamber can be easily constructed with locally available materials. Zero energy cool chamber can keep the temperature 10-15°C cooler than the outside ambient temperature and maintain about 90 per cent relative humidity. It is also known as rural refrigerator. This technology is most effective in hot and dry regions and helps in reducing the post-harvest losses at farm level.

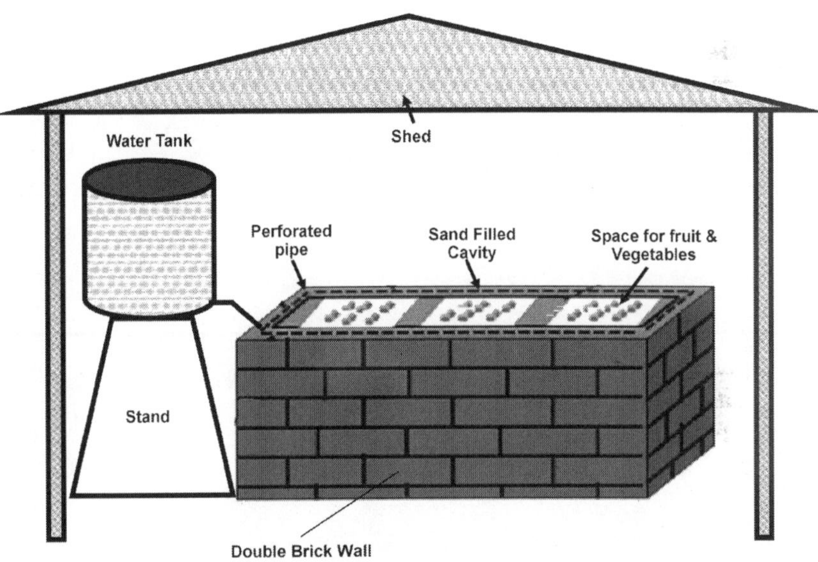

Figure 11.19. Zero Energy Cool Chamber.

The structure is comprised of bricked floor of single layer. Over this, double side wall is erected with spacing of 7.5 cm. Riverbed sand is filled in between the cavity of double wall, and top is covered with bamboo and gunny cloth. Bricks and sand both has great capacity to absorb and retain water which evaporates slowly and steadily depending on the atmospheric temperature and humidity resulting in cooling. The chamber's orientation and shading mechanisms, such as using trees or other structures, shall be such to prevent direct sunlight from heating the chamber. This further aids in maintaining lower temperatures.

The cool chamber is saturated with water for first time thereafter sprinkling of water once in the morning and once in the evening. A drip system with plastic pipes and micro tubes connected to an overhead water source is installed for this purpose. The fruits and vegetables are stored in this chamber by keeping in perforated plastic crates. These crates are covered with a thin polyethylene sheet. These chambers are ideal for storage of perishable for a short period and helpful for small and marginal farmers to store their produce. It can be easily constructed by an unskilled person. Performance evaluation of cool chambers at different locations of the country has been found satisfactory for short term storage.

The size of an evaporative cool chamber can be chosen to meet a range of user storage needs. It is important to select an appropriate size according to the need, to avoid over-building and spending more money than is needed. Punjab Agricultural University, Ludhiana and ICAR–Central Institute of Post-Harvest Engineering and Technology, Ludhiana have designed and developed large capacity evaporative cooled chambers.

11.32 Onion Storage

India is second largest producers of onion in the world second only to China. The crop is grown throughout the country. Most of the onion produced in India comes from the state of Maharashtra, Madhya Pradesh, Karnataka, Andhra Pradesh, Bihar, Gujarat and Haryana. Lack of adequate and appropriate storage facility is one the major constraint which enforces distress sale on farmers.

Onion is a semi perishable vegetable and in India it is grown in three seasons – *kharif* (planted between July-August and harvested in October-December, 20 per cent); late *kharif* (planted between October-November and harvested in January-March, 20 per cent); and *rabi* (planted between December-January and harvested in March-May, 60 per cent). *Kharif* and late *kharif* onion harvest is immediately consumed within one or two months, and therefore does not require storage. However, *rabi* harvest of April-May is in huge quantity creating glut in market. The same crop must continue to meet the consumer demand till the month of October-November every year before the *kharif* crop is harvested and brought to the market. It is therefore vital to successfully store onion for 4-5 months during June-October in order meet the regular supply. It is observed that nearly 30-40 per cent of the crop is lost during storage due to the various reasons in form of physiological weight loss, rotting, sprouting *etc.* The losses occurred during storage are in terms of qualitative as well as in quantitative ways.

Various abiotic factors like temperature, relative humidity affects the health of onions hence their balance is must needed to store the crop with minimum losses. High temperature (above 32°C) and low relative humidity (less than 60 per cent) causes weight losses, while high temperature and high relative humidity (above 70 per cent) shall result in rotting of onions. Sprouting of onions shall occur if onions are stored at low temperature and low relative humidity.

The optimum storage range of temperature and relative humidity for onion is between 25°C to 30°C and 65 per cent to 70 per cent, or 0°C to 5°C and 65 per cent to 70 per cent respectively.

For effective long storage of onion, the parameters essential to be looked after are the bulb size, choice of cultivars, cultivation practices, time of harvest, field curing, removal of tops, drying, grading, packing, storage conditions. Proper ventilation, moisture control, and protection from pests and rodents are critical considerations to ensure the onions remain fresh and of good quality during storage. Additionally, it's important to regularly inspect the stored onions for any signs of spoilage and to remove any affected onions promptly to prevent the spread of decay.

Onion Storage Structures

For onion storage, technology may be either with natural ventilation or with forced ventilation. Although cold storage systems are used in certain countries for onion, this is normally not adopted in India due to poor economics and lack of cold chain facilities required to maintain the quality in the high ambient temperature prevalent in our country. Onion storage in ventilation condition is quite satisfactory when the temperature is maintained between 25°C to 30°C with a relative humidity range of 65 per cent to 70 per cent. This environment reduces the storage losses, which are in the form of physiological loss in weight, rotting and sprouting. The different onion storage structures are as follows:

1. **Naturally Ventilated Structures (*Kandha Chawl*)**: *Kandha Chawl* are the naturally ventilated structure, a scientific onion hut to minimize the losses due to storage. In India, the onions are mostly stored in such structures without any control of temperature and relative humidity. The farmers construct different types of *Kandha Chawl* based on the capacity required. The structure is made of local bamboo. During the storage period, 3 – 4 turnings and screenings of spoiled onions are carried out manually.
2. **Low Cost Thatched Roof Bamboo Storage Structure**: This type of storage structure is usually constructed with bamboo framework having the roof made up with sugarcane leaves and is preferred for 'on-farm' storage of onions. This is a single row storage structure that can be made of 5 to 10 tonnes capacity. The structure is constructed with bamboo rafters. The whole solid bamboos are used for pillars and roof beams. Half split bamboos are used for floor while the sidewalls are made by split bamboos (1/6). The pillars of whole bamboos are erected at 1.5 m distance. The iron angle provide support to all pillars. The bottom ventilation is provided with bricks fixed at the base of all iron angle pillars. The roof is usually made with sugarcane leaves but similar type of grass can be used for this purpose. Inner lining of gunny cloth is provided to check the leakage of rainwater. These

single row onion storage structure should be oriented in the North to South direction (opposite to wind direction) *i.e.*, length facing the East-West direction. This type of storage structure is low cost and easy to construct, but leads up to 40-42 per cent losses of onion during four months of storage. The durability of the structure is low due to use of organic material/bamboo. The temperature and humidity cannot be controlled since it works on natural ventilation mechanism.

Figure 11.20. Naturally Ventilated Onion Storage Structure.

3. **Bottom and Side Ventilated Double Row Storage Structure***:* This has provision ventilated floor made of wooden bantams, central ventilated pathway and extended roof. The framework of the structure is constructed with galvanized iron channels. The floor and sides walls are made with wooden bantam of 2.5 cm thickness and gap of 2.5 cm is kept between the bantams. The roof is constructed with asbestos sheets. The roof is extended to 1 meter to avoid splashes of rain. Shading is used to protect onions from direct sunlight, which can cause excessive drying and affect the quality of the onions. Shading materials like jute cloth or thatch are often used for this purpose. Onions are usually stacked in trays or baskets to allow air circulation between the layers. Proper stacking prevents mechanical damage to the onions and allows for easy monitoring and removal of spoiled or damaged onions. This type of structure has a provision of ventilation from bottom and sides. These structures have length of about 9-15 m

and width of 3.6 m with two rows. The width of each row is 1.2 m and free space of 1.2 m is between the two rows is kept for walking. The length should not be more than 15 m. Double row onion storage structure should be constructed in the direction of wind *i.e.* East-West.

4. **Commercial Onion Storage Structures**: Commercial onion storage facilities, including cold storage units, are used to maintain low temperatures and humidity levels. This helps extend the shelf life of onions and prevents sprouting. In controlled onion storage structures, the onions are stored at 0-5°C and 65-70 per cent RH that leads to much lesser losses as compared to ventilated storage structure. The cost of construction and running cost are very high as energy required to maintain the storage facility in the temperature range of 0-5°C is high. The other problems are condensation and require lot of energy and time. The bulbs start sprouting immediately after they are removed from the cold storage.

Important Points to Remember

1. Bag storage structures are often used to store grains from 25 to 500 tonnes.
2. Lateral pressure in deep bins is calculated by Janssen equation.
3. Grain infesting insects cannot survive at temperatures above 45°C.
4. The optimum moisture content for paddy storage is 12 per cent.
5. The optimum moisture content for one-year duration of groundnut storage is 7 per cent.
6. The recommended safe moisture content for storage of oilseeds is 6 to 8 per cent.
7. Angle of repose of granular materials in general is more than the angle of internal friction.
8. In grain storage, deterioration index of zero corresponds to relative humidity of 30 per cent.
9. The height of stack in CAP grain storage structure should be not more than 18 bags.
10. Bottom of silo should have slope more than/to angle of repose of grain.
11. A bin whose dimensions are such that the plane of rupture meets the grain surface before it strikes the opposite side wall is called shallow bin.
12. The rate of respiration is dependent upon moisture content and temperature of the grain.
13. The moisture content attained by the grain with respect to a set of atmospheric temperature and relative humidity is called equilibrium moisture content of the grain.

14. Stack to stack distance in a bag storage structures should be 2 m and stack to wall distance should be 1 m.
15. The angle of repose is the angle between the base and slope of the cone formed on a free vertical fall of the grain mass to a horizontal plane.
16. The relationship between the ratio of lateral pressure P and vertical pressure V in a grain silo containing granular material with an angle of inter granular friction φ is given by $P/V = (1-\sin\varphi)/(1+\sin\varphi)$.
17. An air tight system in the stored grain is ideal for fumigation.
18. Grains are usually alive; they breathe and produce heat, moisture and carbon dioxide.
19. A bin, in which the place of rupture meets the opposite side before it emerges from the grain, is called deep bin.
20. Grain storage fungi grow most rapidly at 30 to 32°C.
21. Methyl bromide or phosphine is most useful for fumigation.
22. *PUSA* bin is made of mud and plastic film of 700 gauge is used in all the inner sides.
23. In CAP storage structure, cover is rectangular in shape having five sides made from polyethylene film of 1000 gauge.
24. Most common fumigant for storage of cereals is aluminium phosphide.
25. The sprouting of the grain during storage occurs mainly owing to generation of heat as a result of infestation.
26. Hydraulic radius is the ratio of area of cross-section and perimeter of bin.
27. Coefficient of friction between granular materials is equal to the tangent of the angle of internal friction of the material.
28. Aeration is the process of blowing small quantity of air through the bulk of the agricultural produce for cooling and equalizing the temperature.
29. In grain respiration process, CO_2 production of 1 kg corresponds to a dry matter loss of 0.682 kg.
30. The high relative humidity required in a fruit and vegetable storage is favourable to the growth of surface moulds and fungi.
31. Potatoes suffer from internal breakdown known as 'black heart' when stored at the temperature of 0°C.
32. To avoid bruising, the onions should not be stored deeper than 4 meter.
33. Optimum storage conditions for storage of onions are 0°C temperature and 65-70 per cent RH.
34. Most effective method of arresting spoilage of perishables is to lower the temperature of storage environment.
35. The levels of oxygen, carbon dioxide and nitrogen are maintained

in controlled atmosphere storage of fruits and vegetables.

36. Fruits and vegetables are perishable foods.
37. The pre-cooling removes the field heat which ultimately reduces the respiration rate and moisture loss.
38. Moisture migration in stored food grains occurs owing to the changes in temperature.
39. Heat released from fruits and vegetables in storage is due to their respiration.
40. As a thumb rule, for every 10°C rise of temperature, the rate of respiration is doubled.
41. High moisture and warm climate promote mould growth, insect growth and increase rate of respiration of grains.
42. Air tight closure is also known as hermetic closure.
43. During summer, due to moisture migration spoilage of grains takes place at bottom of the bin.
44. During winter, due to moisture migration spoilage of grains takes place at top of the bin.
45. One rat consumes about 10 kg of grain in a year.
46. Phostoxin, Datia and Celphos are the trade name of solid fumigants.
47. Carbon tetra chloride, ethylene dichloride, ethylene dibromide and methyl bromide are liquid fumigants.
48. The stored fodder is known as silage.
49. The cross-section of the trench silo depends upon the size of herd and the length is determined by the number of days the silage is fed in a year.
50. A permanent pit silo is a circular deep well, which is lined all along the side and sealed from the bottom so that water may not rise into it.
51. Moisture content of grains below 13 per cent arrest the growth of most of the microorganisms and mites.
52. One cubic meter of silage weighs 650 kg.
53. Generally, a cow is fed 3 kg of silage per 100 kg of body weight per day.
54. Generally, a cow is fed with 14 to 18 kg of silage ration per day.
55. Generally, a cow is fed about 1.4 kg of silage per 45 kg of her body weight per day.
56. Pit silos have circular cross section.
57. Generally cows should be fed silage at a rate of 3 per cent of body weight per day.

58. Trench silo has trapezoidal cross section.
59. Pit silos are made in areas where the soil is deep and the water table is very low.
60. The body weight of a *Murrah* buffalo on the average is 680 kg.
61. The diameter of pit silo is limited to 6 meter and its depth is kept 2 to 3 times that of the diameter.
62. *Morai* type grain storage structures are popular in South-Eastern India.
63. *Kothar* type grain storage structures are popular in Northern India.
64. Rat proofing cones are provided in grain storage at a height of 1.2 m.

Exercises/Questions

1. Why storage of grain is necessary? What are the requirements of a good storage structure?
2. Explain different types of grain storage structures.
3. List the factors affecting the food grains during storage.
4. Discuss relative advantages and disadvantages of bag and bulk storage.
5. Discuss the criterion to differentiate between a deep and shallow bin.
6. Differentiate between angle of repose and angle of internal friction.
7. Discuss the functional and structural aspects involved in the design of a bin.
8. What are the sources of infestation of the stored grains?
9. Discuss different types of structures for bulk storage of grains.
10. Describe Morai type storage structure with the help of a neat diagram.
11. Describe Kothar type structure with the help of a neat sketch.
12. Describe Bukhari type structure with the help of a neat sketch.
13. Write a short note on Cover and Plinth (CAP) storage. How these structures meet the requirements of safe storage of grains under various agro-climatic zones of India?
14. Discuss various theories used to calculate the grain pressure.
15. Derive Janssen equation for calculating the lateral pressure for design of storage structure for deep bin.
16. Distinguish between Rankine's formula and Janssen's formula in pressure distribution calculation in storage bin.
17. What points will you consider for selection of site of a storage godown?
18. Which points should be considered during bag storage of food grains?
19. What is a silage? Briefly discuss the process of silage making.
20. Describe the different types of silos used for silage making.
21. Explain the technique for the scientific storage of food grains.

22. Write short notes on:
    (a) Janssen's theory for design of bin
    (b) Trench silo
    (c) Rural grain storage
    (d) Deep bins

23. Describe briefly the controlled atmospheric storage system. Also write its merits and demerits.

24. A cylindrical shallow bin is filled with grains having angle of repose of 33°. Calculate the limiting height to diameter ratio of the bin. *(GATE 2017)*

    [**Answer**:1.84]

25 In a circular concrete silo of 2.5 m internal diameter and 8 m height, paddy weighing 800 kg/m³ is loaded. The angle of internal friction of paddy is 28° and that for paddy and concrete is 25°. What will be the maximum lateral pressure (kg/m²) at the bottom of bin section (using Airy's theory)?

    [**Answer**: 3439.4 kg_f/m^2]

26. Angle of internal friction for rice grain is 27°, bulk density of rice at 14 per cent moisture content is 833 kg/m³ and coefficient of friction between rice and concrete wall is 0.5. For a silo of 5 m diameter and 20 m height, what is the ratio between the lateral pressure at the bottom of the silo obtained by Rankine and Janssen formulae? *(GATE 2007)*

    [**Answer**: 3.16]

27. The diameter of a grain storage bin is 4 m and the depth is 16 m. It is completely filled with wheat having bulk density of 800 kg m⁻³. The angle of friction between wheat and wall is 24°. The ratio of lateral and vertical pressure intensity is 0.4. What is the lateral pressure intensity of wheat in kPa on the bin wall at 2 m depth? *(GATE 2009)*

    [**Answer**: 5.28 kPa)

28. A cylindrical silo with 3.0 m diameter and height to diameter ratio of 5:1 is filled with 60 metric ton wheat grains having bulk density of 725 kg m⁻³. The coefficient of friction between grain and silo wall is 0.42 and the ratio of lateral pressure to vertical pressure is 0.5. Calculate the vertical pressure at the bottom of silo in kPa (round off to 2 decimal places). [use 1 kg_f =9.81N] *(GATE 2020)*

    [**Answer**: 24.02 kPa]

29. A RCC cylindrical bin 15 m deep has internal diameter of 5 m and is completely filled with paddy weighing 750 kg/m³. The angle of internal friction for paddy can be taken as 35°, while the angle of

friction between paddy and bin wall is 30°. The ratio of horizontal and vertical pressure intensity k is 0.45. Find the lateral pressure at 3.0 m interval. *(ARS 1994, 2003)*

[**Answers**: 3 m – 753.4 kg_f/m^2, 6 m - 1157.4 kg_f/m^2, 9 m - 1374.1 kg_f/m^2, 12 m - 1490.3 kg_f/m^2, 15 m - 1552.6 kg_f/m^2]

30. Design a bag storage structure for storing 250 tonnes of paddy with a stack of 10 bags in width. Each bag is 85 cm long, 60 cm wide and 30 cm high. Give the complete floor plan of the structure (Godown). Assume other reasonable data wherever necessary, but state them clearly. *(ARS 2011)*

    [**Answer**: 3 stacks of 1200 bags each, overall dimension 24×11×5 m]

31. Design a 5000 tonnes grain warehouse for storing cereal grains. Draw a layout plan for such a godown giving approximately dimensions including inside stacks, compartments, alleys *etc.* *(ARS 1994)*

    [**Answer**: 32 stacks of 1800 bags each, 4 compartments of 8 stacks each, each compartment size 32×28×5.5 m]

32. Design an economical size of silo to store sufficient quantity of silage for a herd of 400 dairy cows having an average body weight of 450 kg each. The cows are fed silage for 200 days in a year at the rate of 3 kg/100 kg of body weight. Assume the density of silage to be 650 kg/cu.m and average 10 cm thick silage is to be removed every day. Also assume losses at 15 per cent. *(ARS 2005)*

    [**Answer**: 3 pits of 6.4 m diameter and 20 m depth]

33. Design a trench silo for a small farm having the following herd. The silage is fed 160 days in a year.

Animal Breed	*Body Weight per Animal (kg)*	*Total Number of Animals*	*Rate of Feeding per 100 kg Body Weight (kg)*
Murrah buffaloes	650	50	4.0
Hariana cows	450	50	3.0
Bullocks	500	10	3.5
Heifers	200	25	3.0

[**Answer**: Bottom width - 6.4 m, top width - 8.9 m, depth - 2.5 m, length - 30 m]

34. Apple having specific heat of 3.5 kJ/kg °C is cooled from the initial temperature of 4°C to the final temperature of -6°C. Given the latent heat of fusion as 335 kJ/kg, what will be the amount of total heat removal (in kJ/kg)?

    [**Answer**: 370 kJ/kg]

35. In a cold store of 30 m×15 m× 15 m size, 4000 tonnes of potato having the specific heat of 3.62 kJ/kg-K and heat of respiration of 20 W/ m^3 is kept at 30°C. Potato is required to be cooled to 2°C in 30 days. Neglecting other sources of heat, what will be the capacity of the refrigeration plant required? *(GATE 2007)*

[**Answer**: 83.3 tonnes]

Chapter 12

Miscellaneous Farm Structures

Miscellaneous farm structures are indispensable for the success of agricultural enterprises. These structures serve a multitude of functions that are essential for the efficient and successful operation of a farm. They contribute to increased productivity, improved product quality, resource protection, and the well-being of both plants and animals. Proper planning, construction, and maintenance of these structures are paramount for the long-term sustainability and profitability of farms. In this chapter, some of the essential structures which are not covered in other chapters are described here.

FARM MACHINERY AND IMPLEMENT SHED

An implement shed is a building used to store farm machinery and equipment. The shed should be large enough to accommodate all of the machinery and equipment available on the farm, and should allow for easy access and manoeuvrability. The structure should be able to withstand harsh weather conditions, as well as heavy equipment and machinery. It should be sturdy, well-ventilated, and secure from trespassers and potential theft. These structures are simple and economical and have less interior finish than their commercial counterparts. Farm machinery is a big investment for any farmer/ organization so it's important that this machinery remains as protected as possible. The storage helps in proper maintenance of farm machines and implements resulting in long life and trouble free operation.

12.1 Construction Details

The construction details for a farm machinery shed are as follows:

1. **Site Selection** - A level and well-drained site shall be selected which is easily accessible for large equipment.
2. **Foundation** – The foundation of the storage structure should be strong and level, and should be able to support the weight of the machinery.
3. **Floor** – Common types of flooring for implement sheds include brick-on-edge, cement-concrete, gravel, and mud. The choice of flooring depends on factors such as the type of machinery being stored, the climate, and the budget. Although comparatively costlier, a concrete floor is durable and easy to clean, making it a good choice for a farm implement shed. Even the ordinary earth floor made of well rammed brick ballast with a thin layer of *morrum* shall serve the purpose. Floor should be located at a higher level than its surroundings. Ramps should be provided to allow movement of machineries over the floor.
4. **Pillars** – Angle iron posts or masonry pillars shall be provided for supporting the corrugated sheet roofing.
5. **Walls** - The walls can be made of bricks, precast panels, or metal siding. They should be strong enough to support the structure and resist damage from exposure to the machines. Generally, side walls of bricks of 20 cm shall be constructed to support the structure.
6. **Roof** – Corrugated metal and asbestos sheet are generally used as roofing material for the shed. A corrugated metal roof is a common choice for a farm implement shed as it is durable, long-lasting and easy to maintain. The asbestos sheet roof is comparatively cheaper.
7. **Door** – If it is desired to have a door, large sliding or hinged doors shall be fitted for easy access for large equipment. The door must be wider than the widest machine on the farm.
8. **Lighting and Electrical** - Adequate lighting and electrical outlets shall be provided in the shed for safety and functionality.
9. **Ventilation** - Proper ventilation is necessary to prevent moisture build-up and to provide a safe and comfortable working environment.
10. **Maintenance** - It should be easy to maintain and keep clean, with a durable roof and walls that can be easily cleaned.

The design of a farm implement shed depends on the specific needs and requirements of the farm, including type and size of the equipment to be stored, the available space, and the budget. An implement storage shed shall feature clear spans, high-studs and extra-wide bays allowing large machinery and implements to be stored away safely. Machinery storage should have as much surface as possible between the interior posts, without being too deep, so that each machine can be taken out easily. Most of the sheds are rectangular, some quite elongated. They could be enclosed, with large gable ends or eaves side doors to admit machinery; or they might be open on one eave side and enclosed on the other three sides. For most purposes, an open-side shed (Figure 12.1) with a well-drained, raised floor will be adequate for machinery storage. In this design, one side of the shed is completely open to allow full access to the shed. The main advantage of this type of shed is that machines do not block access of one another as they are parked side by side. The other three walled sides which may be constructed partly or wholly shall provide enough protection from weather. The cost of this shed is also low as no door is required. An open gable end shed (Figure 12.2) provides wider opening which is useful for wider machinery. By opening both gable ends, one can have full access to the shed and the choice of driving through as well.

Figure 12.1. Open-side Implement Shed.

Figure 12.2. Open Gable End Shed.

12.2 Storage Space Requirement

The storage space requirement for farm machines depends on the size and type of equipment, as well as the available space on the farm. Sufficient aisle space should be provided between equipment to allow for easy movement and manoeuvring. The shed should have enough height to accommodate the tallest piece of equipment. Provide enough room at the entrance and exit points to allow for easy entry and exit of equipment. It is important to consider the future needs of the farm and plan for enough space to accommodate any additional equipment.

A general rule of thumb is to allow for at least 2-3 square meters of floor space per piece of equipment, plus additional space for aisles and maintenance. However, the specific size requirements will vary based on the specific equipment and needs of the farm. The width of the open-side shed shall be 8 to 10 m wide, and the length must be large to accommodate all the machines. Provision of future expansion shall be made. Approximate floor space requirement of the various machines is shown in Table 12.1.

Table 12.1. Space Requirement for Storage of Farm Machines

Sl.No.	*Machine*	*Length (m)*	*Width (m)*	*Area (m^2)*
1.	Tractor	4.0	3.0	12.0
2.	Cultivator			
	(a) Tractor drawn	3.0	3.0	9.0
	(b) Animal drawn	1.5	1.5	2.25
3.	Harrows			
	(a) Tractor drawn	3.0	3.0	9.0
	(b) Animal drawn	1.5	1.5	2.25
4.	Ploughs			
	(a) TD 4 bottom	3.0	5.0	15.0
	(b) Single bottom	3.0	3.0	9.0
	(c) Animal drawn	1.5	3.0	4.5
5.	Seed drills			
	(a) Tractor drawn	2.0	4.0	8.0
	(b) Animal drawn	1.5	1.5	2.25
6.	Planters 4 row type	3.0	5.0	15.0
7.	Sprayers			
	(a) PTO operated	2.0	2.0	4.0
	(b) Engine operated	1.5	1.5	2.25
	(c) Orchard sprayer	2.0	2.5	5.0
8.	Tractor – 4 wheel type	4.0	3.0	12.0

Sl.No.	Machine	Length (m)	Width (m)	Area (m^2)
9.	Trailer			
	(a) 4-wheel	6.0	4.0	24.0
	(b) 2-wheel	4.0	3.0	12.0
10.	Thresher			
	(a) Power operated	8.0	3.0	24.0
	(b) Foot operated	1.0	1.0	1.0
11.	Maize-sheller	4.0	2.0	8.0
12.	Combine harvester	5.0	4.0	20.0
13.	Mowers			
	(a) Mounted type	1.5	3.0	4.5
	(b) Trailed type	2.0	3.0	6.0
14.	Reaper	2.0	1.5	3.0
15.	Chaff-cutter	4.0	2.5	10.0
16.	Water tanker	5.0	4.0	20.0
17.	Potato digger	2.5	1.5	3.75
18.	Baler	5.0	5.0	25.0

The arrangement of machinery stores and workshops depends upon the convenience of their utilization. The store and workshop can be housed in a single building. In which one side of the building is for workshop whereas other side can be used to store the machinery. At the end of the shed, one way is reserved for repair and maintenance and another for tools. The same shed or other can be used for storing seeds, fertilizers, and pesticides.

FARM WORKSHOP

A farm workshop is a repair and maintenance facility inside the farmstead for machines, implements, building *etc.* In addition, farm workshop can be used to store the equipment, supplies and spare parts. It also provides shelter for work during severe weather. The size and design of the farm workshop depends upon the type of repair work done at workshop and the overall size of the farm. The workshop at farm helps in timely completion of repair and maintenance of equipment, and thus help in saving of substantial time and money. The workshop, farm machinery storage structures, fuel and lubricant store, space for tools and spare parts *etc.* shall be combined together. This type of combination is economical in floor space, and will result in a saving of time involved in moving machines and supplies from one to the other.

12.3 Location

The farm workshop shall be situated in a location that is convenient, efficient, and cost-effective for repairing and maintaining farm machinery.

It shall preferably be located close to the work centre of the farm. The area should be well drained. The location should be easily accessible by road and have sufficient parking space for large vehicles and machinery. There should be a good supply of water, electricity, and other necessary resources to support the workshop operations. The size of the workshop should be sufficient to accommodate the repair and maintenance of various types of farm machinery. Adequate ventilation and lighting should be available. The location should be safe for workers and visitors, with proper safety equipment and measures in place. All the current and future needs should be considered while planning and designing the farm workshop.

12.4 Salient Features of a Farm Workshop

The salient features of a farm workshop are as follows:

1. **Durable and Sturdy Flooring**: The flooring should be able to withstand heavy machinery and equipment, as well as resist wear and tear from frequent use. Cement concrete or brick floor is considered to be satisfactory for a farm workshop.
2. **Ample Lighting**: Adequate lighting is important for safety and efficiency, particularly when working with heavy machinery.
3. **Adequate Ventilation**: Proper ventilation is important to ensure a healthy and safe working environment, particularly when working with hazardous materials or chemicals. Provision of adequate number of exhaust fans shall be made to promote air exchange.
4. **Electrical Outlets and Power Supply**: A farm workshop should have enough electrical outlets and power supply to accommodate power tools and other machinery.
5. **Storage Space**: Adequate storage space for tools, equipment, and supplies is crucial for organization and efficiency.
6. **Workspace**: The workshop should have sufficient workspace, such as workbenches, to allow for the repair and maintenance of machinery and equipment.
7. **Safety Features**: Safety features, such as fire extinguishers, emergency exits, and first aid kits, should be readily available in the workshop. It's important to keep the workshop clean, organized, and well-lit to promote safety and efficiency.
8. **Easy Access**: The farm workshop should be easily accessible for vehicles and machinery, with enough room for manoeuvring and parking.
9. **Climate Control**: In some cases, climate control may be necessary to ensure a comfortable working environment, particularly in areas with extreme temperatures.

12.5 Sections of the Workshop

A farm workshop comprises of various sections. The area required by each section shall be taken into account while calculating the overall size. The sections are:

1. **Reception Area** - This area should be well-lit, clean, and welcoming. It should include a desk, chairs, and a waiting area for staff/workers.
2. **Office** - The office for the manager, foreman and other staff should be equipped with a desk, chairs, shelves, and a computer for administrative tasks.
3. **Spare Parts and Tools Storage** - This area should be well-organized, with shelves and racks for storing spare parts and tools.
4. **Work Bays** - There should be several work bays, each equipped with a hydraulic lift, workbench, and welding equipment. In addition, separate carpentry shop, machine shop, fitting shop, tractor repair shop, pump repair shop, electric section, shall be provided. The bays should be well-lit, with enough space to manoeuvre machinery and perform repairs.
5. **Farm Machinery Storage** – This area should be able to store the repaired vehicles and farm machinery.
6. **Cleaning and Servicing Area** - This area should be equipped with cleaning, washing and servicing facilities.
7. **Painting Area** – This area should be equipped with painting equipment as well as adequate ventilation to minimize exposure to fumes and dust.
8. **Fuel and Oil Storage** - This area should be equipped with fuel and oil storage tanks, as well as a dispensing system for filling machinery.
9. **Waste Storage** - This area should be equipped with containers for storing waste materials generated during repairs and maintenance.

It is advised to install the welding system near to main gate to use it inside and outside the workshop easily. The welding section shall be located away from wood working area. The collection of the flammable material like sawdust, shavings and oily rags shall not be allowed inside the workshop to prevent the fire hazard. The separate section shall be for storage of the fuel. Fire extinguisher or two to three buckets of dry sand shall be provided.

It is important to have a comprehensive set of tools and equipment to perform day-to-day repair and maintenance of a wide variety of machinery at a farm workshop. The equipment and machines typically required for repair and maintenance of machinery are workbenches; hydraulic lifts; welding equipment; power tools such as impact wrenches, drill presses, and grinders; hand tools such as wrenches (spanners), pliers, and screwdrivers; measuring instruments such as callipers, micrometres, and dial indicators; cutting tools such as saws, shears, and cutting torches; presses for pressing bearings, bushings, and seals into place; cleaning equipment such as pressure washers, steam cleaners, and degreasers; paint sprayers; storage racks; compressor, battery charger, and computer and software.

12.6 Design of Farm Workshops for different Size of Farms

The design of farm workshops for different sizes of farms should take into account the specific needs and requirements of the farm operation.

1. **For Small Farms**: A small farm workshop may only require a basic design, with a focus on efficiency and functionality. This may include a single work area, adequate lighting and ventilation, and basic storage space. Small farms should have a simple work shelter and store adequate for repair work and the storage of small implements. Under the shelter, a small workbench and vice can also be provided.
2. **For Medium-Sized Farms**: Medium-sized farms may require a more sophisticated design, with multiple work areas, larger storage space, and additional features such as a heating and cooling system, and restroom facilities.
3. **For Large Farms**: Large farms may require a highly organized and efficient workshop, with multiple work areas, extensive storage space, advanced electrical and plumbing systems, and specialized equipment such as overhead cranes. A large commercial farm may need a separate building with extensive equipment for maintaining the farm machinery, tractors and vehicles.

Regardless of the size of the farm, the design of the workshop should prioritize safety, efficiency, and durability. It is also important to consider the type of machinery and equipment that will be stored and used in the workshop, as well as the climate and environmental conditions.

Example 12.1

Draw a floor plan of a farm workshop for a 100 ha mechanized farm having all essential farm equipment.

Solution

1. The workshop should be of a sufficient size to accommodate all the equipment and machinery, as well as provide enough space for repair and maintenance activities.
2. The workshop should be divided into several distinct areas, including a reception area, an office, parts storage, work bays, a cleaning and painting area, fuel and oil storage, and waste storage.

3. The workshop should have access to electricity and water, as well as sufficient lighting and ventilation.

A 100 ha farm shall have the following farm machines and equipment.

1. Tractors – 2 nos.
2. Trailers – 2 wheel and 4 wheel
3. Jeep and trucks
4. Power tillers
5. Electric motors – single phase and three phase
6. Power operated grass cutter
7. Disc ploughs and harrows
8. Cultivators
9. Seed-drill and planters
10. Thresher and grain winnowers and cleaners
11. Combine harvester
12. Diesel engine
13. Centrifugal and submersible pumps

Actual size of the farm machinery shed can be calculated on the basis of the dimensions of each farm machines and equipment.

To cater the different needs, the farm workshop should have following shops:

1. Carpentry and black smithy shop
2. Fitting shop
3. Welding shop
4. Tractor repair shop
5. Pump repair shop
6. Spare-parts storage
7. Raw material storage
8. Fuel storage room
9. Implement shed
10. Foreman office
11. Manager office
12. General office-room

A typical floor plan of the workshop is given below:

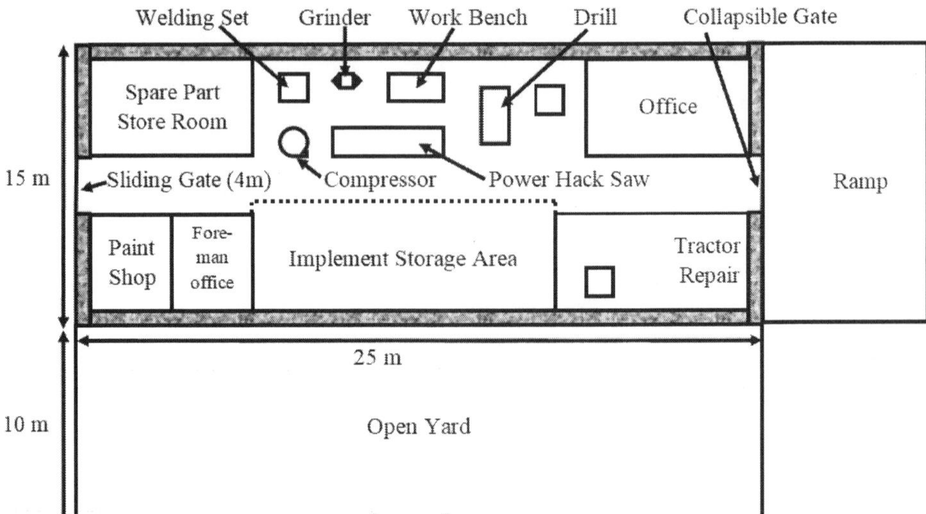

Example 12.2

Draw a typical layout plan of a farm workshop for 1000-hectare farm.

Solution

Large farms shall require a highly organized and efficient workshop, with multiple work bays, and extensive storage space. A farm workshop for such a large size farm is expected to be fully equipped to undertake repairs and over haul of all parts of various farm machines.

Actual size of the farm machinery shed can be calculated on the basis of the number and size of each implement.

The main workshop shall be located at a central position and should have the following sections.

1. Forging section
2. Vulcanising section
3. Welding section
4. Machine shop
5. Carpentry section
6. Electrical section
7. Stores
8. Painting section
9. Administrative office
10. Cleaning and servicing section
11. Sorting section
12. Inspection section
13. Engine section
14. Storage of repaired equipment
15. Yard for vehicles coming to workshop
16. Fuel section
17. Farm machinery storage section
18. Reception area

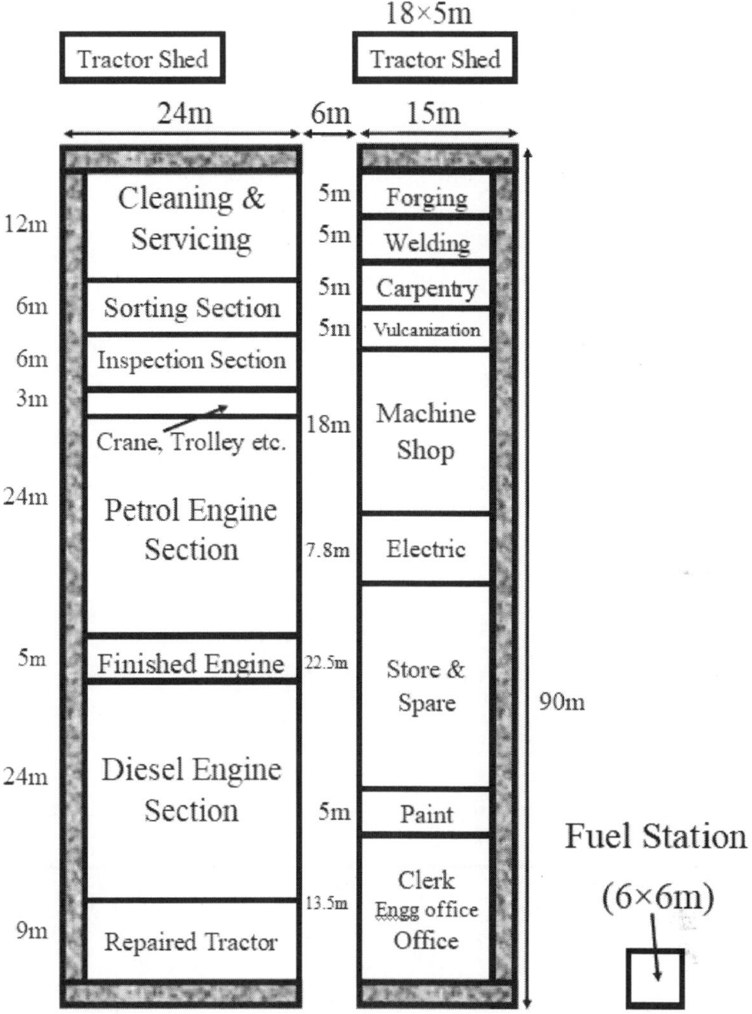

FUEL AND CHEMICAL STORAGE ON FARM

Fuel and chemical storage is a critical aspect of various industries and sectors, ranging from energy production and transportation to manufacturing and agriculture. Proper storage of fuels and chemicals is essential to ensure safety, prevent environmental hazards, and maintain the integrity and quality of these substances. The storage requirements of "hazardous materials" due to their highly flammable or poisonous nature will vary farm to farm. The special storage is required to store the hygroscopic materials like cement and fertilizers. A few basic requirements for safe storage of fuels and chemicals on farm are considered here.

12.7 Storage Considerations

When it comes to fuel and chemical storage, several factors must be taken into consideration to ensure safety and regulatory compliance. These are:

1. **Compatibility**: Different fuels and chemicals have varying properties and compatibility requirements. It is essential to store substances in appropriate containers and tanks that are compatible with the stored material to prevent leaks, reactions, or deterioration.
2. **Containment**: Storage facilities must be designed to contain spills, leaks, and releases effectively. Secondary containment systems, such as dikes or berms, can help prevent environmental contamination in case of accidents or leaks.
3. **Ventilation**: Adequate ventilation is crucial to prevent the build-up of flammable or toxic vapours. Proper airflow can help maintain a safe environment within the storage area and reduce the risk of explosions or health hazards.
4. **Temperature and Humidity**: Some fuels and chemicals are sensitive to temperature and humidity conditions. Storage facilities should be equipped with climate control systems to ensure that stored materials remain within their specified temperature and humidity ranges.
5. **Security**: Fuels and certain chemicals are valuable commodities that can be targets for theft or misuse. Implementing appropriate security measures, such as access control systems, surveillance cameras, and alarms, can help prevent unauthorized access and protect against potential hazards.

12.8 Storage Methods

Fuel and chemical storage methods can vary depending on the type and volume of the material being stored. Here are some common storage methods:

1. **Bulk Storage Tanks** - Large above-ground or underground storage tanks are commonly used for storing fuels and chemicals in significant quantities. These tanks are constructed with materials that are compatible with the stored substance and are designed to withstand internal and external pressures. Regular inspection, maintenance, and compliance with relevant regulations are essential for safe operation.
2. **Drums and Barrels** - Smaller quantities of fuels and chemicals are often stored in drums and barrels made from materials such as steel, plastic, or fibre. These containers should be properly labelled, sealed, and stored in well-ventilated areas away from heat sources or ignition points.
3. **Intermediate Bulk Containers (IBCs)** - IBCs are large containers that typically have a capacity of 250 to 1,000 litres (or more) and are used

for storing and transporting bulk quantities of fuels and chemicals. They are made of durable materials and often feature stackable designs for efficient use of space.

4. **Specialty Storage Solutions** - Certain chemicals and hazardous materials require specialized storage solutions. These may include flammable cabinets for storing flammable liquids, acid cabinets for corrosive substances, or temperature-controlled storage for sensitive materials.

12.9 Location

Some general guidelines on the recommended practices for locating fuel and chemical storage facilities on a farm are:

1. The storage facility should be located at a safe distance from any livestock areas, such as barns, sheds, or grazing fields, to minimize the risk of contamination or accidental exposure to animals.
2. It should also be situated away from water sources, such as ponds, streams, or wells, to prevent any potential water pollution.
3. The storage facility should be easily accessible for farm equipment and emergency vehicles, but it should also be secure to prevent unauthorized access. Fencing, lockable gates, and adequate lighting can help enhance security.
4. If there are residential areas nearby, it is advisable to keep the storage facility at a reasonable distance to minimize any potential risks to the health and safety of people living in those areas.
5. The topography of the farm should also be taken into account. The storage facility should be located on stable ground that is not prone to flooding, erosion, or other environmental risks.
6. Ensure that the storage facility adheres to all relevant government regulations regarding the storage of fuel and chemicals. These regulations may include guidelines on distance requirements, containment measures, fire safety, and spill prevention.

12.10 Storage of Fuel and Chemical on Farm

The storage of fuel and chemical on a farm requires careful consideration and adherence to safety regulations to minimize the risk of accidents, environmental contamination, and health hazards. Here are some guidelines for safely storing fuel and chemicals on a farm:

1. **Separate Storage Areas**: Designate separate storage areas for fuel and chemical to prevent cross-contamination and the potential for chemical reactions. Ideally, these storage areas should be located away from

livestock, crops, and water sources. Aisles, doorways, exits and entry ways must be kept clear.

2. **Fuel Storage**:
   (a) **Fuel Tanks**: Use approved fuel tanks specifically designed for on-site storage. Install tanks on a concrete pad with appropriate secondary containment measures to contain spills or leaks.
   (b) **Fuel Containers**: Quantities of flammable products greater than about 3 litres of cellulose thinner, 10 litres of petrol, 20 litres of kerosene, 50 litres of diesel fuel should be stored in a separate building at least 15 meter from any other building.
   (c) **Ventilation**: Ensure proper ventilation in the fuel storage area to prevent the build-up of flammable vapours.
   (d) **Fire Safety**: Keep fire extinguishers readily available and ensure they are regularly inspected and maintained. Implement fire prevention measures, such as minimizing potential ignition sources and keeping the storage area clear of debris.
3. **Chemical Storage**:
   (a) **Safety Cabinets**: Store chemicals in appropriate safety cabinets made of non-combustible materials. Cabinets should be well-ventilated, lockable, and designed to contain spills.
   (b) **Storage Shelves**: Shelves should be level, stable and secured to the wall or another stable surface. In case of an earthquake, shelves should have raised edge sorrim guards of minimum height of 5 cm to prevent containers from falling. Large bottles/containers should not be stored higher than 60 cm from the floor. Corrosives should be stored on lower shelves.
   (c) **Labelling and Organization**: Clearly label all chemical containers with the contents, hazards, and handling instructions. Store chemicals in an organized manner, grouping them according to compatibility.
   (d) **Spill Containment**: Implement spill containment measures, such as spill pallets or trays, to prevent chemicals from reaching the ground or water sources in case of leaks or spills. Clean up spills and drips immediately.
   (e) **Personal Protective Equipment (PPE)**: Provide and enforce the use of appropriate PPE, such as gloves, goggles, and respirators, when handling chemicals.
4. **Training and Emergency Preparedness**: Ensure that all farm personnel handling fuel and chemicals receive appropriate training on storage procedures, safety protocols, and emergency response. Have an emergency plan in place, including spill response procedures, evacuation routes, and contact information for relevant authorities.

5. **Regular Inspections and Maintenance**: Regularly inspect fuel tanks, safety cabinets, and containment measures for signs of damage or deterioration. Maintain accurate inventory records of fuel and chemical quantities to track usage and identify any discrepancies.

12.11 Storage of Fertilizers and Other Non-hazardous Products

When it comes to the storage of fertilizers and other non-hazardous products, it's important to consider a few key factors to ensure safety and proper handling. Here are some general guidelines:

1. **Suitable Storage Location**: Choose a well-ventilated, dry area away from direct sunlight, heat sources, and ignition points. Ideally, the storage area should be dedicated solely to the storage of fertilizers and similar products. Storage conditions should be as dry as possible as some fertilizers are hygroscopic and easily pick up moisture from humid air or from the ground. Materials which are adversely affected by prolonged exposure to high storage temperatures must be shaded.
2. **Secure Containers**: Store fertilizers in their original, sealed containers whenever possible. If transferring to different containers, ensure they are made of compatible materials and have proper labels indicating the contents. Ensure that the bags are not punctured or otherwise damaged.
3. **Segregation**: Avoid storing fertilizers near incompatible substances, such as flammable materials or organic matter. Different types of fertilizers may have specific storage requirements, so refer to the product labels or safety data sheets for guidance.
4. **Shelving and Stacking**: Use sturdy shelving or pallets to keep containers off the floor and allow for easy access. The roof shall be provided on stacks to protect it from rain. Other water tight cover like polythene can be used. The stacking on the raised platform with plinth allow ventilation and prevent moisture movement from the ground to stacked material. Fertilizers shall be stored away from machines and implement due to their corrosive nature.
5. **Ventilation**: Ensure adequate ventilation in the storage area to prevent the build-up of potentially harmful gases or vapours. This is particularly important for fertilizers that may release ammonia or other gases.
6. **Spill Containment**: Implement measures to contain potential spills.
7. **Fire Safety**: Although fertilizers are generally non-hazardous, some formulations may still pose a fire risk. Keep fire extinguishers nearby and ensure they are suitable for the types of fires that could occur in the storage area.

8. **Accessibility and Labelling**: Arrange the storage area to allow easy access and retrieval of containers. Clearly label all containers with the product name, manufacturer, and any specific handling or storage instructions.
9. **Regular Inspections**: Conduct routine inspections of the storage area to check for any signs of damage, leaks, or deterioration. Replace damaged containers promptly and address any issues that could compromise safety.

12.12 Ventilation

Proper ventilation is crucial when it comes to the storage of fuels and chemicals to ensure safety and prevent potential hazards. The specific requirements for ventilation may vary depending on the type and quantity of the substances being stored. Here are some guidelines for ventilating fuel and chemical storage areas:

1. Familiarize yourself with the specific ventilation requirements for the type of fuel or chemical you are storing. Different substances may have different ventilation needs, so refer to their safety data sheets (SDS) or consult with experts or regulatory agencies for specific guidance.
2. Ensure that the storage area has adequate airflow to prevent the build-up of potentially flammable or toxic vapours. This can be achieved through natural ventilation (*e.g.*, windows, vents) or mechanical ventilation systems (*e.g.*, fans, blowers) depending on the size and nature of the storage area.
3. Store fuels and chemicals in well-ventilated areas that are specifically designed and constructed to minimize the risks associated with storage. Consider locating storage areas away from ignition sources, heat, direct sunlight, or areas prone to sparks or flames.
4. Install appropriate ventilation systems to maintain air circulation. These systems can help remove harmful vapours, fumes, or gases from the storage area. Ensure that the ventilation system is properly designed, regularly inspected, and maintained to ensure its effectiveness.
5. When handling flammable fuels or chemicals, it is important to use explosion-proof electrical equipment and ensure that the ventilation system is designed to prevent the accumulation of flammable concentrations.
6. Regularly monitor the air quality in the storage area to ensure that it meets safety standards. Use appropriate gas detectors or monitoring devices to detect any leaks, spills, or abnormal concentrations of gases or vapours.
7. In the event of a spill, leak, or release of hazardous substances, make sure there are provisions for emergency ventilation. These may include

emergency exhaust systems or procedures to isolate and ventilate the affected area.

8. Familiarize yourself with relevant local, national, and international regulations governing the storage of fuels and chemicals. Adhere to these regulations and ensure that your ventilation systems and storage practices are in compliance.

12.13 Fire Protection

Fire protection for fuel and chemical storage is of paramount importance to ensure the safety of personnel, facilities, and the environment. The design of fire protection system depends on fire characteristics. Here are some essential measures and considerations for fire protection in fuel and chemical storage areas:

1. The fire suppression systems such as sprinkler systems, foam systems, or gas-based suppression systems shall be installed. The type of system depends on the specific hazards present. For example, water extinguishers are suitable to put out fires in ordinary combustibles such as wood, textiles and paper, foam systems are effective for flammable liquids, while gas-based systems like carbon dioxide or clean agents are suitable for areas with sensitive electronics or where water may cause damage. Portable extinguishers and fire blankets shall be provided at strategic points.
2. A reliable fire detection system that includes smoke detectors, heat detectors, and flame detectors shall be installed. These systems can quickly identify the presence of a fire and activate appropriate suppression measures.
3. Ensure proper ventilation in storage areas to prevent the accumulation of flammable vapours. Adequate airflow helps dissipate any released vapours and reduces the risk of ignition.
4. Separate incompatible chemicals and fuels to prevent accidental reactions. Store different substances in dedicated areas or separate storage cabinets to minimize the risk of cross-contamination.
5. Maintain good housekeeping practices to reduce the risk of fire. Keep storage areas clean and organized, promptly remove any combustible materials or debris, and regularly inspect and maintain storage containers and equipment.
6. Develop and communicate an emergency response plan that includes evacuation procedures, fire alarms, and emergency contact information. Conduct regular drills to ensure that personnel are familiar with the procedures and can respond effectively in case of a fire.
7. Provide appropriate training to employees regarding fire safety, including the handling, storage, and disposal of fuels and chemicals.

Educate them on the specific hazards associated with the stored materials and the proper use of fire protection systems.

THRESHING FLOORS

For thousands of years, humans have relied on agriculture to provide food and sustenance. One of the key activities in agriculture is the threshing of crops, which involves separating the edible grain from the inedible chaff or husks that surround it. Threshing is traditionally done on threshing floors, which are constructed specifically for this purpose.

Threshing floors are a common feature of agricultural societies, and they could be found all over the world. They are typically located near fields where crops like wheat, barley, rice, soybean *etc.* are grown. Threshing floors are an essential part of the agricultural process, as they allow farmers to process their crops and prepare them for storage and consumption. In many traditional societies, threshing floors are considered sacred places. They are often the site of religious ceremonies and are seen as a place where the gods could be invoked to bring a good harvest. In some cultures, the threshing floor is also a gathering place for the community, where people would come together to thresh their crops and socialize. Traditionally, threshing floors are used in agricultural societies where people relied on manual labour to cultivate their crops. After the crops are harvested, they are spread out on the threshing floor, which was typically a hard-packed surface made of earth, stone, or wood. Farmers would then beat the grain with flails or sticks, or they would drive animals like horses or oxen over the grain to separate the edible parts from the inedible parts. Threshing floor is not required at places where modern agricultural equipment like combine harvester is used.

Threshing floors are typically located in areas where the wind could blow away the chaff from the grain. In some cases, windbreaks are constructed around the perimeter of the threshing floor to protect the crops from being blown away during threshing. Proper drainage is also important to prevent water from accumulating on the threshing floor, which could damage the crops and the surface. Often threshing floors are created on poor soil unsuitable for agriculture.

The construction of threshing floors varied depending on the region and the materials available. In some areas, threshing floors are simply a hard-packed surface made of earth or clay. In other areas, they are constructed using stones or bricks. Wooden threshing floors are also common in some regions. The surface of the threshing floor has to be durable and hard enough to withstand the threshing process, which could be quite rigorous.

12.14 Types of Threshing Floors

Threshing floors can be classified into different types based on their location, size, and construction. Each type of threshing floor has its own advantages and disadvantages, and the type of threshing floor used depends on a variety of factors, including local agricultural practices, the availability of materials, and the size of the crop being threshed. Here are some common types of threshing floors:

1. **Outdoor Threshing Floor**: This is the most common type of threshing floor, and it is located outdoors. It can be a simple, flat surface made of hard-packed earth or a more elaborate structure made of stone or concrete or wood. Outdoor threshing floors are typically larger in size to accommodate the threshing process. Concrete threshing floors are a popular option due to their durability, strength, and ease of maintenance. They can withstand heavy use and are resistant to weather and erosion. Proper construction is essential to ensure the longevity and effectiveness of the concrete threshing floor. They are typically found in large/commercial farms. Wooden threshing floors are also common in some areas, particularly in rural communities. They can be made of rough-hewn logs or more finished lumber and can be either indoor or outdoor.

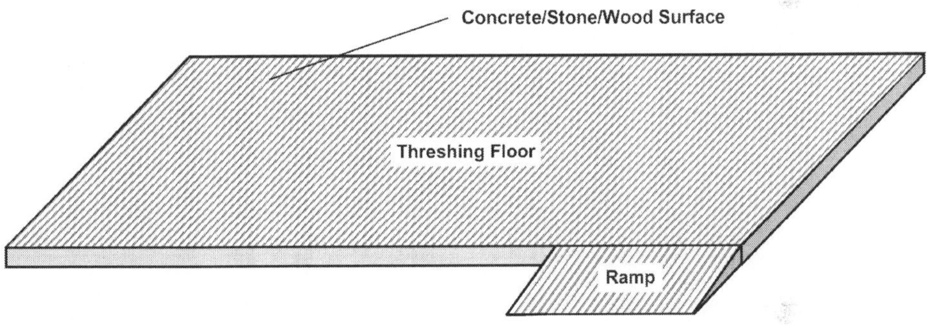

Figure 12.3. Outdoor Threshing Floor made of Concrete.

2. **Indoor Threshing Floor**: Some cultures use indoor threshing floors, which are typically located inside a barn or other agricultural structure. Indoor threshing floors can be smaller in size than outdoor ones, but they still serve the same purpose of separating the grain from the chaff.
3. **Portable Threshing Floor**: In some areas, farmers use portable threshing floors that can be moved from one location to another. These are usually smaller and made of lightweight materials such as woven mats, and they can be folded up and carried to the next field.
4. **Elevated Threshing Floor**: Threshing floors built on an elevated platform can take advantage of the wind to help blow away the chaff. These are typically constructed on a hill or other high location to maximize the wind.

12.15 Method of Construction

The construction details of threshing floors can vary depending on the region and the type of threshing floor. However, here are some common construction details for threshing floors:

1. **Site Preparation**: The site must be cleared of any vegetation, debris or topsoil. The ground must be levelled and compacted to create a flat, even surface. The site should also be properly drained to avoid any water accumulation.
2. **Foundation:** The foundation of the threshing floor must be strong and stable to support the weight of the crops and the people or animals working on it. The foundation can be made of stones, bricks, or concrete.
3. **Surface Material**: The surface of the threshing floor must be hard and durable to withstand the threshing process. The material used for the surface can vary depending on the region, but it can include compacted soil, wooden planks, stone, concrete or a mixture of lime and sand. In some cases, the surface may be covered with straw or other materials to prevent the grain from being damaged during threshing. A concrete mix with a minimum compressive strength of 20 MPa is recommended for the construction of cement-concrete threshing floor.
4. **Borders:** Borders can be constructed around the perimeter of the threshing floor to prevent the crops from spilling over the edge. Borders can be made of stones, logs, or other materials.
5. **Drainage:** Proper drainage is important to prevent water from accumulating on the threshing floor, which can damage the crops and the surface. Depending on the location and the type of threshing floor, drainage can be provided through the use of gutters, trenches, or other means.
6. **Windbreaks**: In areas with high winds, windbreaks may be constructed around the perimeter of the threshing floor to protect the crops from being blown away during threshing. Windbreaks can be made of trees, shrubs, or other materials.

BIOGAS PLANTS

Biogas is produced through anaerobic fermentation of animal excreta and other agricultural wastes. The methane gas generated in a biogas digester is used as smokeless fuel for cooking, for lighting and can also be used as an engine fuel in combination with the liquid fuels. The residual slurry is also useful in the form of excellent manure. Biogas is seen as a clean fuel, and provides a feasible alternative to cooking gas.

12.16 Potential of Biogas in India

It is estimated that about 1300 million tonnes (by wet weight) of cattle dung is produced in India annually which can produce 52,560 million cubic meter of biogas. It has equivalent potential of 33,006 million litres of kerosene, 32,735 million litres of diesel, 37,668 million litres of petrol, 1590 million LPG cylinders and 24,700 million kWh of electric energy.

Assuming 50 per cent of total dung is available for biogas production, the actual availability is estimated as follows:

Total population of cows and buffaloes in India	300 million	
Average dung produced per animal per day	10 kg/day	
Total dung produced per day	3000 million kg/day	
Total dung produced per year	1095 million tones/year	
	Per Day	*Per Year*
Actual available (consider 50 per cent dung available for biogas production)	1500 million kg	547 million tones
Total biogas potential (40 cubic meter gas per tonne)	60 million m^3	21880 million m^3
Equivalent kerosene in litres (coefficient 0.628)	37.68 million litres	13740 million litres
Equivalent petrol in litres (coefficient 0.7167)	43 million litres	15681 million litres
Equivalent electric energy (coefficient 0.47)	28.2 million kWh	10284 million kWh
Equivalent LPG in kg (coefficient 0.43)	25.8 million kg	9408 million kg
Equivalent LPG cylinder (coefficient 1/14.2 = 0.0704)	1.81 million cylinders	662 million cylinders
Available manure (15 per cent dry weight, 1.2 per cent manure of dry weight)	0.4 million tones	148 million tones
Available N, P, K (1.5 per cent N, 1.4 per cent P, 1.5 per cent K)	68 million kg	25 million tones

1 m^3 of biogas = 0.62 litre of kerosene oil = 3.50 kg of wood

= 12.30 kg of cow dung cakes = 1.46 kg of charcoal

= 0.43 kg of LPG = 1.6 kg of coal = 4.7 kWh of electricity

Composition of Biogas

Biogas is a mixture of gases whose composition is as follows:

Methane (CH_4) - 50 to 65 per cent

Carbon dioxide (CO_2) - 30 to 40 per cent

Hydrogen (H_2) - 1 to 5 per cent

Nitrogen (N_2) - 1 per cent

and, traces of hydrogen sulphide (H_2S), oxygen (O_2), and water vapours (H_2O).

Advantages of Biogas

1. Low initial investment
2. Suitable technology for rural areas
3. Domestic use of biogas
4. Lower the dependency on conventional fuel
5. Improvement in the sanitary condition of village
6. Reduction in environment pollution
7. Nitrogen rich manure for the agricultural lands
8. Reduce the health issues like eye and lung disease and drudgery of woman

12.17 Site Selection

Following points may be kept in mind while selecting the site for installation of biogas plant:

1. Site should be at higher elevation than the surroundings to avoid runoff water.
2. The soil should not be too loose and should have sufficient bearing strength (more than 2 kg/cm^2).
3. Should be close to kitchen to minimize the cost of pipeline, and to avoid pressure drop.
4. Should be close to the animal barn for easy handling of manure.
5. Water table should not be very high.
6. The plant should be installed at least 15 meter away from drinking water source to avoid contamination.
7. Adequate supply of water should be there near the plant site.
8. The plant should get clear sunshine during most part of the day. There should not be any tree or building nearby.
9. There should be adequate space for construction of slurry pits for drying purpose.
10. Plant should be at least 1.5 meter away from any wall or foundation.

12.18 Factors that Affect Biogas Production

1. **Temperature:** Optimum temperature inside the digester for biogas production is 35 to 45°C. Microorganisms have optimum activities at this stage. At low temperatures, the gas production is retarded.
2. **pH:** The pH range suitable for gas production is 6.6 to 7.5. Neither acid nor alkaline solution is required for optimum biogas production.

The pH above 8.5 should not be used as it is difficult for the bacteria to survive above this pH.

3. **Carbon to Nitrogen Ratio (C/N ratio):** A carbon to nitrogen (C/N) ratio in the range of 25:1 to 30:1 is found to be optimum for biogas production.
4. **Hydraulic Retention Time (HRT):** The hydraulic retention period of any biogas plant is the time period for which the mixture of cow dung or any organic material with water stays in the digester for production of biogas. HRT for any biogas plant is inversely proportional to temperature. HRT for any biogas plant is different for different areas. HRT for plain areas is less as compared to hilly areas.
5. **Loading Rate**: Loading rate should be optimum. If digester is loaded with too much raw material, acids will accumulate and fermentation will be affected.
6. **Solid Concentration**: The water content should be around 90 per cent of the weight of the total contents. For effective anaerobic fermentation of cow dung, slurry must contain 8 to 9 per cent solid organic matter.

Table 12.2. Plant Capacity and Dung Requirement

Plant Capacity (m^3)	*Suitable for Number of Family Members*	*Daily Dung Required (kg)*	*Approximate Number of Cattle*
1	2 – 3	25	2 – 3
2	4 – 5	50	3 – 4
3	7 – 8	75	5 – 6
4	10 – 11	100	7 – 8
6	14 – 16	150	10 – 12
8	18 – 20	200	16 – 20
10	21 – 24	250	20 – 22

12.19 Components of a Biogas Plant

A biogas plant consists of the following parts:

1. **Digester**: Digester is mostly embedded partly or fully in the ground. It is also called fermentation tank and holds the slurry for a sufficiently long time to complete the digestion.
2. **Gas Holder or Gas Storage Masonry Dome**: Gas holder keeps the biogas for subsequent use.
3. **Mixing Tank**: Mixing tank is used for mixing the dung with water for induction in the digester through inlet.
4. **Inlet Chamber/Pipe**: It is provided for adding the mixture of dung

and water to the digester.

5. **Outlet Chamber/Pipe**: It is provided for taking out the digested portion of the slurry.

Gas outlet pipe, gate valve, gas distribution pipe line, water trap, fittings, gas stove, lamp and appliances are other components of a biogas plant.

12.20 Models of Biogas Plants

There are two models of biogas.

1. Floating-drum type biogas plant (KVIC)
2. Fixed-dome type biogas plants (Janta model and Deenbandhu model)

KVIC Model Biogas Plant

KVIC model biogas plant (Figure 12.4) mainly consists of a digester or pit for fermentation and a floating drum for the collection of gas. The dimensions of digester are 3.5 to 6.5 meter in depth and 1.2 to 1.6 meter in diameter. The digester is divided into two sections by a partition wall. The dung is mixed with water and loaded into the inlet. The loading submerges the partition wall in slurry. The fermented slurry flows out through the outlet pipe and is collected into a compost pit.

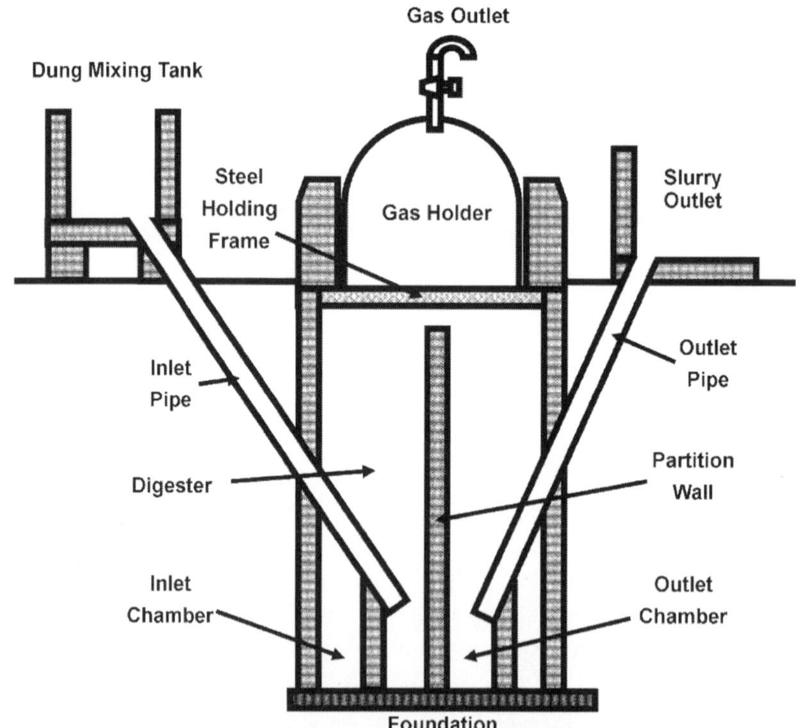

Figure 12.4. KVIC Model Biogas Plant.

The cylindrical gas holder drum is made of mild steel sheets with a concave top supported by angular iron. It fits into the digester like a stopper, sinking into the slurry due to its weight. When gas is generated, it rises and floats freely on the slurry surface to seal it. A central guide pipe prevents the tilting of the gas drum. The gas pressure varies between 7 and 9 cm in the water column. The drum costs about 40 per cent of the plant's total cost and requires periodic maintenance.

Table 12.3. Approximate Material Requirement for the Construction of KVIC Model Biogas Plants

Sl. No.	Size of Plant (m^3)	Bricks (No.)	Cement (Bags)	Brick Ballast (m^3)	Sand (m^3)	Paint (Litres)
1.	1	2460	13	0.60	2.00	2.0
2.	2	2765	17	0.90	2.55	2.5
3.	3	3200	19	0.95	2.90	2.5
4.	4	3730	23	1.25	3.40	3.0
5.	6	4200	27	1.50	4.00	3.0
6.	8	4800	32	1.75	4.25	3.5
7.	10	5500	40	2.00	5.00	4.0

Note: Besides above material, A.C. pipe (10 cm dia), steel drum (gas holder), steel guide frame along with guide pipe will be required.

Janta Model Biogas Plant

The main feature of Janta model (Figure 12.5) is that the digester and gas holder are integrated parts of brick masonry structure. The digester is made of shallow well having a dome-shaped roof on it. The inlet and outlet tanks are connected with the digester through large chutes which are called displacement chambers. The gas pipe is fitted on the top of the masonry dome. There is an opening on the outlet wall of the outlet displacement chamber for the discharge of spent digested slurry. The size of these plants is limited to 15 cubic meters per day. This is a plant where no steel is used, there is no moving part in it and maintenance cost is low. It is cheaper than the KVIC type plants. This model has a higher capacity when compared with KVIC model, hence it can be used as a community biogas plant. This design has longer life than KVIC models.

Table 12.4. Approximate Material requirement for the construction of Janta model biogas plants

Sl. No.	Size of Plant (m^3)	Bricks (No.)	Cement (Bags)	Brick Ballast (m^3)	Stone Ballast (m^3)	Sand (m^3)	Paint (Liters)
1.	2	1500	20	1.00	1.00	2.0	1.5
2.	3	2500	25	1.25	1.25	2.5	2.0
3.	4	3500	30	1.50	1.50	3.5	2.5

Sl. No.	Size of Plant (m^3)	Bricks (No.)	Cement (Bags)	Brick Ballast (m^3)	Stone Ballast (m^3)	Sand (m^3)	Paint (Liters)
4.	6	4000	40	2.00	2.00	4.25	3.0
5.	8	4500	45	2.25	2.25	5.0	3.0
6.	10	6000	65	2.75	2.75	5.5	3.5
7.	15	7500	75	3.50	3.50	6.5	4.0

Note: Besides above material, GI pipe (25 mm dia) with socket, steel for slabs for inlet and outlet chambers, gate valve will be required.

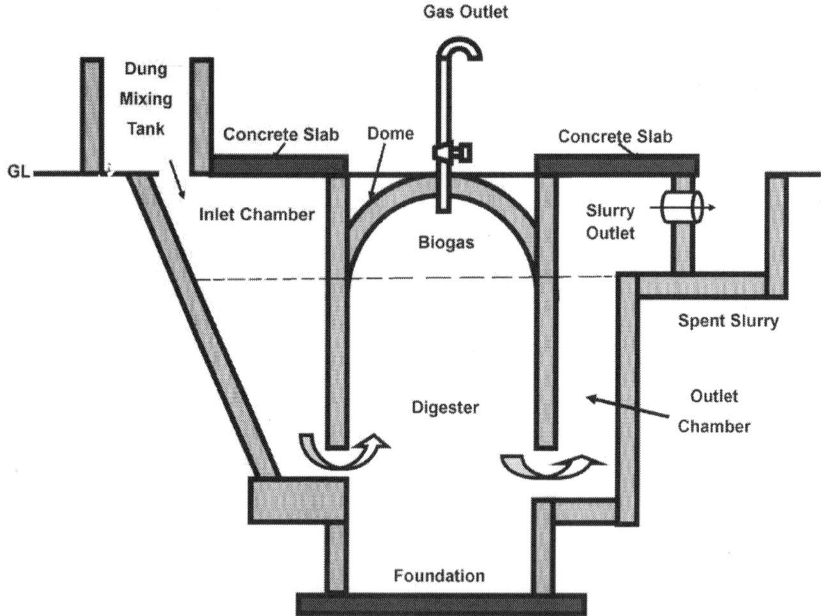

Figure 12.5. Janta Biogas Plant.

Deenbandhu Model Biogas Plant

Deenbandhu model biogas plant (Figure 12.6) was developed by Action for Food Production (AFPRO), New Delhi in 1984. This plant is designed on the principle that the surface area of biogas plants is minimized to reduce their installation cost without sacrificing the efficiency of the plant. It is an improved fixed dome model, where form work is used for construction. Due to shell structure the thickness is considerably reduced. The entire plant is built with brick in cement mortar. Unlike Janta biogas plant it does not require any shuttering, thus reducing the cost of construction. It is much cheaper than KVIC and Janta type plants. The digester, gas storage chamber and the empty space above the slurry are all provided in the spherical shell. All the slurry displaced out of the gas storage chamber is stored in the outlet displacement chamber as there is no displacement space on the inlet side. The inlet is in the form of a pipe which connects the digester with slurry mixing tank. The hydraulic retention time for this design is 40 days for northern plains of the country. For hilly regions the retention time is about 50 days. In colder climates like the lower regions of the Himalaya, the Deenbandhu fixed model is ideal because the digestion chamber is underground providing good insulation against the cold.

Table 12.5. Approximate Material Requirement for the Construction of Deenbandhu Model Biogas Plants

Sl. No.	*Size of Plant (m^3)*	*Bricks (No.)*	*Cement (Bags)*	*Stone Ballast (m^3)*	*Sand (m^3)*	*Paint (Liters)*
1.	2	1000	8	0.45	1.80	1.0
2.	3	1300	10	0.71	2.90	1.5
3.	4	1600	12	0.86	3.50	2.0
4.	5	1900	14	1.00	4.10	2.5
5.	6	2200	15	1.15	5.00	3.0

Note: Besides above material, GI pipe (25 mm dia) with socket, PVC pipe (15 cm dia), steel for slabs for outlet chamber, gate valve will be required.

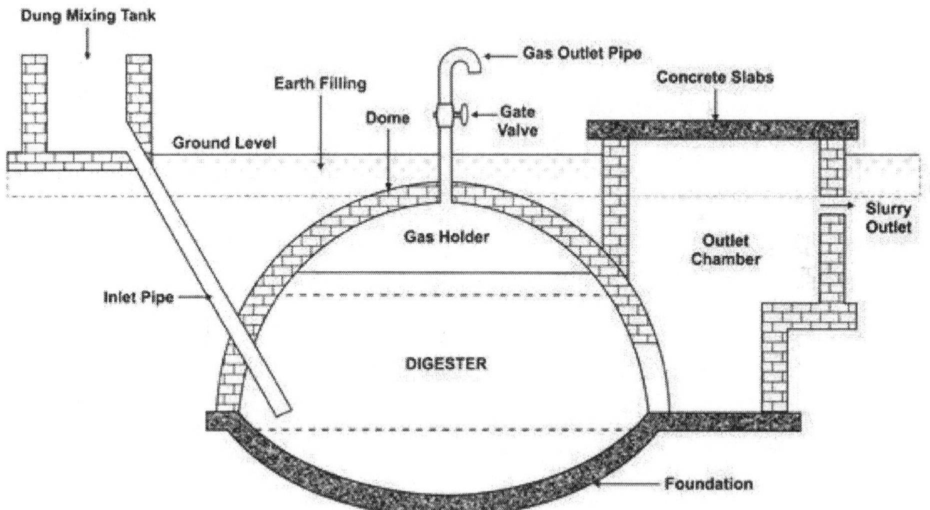

Figure 12.6. Deenbandhu Biogas Plant.

Do's and Dont's for Maintenance of Biogas Plants

Do's

1. Decide the size of biogas plant depending on the availability of feed material.

2. Select the site for biogas plant near to the kitchen. Otherwise, the pressure of the gas would drop at burner. It will also help in reducing the cost of pipe line. Site shall also be nearer to the cattle- shed, as far as possible as it shall be convenient to transport the dung.
3. Install biogas plant in an open space where the plant gets maximum sun rays freely falling during day time, all round the year.
4. Install biogas plant strictly as per the approved drawings and specifications with trained masons/technicians.
5. Do back filling by filling up the gap between the outer surface of the digester and the portion of the earth cutting, firmly compacting with sand/soil and also putting some water while compacting to avoid crack formation in wall.
6. Minimum distance of the source of water supply should be 15 meter from the plant otherwise water may get contaminated.
7. Feed biogas plant with cattle dung and water in 1:1 proportion to make a homogeneous mixture after thorough mixing.
8. Ensure the slurry (mixture of dung and water) is free from soil and straw before feeding in order to maintain gas production.
9. Maintain proper slope in the pipeline and fix water remover (water tap) at the lowest point.
10. Break the scum formed on the surface of the slurry in the digester by rotating the gas holder daily after feeding in case of KVIC model and by inserting a long bamboo throughout and stirring the slurry.
11. Use ISI marked burners and gas lamps to get maximum efficiency.
12. Open the main gas regulator cock only at the time of its actual use.
13. Adjust the air shutter till a blue flame is obtained. At this stage, maximum heat will be obtained.
14 Cover the inlet and outlet tank openings of fixed dome models with wooden plank, stone or reinforced cement concrete (RCC) slab to avoid accidental falling of children/cattle.

Don'ts

1. Do not install much higher capacity plant if you do not have sufficient cattle dung or any other feedstock for the gas production.
2. Do not select a site far-away from cattle shed and also kitchen to avoid operational inconveniences.
3. Do not install biogas plant under a tree or under shade.
4. Do not install biogas plant above ground level.
5. Do not fill soil loosely around the biogas plant.

6. Do not add either more water or less water to make a homogeneous mixture with cattle dung for feeding. Adding more or less water affect the efficient production of gas.
7. Do not allow soil or sand particles to enter the digester along with the slurry.
8. Do not allow any water accumulation in the gas pipeline; otherwise it will disturb the smooth flow of gas.
9. Do not allow formation of scum in the digester; otherwise the production of gas might stop.
10. Do not burn the gas directly at the main gate valve on the dome/gas holder even for the testing purpose as it can be dangerous.
11. Do not keep the outlet tank open since cattle, children may accidentally fall.
12. Do not use burner in the open, otherwise there will be a considerable loss of heat.
13. Do not leave the gas regulator open when the gas is not in use.
14. Do not use the gas if the flame is yellow. Adjust the flame by air regulator till it is blue in colour.
15. Do not make digested slurry pit more than 1.0-meter deep otherwise it will take time to dry.

Some Useful Data

Digester volume = volume of daily charge × retention time

The actual volume should be 10 per cent more.

Gas Requirement

Cooking = 0.3 to 0.4 m^3/person/day

Lighting = 0.10 to 0.12 m^3/hour/100 candle

Engine = 0.45 to 0.54 m^3/hp-h = 0.52 to 0.7 m^3/kWh

Calorific value of biogas = 3500 – 4800 kcal/m^3

Gas per kg of wet dung = 0.04 to 0.05 m^3

Optimum cow dung and water mixing ratio = 1:1

Optimum retention period for cattle dung slurry = 40 to 50 days

Availability of dung

Cow – 10 kg/day; Buffalo – 15 kg/day; Calf – 5 kg/day

IMPROVED COOK STOVES (*CHULHAS*)

Despite the United Nation's continuous effort, only 63 per cent of the world population have access to clean cooking fuels and related technologies. Still, close to 2.6 billion of the world population mainly in developing countries lack access to clean cooking fuels and rely on burning polluting fuels such as wood, animal dung or charcoal as their primary source of energy for cooking and heating. According to International Energy Agency, about 49 per cent people in India have access to clean cooking fuels. It means that 681 million people (51 per cent) still rely on traditional use of biomass, and remain exposed to harmful indoor air pollution caused by traditional cooking fuels. Women are the most sufferers due to inadequate supply of energy.

Wood is the primary source of energy for more than one-third of the world's population who live primarily in rural areas of the developing countries. Wood is the oldest type of fuel which man used for centuries after the discovery of fire. Inadequate attention towards this problem has resulted in heavy deforestation disturbing ecology and environment. Inefficient utilisation of fuel wood, coal, agro-wastes and animal-wastes for cooking has also resulted in serious health problems. Fuel wood consumption has increased 250 per cent since 1960 (the world's population only increased by 90 per cent since 1960). India's current firewood consumption is more than 133 million tonnes; most of it is being used in cooking. To cook 1 kg of food, 1.2 kg of firewood is required. In India, domestic cooking accounts for 77 per cent of the fuel wood consumption and that cooking with this fuel represented more than 35 per cent of all the energy consumed. Dependence of fuel wood is higher in rural areas as compared to urban.

12.21 Effects of Traditional *Chulhas* on Health

Open fires and primitive cook stoves are fuelled by solid materials, including coal and biomass (wood, dung and crop residue). These fuels release harmful particles into the air as they burn. All fires emit carbon dioxide, carbon monoxide, and particulate matter, including white (organic) carbon and black carbon. When inhaled on a regular basis, these particles can cause diseases such as asthma, lung disease and pneumonia. Health effects are mainly on women as they are responsible for cooking. Effect of smoke on women during cooking from traditional *chulhas* (Figure 12.7) is estimated equivalent to smoking of 200 cigarettes. Smoke from cooking activities is so dangerous that it has been called "the killer in the kitchen". The World Health Organization (WHO) estimated that 3.8 million premature deaths occur annually from household air pollution associated with inefficient cooking practices.

Disadvantages of Traditional Cooking Stove

Chulha requires wood for burning. Cooking in a *chulha* is time consuming as one cannot turn on or off the heat level as easily as in the gas stoves. The thermal efficiency of traditional *chulhas* is only 5 to 10 per cent.

Figure 12.7. Traditional *Chulha.*

12.22 Improved Cook Stove (*Chulha*)

Cook stoves are commonly called 'improved' if they are more efficient, emit less emissions or are safer than the traditional cook stoves. Improved cook stove (Figure 12.8) are also known as smokeless *chullah*. Improved cook stoves use biomass (charcoal, wood, agricultural waste or cow dung) and are designed to maximise thermal and fuel efficiency, operate safely and minimise emissions harmful to human health. They consume less fuel, increase convenience, and reduce cooking time as well as emission. The thermal efficiency of improved *chulhas* is 25 to 30 per cent.

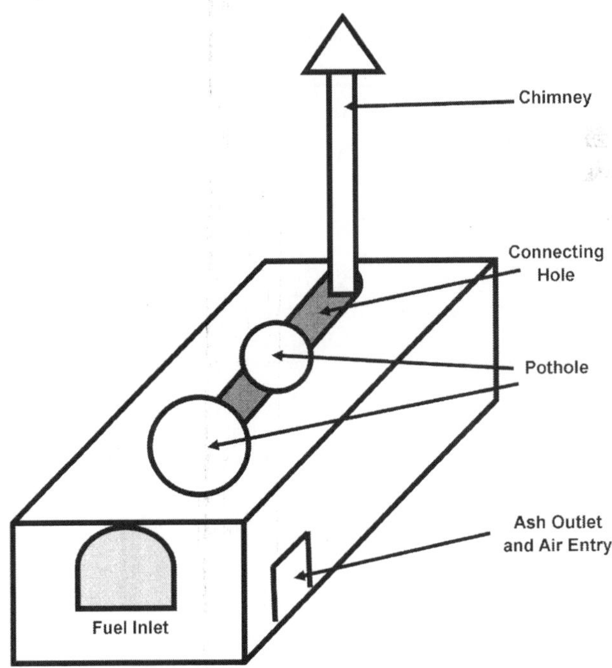

Figure 12.8. Improved Cook Stove.

Design Features

The main features of the improved cook stove are:

1. **A closed hearth** where combustion of the wood takes place. It protects the fire from the vagaries of the wind.
2. **A multi-pot design**. *Chulha* must be capable of carrying more than one pan. It permits a certain amount of recovery from hot gases that would otherwise have gone waste, and
3. **A chimney**. It provides the draught necessary for bringing in air for combustion and removing smoke from the kitchen environment.

Other Requirements

1. **Fire Box**: Fire box must be large enough to accommodate the size and type of fuel, and the pot used. Fuel, oxygen and higher temperatures are required for combustion to take place.
2. **Damper**: Combustion should take place within a closed fire box. Thus, provide damper to the entrance in the front. This can also regulate the flow of the air. A second damper can be used at the entrance of chimney. They should close as tightly as possible.
3. **Cooking Pot**: Cooking pot should have a large surface area exposed to the fire, and a small area exposed to air. Cooking pot should be as close to the heat source as possible to take full advantage of radiant heat, yet not so close that it smoothens the fire. Always cover the cooking pot to prevent heat losses from the contents inside.
4. **Grate**: Grate should be relatively small in size and the surface must always be less than the bottom of the firebox. The properly designed grate will improve the performance of a cooking stove, as air enters from under the grate and passes through the fuel bed. It is generally made of cast-iron.
5. **Baffles**: For maximum efficiency, it is necessary to retain the heat generated for as long as possible and in close proximity of the cooking pots. This can be achieved by slowing down the gas flow by means of obstacles called baffles.
6. **Chimney**: Chimney should be at least two to three feet above the roof, which is very essential in case of thatched roof. It should never be less than six feet high as this is essential to provide an adequate draught. Bends in the chimney should be avoided as far as possible because it creates a pressure loss. Bend, if any, should be made as smooth as possible.
7. **Cowl**: Put a cowl or cap at the top of the chimney. It keeps the sparks from flying out, and also prevents rain from coming in. A good distance between the cap and the chimney top should be kept for the smoke to exit.

A higher efficiency of cooking stove can be achieved by:

1. Better combustion of the fuel by providing an insulated combustion chamber around and above the fire. Use lightweight insulating material for insulating the combustion chamber of the stove in order to minimise heat loss from it and attain a high temperature inside to promote complete combustion.
2. Maximum transfer of heat of combustion from the flame and the hot gases to the cooking pot.
3. Minimum loss of heat to the surroundings.

Advantages of Improved Cooking Stoves

The benefits of improved cooking stoves include health improvements from better indoor air quality, cooking time savings, aesthetic improvements and improved social standing from the use of cleaner stoves, and environmental benefits to society, such as reduced black carbon or greenhouse gas emissions and deforestation. Other benefits of the improved *chulha* is to reduce fuel consumption leading to saving in fuel wood. It reduces the smoke so the women can save time and saving in time leading to more economic activity, employment generation, *etc.*

1. Improved cook stoves for fuel wood save between 25-65 per cent per stove per household when used correctly, while improved charcoal stoves usually save around 25-35 per cent compared to traditional charcoal stoves.
2. Reduction of smoke from the kitchen and lessening of respiratory, eye and other problems associated with smoke inhalation.
3. Check on deforestation and environmental degradation.
4. Reduction in drudgery of women and children from collection of more fuel.
5. Less time spent on gathering wood and hence saving in time. It is estimated that women and children spends more than 374 hours in a year for collecting fuel wood.
6. More cow dung available as fertilizer for agriculture.
7. Reduction in cooking time.
8. Employment opportunity to the local people.

A rural household uses up to 3000 kg of firewood per annum. Improved *chulha* uses 50 per cent less firewood, thus saving 1500 kg of firewood in a year. Saving of 1500 kg of firewood means saving of at least 3 full grown trees. Thus, in a village having 200 families can save 600 trees per annum by adopting improved cook-stoves, which is equivalent to 1.5 acres of forest area. More trees saved means improved environment and reduced pollution.

Types of Improved *Chulhas*

Improved cook stoves can take many shapes. They can be portable or fixed installed in a kitchen. They can be made of different materials such as metal or clay. They come with or without a chimney; and they can have different sizes for households, institutions or small enterprises.

There are a number of approved models in fixed and portable categories for family, community and commercial applications. These are:

1. Mud fixed *chulha* with or without chimney
2. Mud-clad pottery lined fixed *chulha* with or without chimney
3. Portable metallic *chulha*
4. Portable metal-clad ceramic lined *chulha*
5. Portable *chulha* with separate hood chimney system

The name of some of the improved *chulhas* are Sahyog, Thapoly, Dholadhar, Nada, Vikas, Magan, Pusa, Junagarh, Dholadhar, Mohni, Akash, Sohani, Harsha, Sambhu *etc.*

Do's and Don'ts

1. Do plan the cooking in advance to take advantage of graded heat of the various holes.
2. Do not keep any hole uncovered during cooking. Cover them up with a metal sheet, if not required.
3. Do not overstuff the duct with fuel. There must be sufficient space for air to get in.
4. Do not keep the chimney damper wide open to avoid wastage of fuel. Adjust it to the optimum position.
5. Do not locate the chimney far away from the cook stove as it may be difficult to start a draught on a cold day.
6. Use small pieces of wood. Smaller pieces have proportionately greater exposure to air flow and will burn faster. Also small pieces of wood heat up quickly, release gases easily and give off a steady flame.
7. For proper combustion, provide adequate, but not excessive, supply of air.
8. After the fire dies out, dampers can be closed to conserve heat.
9. Clean the chimney at regular intervals, otherwise it would be fire hazard as well as hinder the proper functioning of stove.
10. Do repair the *chulha* as and when any crack is noticed.

Important Points to Remember

1. Common types of flooring for implement sheds include brick-on-edge, cement-concrete, gravel, and mud.
2. A corrugated metal roof is a common choice for a farm implement shed as it is durable, long-lasting and easy to maintain.
3. The width of the open-side implement shed shall be 8 to 10 m wide.
4. Welding section in the farm workshop should preferably be located near the main gate where it can conveniently be used inside or outside the building.
5. Large above-ground or underground storage tanks are commonly used for storing fuels and chemicals in significant quantities.
6. Smaller quantities of fuels and chemicals are often stored in drums and barrels made from materials such as steel, plastic, or fibre.
7. The storage facility for fuels and chemicals should be located at a safe distance from any livestock areas, such as barns, sheds, or grazing fields, to minimize the risk of contamination or accidental exposure to animals.
8. A well-ventilated, dry area away from direct sunlight, heat sources, and ignition points should be chosen for the storage of fertilizers and other non-hazardous products.
9. Proper ventilation is crucial for the storage of fuels and chemicals to ensure safety and prevent potential hazards.
10. The surface of the threshing floor must be hard and durable to withstand the threshing process.
11. Borders can be constructed around the perimeter of the threshing floor to prevent the crops from spilling over the edge.
12. In areas with high winds, windbreaks may be constructed around the perimeter of the threshing floor to protect the crops from being blown away during threshing.
13. Biogas is produced through anaerobic fermentation of animal excreta and other agricultural wastes.
14. One m^3 of biogas is equivalent to 4.7 kWh of electricity.
15. One m^3 of biogas is equivalent to 0.43 kg of LPG.
16. One m^3 of biogas is equivalent to 0.62 litre of kerosene oil.
17. The composition of methane in biogas ranges from 50 to 65 per cent.
18. The biogas plant should be installed at least 15 meter away from drinking water source to avoid contamination.
19. Optimum temperature inside the digester for biogas production is 35 to 45°C.

20. The pH range suitable for gas production is 6.6 to 7.5.
21. A carbon to nitrogen (C/N) ratio in the range of 25:1 to 30:1 is found to be optimum for biogas production.
22. Hydraulic retention time (HRT) for any biogas plant is inversely proportional to temperature.
23. Janta model and Deenbandhu model are fixed-dome type biogas plants.
24. KVIC model is floating-drum type biogas plant.
25. Janta model gas plants are cheaper than the KVIC type plants.
26. Deenbandhu model biogas plant is cheaper than the KVIC type plants.
27. Calorific value of biogas is 3500 – 4800 kcal/m^3.
28. The thermal efficiency of improved *chulhas* is 25 to 30 per cent.

Exercises/Questions

1. Describe the construction details of a farm machinery shed.
2. What should be the best location of a farm workshop?
3. What are the different sections of the farm workshop? Discuss them briefly.
4. What are the storage considerations for a fuel and chemical storage?
5. Discuss the different storage methods for storing fuels and chemicals.
6. What points should be considered while selecting a site for fuel and chemical storage?
7. What is the purpose of a threshing floor?
8. What are the different types of threshing floors?
9. Discuss the construction details of a threshing floor.
10. Discuss the potential of biogas in India.
11. What points should be considered while selecting a site for a biogas plant?
12. What are the factors that affect the biogas production?
13. List down the main components of a biogas plant.
14. Write the composition and calorific value of biogas.
15. How does a KVIC biogas plant operates? Explain with the help of a figure.
16. Describe the construction and working of *Janta* biogas plant.
17. Describe the construction and working of *Deenbandhu* biogas plant.
18. What are the methods to maintain biogas production in a biogas plant and explain them in brief?
19. What are the effects of traditional *chulhas* on health of the people?

20. Describe the design features of an improved cook stove.
21. What are the advantages of the improved cook stoves?
22. Write short notes on:
    (a) Storage of fertilizers and other non-hazardous products
    (b) Ventilation requirement for fuel and chemical storage
    (c) Fire protection for fuel and chemical storage
    (d) Threshing floors
    (e) Improved cook stoves
    (f) Fuel and chemical storage

Chapter 13

Rural Roads and their Construction

Farm roads are the arteries that connect remote villages to the broader transportation network in the country. They play a vital role in connecting communities and driving economic development in the countryside. They enable farmers to transport their produce to markets efficiently, reducing post-harvest losses and ensuring timely access to consumers. These roads also facilitate the transportation of inputs such as seeds, fertilizers, and machinery to farms, enhancing agricultural productivity. Furthermore, farm roads provide access to healthcare, education, and other social amenities for rural communities.

Many farm roads in India lack proper construction, making them prone to damage during monsoons and causing inconvenience to farmers. Insufficient drainage systems and inadequate maintenance contribute to deteriorating road conditions. In some regions, remote villages still lack access to farm roads, resulting in isolation and limited market opportunities for farmers. The absence of proper connectivity hampers the overall economic development of these regions. In areas with heavy rainfall or extreme weather conditions, farm roads may become inaccessible during certain seasons, disrupting the transportation of agricultural produce and essential services.

Recognizing the critical role of farm roads in rural development, the Indian government as well as State governments have implemented various initiatives to improve rural connectivity and address the challenges faced. Some notable initiatives include *Pradhan Mantri Gram Sadak Yojana* (PMGSY), *Bharatmala Pariyojana, etc.*

13.1 Types of Farm Roads

Farm roads can vary in design and purpose based on the specific needs of the farm and the surrounding landscape. Here are some common types of farm roads:

From the service point of view, the farm roads shall be grouped into the following two classes:

1. Roads serving farmstead, and
2. Roads serving field

Roads can be classified into flexible or rigid road on the basis of construction and materials used. The flexible roads are made up of layers of various materials such as lime stone, gravel, *kanker*, red earth *etc.* while the rigid roads are made up of cement concrete.

Farmstead Roads

These roads provide access to different areas of the farm, connecting various farm buildings, fields, and production areas. They are usually designed for vehicles and equipment to move easily between different parts of the farm. Farmstead roads typically carry heavier and denser traffic than field roads. Hence, farmstead roads shall have a smooth but hard surface. These roads require minimal maintenance and facilitate easy cleaning.

Field Roads

These roads are built within fields to facilitate the movement of farm machinery and vehicles for planting, harvesting, and other field operations. They are often unpaved or minimally paved and are designed to withstand heavy loads and frequent traffic. Field roads should have sufficient hardness and not necessarily smoothness. Hence all that is required is a good hard base with a surface of gravel of broken stone. The base may be constructed using rubble, broken brick, laterite or any other cheap material available in the locality.

13.2 Requirements of a Good Road

Good roads typically have several key requirements such as level, smooth, hard and dry to ensure safety, efficiency, and durability. The road requirements may vary based on the type of road (urban, rural, highway) and the local conditions and regulations. Rural roads should provide access to various communities, schools, healthcare facilities, and economic centres. They should be well-connected to the broader road network, enabling efficient transportation of goods and services. Some important factors that contribute to the quality of a road are listed below:

1. **Alignment:** The alignment of a road refers to its horizontal and vertical curves. Roads should be properly aligned to provide good visibility, minimize sharp turns, and maintain a comfortable driving experience.

Easy plying and good visibility on straight roads prevent accidents. Curves should be smooth and easy to navigate, as side thrusts can cause vehicles to slip or skid and overturn. To maintain the equilibrium of vehicles on curves of roads, the outside edge of the road is kept at a higher level, known as super elevation. This inclination reduces centrifugal force on a running vehicle. The super elevation is expressed as a fraction of the difference in level between the outer and inner edges of the road to its width. This helps to maintain vehicle equilibrium and prevent accidents. The generally adopted super elevation for the various kinds of roads is given in the Table 13.1.

Table 13.1. Super Elevation for Various Kinds of Roads

Sl.No.	*Kind of Roads*	*Camber*
1.	Water bound macadam road	1 in 48
2.	Tar or bitumen	1 in 60
3.	Cement concrete	1 in 72

2. **Gradient**: Roads need to have appropriate gradients (slopes) to allow for efficient drainage of rainwater and prevent waterlogging. Gradient is the slope (rise or fall) of a road along its longitudinal section. It is expressed as a ratio of the difference in level between two points to their longitudinal distance. It is also expressed as a percentage of slope (*i.e.*) the rise or fall for 100 metres distance. In road making, a level road would be ideal from the attractive and efficiency point of view. Gentle curves and gradients are preferable for better visibility and reduced stress on vehicles.
3. **Smoothness**: The surface of a road should be smooth, well graded, and free from potholes and major cracks. It should provide good traction for vehicles and minimize vibrations and discomfort for passengers. Regular maintenance and repair are essential to ensure the road surface remains in good condition. A properly constructed and maintained surface reduces wear and tear on vehicles and provides a comfortable ride for users. Secondly, it helps to drain off water since on smooth water proof surface water will not stick but will be shed to ditches at the sides.
4. **Drainage**: Effective drainage is crucial to prevent water from accumulating on the road surface. Proper design and construction of drainage systems, including adequate provision of gutters, culverts, and storm water drains, help prevent water-related damage and erosion, and improve road safety. Roads are designed with sufficient camber to facilitate surface drainage and easy traffic. Steep camber can cause the slipping of high-speed vehicles and erosion on berms.

Camber is the rise of a road's centre relative to the edge, typically resulting in a flat contour at the central half and slightly steeper at the edges. The camber adopted for different kinds of roads are given in the Table 13.2.

Table 13.2. Values of Camber for Various Kind of Roads

Sl.No.	Kind of Roads	Camber
1.	Earthen roads	1 in 20 to 1 in 24
2.	Water bound macadam road	1 in 30 to 1 in 48
3.	Tar or bitumen	1 in 36 to 1 in 48
4.	Cement concrete	1 in 60 to 1 in 72

5. **Strong Foundation**: A good road requires a strong foundation to support the weight of vehicles and prevent structural failures. The roadbed should be appropriately prepared, which may involve compacting the soil, adding stabilization materials, or constructing a suitable sub-base layer.
6. **Width of Roads**: Roads should have a sufficient width to accommodate the expected volume of traffic. This allows vehicles to pass each other safely and reduces the chances of collisions. If berms (space from the road edge to the edge of the drain) known as shoulders are provided, 2.4 m to 3.0 m (8 feet to 10 feet) wide roads are sufficient for running through fields. However, for the farmstead still wider roads are necessary, since vehicles are housed here and the traffic will be denser. For long vehicles and trailers, the width should be increased on curves to negotiate. The increase in width depends on the radius of curves; 1 m for 30 m radius and 2.5 m for 18 m radius. According to the Indian Road Congress (IRC) guidelines, the recommended width for a single-lane rural road with light traffic is typically around 3.75 meters (12.3 feet). For two-lane rural roads, the recommended width is usually around 7.0 meters (23 feet) to accommodate two vehicles traveling in opposite directions. However, it's important to note that these are recommended standards, and the actual width of rural roads can vary based on local conditions and available resources. In some cases, particularly in more remote and underdeveloped areas, rural roads may be narrower than the recommended standards due to various factors such as budget limitations, land constraints, or lack of proper maintenance. These narrower roads can range from 2.5 meters (8.2 feet) or even less, making it challenging for vehicles to pass each other comfortably.
7. **Safety Features**: Roads should be equipped with necessary safety features, such as road signs, guardrails, reflective markers, and road

markings. These elements help guide and regulate traffic, enhance visibility, and reduce the risk of accidents.

8. **Adequate Lighting**: Appropriate lighting is essential, especially for roads used during night time. Streetlights placed at regular intervals help improve visibility and safety for drivers and pedestrians.
9. **Maintenance and Repair**: Regular maintenance and timely repairs are crucial for the longevity and quality of a road. This includes activities like filling potholes, resurfacing damaged sections, maintaining proper road markings, clearing debris, and addressing any structural issues promptly.
10. **Environmental Considerations**: Roads should be designed and constructed while considering environmental factors. This can include measures like minimizing the impact on natural habitats, implementing erosion control measures, and incorporating eco-friendly materials, employing energy-efficient techniques, and minimizing environmental impacts during construction and maintenance.

13.3 Composition of a Road Structure

Roads typically consist of several layers, each serving a specific purpose in providing a durable and functional transportation surface. A road structure can vary based on design standards and regional practices, and is composed of following layers (Figure 13.1) from bottom to top:

1. **Sub-Soil or Road Bed**: The ground beneath the road can be natural or prepared after top soil removal. It must be strong enough to resist vehicle movement and distortion. It forms the lowest layer and provides support for the entire road structure.
2. **Subgrade**: The subgrade is a compacted layer of granular permeable material. It prevents water from entering the road foundation. It also supports the foundation and maintains its level. In favourable natural conditions, the subsoil and subgrade can be combined to form the base and foundation courses.
3. **Foundation Course or Sub-Base**: The sub-base is a layer of granular or stabilized material placed on top of the subgrade. It helps distribute the load from traffic and provides additional support to the road, and offers resistance to the movement or distortion due to moving loads to a maximum extent. This is also known as 'soling' or 'bottoming'. Common sub-base materials include crushed stone, gravel, or recycled materials.
4. **Base Course**: The base course lies above the sub-base and is composed of a stronger and more durable material. It serves as a load-spreading layer, providing further support to the road. Typical materials used for

the base course include crushed stone, recycled concrete, or asphalt-treated base.

5. **Binder Course:** The binder course, also known as the intermediate course or binder layer, is made up of asphalt concrete (commonly referred to as asphalt). It is designed to provide strength, stability, and flexibility to the road surface, ensuring it can withstand the stresses from traffic loads.
6. **Wearing Course or Surface Course**: The surface course is the topmost layer of the road and is designed to provide a smooth and skid-resistant driving surface. It also protects the base from abrasion and from the effects of rain. It is composed of a thin layer of high-quality asphalt concrete or other materials like concrete, depending on the specific road design and requirements. In some road designs, additional layers may be included, such as a wearing course or a special surface treatment for specific purposes like noise reduction or water drainage.

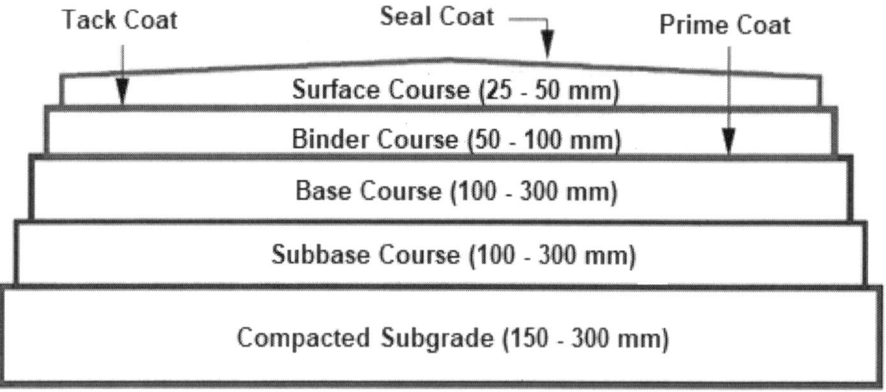

Figure 13.1. Typical Cross-Section of a Flexible Pavement.

13.4 Components of Road Structures

Road structures consist of various components and elements that work together to create a functional and safe roadway. Some of the key components (Figure 13.2) of road structures are listed below. The specific features and designs may vary depending on the type of road, location, and local regulations.

1. **Roadway**: It is the main driving surface where vehicles travel. It is typically made of asphalt or concrete.
2. **Shoulders**: These are the areas on either side of the roadway, designed to provide additional space for emergency stopping or manoeuvring.

3. **Lane Markings**: Lines painted on the road to delineate lanes, including solid lines, broken lines, and arrows indicating traffic flow.

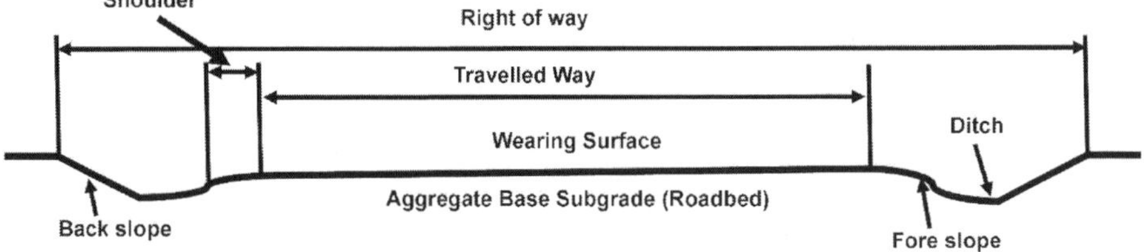

Figure 13.2. Typical Gravel Road Section.

4. **Curbs**: These are raised edges of the road that help contain and channel water runoff and define the boundary between the roadway and the adjacent areas.
5. **Sidewalks**: These are paved pathways alongside the road for pedestrians to walk on, separated from the roadway by curbs or other barriers.
6. **Medians**: These are dividers between opposing lanes of traffic, often with barriers or vegetation to prevent vehicles from crossing over.
7. **Bridges**: These are structures that carry the road over obstacles such as rivers, valleys, or other roads.
8. **Culverts:** Culverts are necessary at field gateways and road junctions to discharge drain water through pipes. The minimum diameter of pipes shall be 22.5 cm. The size of the pipes depends on the road's length and the amount of water to be discharged. The culverts also allow passage over narrow streams, drains, or small waterways.
9. **Drainage Systems**: Channels, gutters, and catch basins designed to collect and divert rainwater away from the road surface. Water stagnation on earthen roads leads to ruts, pot holes, and erosion. The paved roads are also susceptible to erosion due to stagnation. The eroded road with ruts and potholes makes it unpleasant. Side ditches are necessary for water removal and intercepting water from adjoining lands. Side ditches can be constructed on one or both sides of the road. These drains should be located 1.2 metres from the road's edge and 0.3 metres below the formation, with the verge falling from the road to the ditch. The ditches should have a sufficient cross-sectional area to discharge the entire water collected in them. This prevents water from accumulating on the road and makes it more pleasant.
10. **Traffic Signs**: Signage placed alongside or above the road to provide information, warnings, and regulations for drivers.

11. **Street Lighting**: Illumination provided by streetlights to enhance visibility during night time driving.
12. **Traffic Calming Measures**: Speed bumps, rumble strips, and other features designed to reduce vehicle speeds and increase safety.

13.5 Detailed Specification of Roads

Earth Filling

Earth filling of roads, also known as embankment, is a crucial process in road construction that involves the placement of compacted soil or other suitable materials to elevate the road surface. This technique is employed to bridge gaps, level uneven terrain, or create a stable foundation for roadways. Earth filling plays a vital role in ensuring smooth and safe transportation, as it helps to eliminate steep slopes, reduce the impact of natural obstacles such as rivers or valleys, and maintain a consistent road gradient.

Before beginning earthwork, the ground must be cleared of trees, jungle, and roots. An embankment should be created from borrow pits on either side of roads. The earthwork shall be laid in layers of 15 to 23 cm and consolidated by rollers. Large clods should be broken and settlement allowances made equal to 10 per cent of the embankment's height. Side slopes should be 4 to 1, except in high embankments, and 2 to 1 in good soil.

Soling Coat

The soling layer acts as a transition between the roadbed and the subsequent layers, such as the base course and the wearing surface. Soling refers to the placement of a durable and stable layer of material on the roadbed, serving as a foundation for the upper layers. The primary purpose of soling is to provide strength, stability, and load-bearing capacity to the road surface. Soling must be provided under the wearing coat, except on hard natural surfaces like rock. The width of soling should be 30 cm more than the proposed carriageway width. If brick on edge is provided, the width of soling should be the same as the carriageway width.

During the soling process, various types of materials are used, depending on the specific requirements of the road and local conditions. Common materials include crushed stone, gravel, or well-graded aggregates. These materials are carefully selected and compacted to create a solid base layer. The thickness of the soling layer may vary based on factors such as the expected traffic load and the underlying soil conditions.

Collection of Soling

Burnt or over-burnt bricks or stones shall be used for the soling coat. Avoid using bats due to their smaller size than bricks. They should be stacked parallel to the road's centre line and clear of the formation width. The stone should be hard, durable, and tough, while the *kankar* should be tough and heavy with a bluish fracture. The weathering of *kankar* and stones for one month is necessary before their use on the roadside.

Consolidation of Soling

The soling should be carefully laid. It shall be hand-packed with interstices filled with smaller pieces of the same material. The subgrade should be thoroughly levelled to ensure it is hard and consolidated. The sub-grade shall be free of soft pots or depressions. After laying the soling, spread the earth or sand over it to a thickness of 2.5 cm to fill up the joints.

Wearing Coat

The soling surface must be cleaned and cambered to create a finished road. Two bunds of clay puddle, 23 cm wide and 15 cm deep, should be made along the outer edges of the metalling. The bunds prevent new metal from spreading and retain water used in consolidation. Stone metal should be spread over the surface, carefully packed, with larger pieces placed below and smaller pieces in the interstices. Templates should be used at a distance of approximately 7.5 m from one another.

The metal should be compacted using a road roller from the edges to the centre. The metal shall be well compacted without any appreciable wave in front of the roller. The metal shall be thoroughly watered and continued until the consolidation is complete. After the consolidation, the binding material, such as fine *bajri*, shall be spread over the metal. The water shall be added with *bajri* to form slurry. The slurry shall be grouted into interstices.

The template's camber shall be 1 in 60 for painted roads, 1 in 72 for cement concrete surfaces, and 1 in 48 for water-bound roads. For example, a 1 in 72 camber would require a 1 cm height above the outer edges for each 1.44 m road width. To ensure good consolidation, a loaded cart should not leave an indentation when passing over the finished work, and a metal piece of wall nut size should be crushed when a roller is passed over it.

Surface Painting

The water bound road shall be rendered free from all ruts and pot holes or other defects. The surface shall be swept clean and free from dust, dirt or other deleterious matter by hand brushing with wire brushes, and finally by dusting with gunny bags. Only that area of the surface shall be broomed and brushed which is to be painted on the same day.

Quantities of materials for various coats shall be as follows:

Coat	*Binder (kg/m² of road area)*	*Grit (m³/100 m² of road area)*
First coat	1.74	1.23
Second coat	0.90	0.75
Subsequent or renewal coat	0.80 to 0.90	0.60 to 0.75

Application of Binder (Bitumen): The bitumen shall be spread in a rate of uniform thickness by means of a pressure sprayer. The standard grade for surface painting which has the widest use of any bitumen in India is MAXPHALT 80/100 penetration. The quantity of bitumen for first coat shall be 1.74 kg/m^2 of road area. Immediately following the application of the bitumen, stone chippings (*bajri* or grit) which should be in a perfectly dry condition having gauge 6 to 20 mm shall be carefully spread in uniform layer. Any excess chippings in one place shall be brushed into hungry spots with a *Pilchhi* broom. The quantity of *bajri* shall be 1.23 m^3/100 m^2 of road area for first coat. Immediately after gritting and completion of light brooming, the whole area shall be rolled with a roller weighing 8 to 10 ton, and rolling shall be continued until the chippings are thoroughly embedded in the bituminous surface.

The second coat of paint shall be applied as soon as the first coat has reached its optimum condition. This shall be given when paint has hardened, the surface has become smooth, and all loose grit has been absorbed. The same process and precautions shall be followed as that of the first coat except the quantity of binder and grit. The grade of the *bajri* for second coat shall be 6 to 10 mm.

13.6 Classification of Roads on the Basis of Materials Used for Construction

Roads can be classified into various categories based on the materials used in their construction. Here are some common classifications:

Flexible Pavement

1. **Bituminous Roads**: These roads are constructed using bitumen (a sticky black substance) as the binder material. Bituminous roads are flexible and have a smooth surface. They are suitable for rubber tyred vehicles of both slow and high speed.
2. **Asphalt Roads**: Asphalt is a mixture of aggregates (such as stone, sand, and gravel) and asphalt binder. Asphalt roads are commonly used for high-traffic roads and highways.

Rigid Pavement (Concrete roads)

Concrete roads are constructed using cement, aggregates, and water. They have a rigid structure and are known for their durability and strength. They are suited for all kinds of traffic.

Gravel Roads

Gravel roads are constructed using layers of compacted gravel. They are common in rural areas and are relatively less expensive to build compared to paved roads. These roads are better than earthen roads and are stable for light traffic.

Earthen Roads

Earthen roads are the simplest type of roads, consisting of a compacted layer of soil or earth. They are often found in rural or remote areas and are generally less durable than other types of roads. They are suitable for low speed vehicles such as bullock drawn carts.

Water Bound Macadam Roads

The road having its wearing surface consists of clean, crushed aggregates, mechanically interlocked by rolling and bound together with filler material and water laid on a prepared base course is called WBM road. They are less common today but have been influential in the development of modern road construction techniques. They are suited for heavy traffic of low speed vehicles.

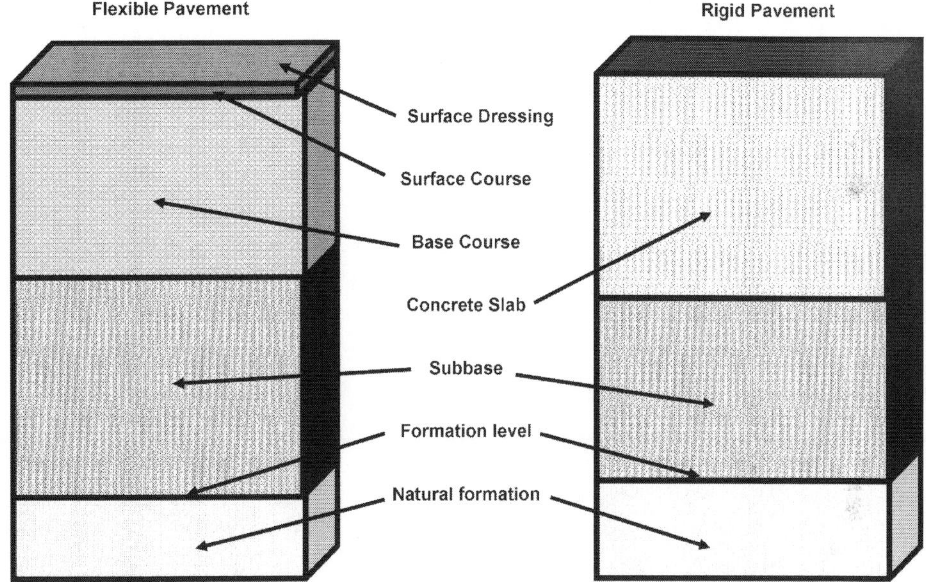

Figure 13.3. Difference between Flexible Pavement and Rigid Pavement.

13.7 Construction of Roads

13.7.1 Earthen Road

Constructing an earthen road involves the process of preparing the roadbed, shaping the road surface, and ensuring proper drainage. Soils with up to 30 per cent sand and 70 per cent clay content can create first-class un-metalled roads. Properly maintained un-metalled roads offer better resilience and cost-effectiveness compared to many other road types.

The general steps involved in the construction of an earthen road are as follows:

1. **Survey and Design**: Conduct a survey of the area where the road is to be constructed. This includes assessing the terrain, soil conditions, and determining the alignment and gradient of the road. Design the road layout accordingly, considering factors such as traffic volume and load-bearing capacity.
2. **Clearing and Grubbing**: Remove any vegetation, trees, rocks, or debris from the road alignment after peg marking. Clearing the path ensures a clean surface for construction.
3. **Earthwork**: Prepare the roadbed by cutting or filling the soil, depending on the topography. Cut the soil to lower elevations and fill the soil in areas where the ground is lower than the desired road level. All the clods and lumps should be crushed and spread. Ensure proper compaction of the soil to achieve stability. The soil is spread over uniformly to the required width in layers not exceeding 15 cm thick.
4. **Subgrade Preparation**: Compact the subgrade, which is the natural soil or compacted fill on which the road will be constructed. Red earth or clayey soil mixed with sand is ideal for earthen roads. The soils shall be spread in layers and consolidated with a roller using water. Use heavy machinery like rollers or compactors to achieve the desired density and strength. This step is crucial to prevent settling or deformation of the road in the future.
5. **Shaping and Grading**: Shape the road surface according to the designed alignment and cross-section. Use graders or other earth-moving equipment to achieve the desired profile, including the road crown (higher centre and sloping sides for proper drainage). A camber of 1 in 24 is given for earthen roads.
6. **Compaction**: Compact the road surface using compactors or rollers to increase the density and strength of the earthen material. This helps to minimize future erosion and maintain a smooth and stable road surface.
7. **Surface Treatment**: Apply surface treatments to protect the road surface from erosion and improve durability. This may involve applying a layer of gravel or other suitable materials, depending on the specific requirements and budget.
8. **Drainage**: Install proper drainage systems to ensure the efficient flow of water off the road. Construct ditches or side slopes to carry away surface water and prevent water accumulation on the road surface, which can cause erosion and damage.
9. **Maintenance**: Regularly inspect and maintain the earthen road to address any issues like erosion, potholes, or drainage problems. This includes periodic grading, compaction, and repair of surface damage.

13.7.2 Gravel Road

Gravel roads are an improved form of unsurfaced earthen roads. It is made by using well-graded gravel of 2–5 cm in size. The soil acts as a binding material among the gravels. First three steps as listed for earthen road are also applicable for gravel road. Thereafter, the soil subgrade of total thickness 300 mm is constructed by compacting the soil in suitable layers at its optimum moisture content. After sub-grade gets sufficiently dry, the gravel base course of specified thickness (100-200 mm) is constructed using the selected gravel or crushed stone in suitable layers. The material is compacted using a roller to enhance its stability and load-bearing capacity. A moderate sprinkling of water will be of help to aid compaction. Then, a layer of gravel or crushed stone on top of the compacted base layer is applied. The compaction process is repeated multiple times to achieve the desired level of compaction. The camber given for gravel road is 1 in 28 to 1 in 32. Maintain the gravel road by grading the surface to remove potholes, ruts, and other irregularities at regular interval. Additional gravel to maintain the road's integrity shall be added as and when needed.

Advantages of a gravel road over an earthen road are:

1. The construction is easy.
2. Gravel road is economical.
3. The surface is smooth, and it is easy for driving on road.
4. It is suited even for heavy traffic when the thickness is suitably designed, and
5. It is suitable in all weather.

13.7.3 Water Bound Macadam Road

Macadam road construction refers to the process of building a road using layers of compacted crushed stone or gravel. The pavement base course is made of crushed or broken aggregate mechanically interlocked by rolling, and the void filled with screening and binding material with the assistance of water. The macadam road construction method was developed by Scottish engineer John Loudon McAdam in the 19th century and has since evolved, but the basic principles remain the same. It is the most common type of road in India, and is used in certain situations, such as low-traffic rural roads or temporary construction access roads.

Water-bound macadam roads are smoother and more resilient than earth and gravel roads. These roads offer good traction, better wheel hold, and help in avoiding slipping or skidding. These roads are not slippery even when wet, making them a safer option for vehicles. The steps involved in constructing a macadam road are:

1. **Clearing and Grading**: The area where the road will be constructed is cleared of vegetation, rocks, and other obstructions. The surface is then graded to ensure proper drainage and a smooth foundation.
2. **Subgrade Preparation**: The subgrade is the natural soil or compacted earth beneath the road. It is prepared by removing any weak or unsuitable material and ensuring a stable base for the road. On this, a layer of sand for the designed thickness to a maximum of 10 cm is spread evenly and consolidated well.
3. **Base Course**: A layer of larger-sized crushed stone or gravel, known as the base course, is placed on top of the subgrade. This layer provides stability and helps distribute the load of traffic. It is compacted using heavy machinery.
4. **Binder Course**: A layer of smaller-sized crushed stone or gravel, called the binder course, is laid on top of the base course. The binder course further enhances stability and helps prevent the migration of the upper layers.
5. **Surface Course**: The final layer, known as the surface course, is composed of smaller stones or gravel. It provides a smooth and durable driving surface. The surface course is compacted and shaped to the desired profile.
6. **Compaction**: Throughout the construction process, each layer is thoroughly compacted using rollers or other compaction equipment. This ensures proper bonding between the particles and increases the road's strength and durability.

13.7.4 Bitumen Road

Bitumen or tar roads' top surfaces are made of broken metal binded by bitumen or tar. Old surfaces with a strong, water-bound macadam crust serve as a foundation. Hard-broken stone and bitumen are used for bitumen roads. The granite broken stones are the best for bitumen roads. Bitumen coats aggregates and provides adhesive properties to prevent disintegration and protect water content. Constructing a bitumen road involves several steps and processes. These are:

1. **Site Preparation**: Construction site shall be cleared of any vegetation, rocks, or debris. Then, excavate the roadbed to the required depth, taking into account any necessary slopes or curves. The subgrade soil shall be compacted to ensure stability and prevent future settlement.
2. **Base Course Construction**: A layer of granular material such as crushed stone or gravel is placed on top of the compacted subgrade. The base course is compacted using heavy machinery like rollers to achieve the desired density to the required thickness and smoothness specifications.

3. **Binder Course Construction**: A binder course consists of a mixture of bitumen and aggregates is applied. It serves as an intermediate layer between the base and surface courses. The mixture is spread and compacted using specialized paving equipment to achieve the desired thickness and density. The surface should be swept, clean and kept free from dust before applying the bituminous material.
4. **Surface Course Construction**: The final layer, known as the surface course or wearing course, is applied. The surface course also consists of a bituminous mixture, but it usually contains finer aggregates for a smoother finish. The mixture is spread and compacted using pavers and rollers to achieve the desired thickness, texture, and compaction.
5. **Surface Treatment**: Surface treatments like seal coats or micro surfacing to enhance the road's durability, skid resistance, and waterproofing is applied. These treatments help protect the surface from oxidation, aging, and moisture damage.

13.7.5 Cement Concrete Road

The materials required for cement concrete roads are cement, sand, hard and durable broken stone metal of 2.5 to 5 cm size and water. The slab is constructed either by alternate bay method (Figure 13.4) by dividing the carriage-way into suitable strips or by the continuous slab method.

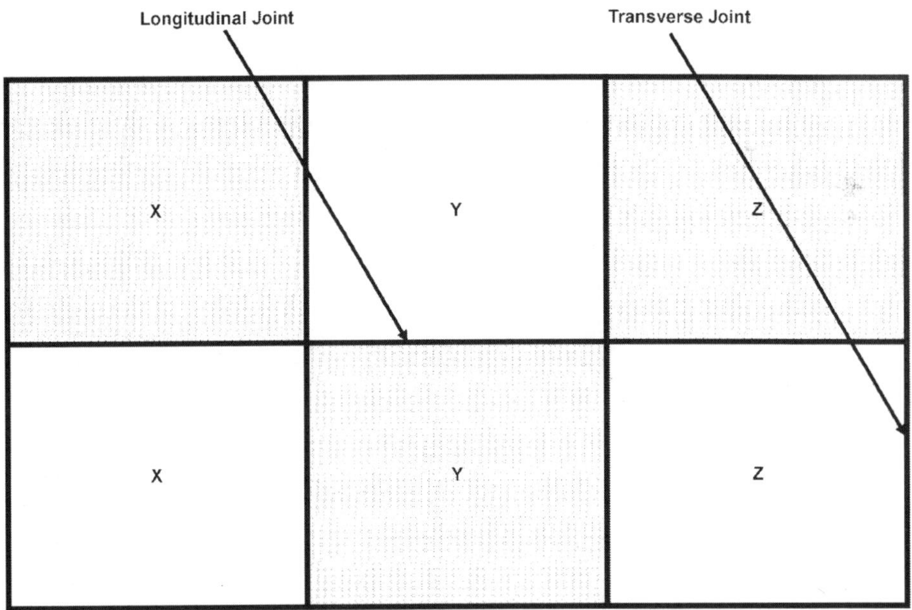

Figure 13.4. Alternate Bay Method.

The subgrade is prepared and checked at least two days in advance of correcting. The subgrade or sub-base is kept in moist condition at the time when the cement concrete is placed. The steel or wooden forms are placed. These have minimum base width of 10 cm for slab thickness of 20 cm and minimum base width of 15 cm for slab over 20 cm. The length of steel forms shall preferably be equal to the length of the slabs to be laid, but shall in no case be less than 6 m in length. Material is batched and mixed as per design, then it is transported to the site and placing of concrete (1:2:4) is carried out. Spreading is done uniformly. Concrete shall be handled from the place of mixing to the place of final deposit as rapidly as practicable by method which will prevent the segregation, contamination, loss or setting of the ingredients. The concrete shall be deposited within 30 minutes after mixing. Needle vibrator is employed in lieu of rodding and splicing of the concrete. The surface of pavement is compacted either by means of power driven finishing machine or by a vibrating hand screed. At small width or corner hand consolidation and finishing may be adopted. The concrete is further compacted by means of longitudinal float. Just before the concrete becomes hard, the surface is belted with two ply canvas belt. After belting pavement is given broom finish with fibre broom brush. Finally curing of newly laid pavement surface is done. For initial curing burlap or cotton mats saturated with water are placed in connect with the surface. In initial curing edges of the slab are banked with soil berm. The soil is thoroughly kept saturated with water for 14 days.

13.8 Use of Geo-Textile for Rural Roads

Geo-textiles can be highly beneficial in the construction and maintenance of rural roads. The use of geo-textiles in rural road construction can enhance durability, improve drainage, control erosion, and increase the load-bearing capacity of the road. Geo-textiles have the ability to separate, filter, reinforce, protect, or drain. They are typically made from polypropylene or polyester, and come in two basic forms woven and nonwoven. It is important to select the appropriate type and specification of geo-textile based on the specific requirements and conditions of the project.

Some of the common uses of geo-textiles in rural road applications are:

1. **Soil Stabilization**: Geo-textiles are used to reinforce and stabilize the soil subgrade beneath the road surface. They prevent the lateral movement and displacement of soil particles, increasing the overall strength and load-bearing capacity of the road.
2. **Separation:** Geo-textiles are used as a separation layer between different soil layers or materials. For example, they can be placed between the subgrade and the base course to prevent the mixing of fine-grained soil with coarser aggregate materials, ensuring proper drainage and maintaining the structural integrity of the road.

3. **Filtration and Drainage**: Geo-textiles act as filters to prevent the migration of fine particles from the subgrade into the base or drainage layers. They allow water to pass through while preventing the clogging of underlying layers, thereby improving drainage and reducing the potential for water-related damage, such as erosion and weakening of the road structure.
4. **Erosion Control**: In areas prone to erosion, geo-textiles can be used to stabilize slopes and embankments along the road. They help retain soil particles and prevent erosion by providing a barrier against the erosive forces of water and wind.
5. **Pavement Overlay**: Geo-textiles can also be used in pavement overlays to improve the performance and longevity of existing rural roads. Placing a geo-textile between the existing pavement and the new overlay helps to reduce reflective cracking, moisture intrusion, and deformation caused by differential settlement.
6. **Dust Control**: Geo-textiles can be utilized as a surface treatment on unpaved roads to reduce dust generation. They form a temporary or permanent barrier that stabilizes the surface and minimizes the generation of airborne dust particles, thereby improving air quality and visibility.

13.9 Erosion of Earth Roads

The erosion of earth roads is a natural process that occurs when the soil and other materials that make up the road surface are gradually worn away by various factors such as water, wind, and traffic. The severity of erosion on earth roads can vary depending on factors such as climate, terrain, soil type, and the level of road construction and maintenance. Addressing erosion requires a comprehensive approach that considers these factors and implements appropriate measures to minimize its impact. Here are some key points about the erosion of earth roads:

1. **Water Erosion**: Water is one of the primary causes of erosion on earth roads. When it rains or when there is poor drainage, the water can accumulate on the road surface and flow over it, leading to the removal of soil particles. This process is known as sheet erosion. Water can also create ruts and gullies by forming channels that transport soil away from the road.
2. **Wind Erosion**: In areas with dry and loose soil, wind erosion can be a significant factor. Strong winds can blow away the top layer of soil on earth roads, gradually wearing down the road surface over time. This is particularly common in arid or desert regions where there is limited vegetation to hold the soil in place.

3. **Traffic Erosion**: The movement of vehicles on earth roads contributes to their erosion. The weight and friction generated by vehicles can loosen and displace the soil, causing it to break apart. Heavy traffic can accelerate the erosion process, especially if the road is not properly designed or maintained to withstand the load.
4. **Lack of Vegetation**: Vegetation plays a crucial role in preventing erosion. The roots of plants bind the soil together, creating a stable surface that resists erosion. In areas where vegetation is sparse or absent, such as construction sites or newly cleared areas, earth roads are more susceptible to erosion.
5. **Maintenance and Mitigation**: Regular maintenance and appropriate mitigation measures can help minimize erosion on earth roads. This includes maintaining proper drainage systems to prevent water from accumulating on the road surface, implementing erosion control techniques like the use of gravel or stabilizing agents, and establishing vegetation cover along road embankments and slopes.

13.10 Road Cost Estimates

Road cost estimates are crucial in planning and implementing transportation infrastructure projects. These estimates encompass a wide range of factors and considerations to determine the financial resources required for road construction, maintenance, and improvement. They involve various elements such as land acquisition, engineering design, labor, materials, equipment, and associated permits. Additionally, costs are influenced by factors like the type and length of the road, terrain conditions, environmental impacts, and local regulations. Advanced techniques such as cost indexing, value engineering, and life-cycle analysis are employed to accurately project costs over the long term. Road cost estimates play a vital role in budgeting, resource allocation, and ensuring the efficient and sustainable development of road networks, thereby facilitating safe and reliable transportation for communities.

Some common approaches used in estimating road costs are:

1. **Historical Data**: One method is to gather data from past road construction projects of similar nature and scale. By analysing historical data, including project size, materials used, labour costs, equipment requirements, and other relevant factors, one can make estimates based on previous project costs.
2. **Cost per Unit Length**: Another method involves calculating the cost per unit length of road construction. This approach requires determining the unit cost for specific elements of road construction, such as excavation, sub-base preparation, pavement layers, drainage systems, signage, lighting, and landscaping. The unit costs are then multiplied by the length of the road to estimate the total cost.

3. **Cost Estimating Software**: Various software tools and applications are available to assist in road cost estimation. These tools often incorporate databases of material and labour costs, allowing one to input project-specific information and generate accurate estimates based on current market rates.
4. **Parametric Estimating**: Parametric estimating involves developing mathematical models or algorithms based on known relationships between project characteristics and costs. For example, the cost of a road project may be estimated based on factors such as traffic volume, road width, soil conditions, and terrain. These relationships are established using historical data or expert judgment.
5. **Comparative Estimating**: This method involves comparing the cost of a proposed road project to similar completed projects. By identifying key similarities and differences, adjustments can be made to account for variations in design, scope, location, and other factors.

13.11 Pot-Holes Repairs

Pot-hole repairs are an essential part of maintaining the infrastructure of roads. Pot holes, which are depressions or cavities in the surface of the road, can be caused by a variety of factors such as heavy traffic, harsh weather conditions, or poor construction materials. Repairing pot holes is crucial to ensure smooth and safe driving conditions for motorists.

Before starting the repairs, the area around the pot hole is prepared. This may involve removing loose debris, cleaning the surface, and ensuring proper drainage to prevent future damage. The damaged area is cut out the full depth of the metal in a rectangular form and shaped to create a clean, defined edge. This helps ensure proper adhesion of the repair material and increases the longevity of the fix. Depending on the size and severity of the pot hole, suitable material, such as asphalt or concrete may be used for filling. The filling is similar to the new road construction. The repair material is carefully applied, levelled, and compacted to create a smooth surface. Compaction is crucial to achieve a solid and durable repair. It involves using specialized equipment, such as rollers or compactors, to compact the repair material and remove any air pockets. Where the pot holes are too many, it is better to relay the whole surface.

13.12 Relaying of Wearing Coat

The relaying of wearing coats on roads is an essential process in maintaining and improving the quality and longevity of road surfaces. A wearing coat, commonly known as an asphalt overlay, refers to a layer of fresh asphalt that is applied over the existing road pavement. This technique is typically employed when the road surface has experienced significant wear and tear, resulting in cracks, potholes, or general deterioration.

The relaying process involves picking up the old metalled surface to a depth of 5–10 cm. It ensures evenness and proper bedding for new materials. This process of relaying is similar to the construction of new metalled roads.

Relaying wearing coats offers several benefits. Firstly, it enhances the structural integrity of the road by providing a new, durable surface that can withstand heavy traffic loads and adverse weather conditions. It also improves the ride quality for motorists, reducing the impact of bumps, cracks, and other surface irregularities. Additionally, a well-maintained road surface contributes to enhanced road safety, as it reduces the risk of accidents caused by uneven surfaces or potholes. Furthermore, relaying wearing coats extends the lifespan of roads by protecting the underlying pavement from further degradation. By sealing cracks and preventing water infiltration, the wearing coat helps to prevent further damage caused by freeze-thaw cycles and moisture-related issues.

13.13 Repair and Maintenance of Roads

Road repair and maintenance are crucial for ensuring safe and efficient transportation. Regular upkeep helps prolong the lifespan of roads, improves driving conditions, and reduces accidents. Some key aspects of road repair and maintenance are:

1. **Inspections and Assessments**: Regular inspections shall be conducted to identify road damage, potholes, cracks, and other issues. These assessments help determine the scope and priority of repairs.
2. **Routine Maintenance**: Routine maintenance activities include cleaning, debris removal, sweeping, and vegetation control along roadsides. Regular inspections of road signs, markings, and traffic signals are also essential.
3. **Pothole Patching**: Potholes are a common road issue caused by water seepage and freeze-thaw cycles. Patching involves filling potholes with suitable materials, such as asphalt or concrete, to restore a smooth and safe driving surface.
4. **Crack Sealing**: Cracks on the road surface allow water penetration, leading to further damage. Crack sealing involves filling cracks with specialized sealants to prevent water infiltration and subsequent deterioration.
5. **Resurfacing**: Over time, the top layer of roads can deteriorate due to traffic, weather conditions, and aging. Resurfacing involves applying a new layer of asphalt or concrete to restore the road's smoothness, skid resistance, and structural integrity.
6. **Rehabilitation and Reconstruction**: When roads experience extensive damage or reach the end of their lifespan, rehabilitation or reconstruction may be necessary. This involves more extensive repairs,

including base layer reconstruction, drainage improvements, and full-depth pavement replacement.

7. **Drainage Maintenance**: Proper drainage is essential to prevent water accumulation on roads. Regular maintenance of ditches, culverts, and storm water drains ensures efficient water flow and minimizes erosion and pavement damage.

Important Points to Remember

1. Farm roads are the arteries that connect remote villages to the broader transportation network in the country.
2. According to the construction and materials used, the roads may be grouped into the flexible road and the rigid road.
3. The flexible roads are made up of layers of various materials such as lime stone, gravel, *kanker*, red earth *etc.* while the rigid roads are made up of cement concrete.
4. Field roads are often unpaved or minimally paved.
5. To avoid slipping or skidding and overturning and to maintain equilibrium of the vehicle moving on a curve, the outside edge of the road is kept at a higher level to counter act the centrifugal force. This rise is known as super elevation.
6. Super elevation is the inclination given to the cross section of road on curves in order to reduce the centrifugal force on a running vehicle.
7. Camber is the rise of the centre of roads with reference to the edge.
8. If berms known as shoulders are provided, 2.4 m to 3.0 m (8 feet to 10 feet) wide roads are sufficient for running through fields.
9. According to the Indian Road Congress (IRC) guidelines, the recommended width for a single-lane rural road with light traffic is typically around 3.75 meters (12.3 feet). For two-lane rural roads, the recommended width is usually around 7.0 meters (23 feet) to accommodate two vehicles traveling in opposite directions.
10. Sub-soil forms the lowest layer and provides support for the entire road structure.
11. Roadway is the main driving surface where vehicles travel.
12. Shoulders are the areas on either side of the roadway, designed to provide additional space for emergency stopping or manoeuvring.
13. The road having its wearing surface consists of clean, crushed aggregates, mechanically interlocked by rolling and bound together with filler material and water laid on a prepared base course is called Water Bound Macadam (WBM) road. They are suited for heavy traffic of low speed vehicles. They afford better hold for wheels so that

slipping or skidding of vehicles is avoided. They are not slippery even when they are wet.

14. The camber given for gravel road is 1 in 28 to 1 in 32.
15. The cement concrete road is constructed either by alternate bay method by dividing the carriage-way into suitable strips or by the continuous slab method.
16. Water is one of the primary causes of erosion on earth roads. In areas with dry and loose soil, wind erosion can be a significant factor.

Exercises/Questions

1. Discuss different types of farm roads.
2. What are the requirements of a good road?
3. Describe the composition (different layers) of a road structure. Also draw the cross-section of a flexible pavement.
4. Define roadway, shoulders, sidewalks, median, and culvert.
5. Give the detailed specification of a road.
6. What is the difference between flexible pavement and rigid pavement?
7. Discuss the use of geo-textile for rural roads.
8. Discuss the steps for the construction of (a) Earthen road, (b) Water bound macadam road, (c) Gravel road, (d) Bitumen road, (e) Cement concrete road.
9. Write short notes on:
   (a) Drainage system for road
   (b) Erosion of earth roads
   (c) Pot-hole repairs
   (d) Relaying of wearing coat
   (e) Repair and maintenance of roads
   (f) Farm roads

Chapter 14

Aquaculture

14.1 Inland Fish Farming

The commercial rearing of the aquatic animals for the human consumption using the inland water resources either natural or man-made other than the sea is known as the inland fish farming. Inland fish farming is also known as the aquaculture. The aquatic animals include fishes, shrimp, prawn, crabs *etc.* The inland water resources are the rivers, streams, lakes, rice-fields, water reservoirs, and ponds. On the basis of salinity, aquaculture is divided into three categories *i.e.* freshwater (<0.5 ppt), brackish (0.5 to 30 ppt) water and marine water (>30ppt). The inland fish farming is for food, income generation and employment. The inland aquaculture in India contributes about 25 per cent of total fish production in world. The habitat alteration and fragmentation, altered flow, water and soil pollution, exotic species *etc.* are the major challenges faced during the inland fish farming. Aquaculture is also known as culture based fisheries. The stock material like post larvae, fry, fingerlings, *etc.* from the nurseries or hatcheries is reared in the water bodies.

Fish seeds are fertilized fish eggs. They are used to seed new ponds in fisheries. Four stages of fish seed production include (Figure 14.1):

1. Egg to spawn production for 3 days
2. Spawn to fry nursing for 15-20 days
3. Fry to fingerling rearing for 60-90 days
4. Fingerling to yearling rearing for 8-9 months

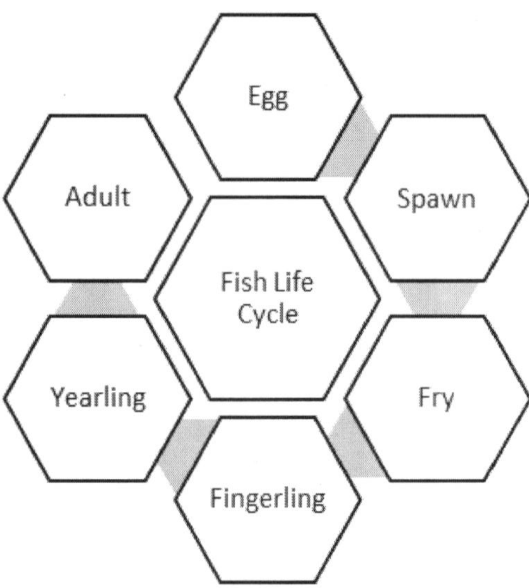

Figure 14.1. The Life Cycle of the Fish.

14.1.1 Consideration for Inland Fish Farming

1. **Water Supply**: Although, fish farming is non consumptive of water but the intensive fish farming requires supply of fresh water for the growth and well-being of the aquatic animals.
2. **Skill Development**: The aquatic animals are susceptive to environment changes, feed composition, water quality of water bodies *etc.* therefore requires continuous monitoring and supervision. The skilled personnel can take the corrective actions accordingly.
3. **Water Bodies**: The water bodies used for the aquaculture are generally common for peoples living in the rural areas. It can be used for washing of cloths, drainage of rainwater *etc.* that pollute the water, hence it is not suitable for intensive fish farming.
4. **Market and Infrastructure**: After harvesting of the fishes, it is necessary to maintain the cold chain to avoid spoilage. Dead fishes are kept in ice whereas live fishes are kept in water before their sale in market. Distant places from the market need infrastructure for cold chain along with the roads, trucks, cold store *etc.*
5. **Synergistic Interactions**: Incompatible fish species instable the food chain and that reduce the food resources and deteriorate the environment. It is also depending upon the degree of intensification of fish inside the pond. The extensive fish production is comparatively more sustainable than the intensive system.

14.2 Site Selection for Aquaculture

1. The site for the fish production should have year round uninterrupted supply of the quality water. Prefer water supply and drainage of pond by the gravity due to economics.
2. In the case of runoff operated ponds, pond should be in 1 hectare for the catchment area of 10-15 hectare.
3. Flood prone and poor rainfall areas, location with radio towers and high voltage electricity poles, industrial zones are not recommended for the construction of pond.
4. The culture and tradition of location need to be evaluated due to their involvement in the supply chain of the fish from pond to market/ consumer.
5. The location should have facility of electricity, road, transportation, labour and medical facilities.
6. The suitability of soil and water of the location need to be judged before construction of the pond.
7. The site should be protected by the rules and regulation for aquaculture and it should be free of legal issues.
8. The site should have accessibility to market for local, national and international business along with the supply for consumables required for the fish farming.
9. The bottom of the pond should be above the level of water table. The level of water table decides whether to dig the pond under the soil or construct pond over the soil.
10. The pond should be constructed in the open area for sufficient availability of sunlight. Sunlight keep a healthy natural ecosystem inside the pond by photosynthesis.

Major Breeds of the Indian Major Carp (IMC) Fishes

Sl.No.	*Category*	*Name of Breed*	*Biological Name*
1.	Freshwater	Catla	*Labeo catla*
2.		Rohu	*Labeo rohita*
3.		Mrigala	*Cirrhinus mrigala*

14.3 Fish Physiology and Micro-Climatic Considerations

Fish physiology include the respiration, digestion, reproduction, excretion, blood circulation and osmoregulation. The survival, reproduction and production of fish depends upon the micro climate of the pond. The deterioration in the micro climate also develops the diseases in the fishes. The major component of the pond system is water that affect the basic survival physiological functions of the fish. The pH, dissolved oxygen, hardness and temperature are comparatively important parameters need to be considered in water quality. The dissolved oxygen should be 4-6 mg/L, 7-8 pH, <5-10 mg/L CO_2, <0.05 ppm ammonia is preferred for intensive fish farming. The total soluble solids of the water should be less than 200 mg/L. Acidity in water affect the osmoregulation that imbalance the ion transport by the gills of fish. Whereas, the alkalinity supports the growth of aquatic animals and plants that support the fish farming by synergic interaction. The algae consume carbon dioxide and produce oxygen for sustainable ecosystem inside pond. In addition, alkalinity helps in maintaining the pH above 7 by buffering capacity, prevent leaching of heavy metals, facilitate utilization of the micro elements for disease treatment. The higher concentration of ammonia represents the pollution inside the pond water therefore 0.01 mg/L is highly desirable for the fish production. High level of carbon dioxide create hypoxia in fish due to low oxygen in the blood. The high temperature of the pond reduces the availability of oxygen for the fish and increase the toxicity of heavy metals. Low temperature of water restricts the growth of fish due to low feeding and minimum activity. The hard water is more beneficial for the fish farming that provide the calcium used for osmoregulation. The fish farming in hard water prevent diseases of pancreases, kidney, viral and bacterial infections.

14.4 Hatcheries

Hatcheries are the separated place where breeding of fish, fertilization, incubation and hatching of eggs and rearing of the initial stages of fishes take place under controlled condition. Hatcheries should have provision of the water quality management and the disease prevention. The fishes are kept inside the hatcheries until its fry life stage. The type of hatcheries depends upon the type of fishes and the type of water used for fishes.

14.5 Design of Fish Rearing Structures

Rectangular shape of ponds is preferred over the circular ponds. As the rectangular ponds prevent escape of fishes that help for easy harvesting of the fishes. The ratio of length to breadth should be 3:1. Although, the breadth of the pond should be between 30-50m. Fish rearing ponds are of three type *i.e.* Nursery pond, Rearing pond and Stocking pond. The nursery pond is constructed for seed rearing. The area of the nursery pond is 100-500 m^2 and depth of the pond is in between 1–1.5 m. The area of the rearing pond is 500-1000 m^2 and depth of the pond is in between 1.5–2 m. The rearing pond is used for fry and fingerlings for 2-3 months.

To construct a pond, remove the 20 cm of the top soil of the pond site. This fertile soil will be used to grow vegetation over the top side and outer side of the dyke. The pond floor should have slope of 2-4 per cent for drainage. Dig out the 1-1.5 m on the outlet side whereas depth on inlet side should be kept between 0.5 to 0.75 m. The soil should be used for the development of dyke. The digging should be done with gradual slope from inlet to outlet. The soil should be compacted after every 20 cm to prevent damage to dyke due to wave action of pond water. The compaction makes the dyke solid and water proof. The dyke should be elevated to 30 cm for small pond and 50 cm in big pond from the surface of pond water. After excavation of pond, use the fertile soil over the dyke to grow vegetation that helps to keep dyke intact for longer time (Figure 14.2).

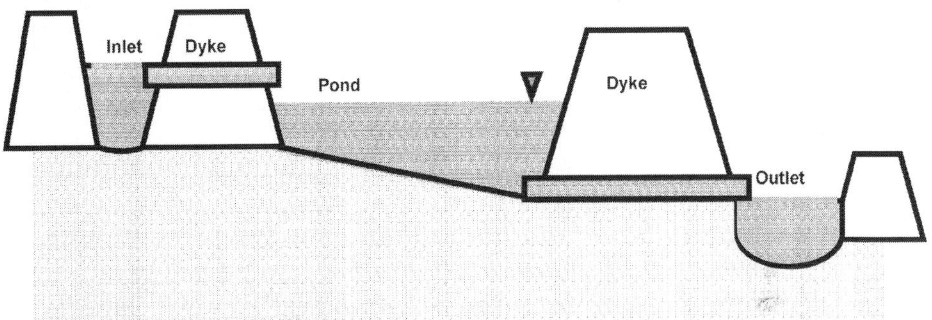

Figure 14.2. The Excavated Pond with Inlet and Outlet Structures.

Inlet and Outlet Structures

The inlet and the outlet system in the pond should be in opposite direction. The outlet system is generally made of plastic or metal pipe; the bamboo can be used as per the local availability. A ditch of 50 cm at the outside of the pond is required for the drainage of the pond water. A sluice made of brick or concrete can be used for outlet of pond water. Although, it is costly over the pipes (Figure 14.3). Wooden plank is used inside the sluice to control the flow. A monk is similar structure like sluice made of brick and concrete. It is also work as drainage system for the overflow of pond water. It is used inside the big pond (Figure 14.4).

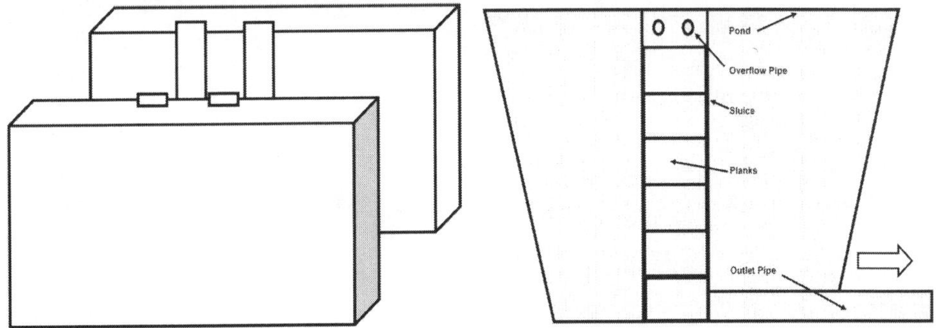

Figure 14.3. Sluice Used for Inlet and Outlet Structures in Pond.

Figure 14.4. Monk Used for the Outlet Structures in Big Ponds.

The bottom of inlet structure should be 10 cm above over the surface of pond water. It incorporates the oxygen to the pond water while filling of pond. The inlet may be made of metal or plastic pipes, bamboo and brick and concrete. An inlet sluice like outlet can be used as outlet structure. The screen should be provided in the inlet and outlet structures to avoid pollution of pond water, to keep small fishes inside the pond, to keep predators away from the pond *etc.* The inlet channel should be constructed 3-4 m away from the water sources.

14.6 Aeration

The fish absorb the dissolved oxygen of water by using gills and produce carbon dioxide. In the case of extensive fish farming, the oxygen is replenishing by the photosynthesis of aquatic plants in sustainable manner. Aeration improve the water quality, increase the availability of dissolved oxygen, regulate the temperature of water and reduce the growth of aquatic plants. Whereas in the case of intensive and semi-intensive system, the concentration of dissolved oxygen depleted with the passage of time. Therefore, external aeration system is required for the supply of oxygen inside the water for survival of fishes. Aeration depends upon the surface area, water turbulence and status of dissolved oxygen in pond. The aeration system should be installed in the middle of longest side of pond. The aerator should be installed 3-4 m away from the embankment of pond to avoid soil erosion. There are three type of aeration systems *i.e.* diffuser, surface and submersible used in fish farming.

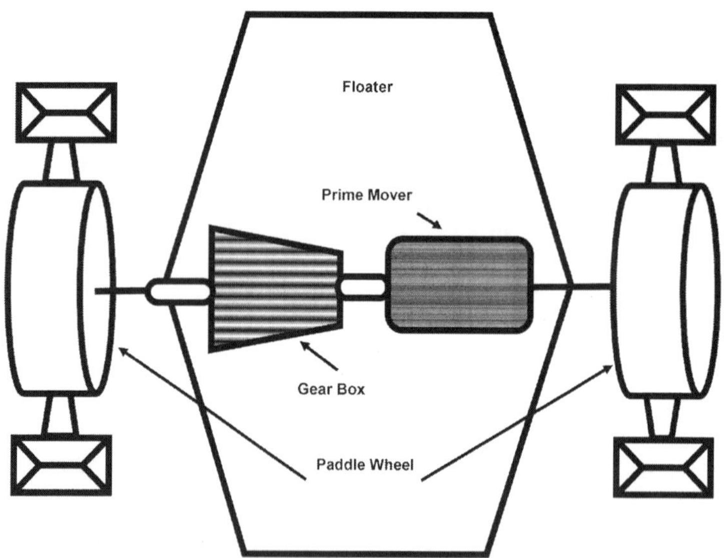

Figure 14.5. Paddle Wheel Aerator for the Fish Pond

The paddle wheel aeration system or side winders is a surface system which is powered by the tractor PTO, engine, electricity motor, solar power from outside or inside the pond. The float is equipped with engine/motor and paddle that move on water surface and incorporate the oxygen in the water. The tractor/engine/motor operated paddle wheel are operated from the bank of the pond on 20-30 rpm (Figure 14.5). Propeller-aspirator-pumps operates from the surface in which air is inserted in water by using a hollow pipe that have one end open in air and other inside the water. The paddle wheel has highest oxygen transfer rate and aeration efficiency followed by the propeller, vertical and diffuser pump.

Fountain aerator is a vertical submersible pump that throw the water above the water surface in a vertical direction. The creation of the fountain increases the surface area of water that absorb the oxygen of air. It is not effective if the depth of pond is above the 6 feet. It is also enhancing the cosmetics of the pond by its fountain type of vertical water flow (Figure 14.6).

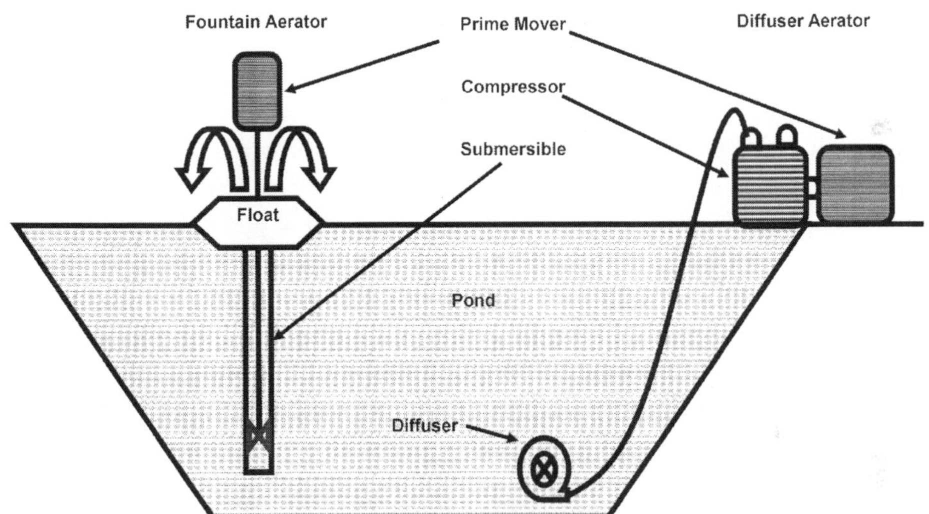

Figure 14.6. Fountain and Diffuser Aerator.

Diffuser type of aeration system is the most effective way of aeration in deep ponds than fountain and paddlewheel. The compressed air is fed in the water near to the bottom of the pond. The diffuser type of system releases low pressure and high volume air inside the water that incorporate oxygen through bubbles. The tiny bubbles are more desirable for more oxygen incorporation.

14.7 Recirculation of Water

Recirculation of the water is required to prevent the thermal and chemical stratification inside the pond. The recirculation device create turbulence that maintains the similar oxygenated condition throughout the pond. During the day time, the pond water on surface is supersaturated whereas the water in deep is oxygen deficit. Recirculation mix the surface and sub-surface water and help the surface and submersible aeration system to prevent oxygen super saturation. The recirculation of water also promotes growth of the aquatic plants that add oxygen in the water by the photosynthesis. An air lift pump sucks the subsurface water and release water over the surface of pond. This pump may be equipped with the diffuser to insert the bubble for aeration. A ballast in the bottom and float on the top of air lift pump is required to keep it stable inside the pond. The air lift pump hinders the fish harvesting operation, therefore it is useful in the small ponds. In the big ponds, the turbine impeller water recirculation system is used at bottom of pond on 60 rpm. It is equipped with submersible pump and a cart that helps its movement inside the pond to target the dead spot for recirculation.

Water Exchange

The water exchange dilutes the pond water that flushes the nutrients, prevent excessive algae production, remove the metabolic waste like ammonia, reduce the salinity during the dry season. It is also treated as substitute of aeration that maintain the oxygen concentration in pond water. Generally, the pond water is exchanged by the pumping the fresh water in the pond water. This is the less effective way of water exchange. Although, its effectiveness can be increased by the incorporation of fresh water on the surface and drain of pond water should be performed from the bottom of pond. The pond water which to be replaced should be drain out first before filling the fresh water. The rate of water exchange depends upon the quality of pond water. The pond water should be replaced with the good quality water. Water exchange pollute the water resources outside of the pond. The water of bore well is deficit of the oxygen therefore aeration of the ground water is mandatory before the water exchange. The average water exchange rate of 5-10 per cent per day is applied in the case of fish farming using the brackish water.

Water Reuse Systems and its Design

High density of fish in Intensive aquaculture pollute the pond water at higher rate by their excreta, unused feed, pond debris, chemical run off due to rainfall *etc.* This pollute water need to exchange or replace to ensure the healthy environment for the fish rearing. The replacement of pond water is an energy intensive and deplete the natural resources of water like ground water, river, pond *etc.* Therefore, the installation of water reuse system helps to clean and disinfect the pond water. It decreases the requirement of new water in the fish pond.

The water reuse system consists of water pump and water filtration system. The water filtration system has physical, chemical and biological ways to treat the polluted pond water. The water reuse system can work in batches or in a continuous manner (Figure 14.7). The major disadvantage of the water reduce system is the associated high installation cost of the water filtration system and the running cost of the filters used for the purification of the water. The sensors for the monitoring of the degree of pollution need to be installed in the pond water. The water is also inspected for degree of purification after the filtration before adding into the fish pond. The centrifugal type of pump is generally utilized for the movement of the pond water. The submersible pump can be utilised but it is costly and it produces the heat during operation inside the pond water that increase the temperature of pond water. Physical filtration system includes the use of centrifuge, screen, sedimentation tank *etc.* Biological filtration uses beneficial bacteria that consume the organic matter. In the case of chemical filtration system, ozone is added to improve the microbiological quality of the pond water.

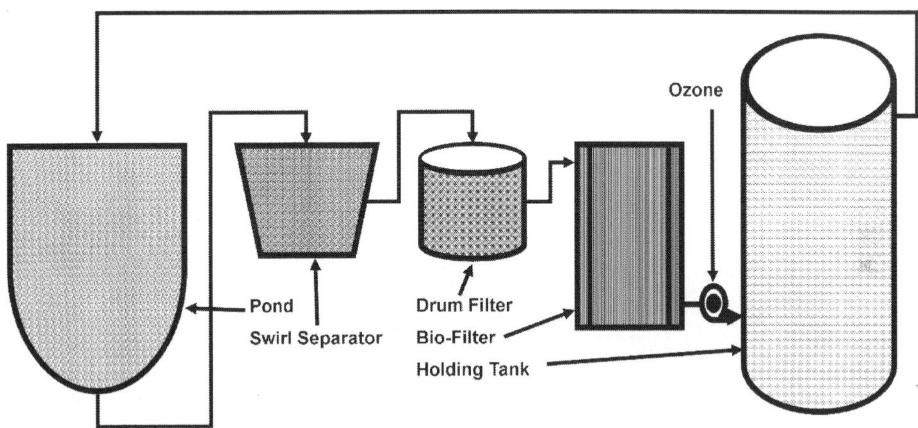

Figure 14.7. A Water Reuse System for the Fish Pond.

14.8 Water Treatment Plants

Research suggested that the 10 per cent of the pond water need to be replaced to achieve the sustainability in the fish production. This replaced pond water is polluted by the dissolved and particulate organic matter, nutrients like phosphorus and nitrogen *etc.* due to faeces of aquatic animals and the uneaten food. This waste water may deteriorate the water bodies if not disposed properly by scientific methods. On the other hand, this waste water can be reused/recycle by using the water treatment plants. The physical, chemical and biological methods are used to treat the waste water of fish pond. The waste water is treated in three stages *i.e.* Primary, Secondary and Tertiary treatment (Figure 14.8). Primary treatment is about the removal of solids by using the physical methods *i.e.* screen, sieve and settling tank. The primary stage removes 90 per cent of the suspended solids and reduce the 50 per cent of the BOD (biological oxygen demand) for water treatment. The waste water stays for hours in the settling tank and lighter impurities like grease form scum on the top whereas the dense material like sand form sludge in the bottom of tank.

The secondary stage uses the aerobic and anaerobic way to digest and speared the organic matter. The bacteria in the presence of oxygen break down the organic matter. The secondary stage removes 90-95 per cent of pollutants and reduce the BOD by 85-90 per cent. The activated sludge process, trickling filter, aerated lagoons and oxidation ponds *etc.* are the different biological methods used in the secondary stage of waste water treatment. The activated sludge is basically the culture of microbes that is added in the waste water for degradation of organic impurities. The aerator need to be installed in the settling tank to provide oxygen to microbes. The tricking filter sprinkle the waste water over the filter that have high surface area to volume ratio. This filter increases the availability of the oxygen to microbes. Aerated lagoon is generally a pond that have installed aeration system to incorporate the oxygen into the waste water to increase the activity of the microbes. Oxidation pond is the advanced version of the aerated lagoon that utilize the interaction of sunlight, microbes and algae for the treatment. The algae used the sunlight and in-organic compounds for its growth and produce oxygen. This oxygen is consumed by the microbes present in the waste water for its treatment. The anaerobic treatment used the methanogenic bacteria to produce the methane gas by breakdown of the organic compounds.

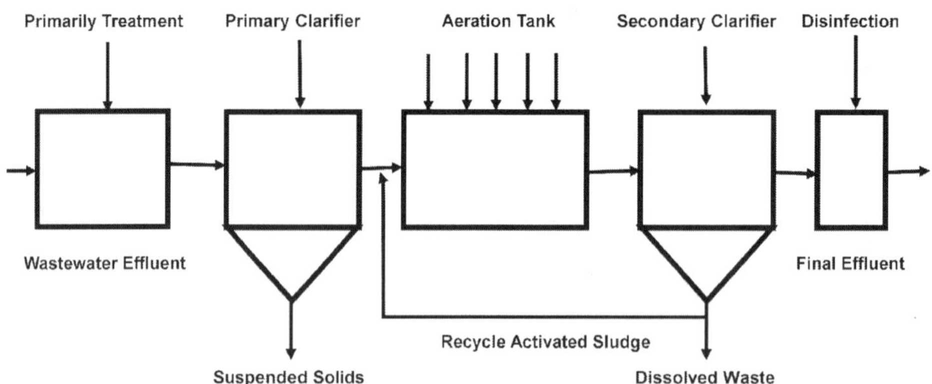

Figure 14.8. Water Treatment System for the Fish Pond.

Tertiary treatment stage is about improvement of the quality of the treated water by removing the colour, nutrients, metals and microbes from the waste water. Alum flocculate and coagulate the impurities that is easily removed by the filtration system. Chlorine is used for the disinfection of the pond water. Sodium bi-sulphite need to be added in chlorinated water as the chlorinated water is harmful for the aquatic life.

14.9 Feeding Systems

Fish is reared in three type of pond system *i.e.* extensive, semi-intensive and intensive. The extensive system is self-sustainable as the fish feed on the natural food available in the pond made by photosynthesis. Whereas in the case of semi-intensive system, supplementary food need to provide for growth and health of fish. The density of the fish in the intensive system is highest therefore complete food has to be provided to the fish. Therefore, the fish food categorised under three categories *i.e.* natural, supplementary and complete food. The natural food includes insect, worm, aquatic plants *etc.* The agricultural by-products (broken, bran, oil cakes), slaughterhouse waste, kitchen waste, left-over food *etc.* are the supplementary food. The complete food is designed as per the nutrition, health, feeding habit, age, type of fish that helps in growth and easy to digest. The complete food is costly than the supplementary food. The quantity of the feed depends upon the Feed Conversion Ratio (FCR) that means the weight of feed provided to fish per kg of produced fish. The number of meals in a day depends upon the size of fish. Smaller the fish size requires frequent feed up to 8 times in a day. Whereas, the mature fish(40g) need minimum of 2 feed in a day. The feeding of the fish need to be monitored for exact determination of feed quantity otherwise higher quantity of feed go waste that will also pollute the pond water. The morning time is the best time to feed the fish. Moreover, in the afternoon before the sunset, fish may be fed two times in a day. The feed can be placed on the raised pond bed, floating frames, submerged feed tray, demand feeder, IOT based automatic feeder as per the available investment or size of pond. The feeder can be classified as manual, automatic and demand feeder (Figure 14.9). Manual feed is costly, time and labour consuming. Automatic feeder equipped with timer whereas demand feeder has trigger that release the fish food in the pond (Figure 14.10).

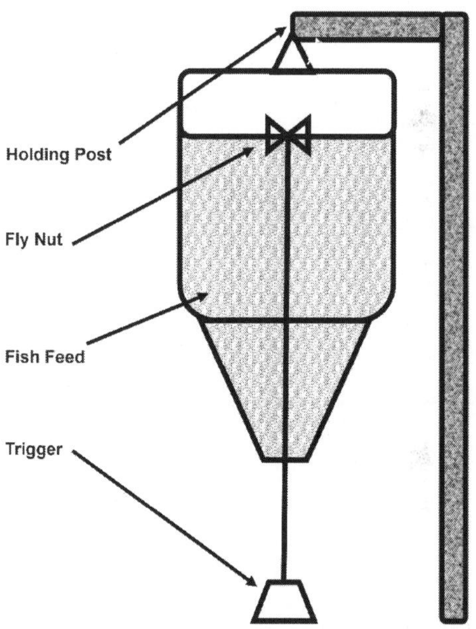

Figure 14.9. Demand Feeder for the Fish Pond.

The commercial and complete fish food is either extruded or pelleted. The extruded food float over the water whereas the pellet food sink in the pond water. Extruded food are comparatively costly than the pellet food. The selection of the pellet or extruded food depends upon the preference of fish. Extruded food is more prefer to use due to its floating nature that helps in fish monitoring. Feed acceptability, palatability, and digestibility are the parameters that decide the feed efficiency for fishes. Generally, fish need feed 1-5 per cent of their body weight in a day. Feed should be not kept longer than the 90-100 days. Feed stock should have kept away from environment of high temperature and high humidity. Fish should be fed with the 80 per cent of the required feed and it should be finished within 5-10 min.

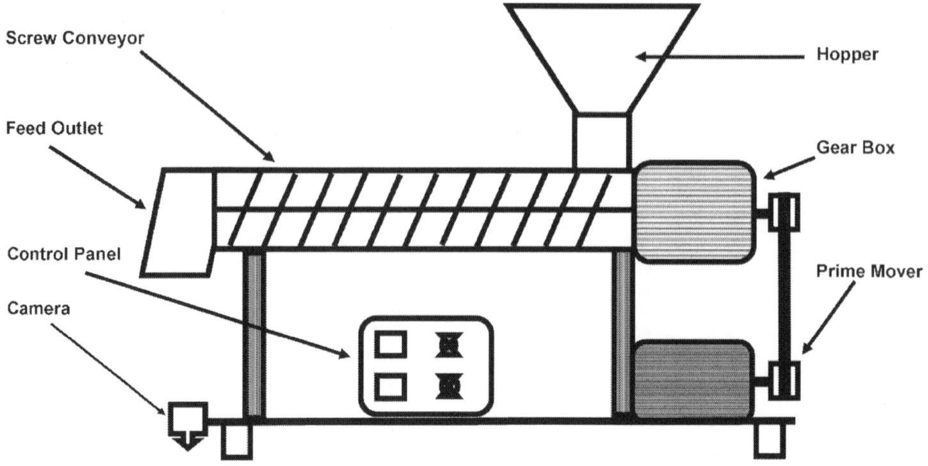

Figure 14.10. Image Based Automatic Demand Feeder for the Fish Pond.

14.10 Sterilization and Disinfection of Ponds

The disinfection is the elimination of the harmful microbes whereas the sterilization is about killing of all the microbes inside the pond and other equipment. Disinfection and sterilization of pond ensure the healthy and disease free ambience for the aquatic animals. Disinfection control the amount of harmful microbes and prevent the infectious diseases among fish. There is four ways of disinfecting the ponds that are physical, chemical, mechanical and radiation. Ozone, hydrogen peroxide, alcohol, lime, sodium hypochlorite *etc.* are used under the chemical disinfection. Hydrated lime increase the pH of water to 11 for one hour that is enough to disinfect the pond water. 1 per cent NaOH, 200-250 ppm iodine, 50 ppm formaldehyde used to disinfect the fish net, containers, boots *etc.* Physical methods include sunlight and UV light to control microbes (Figure 14.11). The removal of the particles with adhere microbes from the pond water using the mechanical sieves. The electromagnetic rays *i.e.* Gamma rays are treat the water and kill microbes. In addition, the quarantine or removal of the diseased/infected fish is also important to keep the load of the harmful microbes under control. The water reuse or recycle system should be equipped with suitable sterilizing or disinfection modules to reduce the microbes. The use of river water should be avoided to direct refilling of ponds. It may deliver new parasites that cause diseases in fish. The birds also transmit diseases by their droppings in the pond water. The fish or other living organism of other pond also transported by the birds that may carry the pathogens. The equipment like fish net need to be dried under sun or disinfected before use in the pond water. Moreover, the effectiveness of disinfectant depends upon the type of disinfectant and exposure time to the disinfectant.

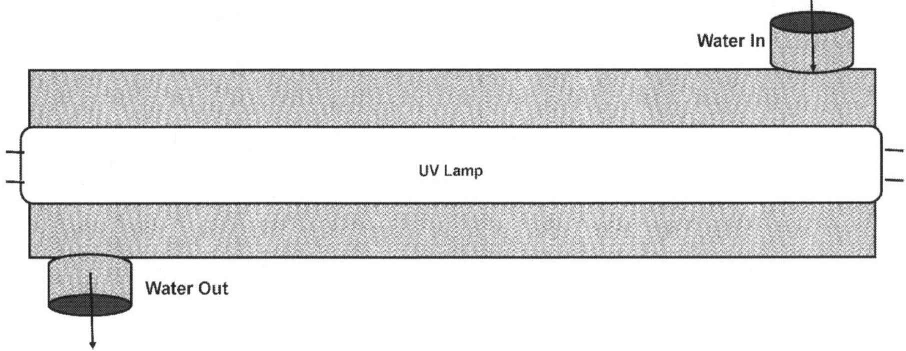

Figure 14.11. UV Lamp for the Sterilization of the Pond Water.

Important Points to Remember

1. The fry took 30-60 days to change into the fingerlings.
2. Mahua oil cake can kill fishes in the pond within hour due to 4-6 per cent saponia.
3. The area of rearing pond should be 600-100 m^2.
4. Management and rearing of fishes only is known as pisciculture.
5. Management and rearing of aquatic animals is known as fishery.
6. Salting, canning, chilling *etc.* can be used for preservation of fishes.
7. Bio filtration in recirculation system decrease the toxicity of ammonia.
8. ICAR-Central Marine Fisheries Research Institute is located at Kochi, Kerala, India.
9. Blue revolution is about Integrated Development and Management of Fisheries.
10. National Fisheries Development Board (NFDB) is located at Hyderabad
11. Pradhan Mantri Matsya Sampada Yojana (PMMSY) was started in 2020.
12. India stands second in culture fisheries production after China.
13. Catla, Rohu and Mrigal are the three Indian Major Carps.
14. CCRA is the abbreviated form of Code of Conduct for Responsible Aquaculture.
15. Andhra Pradesh is the top producer of the fish in India.

Exercises/Questions

1. List various types of aeration systems used in fish pond. Explain in detail the construction and working of paddle wheel aerator?
2. Discuss various types of inlet and outlet structures used for fish pond.
3. Write about the disinfection and sterilization of pond water?
4. Discuss about the difference between the water reuse and water exchange system?
5. What should be the consideration for the fish pond? Discuss about its location and size. Give the construction details of the fish pond.
6. Write short notes on:
   (a) Indian major carps
   (b) Demand feeder
   (c) FCR
   (d) Hatchery
   (e) Life stages of fish

Chapter 15

Cost and Material Estimation

Estimation is a computation of the quantities of various materials required, labour involved, and expenditure likely to be incurred in the construction of a work. It also prevents misuse (or pilferage) of building materials at site thus results in overall economy. In case the funds available are less than the estimated cost, the work is done in part or by reducing it or specifications are altered. It assists in maintaining the standard of construction by careful supervision and selecting proper specifications. Thus, the estimate is the probable cost of work and is determined theoretically by mathematical calculations based on plans and drawings and current rates. Accurate cost estimation is vital to ensure that the project is financially viable and within budget. It involves estimating the expenses associated with labour, materials, equipment, and other project-related activities. Estimating the cost and required materials involves a systematic approach that takes into account various factors such as project scope, design, specifications, and market conditions.

The following requirement are necessary for preparing an estimate:

1. Drawings like plan, elevation and sections of the proposed structure
2. Detailed specifications about workmanship and properties of materials *etc.*
3. Standard Schedule of Rates (SSR) of the current year

Why Estimation and Costing?

An estimate provides a cost estimate that determine feasibility of the project to be completed within available funds. It also indicates the time required for completion of project. It helps to invite tenders and quotations, arranging contracts, controlling expenditure during work execution. It decides whether the proposed plan matches available funds.

This chapter will delve into the essential aspects of cost and material estimation, providing an overview of the methodologies and considerations involved.

15.1 Types of Estimates

There are different methods of cost estimation. The estimates may be preliminary or detailed. The preliminary or rough cost or approximate estimate is the method to find an approximate cost in a short time, while the detailed estimate is the most accurate estimate.

15.1.1 Preliminary or Rough Cost or Approximate Estimate

1. **Cost of Comparable Structure:** It involves comparison of the proposed structure on a total cost basis with recently completed nearby construction of similar character and size for which actual costs are known.
2. **Plinth Area Estimate:** This method is commonly used for approximate estimating the cost of a structure. If the cost per unit area for a particular type of structure is known, then the total cost can be estimated by multiplying the cost per unit area with the total proposed area of construction. Plinth area is calculated by measuring the length and width of the building at the ground level. This is the area within the external walls of the building at the ground floor level, including balconies and projections. Exclude open areas like courtyards and voids. It's important to note that the plinth area method provides an approximate cost estimation and may not account for specific project requirements or variations in material prices.
3. **Cubic Rate Estimate for Building:** This is another method of approximate estimate in which the estimate is prepared on the basis of the overall volume or cubic content of the building. This method is generally used when height of the structure is more than the conventional height. The cube rate being deduced from the cost of the similar building having similar specifications and construction in the neighbourhood. If the cost per cubic for a particular type of structure is known, then the total cost can be estimated by multiplying the cost per cubic with the total volume of construction (length × breadth × height). If the building has multiple floors, calculate the volume for each floor separately and then sum them up. Estimate using the cube rate is accurate as compared to plinth area estimate. It's important to note that the cubical content method provides a rough estimate and is not as accurate as more detailed cost estimation methods.

4. **Cost per Unit:** For many structures, such as livestock, school, hospitals, roads *etc.*, the cost per unit is considered for rough cost estimate. For a livestock structure, the actual construction cost may be expressed as cost per animal housed. If the cost of structure for housing of one animal is known, then the cost of structure for housing any number of animals can be calculated by multiplying the number of animals with the cost per animal unit. Similarly, for a school, hospital and road, cost per class room, cost per bed and cost per km respectively is considered. The cost worked out by this method is approximate and depends upon shape, size, kind of construction and locality.

15.1.2 Detailed Estimates or Item Rate Estimate

This is the most accurate estimate and in this case quantities of each item of work are prepared and the cost is worked out accordingly. The dimensions of length, breadth, and height of each item such as earth work, brick work, R.C.C., plastering, *etc.* are taken out correctly from the drawings. Quantities of each item are then calculated and estimate is done. An amount equivalent to 3 to 5 per cent of the estimated cost is added as contingencies for unforeseen items.

Various Forms

Form I: Form for Details of Measurements and Calculation of Quantities

A commonly and widely used form for entering all the measurements of a building, sub-head wise to calculate the quantities is given below. The quantities are worked out on the basis of drawings and plan of the building.

Sl. No.	*Description of Work*	*No.*	*Measurement*			*Quantity*	*Remarks*
			Length (L)	*Breadth (B)*	*Height/ Depth (H/D)*		

Form II: Form for Abstract of Cost

The total estimate of the work is calculated after calculation of quantity of each sub-head of work, and following form is used for this purpose:

Sl.No.	*Sub-Head of Work*	*Quantity*	*Unit*	*Rate*	*Amount*
I	Earth work				
II	Concrete work				
III	Brick work				
IV	Roofing				
V	Flooring				

Sl.No.	Sub-Head of Work	Quantity	Unit	Rate	Amount
VI	Wood work				
VII	Steel work				
VIII	Finishing				
IX	White washing				
X	Miscellaneous				
	TOTAL AMOUNT				

Form III: Material Statement Form

The estimate is also accompanied with a statement of materials required for the completion of the particular work. A statement w.r.t. this is prepared in advance for arranging materials so that no difficulty should arise later on during the execution of work. The form used for such a statement is called Material Statement Form.

Sl. No.	Parti-cular of Work	Materials									
		Cem-ent	Coarse Sand	Fine Sand	Bajri	Bricks	Lime	Surkhi	Brick Ballast	Steel	and so on

Example 15.1

Estimate a 500 m^2 residential building's floor area if a similar building in the same location costs Rs. 1,60,00,000 for 800 m^2 construction.

Solution

$$\text{Plinth area rate} = \frac{\text{Cost of building}}{\text{Plinth area}} = \frac{1{,}60{,}00{,}000}{800} = \text{Rs. } 20000/\text{- per } m^2$$

Approximate cost of new residential building = Plinth area × rate

$$= 500 \times 20000 = \text{Rs. } 1{,}00{,}00{,}000/\text{-}$$

Example 15.2

A 200 cows dairy shed is constructed in Hisar at the cost of construction of Rs 50,00,000/-. Find the approximate estimate for a dairy shed of 60 cows in the similar locality using unit method.

Solution

$$\text{Rate per animal} = \frac{\text{Cost of building}}{\text{Number of animals}} = \frac{50{,}00{,}000}{200} = \text{Rs. } 25{,}000 \text{ per animal}$$

Approximate cost of 60 cows' dairy shed = Number of animals × rate per animal

$$= 60 \times 25{,}000 = \text{Rs. } 15{,}00{,}000/-$$

Example 15.3

A warehouse of 250 m^2 is constructed at the cost of construction of Rs 40,00,000/-. Height of the warehouse is 5 meter above ground. Find the approximate estimate for a warehouse of 4-meter height and 200 m^2 plinth area in the similar locality using cube rate estimate.

Solution

$$\text{Rate of construction on cube rate basis} = \frac{\text{Cost of construction}}{\text{Volume of building}} = \frac{40{,}00{,}000}{250 \times 5}$$

$$= \text{Rs. } 3200 \text{ per } m^3$$

Approximate estimate of new building = Volume of new building × rate/m^3

$$= (200 \times 4) \times 3200 = \text{Rs. } 25{,}60{,}000/-$$

15.2 Items of Work

The estimator must be well known to all items of works and market prevailing rates of each item. For detailed cost estimation, the whole construction work is distributed in different items of work for ease in calculation. The main items of work are as follows:

1. **Earth Work:** Earth work is required to be done in excavation as well as in filling, and should be taken out separately under different types. It is estimated in cubic meters (length × breadth × depth or height).
2. **Concrete in Foundation:** It is measured in cubic meter (length × breadth × thickness). The length and width are easily known from excavation data. The thickness usually varies from 20 cm to 45 cm. The concrete in foundation may be 1:4:8 or 1:5:10.
3. **Masonry or Brick Work:** Masonry is computed in cubic meter (length × breadth × height). Foundation and plinth masonry is taken as one item and masonry for super structure is taken as separate item. In buildings having more than one floor, the masonry for superstructure for each floor is computed separately. The partitioned wall and honeycomb brick work are taken as separate item, and measured in square meters. Different types of masonry are taken under separate items.

4. **Damp Proof Course (DPC):** It is provided at plinth level and shall extend to the full width of superstructure wall except in the case of outer walls where it shall be continued up to outside face of plinth. The DPC shall not be carried across doorways and verandah openings. It is measured in square meter (length × breadth) with specified thickness. In general, 2.5 cm thick cement concrete (1:1.5:3) with water proofing material is used as DPC.
5. **Reinforced Cement Concrete (RCC) or Reinforced Brick (RB) Work:** RCC and RB work may be in foundations, columns, lintels, beams, floor slabs or roof. The estimate is calculated in cubic meters, considering length, breadth, and thickness without steel reinforcements. Steel quantity is measured in kg, quintals, or metric ton as separate item. If detailed drawings are unavailable, 0.8 to 3 per cent of concrete can be used as steel quantity, multiplied by density.
6. **Lintel:** Lintels are made in RCC, and quantity is calculated in cubic meter. Length of lintel is taken equal to the size of opening plus 150 mm each on both sides.
7. **Flooring:** The baseline concrete and floor finishing of cement concrete or stone or marble or mosaic *etc.* are usually taken as one job and quantity is calculated in square meter by multiplying the length and breadth. For upper floors, bed of RCC is taken in cubic meter, and floor finishing is taken separately in square meter.
8. **Roof:** Flat roofs are calculated in cubic meter. In case of tiled, galvanized iron sheet or asbestos cement sheet roofing, the roof coverings are measured in square meter. Tiles on hip and valley are measured in running meter.
9. **Plastering and Pointing:** Plastering is measured in square meters with a specified thickness. The entire face of the wall on both sides shall be taken into account for the calculation of plastering. Deductions are made for openings, and external and internal plastering are calculated separately under different items. Wall pointing is also calculated in square meters with a similar deduction as plastering.
10. **Pillars:** Pillars are taken separately in cubic meter for their net volume (length × breadth × height). It is of different shapes. Quantities are calculated by correct geometrical measurements by simple mensuration methods. Plastering works in pillars is calculated in square meter by multiplying the perimeter by the height.
11. **Doors and Windows:** Door and window frame work is computed in cubic meter. Shutters for doors or windows are computed in square meter by multiplying the breadth by the height of the shutters. Wooden beams, posts, roof truss *etc.* are computed in cubic meters by taking into account their finished sizes.

12. **Iron Work:** This is computed by weight (kg or quintal).
13. **White Washing, Colour Washing and Distempering:** Similar to plastering, the quantities are also measured in square meters. The number of coats is considered a separate item. The rates for the number of coats are also specified.
14. **Painting:** The cost of painting or varnishing doors and windows is calculated in square meters.

15.3 Units of Measurements and Units of Payments

The units of different works depend on their nature, size and shape. In general, the units of items of work are based on the following principle:

1. Mass, voluminous and thick works shall be taken as cubic unit or volume.
2. Shallow, thin and surface works shall be taken as square unit or area.
3. Long and thin work shall be taken in linear or running unit.
4. Piece work, job work, *etc.* shall be taken in number.

The units of payments and measurements of various items of work in metric system. Whereas, earthwork is measured in cubic meter but payment is made per 100 cubic meter (per per cent cu m).

The measurement of work for different items is done in the units as decided by Bureau of Indian Standards (BIS). A booklet containing rates of various engineering items for preparation of detailed estimates is known as Standard Schedule of Rates (SSR). Rates approved for payment (rates of materials, daily wages of labour, carriage expenditure and tables for quantities of various materials required in construction) is mentioned in SSR and a premium is given above these rates. The premium rates are updated periodically. Each state has its own SSR. The unit of measurement and unit of payment of some of the important items of works are presented in Table 15.1.

Table 15.1. Units of Measurements and Payments for Various Items of Works and Materials

Sl. No.	*Particulars of Items*	*Units of Measurement*	*Units of Payment*
	Sub Head I: Earth work		
1	Earth work in excavation in foundation	cum	per % cum
2	Earth filling in plinth	cum	per % cum
3	Cutting of trees (girth specified)	No.	per No.
	Note: *For earthwork, normal lead is 30 m, and normal lift is 1.5 m.*		

Sl. No.	Particulars of Items	Units of Measurement	Units of Payment
Sub Head II: Concrete work			
1	Lime concrete in foundation	cum	per cum
2	Cement concrete in foundation	cum	per cum
3	Plain cement concrete	cum	per cum
4	Reinforced cement concrete (RCC)	cum	per cum
5	Damp proof course (DPC) - Cement concrete, rich cement mortar, asphalt *etc.* (thickness specified)	sqm	per sqm
6	RCC in shelves or jali work *etc.* (thickness specified)	sqm	per sqm
7	Reinforced brick work	cum	per cum
Sub Head III: Brick work			
1	Ist class burnt brick in cement mortar (1:6) in foundations and plinth	cum	per cum
2	Ist class burnt brick in cement mortar (1:6) in suoerstructure	cum	per cum
3	Ist class burnt brick in cement mortar (1:7) in arches	cum	per cum
4	Sun-dried bricks in mud mortar in superstructure	cum	per cum
5	Half brick work with or without reinforcement	sqm	per sqm
6	Random rubble masonry in mud, mortar, lime mortar or cement mortar (ratio specified)	cum	per cum
7	Stone slabs in roof, shelves, stone chujjas, sun shade *etc.* (thickness specified)	sqm	per sqm
Sub Head IV: Roofing			
1	Reinforced cement concrete slab	cum	per cum
2	Reinforced brick slab roofing	cum	per cum
3	Corrugated galvanized iron roofing, Asbestos cement sheet roof	sqm	per sqm
4	Tiled roof -Allahabad tile, Mangalore tile *etc.* including battens	sqm	per sqm
5	Thatch roofing including bamboo frame (thickness specified)	sqm	per sqm
Sub Head V: Flooring			
1	Mud flooring finished	sqm	per sqm
2	Brick on edge floor over lime concrete	sqm	per sqm
3	Flat brick flooring laid in cement mortar over lime concrete and thick sand and cement pointed	sqm	per sqm

Sl. No.	Particulars of Items	Units of Measurement	Units of Payment
4	Conglomerate floor of cement concrete laid over cement concrete	sqm	per sqm
5	Mosaic or terrazzo floor over cement concrete and sand	sqm	per sqm
6	Providing and fixing glass or aluminium strip in floors	m	per m
	Sub Head VI: Wood work		
1	Wood work (name of wood specified) in door and window frame or chowkhat, rafters, beams, roof trusses, *etc.*	cum	per cum
2	Wooden doors and windows shutters, panelled or battened, or panelled and glazed or glazed or wire gauged *etc.* (thickness specified), complete in all respects and fixed in position	sqm	per sqm
3	Wire gauged fixed on chowkhats	sqm	per sqm
	Sub Head VII: Steel work		
1	Rolled steel beams including hoisting, fixing in position and painting 3 coats.	quintal	per quintal
2	Mild steel bars, flats and angles fixed in position	quintal	per quintal
3	Expanded metal (size specified) or iron grill work	sqm	per sqm
4	Steel doors and windows	sqm	per sqm
5	Barbed wire fencing including fixing to post after stretching	m	per m
6	Collapsible steel shutter complete with fittings	sqm	per sqm
7	GI sheets	sqm	per sqm
8	Mild steel bars reinforcement in RCC or RB work	quintal	per quintal
	Sub Head VIII: Finishing (Plastering, Pointing *etc.*)		
1	Cement plaster	sqm	per sqm
2	Lime plaster with stone lime and surkhi mortar	sqm	per sqm
3	Mud plaster with g*obri* lapping (cow dung wash)	sqm	per sqm
4	White washing or colour washing (no. of coats specified)	sqm	per sqm
5	Lime or cement pointing deep	sqm	per sqm
6	Cement pointing flush	sqm	per sqm
	Sub Head IX: Painting		
1	Painting, varnishing doors and windows *etc.* (no. of coats specified)	sqm	per sqm

Sl. No.	*Particulars of Items*	*Units of Measurement*	*Units of Payment*
2	Polishing wood work (no. of coats specified)	sqm	per sqm
3	Distempering (no. of coats specified)	sqm	per sqm
	MISCELLANEOUS ITEMS		
1	Pipe-rainwater, sanitary, water pipe *etc.* (diameter specified)	m	per m
2	Glass panes supply (thickness specified)	sq m	per sq m
3	Wiring in the building for all types of points *i.e.*, for electric bulbs, plug, fan point, power plugs, *etc.*	point	per point
4	Water closet (WC), wash hand basin, man hole *etc.* (size specified)	No.	per no.
	SUPPLY OF MATERIALS		
1	Ordinary cement	bag of 50 kg	per bag
2	Coloured cement	bag or kg	per bag or per kg
3	Bricks	1000 nos.	per 1000 nos.
4	Sand, surkhi, bajri, brick ballast, stone ballast *etc.*	cu m	per cu m
5	Lime	quintal	per quintal
6	Timber	cum	per cum
7	Commercial plywood, glass panes	sqm	per sqm
8	Bitumen	tonne	per tonne
9	G.I. (galvanized iron) sheet	tonne	per tonne
10	A.C. sheets (measured flat)	sqm	per sqm
11	Mild steel	quintal	per quintal
12	Marble chips of any colour	quintal	per quintal
13	Varnish, paint, oil *etc.*	liter	per liter
14	Distemper	quintal (kg)	per quintal (kg)
15	Switches, plugs, electric bulbs, call bell, *etc.*	no.	per no.
16	Water proofing compound such as snowcem *etc.*	kg	per kg
17	Wire gauze	sqm	per sqm

Rules for Measurement

In general, dimensions should be measured to the nearest 1 cm, areas should be worked out to the nearest 0.01 square meter and cubic contents should be worked out to the nearest to 0.01 cubic meter. Thickness of slabs, partitions *etc.* and sectional dimensions of columns, pillars, beams, *etc.* should be taken to the nearest half centimetre (5 mm).

15.4 Approximate Sizes and Dimensions of Various Works

The approximate sizes and dimensions of various works in MKS and FPS units are as follows:

Sl.No.	Work	Size in MKS Units	Size in FPS Units
1.	Size of doors	1.20 m × 2.10 m	4'0"×6'6.75"
		1.10 m × 2.00 m	3'6"×6'6"
		1.00 m × 1.90 m	3'3"×6'3"
		0.90 m × 1.80 m	3'0"×6'0"
		0.75 m × 1.80 m	2'6"×6'0"
2.	Size of windows	1.00 m × 1.50 m	3'3"×5'0"
		0.90 m × 1.20 m	3'0"× 4'0"
3.	Size of *chowkhats* section for doors and windows	12 cm × 7.5 cm	5"×3"
		10 cm × 10 cm	4"×4"
		10 cm × 7.5 cm	4"×3"
		7.5 cm × 7.5 cm	3"×3"
4.	Thickness of doors and windows, shutter or leaves	2.5 cm/3.0 cm/4.0 cm/4.5 cm	1"/1.25"/1.5"/1.75"
5.	Thickness of lime concrete in foundation	15 cm/20 cm/25 cm/30 cm/40 cm/45 cm	6"/8"/10"/12"/16"/18"
6.	Thickness of RCC slab	7.5 cm/10 cm/12.5 cm/15 cm	3"/4"/5"/6"
7.	Thickness of plastering	10 mm/12 mm/15 mm/20 mm	3/8"/1/2"/5/8"/3/4"
8.	Thickness of DPC	2 cm/2.5 cm/4 cm	3/4"/1"/1.5"
9.	Thickness of C.C. floor	2 cm/2.5 cm/4 cm	3/4"/1"/1.5"
10.	Height of plinth	45 cm/60 cm/75 cm/90 cm	1'6"/2'0"/2'6"/3'0"
11.	Height of building	2.80 m/3.00 m/3.30 m/3.60 m/3.90 m/4.20 m	9'3"/10'0"/11'0"/12'0"/13'0"/14'0"
12.	Size of rooms	3.00 m × 3.00 m	10'×10'
		3.00 m × 3.60 m	10'×12'
		3.60 m × 3.60 m	12'×12'
		3.60 m × 4.20 m	12'×14'
		3.60 m × 4.80 m	12'×16'

15.5 Estimation of Quantity of Construction Materials

Some important data pertaining to building construction is as follows:

1.	Metric size of brick	19×9×9 cm
2.	Nominal size of brick with mortar	20×10×10 cm
3.	Number of bricks required for laying of 1 m^3 brick masonry	500
4.	Quantity of dry mortar required for 1 m^3 brick masonry	0.30 cum
5.	Quantity of wet mortar required for 1 m^3 brick masonry	0.25 cum
6.	One bag of cement	50 kg = 0.03472 cum
7.	Quantity of dry ingredients for 1 m^3 of finished wet concrete	1.55 cum
8.	One cum of unslaked lime	580 to 600 kg
9.	Quantity of undressed stones required for laying 1 cum of stone masonry (fine ashler)	1.33 cum
10.	Quantity of undressed stones required for laying 1 cum of rubble masonry	1.10 cum+20 per cent through stones
11.	Quantity of dry mortar required for 1 m^3 fine ashlar	0.12 to 0.20 cum
12.	Quantity of dry mortar required for 1 m^3 rubble masonry	0.35 to 0.40 cum
13.	Quantity of unslaked lime for whitewashing 100 sqm surface (one coat)	15 kg
14.	Quantity of unslaked lime for whitewashing 100 sqm surface (2 coats)	25 kg
15.	Quantity of unslaked lime for whitewashing 100 sqm surface (3 coats)	35 kg
16.	Quantity of unslaked lime for colour washing 100 sqm surface (one coat)	10 kg
17.	Quantity of unslaked lime for colour washing 100 sqm surface (2 coats)	20 kg
18.	Weight of 1 cum of mild steel	7850 kg
19.	Number of bricks required for the preparation of 2.83 cum brick ballast	1000 numbers
20.	Quantity of distemper for 100 sqm surface new work (2 coats)	17.1 liters or kg
21.	Bitumen required for DPC for one coat of 100 sqm surface	100 kg
22.	Water proofer for DPC per bag	1 kg
23.	G.C.I. sheets for 100 sqm roof	128 sqm
24.	A.C. sheets for 100 sqm roof	115 sqm
25.	Bricks required for R.B. work per cum	420 numbers

1. Materials required for 1:2:4 cement concrete

In order to get one cubic meter of compacted dense concrete, 1.5 to 1.6 times of dry volume of the materials are required.

Calculations

Let the compacted quantity of mix required be 1 cum

Thus, quantity of dry material required = 1.55 cum

Volume of cement = $\frac{\text{ratio of cement} \times 1.55}{\text{total volume of mix}} = \frac{1 \times 1.55}{1+2+4} = \frac{1.55}{7}$ = 0.221 cum

Volume of one bag of cement = 0.03472 cum

Therefore, cement in bags = $\frac{0.221}{0.03472}$ = 6.4 bags

Volume of coarse sand = 0.221 × 2 = 0.442 cum

Volume of coarse aggregate (bajri) = 0.221 × 4 = 0.884 cum

Similarly, any ratio of cement concrete can be worked out.

2. Materials required for brick masonry laid in 1:7 cement mortar

It is found that about 30 per cent of the mortar is consumed in brick masonry.

Let the quantity of brick masonry be 1 cum

Normal size of one brick = 20 cm × 10 cm × 10 cm (including mortar)

Thus, volume of each brick = $\frac{20 \times 10 \times 10}{100 \times 100 \times 100}$ = 0.002 cum

Therefore, number of bricks required = $\frac{1}{0.002}$ = 500

Cement mortar required (30 per cent of 1 cum) = 0.30 cum

Cement required = $\frac{0.30}{1+7} = \frac{0.30}{8}$ = 0.0375 cum

Therefore, cement required in bags = $\frac{0.0375}{0.03472}$ = 1.08 bags

Fine sand required = 0.0375 × 7 = 0.26 cum

3. Materials required for 10 cm thick brick masonry laid in 1:4 cement mortar

It is found that about 30 per cent of the mortar is consumed in brick masonry

Let quantity of brick masonry be 1 sq. m

Volume of brick masonry = $\frac{1 \times 10}{100}$ = 0.1 cum

Nominal size of bricks and volume = 20 × 10 × 10 cm =0.002 cum

No. of bricks required = $\frac{0.1}{0.002}$ = 50 Nos.

Cement mortar required = $\frac{0.1 \times 30}{100} = 0.03$ cum (30 per cent)

Cement required = $\frac{0.03}{1+4} = \frac{0.03}{5} = 0.006$ cum

Cement required in bags = $\frac{0.006}{0.03472} = 0.173$ bags

Fine sand required = 0.006 × 4 = 0.024 cum

4. Material required for stone masonry laid in cement mortar

In practice it is found that the requirement of stone varies from 110 per cent to 133 per cent depending upon the quantities of dressing required. Generally, 35 per cent to 45 per cent of dry mortar is considered sufficient for rubble stone masonry and 12 to 20 per cent for Ashlar stone masonry.

Material required for rubble masonry laid in cement mortar 1: 6

Let quantity of stone masonry be 1 cum

Quantity of undressed stone = $\frac{1 \times 110}{100} = 1.1$ cum (taken 110%)

Quantity of mortar required = $\frac{1 \times 35}{100} = 0.35$ cum (mortar taken as 35 per cent)

Quantity of Cement required = $\frac{0.35 \times 1}{7} = 0.050$ cum

Cement in bags = $\frac{0.050 \times 6}{0.03472} = 1.44$ bags

Quantity of fine sand required = 0.050 × 6 = 0.30 cum

5. Materials required for 12.5 mm thick cement plaster in 1:4 cement mortar

Cement plaster is done on brick walls so as to make them immune from absorption of water. Generally, 12.5 mm thick plaster layer is considered to be quite effective. The quantities of dry mortar required is 50 to 60 per cent more than wet mortar though the reduction in volume of the wet mortar is hardly 20 per cent but this higher percentage of 50 to 60 per cent is adopted to compensate for wastage and for covering the irregularities in the surface to be plastered.

Let the area to be plastered be 1 sq. m

Thickness of plaster = 12.5 mm

Thus, volume of the mortar = $\frac{1 \times 12.5}{1000}$ = 0.0125 cum

Volume of dry mortar (1.55 times) = 0.0125 × 1.55 = 0.0194 cum

Therefore, cement required = $\frac{0.0194}{1+4}$ = 0.004 cum

Cement required (in bags) = $\frac{0.004}{0.03472}$ = 0.12 bag (approx.)

Fine sand required = 0.004 × 4 = 0.016 cum

6. Materials required for pointing in cement mortar 1:2 for 1 m².

Area to be pointed = 1 sq. m

Area of 50 bricks of metric size $\frac{50 \times (19 \times 9)}{100 \times 100}$ = 0.85 sqm

[In one sq. m, no. of bricks of size 20×10×10 cm(including mortar) required = $\frac{1 \times 100 \times 100}{20 \times 10}$ = 50]

Area covered by joints = 1.00 – 0.85 = 0.15 sq. m

Average thickness l3mm depth of joint +7 mm(for any irregularities) = 20 mm.

Volume of wet mortar = $\frac{0.15 \times 20}{1000}$ = 0.0030 cum

Volume of dry mortar (including losses) = $\frac{0.0030 \times 155}{100}$ = 00465 cum

Sum of ratio of mortar = 1 + 2 = 3

Cement required = $\frac{0.00465}{3}$ = 0.00 155 cum

Therefore, cement in bags = $\frac{0.00155}{0.03472}$ = 0.045 bag

Fine sand required = 0.00155 × 2 = 0.003 cum

15.6 Split Up of Cost of Building Works

Sl.No.	Work	Cost
1	Cost of labour	35 to 40 per cent of the whole cost
2	Cost of materials	60 to 65 per cent of the whole cost
3	Cost of foundation and plinth	18 to 20 per cent of the whole cost
4	Cost of superstructure	80 to 90 per cent of the whole cost
5	Cost of second storey	85 to 90 per cent of the ground floor
6	Earth work in excavation and filling	0.5 to 1 per cent of the whole cost
7	Concrete in foundation	5 to 6 per cent of the whole cost
8	Damp proof course	1 to 1.5 per cent of the whole cost
9	Brick work	30 to 35 per cent of the whole cost
10	Roofing	20 to 25 per cent of the whole cost
11	Flooring	5 to 6 per cent of the whole cost
12	Doors and windows	16 to 20 per cent of the whole cost
13	Plastering and pointing	10 to 12 per cent of the whole cost
14	White washing, colour washing, painting *etc.*	2 to 3 per cent of the whole cost
15	Miscellaneous	6 per cent of the whole cost
16	Cost of sanitary and water supply installation	8 to 10 per cent of the whole cost
17	Electric installation	8 to 10 per cent of the whole cost
18	Cost of bricks	22 to 25 per cent of the whole cost
19	Cost of cement	12 to 15 per cent of the whole cost
20	Cost of steel bars	10 to 12 per cent of the whole cost
21	Cost of timber	10 to 12 per cent of the whole cost

Important Points to Remember

1. There are mainly two methods of cost estimation. The estimates may be preliminary or detailed.
2. Cubic rate estimate for building is generally used when height of the structure is more than the conventional height.
3. Detailed estimates or item rate estimate is the most accurate estimate and in this case quantities of each item of work are prepared and the cost is worked out accordingly.
4. An amount equivalent to 3 to 5 per cent of the estimated cost is added as contingencies for unforeseen items.
5. Masonry or brick work is computed in cubic meter, while the partitioned wall and honeycomb brick work are taken as separate item, and measured in square meters.

6. In general, 2.5 cm thick cement concrete (1:1.5:3) with water proofing material is used as DPC.
7. Quantity of steel/iron work is measured on weight basis.
8. Plywood is specified by thickness.
9. Painting work is specified by area of the painted surface.
10. One cubic meter of mild steel weighs about 7850 kg.
11. The measurement of grills is usually taken in terms of weight.
12. Door and window frame work is computed in terms of volume.
13. Shutters for doors or windows are computed in square meter by multiplying the breadth by the height of the shutters.
14. A booklet containing rates of various engineering items for preparation of detailed estimates is known as Standard Schedule of Rates (SSR).
15. In order to get one cubic meter of compacted dense concrete, 1.5 to 1.6 times of dry volume of the materials are required.
16. The quantity of mortar required in brick masonry is about 30 per cent.
17. The requirement of stone varies from 110 per cent to 133 per cent depending upon the quantities of dressing required. Generally, 35 per cent to 45 per cent of dry mortar is considered sufficient for rubble stone masonry and 12 to 20 per cent for Ashlar stone masonry.
18. Generally, 12.5 mm thick cement plaster layer is considered to be quite effective. The quantities of dry mortar required is 50 to 60 per cent more than wet mortar.
19. Number of bricks required for laying of 1 m^3 brick masonry is 500.
20. Quantity of un-slaked lime required for whitewashing 100 sq. m surface (2 coats) is 25 kg.
21. G.C.I. sheets required for 100 sq. m roof is 128 sq. m.
22. A.C. sheets required for 100 sq. m roof is 115 sq. m.

Exercises/Questions

1. Explain any two methods of calculating rough estimate of a structure.
2. Name and explain briefly different types of estimates.
3. Write down the uses of approximate estimate.
4. What do you mean by detailed estimate? Explain in brief.
5. Write short note on detailed estimate.
6. What are the major items included in detailed estimate?
7. Write down the units of measurement of the following items.
   (a) Earth work (b) Brick work in foundation
   (c) Wire fencing (d) Marble flooring

(e) Glass panes (f) Painting on doors
(g) Door frames (h) Laying of C.I. pipes
(i) Half brick work (j) Mosaic flooring
(k) Sanitary fittings (l) Rain water pipe
(m) Honey comb brickwork (n) D.P.C.
(o) Supply of bricks (p) Wash basin

8. Calculate the materials required for 1:8:16 cement concrete for 1 m^3.

   [**Ans:** Cement in bags =1.8, fine sand = 0.50 cum, coarse aggregate =1.00 cum]

9. Calculate the materials required for 1:4:8 cement concrete for 1 m^3.

   [**Ans:** Cement in bags =3.46, coarse sand = 0.48 cum, coarse aggregate = 0.96 cum]

10. Calculate the material required for brick masonry laid in 1:5 cement mortar for 1 m^3.

    [**Ans:** Cement = 1.44 bags, fine sand = 0.25 cum, bricks = 500 nos.]

11. Calculate the material required for 12.5 thick cement plaster in 1:6 cement mortar for 1 m^2.

    [**Ans:** Cement required = 0.08 bag, fine sand required =0.0166 cum]

Environmental Control

Chapter 16

Environmental Control in Agricultural Structures

The primary rationale behind the construction of agricultural structures is to alter the prevailing natural climate to some extent. In an ideal scenario with a consistently perfect natural climate, the confinement of livestock and poultry would require nothing more than a simple fence. Grain and other biological products could be stored outdoors without any degradation in quality, and there would be no necessity for protective measures for plant growth. Unfortunately, there are scarce locations on earth where the natural climate consistently provides optimal conditions for livestock, stored products, and plant growth. In certain places, temperatures can be excessively cold during one season and overly hot during another, leaving only brief intervals in between that are truly conducive to ideal conditions.

In tropical countries, temperature, humidity, and radiation intensities generally remain relatively stable, but they often reach levels that are too high for optimal livestock and poultry production. The primary requirement for climatic modification in tropical regions is the installation of effective shading. On the other hand, cold climates necessitate completely enclosed, insulated, and mechanically ventilated structures. Additionally, stored products require distinct structural and environmental protection measures in hot climates compared to cold weather. Thus, modification of environment is necessary for successful production of many animals and plants, and storage of grain and other biological products in various areas of the world.

16.1 Climatic Zones

Climate refers to the long-term patterns of temperature, humidity, wind, precipitation, and other atmospheric conditions in a particular region. It represents the average weather conditions, including variations and extremes, over an extended period, typically 30 years or more. Climate plays a crucial role in shaping the natural environment, ecosystems, and human activities in a given area. It is distinct from weather, which refers to the short-term and day-to-day variations in atmospheric conditions.

The most common system used to classify climatic zones is the Köeppen classification system. This climate classification scheme was developed by Wladimir Peter Koeppen in 1884. He recognized a close relationship between the distribution of vegetation and climate. The categories are based on the data of annual and monthly averages of temperature and precipitation. He selected specific values of temperature and precipitation and related them to the distribution of vegetation and used these values for classifying the climates. The Koeppen climate classification system recognizes five major climatic types on earth namely, tropical, dry, temperate, continental and polar. These climate types are general categories, and within each type, there can be variations and subcategories based on specific geographical and regional factors.

(i) **Tropical Climate:** High temperatures throughout the year, typically with little variation. The average temperature of the coldest month is 18°C or higher. Tropical wet climate (high temperature and heavy rainfall throughout the year) is found in the western parts of the Western Ghats and the north-eastern states. While tropical wet and dry climate (distinct wet and dry seasons with a monsoonal influence) is predominant in most of the country, including the central and northern regions.

(ii) **Dry or Arid Climate:** Low rainfall and high temperatures, leading to dry conditions. Potential evaporation exceeds precipitation in the dry climates. Arid climate is found in the north-western parts of India, including Rajasthan and parts of Gujarat.

(iii) **Temperate Climate:** Moderate mean annual temperatures with distinct seasons, including warm summers and cool winters. Warm temperate climate has average monthly temperatures above –3°C but below 18°C in the coldest month. Whereas, cool temperate climate is one where average temperature of the warmest month is above 10°C, and average temperature of the coldest month is under –3°C. The northern India lying above the tropic of cancer usually experiences temperate climates.

(iv) **Continental Climate:** Large temperature variations between summer and winter, often with cold winters. The northern regions of the country have a continental climate with cold winters and hot summers.

(v **Polar Climate:** Cool summers and extremely cold winters. Every month a polar climate has an average temperature of less than 0°C. The June, the warmest month, seldom ever exceeds 10°C. Parts of Jammu and Kashmir, Himachal Pradesh, and Uttaranchal have polar climates.

The another classification of climatic zones was done by Bansal and Minke (1988). They had carried out detailed studies and reported that India can be categorized into six climatic zones: hot and dry, warm and humid, moderate, cold and cloudy, cold and sunny, and composite. The criteria of classification are outlined in Table 16.1. A place is categorized into one of the first five climatic zones if the specified conditions persist for more than six months. In cases where none of the defined categories can be identified for six months or longer, the place is designated as composite climatic zone.

Table 16.1. Climatic Zone Classification Criteria (Bansal and Minke 1988)

Sl. No.	*Climatic Zone*	*Mean Monthly Temperature (°C)*	*Relative Humidity (per cent)*	*Precipitation (mm)*	*Number of Clear Days*
1.	Hot and Dry	>30	<55	<5	>20
2.	Warm and Humid	>30	>55	>5	<20
3.	Moderate	25-30	<75	<5	>20
4.	Cold and Cloudy	<25	>55	>5	<20
5.	Cold and Sunny	<25	<55	<5	>20
6.	Composite	Any of the season does not fall for more than six month			

As per the National Building Code of India 2005, the climate of India has been categorized into five major climatic zones (hot-dry, warm- humid, temperate, cold, and composite). The revised classification isn't significantly distinct from the previous one, except for the grouping of 'cold and cloudy' and 'cold and sunny' climates under the cold climate category, and the moderate climate designation has been renamed as temperate climate (Bureau of Indian Standards, 2005).

16.2 Environmental Factors

Environment can be understood as all external conditions that might affect animals, birds, plants, or stored biological products. For animal and bird housing, these environment factors can be broadly categorized into social, physical, and thermal aspects. Considering and managing these social, physical, and thermal factors is critical in designing and maintaining animal and bird housing facilities. This holistic approach ensures a conducive environment for the inhabitants, promoting their well-being and allowing them to express natural behaviours.

16.2.1 Social Factors

The social factors pertain to animals and birds are space, group dynamics, and environmental enrichment.

1. **Space Requirements:** Adequate space is crucial to prevent overcrowding, which can lead to stress, aggression, and the spread of diseases among animals. Birds also require sufficient space to exhibit natural behaviours such as flying, perching, and nesting.
2. **Group Dynamics:** Consideration of social hierarchies and compatibility within groups is essential to minimize conflicts among animals or birds. Group size and composition can influence the overall social dynamics of the population.
3. **Environmental Enrichment:** Provision of environmental enrichment, such as toys, structures, or substrates, helps prevent boredom and encourages natural behaviours. Enrichment can vary based on the species and individual needs.

16.2.2 Physical Factors

The physical factors pertain to surroundings of animals and birds. These are lighting, noise levels, flooring, pens, cages, stalls, and shelter.

1. **Lighting:** Natural or artificial lighting influences circadian rhythms, behaviour, and reproductive cycles in animals and birds. Providing appropriate light levels is crucial for the overall well-being of the inhabitants.
2. **Noise Levels:** Animals, especially those in close confinement, can be sensitive to noise. Excessive noise levels can cause stress and disrupt normal behaviours.
3. **Flooring or Substrate:** The type of flooring or substrate provided should be suitable for the species. For example, proper bedding is essential for poultry, while comfortable flooring is crucial for the health of livestock.
4. **Pens, Cages and Stalls:** The shape, size, and material of construction of pens, cages, and stalls are crucial for the comfort of animals and birds.
5. **Shelter:** Adequate shelter is necessary to protect animals and birds from extreme weather conditions, providing shade, and ensuring a sense of security. Proper fencing and enclosures prevent escape and protect animals from potential predators.

16.2.3 Thermal Factors

The thermal factors are temperature, humidity, and air movement.

1. **Temperature:** Maintaining an optimal temperature range is crucial for animals and birds' health. Extremes in temperature can lead to stress, reduced productivity, and health issues. In colder climates, heating systems may be required to prevent hypothermia, while cooling mechanisms such as fans or misters may be needed in hotter conditions. Broilers, for example, may require additional heat during the early stages of development.
2. **Humidity:** Proper humidity levels are essential, as excessive humidity can lead to respiratory issues and promote the growth of harmful microorganisms.
3. **Air movement:** Adequate airflow and ventilation are critical for maintaining air quality, preventing the build-up of harmful gases, and preventing respiratory issues.

Similarly, different plants have temperature preferences for optimal growth. Frost or extreme heat can damage or kill plants. Plants also require light for photosynthesis. Light intensity and duration influence growth and flowering.

16.3 Shelter Design

The foremost priority of plant and animal producers is to optimize production while minimizing costs. Shelters or structures play a crucial role in this objective by serving as a means to alter natural climatic thermal factors, creating conditions that are more conducive to optimal growth. In cases where only minimal environmental control is necessary, a shelter functions by mitigating the extremes of the natural climate, thereby diminishing the peak stresses experienced by animals or plants. Shelters shield animals and plants from extreme weather elements such as rain, wind, snow, and excessive heat, ensuring they can thrive despite varying climatic conditions.

The shelters can be categorized as shade-type, open-type, and environmentally controlled structures. Structures of the shade type serve as a shield against intense radiation, offering protection from the sun's powerful rays. In contrast, open-type structures find utility in colder climates, offering defence against rain, wind, and snow. Environmentally controlled structures, on the other hand, are enclosed shelters designed to modify various environmental factors such as temperature, humidity, air velocity, and radiation. In certain instances, these environmentally controlled structures must also provide control of light intensity and duration.

The thermal environment of these structures may be controlled or altered to a greater extent by manipulation of different structural and management parameters. The structural and management parameters of different types of shelters are listed in Table 16.2.

Table 16.2. List of Structural and Management Parameters of different Types of Shelters

Shade-Type	*Open-type/Partially Enclosed*	*Enclosed Structure*
Location	Location	Location
Orientation of the structure	Orientation of the structure	Orientation of the structure
Physical dimensions of space	Physical dimensions of space	Physical dimensions of space
Construction materials	Construction materials	Construction materials
Colour of walls and roof	Colour of walls and roof	Colour of walls and roof
Surrounding conditions	Surrounding conditions	Surrounding conditions
	Wind breaks	Mechanical ventilation
	Side-walls openings	Air inlets and distribution
	Roof openings	Evaporative cooling
	Animal and manure management	Animal/bird and manure management
		Solar radiation

16.4 Environmental Indices

16.4.1 Thermal Humidity Index (THI)

Temperature Humidity Index, also known as the heat index or the apparent temperature, is a measure that accounts for the combined effects of environmental temperature and relative humidity and is a useful and easy way to assess the risk of heat stress. Research has identified THI values above which heat stress begins. These THI values vary between the different livestock species of interest and within the same species, they vary for the different classes of animals. It is used to gauge the level of discomfort or danger associated with exposure to heat and humidity. By incorporating both temperature and humidity, the index provides a more accurate representation of the perceived temperature and its impact on human and animal comfort. THI is commonly used for evaluating heat stress in animals.

Heat stress is caused by a combination of temperature, relative humidity, solar radiation, air movement, and precipitation. The majority of studies on heat stress in livestock focus on the two main environmental stressors: temperature and relative humidity. Researchers have developed different formulas for the calculation of THI. It can be calculated easily from dry bulb temperature and relative humidity and both these measures are routinely collected. The formula for the calculation of THI is:

$$THI = 1.0 \times T_{db} + 0.36 \times T_{dp} + 41.2$$

where, T_{db} and T_{dp} are the dry-bulb temperature and the dew point temperature, respectively in °C. Dew point can be determined with the help of psychrometric chart.

Another simple and most common formula to calculate the THI is

$$THI = 0.8 \times T_{db} + RH \times (T_{db} - 14.4) + 46.4$$

where, RH=relative humidity expressed as a proportion *i.e.* 70 per cent humidity is expressed as 0.70.

When the THI index reaches 70, 10 per cent of the people will feel uncomfortable, and at a THI of 79, 100 per cent will feel uncomfortable. Milk production declines one standard deviation or more between a THI of 76 and 77. Environmentally controlled summer shelters for dairy cows should ideally provide a THI of 75 or lower. Figure 16.1 shows the THI plotted on psychrometric chart.

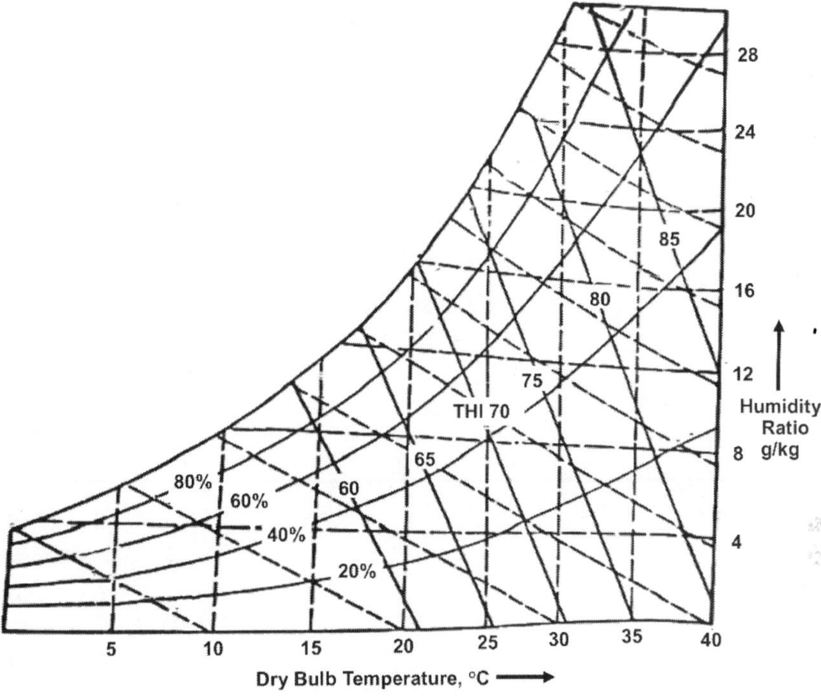

Figure 16.1. THI Superimposed on a Psychometric Chart.

The initial studies conducted in the 1950s at the University of Missouri indicated a stress threshold of a 71 THI, so animals were experiencing heat stress at a THI of 72 and greater. The levels of stress were separated into:

1. Mild - 72 to 79 THI
2. Moderate - 80 to 89 THI
3. Severe - 90 THI or greater

In India, especially during the summer months, the THI can rise to significant levels, indicating the potential for heat stress and discomfort. Higher THI values imply a greater level of heat stress, while lower values indicate more comfortable conditions. Heat stress can have adverse effects on human and animal health, leading to heat-related illnesses, dehydration, and other related issues. Table 16.3 shows the temperature humidity index for various combinations of temperature and relative humidity, and varying levels of risk for animals.

Table 16.3. Temperature Humidity Index (THI) at Specific Temperatures and Relative Humidity

Temperature (°C)	*Relative humidity (per cent)*							
	40	*50*	*60*	*70*	*80*	*90*	*100*	
18.9	63	64	64	65	65	66	66	**No stress**
20.0	65	65	66	66	67	68	68	
21.1	66	67	67	68	69	70	70	
22.2	67	68	69	70	71	72	72	**Mild stress**
23.3	69	70	71	71	72	74	74	
24.4	70	71	72	73	74	75	76	
25.6	71	73	74	75	76	77	78	
26.7	73	74	75	76	78	79	80	**Severe stress**
27.8	74	75	77	78	79	81	82	
28.9	75	77	78	80	81	83	84	
30.0	77	78	80	81	83	84	86	
31.1	78	80	81	83	85	86	88	
32.2	79	81	83	85	86	88	90	**Very severe stress**
33.3	80	82	84	86	87	90	92	
34.4	82	84	86	88	90	92	94	
35.6	83	86	88	90	92	94	96	
36.7	85	87	89	91	94	96	98	
37.8	86	88	91	93	95	98	100	**Dead cattle**
38.9	87	90	92	95	97	99	102	
40.0	88	91	94	96	99	101	104	
41.1	90	93	95	98	101	103	106	
42.2	92	94	97	100	102	105	108	
43.3	93	96	98	101	104	106	110	

Effects of Thermal Humidity Index on Milk Production

The thermal humidity index (THI) is a measurement used to assess the combined impact of temperature and humidity on livestock, including dairy cows. High THI values indicate conditions that can be stressful for animals, potentially leading to decreased milk production. Here are some effects of the thermal humidity index on milk production:

1. **Heat Stress**: High THI values can cause heat stress in dairy cows. Heat stress negatively affects their comfort, appetite, and overall well-being. When cows experience heat stress, they divert energy away from milk production to regulate their body temperature, resulting in reduced milk yield.
2. **Reduced Feed Intake**: Heat and humidity can reduce a cow's appetite, leading to decreased feed intake. Cows may eat less during hot and humid weather, which results in fewer nutrients available for milk production.
3. **Altered Rumen Function**: Heat stress can disrupt the rumen function in dairy cows. The rumen is responsible for digesting and fermenting feed, and any disturbances in its function can lead to decreased milk production. Heat stress can alter the microbial population in the rumen, affecting nutrient utilization and overall cow performance.
4. **Increased Respiration Rate**: When cows are exposed to high temperatures and humidity, their respiration rate increases as a cooling mechanism. The increased respiration rate can result in the loss of carbon dioxide and water from the cow's body, potentially leading to dehydration. Dehydration negatively affects milk production and overall cow health.
5. **Impaired Reproductive Performance**: High THI values can have a negative impact on the reproductive performance of dairy cows. Heat stress can lead to reduced fertility, longer calving intervals, and decreased conception rates, which ultimately affects milk production.

To mitigate the negative effects of high THI on milk production, dairy farmers implement various management practices. These may include providing shade, ventilation, cooling systems, and access to fresh, cool water for cows. Additionally, adjusting feeding strategies and using heat stress mitigation techniques, such as evaporative cooling or fans, can help alleviate the negative impact of heat stress on milk production.

16.4.2 Daily Range

The daily range of temperature is the difference between the highest and lowest temperatures recorded in a given day. It can have significant implications for various aspects of the environment, including plant and animal behaviour, energy consumption, and human comfort. For example, in a desert climate, the daily temperature range can be quite large, with scorching heat during the day and a significant drop in temperature at night. In contrast, coastal regions often have smaller daily temperature ranges due to the moderating influence of the ocean.

16.4.3 Wet Bulb Depression

Wet bulb depression, also known as wet bulb temperature depression or wet bulb deficit, is a meteorological term that refers to the difference between the dry bulb temperature and the wet bulb temperature in a given environment. The wet bulb depression is calculated by subtracting the wet bulb temperature from the dry bulb temperature. It provides an indication of how much evaporative cooling potential exists in the atmosphere. A higher wet bulb depression suggests a greater capacity for evaporation and thus a greater potential for cooling.

Wet bulb depression is particularly relevant in assessing the risk of heat stress or heat-related illnesses, as it is an indicator of the cooling potential of the surrounding air. Higher wet bulb depression values generally imply a more comfortable or favourable environment for human comfort, while lower values indicate less evaporative cooling potential and increased heat stress.

16.4.4 Effective Temperature

Effective temperature typically refers to a calculated value that represents the combined impact of various environmental factors, such as temperature, humidity, air movement, and sometimes radiant heat, on the perceived comfort or discomfort experienced by humans or animals. In the context of human comfort, effective temperature is often used in occupational health and safety or building design to assess the thermal comfort of individuals. It takes into account factors like air temperature, humidity, air velocity, and radiant heat exchange to provide a single value that reflects how the environment feels to a person. For animals, particularly livestock, effective temperature serves a similar purpose in assessing the comfort level within housing facilities or outdoor environments. It considers factors relevant to the species, such as coat type, metabolic rate, and behaviour, to determine the thermal conditions that are conducive to their well-being and productivity.

16.4.5 Black Globe Thermometer

A black globe thermometer (Figure 16.2) is a type of instrument used to measure the combined effects of air temperature, humidity, and radiant heat on the thermal comfort of humans or animals in various environments. It consists of a hollow, black-coloured sphere (the globe) with temperature sensors inside. When exposed to the environment, the globe absorbs radiant heat from surrounding surfaces and emits heat due to its own temperature. The rate of heat exchange between the globe and the surrounding air depends on factors like air temperature, humidity, air movement, and radiant heat sources. By measuring the temperature inside the globe, along with air temperature and humidity, one can assess the thermal environment and evaluate heat stress risks for occupants or animals. Black globe thermometers are commonly used in various applications, including occupational health and safety, agriculture, and sports science.

Figure 16.2. Black Globe Thermometer

16.4.6 Mean Radiant Temperature (MRT)

Mean radiant temperature is a crucial concept in thermal comfort and building physics. It refers to the average temperature of all surfaces surrounding an individual, as perceived by their radiant heat exchange. Unlike air temperature, which is typically measured with a thermometer, MRT considers the combined effect of radiant heat exchange between a livestock/person and surrounding surfaces, such as walls, ceilings, floors, and other objects. MRT plays a significant role in livestock/human comfort because the bodies not only exchange heat with the surrounding air but also with nearby surfaces via radiation. For example, if you're standing next to a cold window on a sunny day, even if the air temperature is warm, you may feel chilly because your body is losing heat to the cold window surface through radiation.

The mean radiant temperature is a function of surface temperatures, emissivity of surfaces, solar radiation, geometry of the shelter surrounding the animals (view factors), air temperature, and air velocity. By managing these factors effectively, one can optimize thermal comfort and energy efficiency in buildings. The mean radiant temperature for humans or animals can vary based on their body's shape factor with respect to the various surrounding radiating sources and convective properties. However, the black globe is commonly employed in experiments to represent a person or animal.

The equation for MRT is expressed as

$$T_m = 100\left[2.50\sqrt{v}(T_g - T_a) + \left(\frac{T_g + 273}{100}\right)^4\right]^{1/4}$$

where,

T_m = mean radiant temperature in K,

T_g = black globe temperature in °C,

T_a = air temperature in °C, and

v = air velocity in m/s.

16.4.7 Black Globe Humidity Index (BGHI)/Wet Bulb Globe Temperature (WBGT) Index

The THI does not include the effect of radiant heat load on the human body, animals and birds.

To address this limitation, a comprehensive thermal index suitable for both hot and cold environments has been developed, known as black globe humidity index (BGHI). The BGHI takes into account the combined effects of temperature, humidity, and radiant heat on the human body, animals and birds. It is commonly used to assess the risk of heat stress. The index takes into account three components:

1. **Wet Bulb Temperature**: This measures the cooling effect of evaporation and is influenced by both temperature and humidity.
2. **Dry Bulb Temperature**: This is the normal air temperature measured by a regular thermometer.
3. **Globe Temperature**: This represents the combined effects of air temperature and radiant heat from the sun or other sources.

The equation to calculate the BGHI is as follows:

$$BGHI = 0.7T_{wb} + 0.2T_g + 0.1T_d$$

OR

$$BGHI = T_g + 0.36\,(T_{dp}) + 41.5$$

where,

T_{wb} = wet bulb temperature in °C,

T_g = black globe temperature in °C,

T_d = dry bulb temperature in °C, and

T_{dp} = dew point temperature in °C.

The black globe temperature accounts for the combined effects of air temperature and radiant heat from the sun or other sources. In comparison, the BGHI proves to be significantly more accurate and representative than the THI, particularly in tropical climates with intense thermal radiation.

16.4.8 Degree Days

Degree days are a measure used in climate analysis and energy management to quantify the amount of heating or cooling required in a specific location over a given period of time. Degree days are calculated as the difference between the daily mean temperature and the reference temperature. A "degree day" records how hot or how cold it has been over a 24-hour period.

In heating degree day (HDD), the reference temperature is typically set at 18°C. It represents a temperature at which no heating is required. HDD is used to estimate the amount of energy needed to heat a building during a certain period. By subtracting the average daily temperature from the reference temperature and summing up these differences over a specific time frame, the total heating degree days are determined. For example, if the mean temperature for a day is 12 degrees Celsius, there are 6 HDD for that day (18°C - 12°C = 6°C). If the daily mean temperature is above 18°C, the HDD value for that day is set to zero.

In cooling degree day (CDD), the reference temperature is usually set at 18°C as well. It represents a temperature at which no cooling is required. CDD is employed to estimate the amount of energy required for air conditioning during a particular time period. The average daily temperature is subtracted from the reference temperature, and the resulting differences are summed up to obtain the total cooling degree days. If the average daily temperature is 25°C, then the difference from 25°C to 18°C is 7°C. Therefore, there are 7 Cooling Degree Day. In other words, you have had to cool the building by 7°C that day to reach 18°C.

Mean daily temperature (MDT) =

$$\frac{(\text{daily high temperature} + \text{daily low temperature})}{2}$$

$$\text{HDD} = 18°\text{C} - \text{MDT}$$

$$\text{CDD} = \text{MDT} - 18°\text{C}$$

By summing up the HDD or CDD over a specific period, typically a month or a heating/cooling season, the heating or cooling requirements for that period can be determined. The calculation of heating load using degree days involves a straightforward equation. The heating load is the product of the total degree days and the heating degree day factor (HDDF) for a particular location and time period. The formula is as follows:

Heating load = Total Heating Degree Days × Heating Degree Day Factor

where,

Total Heating Degree Days = Σ (Base Temperature - Daily Mean Temperature)

The "Σ" symbol denotes summation, which means to add up the differences between the base temperature and the daily mean temperature for the desired time period.

The Heating Degree Day Factor (HDDF) is a factor that represents the heat loss rate of a building per degree day (W/K). It takes into account factors such as the building's insulation, efficiency of heating systems, and other local climate considerations. HDDF is specific to each building and location and is usually provided by heating engineers or energy experts.

This calculation provides an estimate of the heating load, and actual heating requirements may differ due to factors like internal heat gains and fluctuations in weather conditions.

For cooling load calculations, a similar process can be followed using Cooling Degree Days (CDD) and the Cooling Degree Day Factor (CDDF), but the specific equations would differ based on the region's climate and cooling systems.

An estimate of an annual heating energy consumption of a building in watt-hours would be the heat loss coefficient (W/K), multiplied by the number of degree days multiplied by 24 (to convert from days to hours). Dividing by 1000 then gives the result in kilowatt-hours (kWh).

16.4.9 Basic Solar Earth Angles

The basic solar earth angles refer to the angles formed between the Sun, the Earth, and various reference points on the Earth's surface. These angles play a crucial role in determining the intensity and distribution of solar radiation received by different locations on the Earth.

Basic Angles

1. **Latitude** (ϕ) of a location is the angle made by the radial line joining the location to the centre of the earth with the projection of the line on the equatorial plane. It is the angular distance north or south of the equator measure from the centre of the earth.
2. The **hour angle** (ω) is the angle through which the earth must turn to bring the meridian of a point directly in line with the sun's rays. At solar noon the hour angle is zero.

   One hour = 15°
3. **Declination** (δ) is the angle made by the line joining the centres of the sun and the earth with its projection on the equatorial plane. The declination angle varies from a maximum value of $+23.45^{\circ}$ on June 21 to a minimum value of -23.45° on December 21. It is zero on two equinox days of March 21 and September 22.

Derived Solar Angles

1. **Solar Zenith Angle** (θ_z): The solar zenith angle is the angle between the vertical line (perpendicular to the Earth's surface) and the line connecting the Sun to a specific location on Earth. It is measured from directly overhead (zenith) to the position of the Sun. The solar zenith angle varies throughout the day and depends on the latitude and the time of year. At solar noon, when the Sun is highest in the sky, the solar zenith angle is 0° at the location directly beneath the Sun.
2. **Solar Altitude Angle** (θ_a): The solar altitude angle is the complement of the solar zenith angle, representing the height of the Sun above the horizon. It is the angle between the horizontal plane and the line connecting the Sun to a specific location on Earth. The solar altitude angle ranges from 0° (Sun on the horizon) to 90° (Sun directly overhead at the zenith). Higher solar altitude angles generally result in more intense solar radiation.
3. **Solar Azimuth Angle** (θ_{az}): Surface azimuth angle (is the angle made in the horizontal plane between the line due south and the projection of the line of sight of the sun on the horizontal plane. Thus, it gives the direction of the shadow cast in the horizontal plane by a vertical rod. It is the angle of deviation of the normal to the surface from the local meridian, the zero point being south, east positive and west negative. It can vary from -180° to +180°. The angle is positive if the normal is east of south and negative if west of south.

These angles vary based on the time of day, the observer's latitude, and the date.

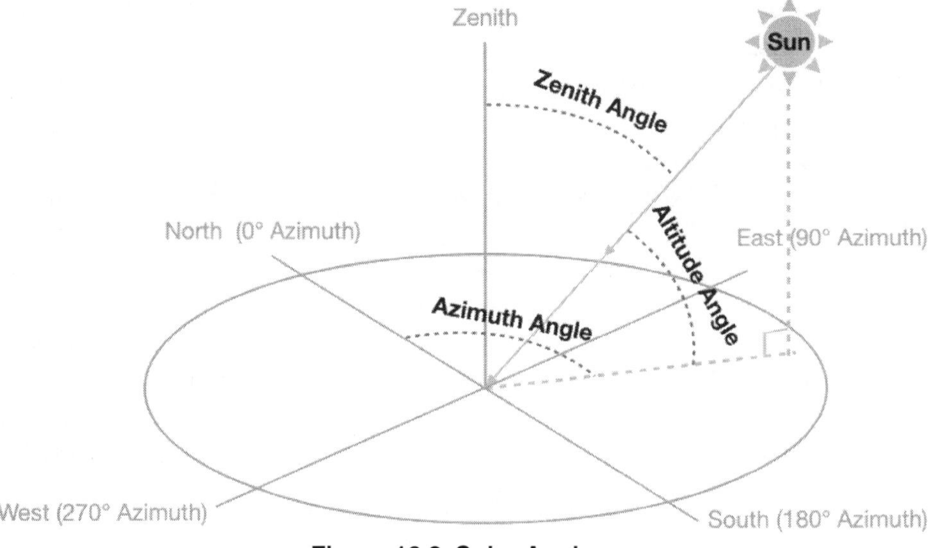

Figure 16.3. Solar Angles.

16.4.10 Sol-Air Temperature

The term "sol-air temperature" refers to the temperature of the outdoor air, taking into account both solar radiation and air temperature. It is an important parameter in building energy calculations and is used to evaluate the heat gains or losses through building envelopes (such as walls, windows, and roofs) due to solar radiation.

Sol-air temperature is typically calculated by combining the effects of the outdoor air temperature and the solar radiation incident on the building's surfaces. It takes into consideration factors such as solar angle, solar intensity, and the thermal properties of the building materials. It is particularly relevant for assessing the cooling load on a building in warmer climates or during periods of high solar exposure. Sol-air temperature may be described as

$$T_e = T_o + \frac{\alpha I}{h_c}$$

where

T_e = sol-air temperature in °C,

T_o = outside air temperature in °C,

α = solar absorptivity of the outside roof surface,

I = the combined incident solar radiation upon the surface in W/m^2, and

h_c = the convective film coefficient of the outside roof surface in W/m^2-°C.

Important Points to Remember

1. The primary rationale behind the construction of agricultural structures is to alter the prevailing natural climate to some extent.
2. The Koeppen climate classification system recognizes five major climatic types on earth namely, tropical, dry, temperate, continental and polar.
3. As per the National Building Code of India 2005, the climate of India has been categorized into five major climatic zones (hot-dry, warm-humid, temperate, cold, and composite).
4. Environment can be understood as all external conditions that might affect animals, birds, plants, or stored biological products.
5. For animal and bird housing, these environment factors can be broadly categorized into social, physical, and thermal aspects.
6. The social factors pertain to animals and birds are space, group dynamics, and environmental enrichment.
7. The physical factors pertain to surroundings of animals and birds. These are lighting, noise levels, flooring, pens, cages, stalls, and shelter.

8. The thermal factors are temperature, humidity, and air movement.
9. The shelters can be categorized as shade-type, open-type, and environmentally controlled structures.
10. Milk production declines one standard deviation or more between a THI of 76 and 77.
11. Environmentally controlled summer shelters for dairy cows should ideally provide a THI of less than 75.
12. When the THI index reaches 70, 10 per cent the people will feel uncomfortable.
13. When the THI index reaches 79, 100 per cent of the people will feel uncomfortable.
14. Animals experience heat stress at a THI of 72 and greater.
15. THI can be calculated using the relationship $THI = 1.0\ T_{db} + 0.36\ T_{dp} + 41.2$ or $THI = 0.8T_{db} + RH \times (T_{db} - 14.4) + 46.4$
16. Wet bulb depression is an indicator of the cooling potential of the surrounding air.
17. The black globe humidity index (BGHI) or wet-bulb globe temperature (WBGT) index is a measure of the combined effects of temperature, humidity, and radiant heat on the human body, animals and birds.
18. Sol-air temperature refers to the temperature of the outdoor air, taking into account both solar radiation and air temperature.
19. Temperature Humidity Index (THI) is commonly used for evaluating heat stress in animals.
20. Higher THI values imply a greater level of heat stress, while lower values indicate more comfortable conditions.

Exercises/Questions

1. Discuss the importance of environmental control in agricultural structures.
2. Describe the social, physical, and thermal factors for animal and bird housing.
3. Discuss the classification of climatic zones.
4. Based on the climate data of your city/state, identify the zone in which type of climate your area falls?
5. What are the different types of shelters for animals?
6. What are the effects of thermal humidity index on milk production of dairy cows?
7. Define degree days. Differentiate between heating degree day and cooling degree day.

8. Write short notes on the following:
   (a) Temperature humidity index (b) Black globe humidity index
   (c) Sol-air temperature (d) Wet-bulb depression

Chapter 17

Psychrometrics

Psychrometrics, the study of the properties of air and water vapour mixtures, is highly relevant for the environmental control of farm buildings. Proper environmental control is essential to create optimal conditions for the health, comfort, and productivity of livestock, as well as for crop growth and storage. Psychrometrics provides valuable insights into the relationship between temperature, humidity, and air properties, enabling farmers to create and maintain optimal environmental conditions in farm buildings for livestock health, productivity, and crop management.

Psychrometrics principles are widely used in HVAC (Heating, Ventilation, and Air Conditioning) systems design, drying processes, and other applications where understanding the behaviour of moist air is essential. Psychrometrics calculations help to optimize indoor environments for comfort, energy efficiency, and various industrial processes.

17.1 Composition of Atmospheric Air

Atmospheric air is made by mixing the several gases. Nitrogen is the most abundant component of atmospheric air, making up approximately 78 per cent of the atmosphere. Oxygen is the second most abundant component, constituting about 21 per cent of the atmosphere. It is essential for supporting life and is necessary for the process of respiration. Argon is a noble gas and makes up approximately 0.93 per cent of the atmosphere. It is chemically inert and does not participate in most chemical reactions.

In addition to these major components, the atmosphere also contains trace amounts of several other gases, including carbon dioxide, methane, ammonia, hydrogen, ozone, *etc.* Apart from these gases, the atmosphere also contains varying amounts of water vapour, dust particles, and other airborne substances that contribute to its overall composition. The composition of the atmosphere can vary slightly depending on factors such as altitude, location, and prevailing environmental conditions. The values provided here are approximate and represent average atmospheric composition at sea level.

Carbon dioxide is a greenhouse gas and plays a crucial role in the Earth's climate system. It makes up about 0.04 per cent (or 400 parts per million) of the atmosphere. The amount of carbon dioxide varies more than any of the other gases, particularly in livestock buildings. Methane is another greenhouse gas, albeit present in trace amounts (about 1.9 parts per million). It is released from various natural and human/animal related sources. Ammonia exists naturally in the air at levels between 1 and 5 parts in a billion parts of air. Methane and ammonia may be quite prevalent in livestock and poultry buildings. In poorly ventilated buildings ammonia concentration in the air can reach a level detrimental to birds and animals.

17.2 Thermodynamic Properties of Moist Air

Moist air is a mixture of dry air and water vapour. The thermodynamic properties of moist air are important in various engineering and scientific applications, such as heating, ventilation, and air-conditioning. Moist air may contain various amounts of water vapour ranging from zero (dry air) to saturation, which varies with air temperature. Moist air is saturated when it will co-exist in neutral equilibrium with the liquid or solid phase present on a flat surface *i.e.*, the air will absorb no moisture. The application of the perfect gas law and Dalton's law to moist air develops several fundamental relationships.

Perfect Gas Law

The perfect gas law, also known as the ideal gas law, describes the behaviour of an ideal gas under various conditions. It relates the pressure, volume, temperature, and amount of gas in a system. The ideal gas law equation is given by:

$$\frac{PV}{T} = \text{Constant}$$

or, $PV = n\,R\,T$

where,

P = pressure of the gas in Pascal,

V = volume of the gas in cubic meter,

n = amount of gas present, usually measured in moles,

T = absolute temperature of the gas, measured in kelvin (K), and

R = universal gas constant, which is the same for all gases = 8.314 kJ/kg-mol-K

This equation shows the relationship between the pressure, volume, temperature, and amount of gas in a system. It implies that, for a given amount of gas, increasing the temperature or decreasing the volume will result in an increase in pressure, and vice versa. The thermodynamic state of a gas can be determined if any two of the three variables are known.

The ideal gas law assumes that the gas behaves perfectly and that there are no intermolecular forces or molecular volume. In reality, real gases deviate from ideal behaviour at high pressures and low temperatures due to factors like molecular interactions and non-zero molecular volumes. However, the ideal gas law is still useful in many situations and provides a good approximation for the behaviour of gases under ordinary conditions.

Dalton's Law

Dalton's law, also known as Dalton's law of partial pressures, describes the behaviour of ideal gases in a mixture. It states that the total pressure exerted by a mixture of non-reacting gases is equal to the sum of the partial pressures of the individual gases in the mixture. For a mixture of air and water vapour, the total pressure as per Dalton's law is

$$p_t = p_a + p_w$$

where, p_t is the total pressure of the gas mixture, p_a and p_w are the partial pressures of the dry air and water vapour in the mixture.

Humidity Ratio or Specific humidity or Humidity or Moisture Content (ω)

The ratio of the mass of water vapour to the mass of moist air in a given volume is known as the humidity ratio. It depends only on the partial pressure p_v of water vapour in the air and on the total pressure, p (101.325 kPa)

$$\omega = \frac{m_v}{m_a}$$

where,

m_v = mass of water vapour in mixture

m_a = mass of dry air in mixture (always 1 kg on the psychrometric chart)

The perfect gas relation for air and water vapour for a given volume V of moist air provides

$$m_v = \frac{p_v V}{R_v T}$$

and the mass of air is,

$$m_a = \frac{(p - p_v)V}{R_a T}$$

where,

p is the total pressure of the moist air (also known as atmospheric pressure or barometric pressure = 101.325 kPa or 1 atm abs or 760 mm Hg or 101.3 kN/m²).

R_a and R_v are the gas constant for dry air and water vapour respectively.

Gas constant for dry air, R_a = 8314/28.96 = 287.2 J/kg-K = 0.287 kJ/kg-K

Gas constant for water vapour, R_v = 8314/18 = 461 J/kg-K = 0.461 kJ/kg-K

(Molecular weight of water and air are 18 and 28.96 respectively)

Thus,

$$\text{Humidity ratio } (\omega) = \frac{m_v}{m_a} = \frac{p_v}{p - p_v} \times \frac{R_a}{R_v} = 0.622 \times \frac{p_v}{p - p_v}$$

Saturation Humidity

Saturation humidity is known as humidity of saturated air or maximum humidity.

$$\omega_s = 0.622\left[\frac{p_s}{p - p_s}\right]$$

where,

p_s is partial pressure of saturated air at dry-bulb temperature

Degree of Saturation (μ) or Percentage Humidity

It is the ratio of the percentage weight of water vapour present in moist air to the percentage weight of water vapour present in saturated air at the same temperature. It is also defined as the ratio of the actual humidity of air to the humidity of saturated air at the same temperature and pressure. It is also known as percentage humidity. Thus,

$$\mu = \frac{\omega}{\omega_s} \times 100 = \frac{p_v}{p_s}\left(\frac{p - p_s}{p - p_v}\right) \times 100$$

Relative Humidity (ϕ): It is the ratio of the actual partial pressure of water vapour present in moist air to the saturation vapour pressure at the same temperature., *i.e.*,

$$\phi = \frac{\text{existing partial pressure of water vapour}}{\text{saturation pressure of pure water vapour at the same temperature}} = \frac{p_v}{p_s} \times 100$$

Substituting the value of ϕ from the above equation

$$\mu = \phi\left(\frac{p-p_s}{p-p_v}\right)\times 100$$

Note: The difference between the relative humidity and degree of saturation is not very large. The RH and degree of saturation are zero for dry air and unity for saturated moist air. The above equation shows that the degree of saturation being less than relative humidity.

Specific Volume

The specific volume is the volume per unit mass of dry air at the state of the mixture. It is usually expressed in m^3/kg dry air. The specific volume of an air-vapour mixture is an important quantity in environmental calculations for agricultural buildings because most air movement equipment such as fans *etc.* are rated by volume of air. Thus, specific volume value is used for conversion of units from volume of air to mass and vice-versa.

Specific volume, $\upsilon = \upsilon_a + \omega \upsilon_v$

From an ideal gas law:

$$p_a V = \frac{m_a RT}{M_a}$$

Thus, the **specific volume of dry air** can be found as:

$$\upsilon_a = \frac{V}{m_a} = \frac{RT}{p_a M_a} = \frac{R_a T}{p_a}$$

$$or,\ \upsilon_a = \frac{R_a T}{(p - p_v)}$$

The saturated or superheated vapour follows ideal gas laws below 66°C. Thus, **specific volume of water vapour** can be found as:

$$\upsilon_v = \frac{V}{m_v} = \frac{R_v T}{p_v}$$

Also, using the ideal gas law,

$$\upsilon = \frac{22.41}{273} T\left(\frac{1}{29} + \frac{1}{18}\omega\right) \quad when\ \text{T is temperature in K}$$

$$\upsilon = (0.082T + 22.4)\left(\frac{1}{29} + \frac{1}{18}\omega\right) \quad when\ \text{T is temperature in } ^\circ\text{C}$$

Enthalpy

Enthalpy is the total energy content of an air-water vapour mixture. The enthalpy of moist air consists of the enthalpy of dry air plus the enthalpy of the water vapour. A mixture of dry air and liquid water at 0°C is considered to have an enthalpy of zero (datum point for calculations).

Specific Heat of Dry Air (c_{pa}): The specific heat of dry air in -40 to 60°C ranges from 0.997 to 1.022 kJ/kg-K at 1 atm, with an average value of 1.005 kJ/kg-K for most calculations.

The **enthalpy of the dry air** is the energy which would have to be added to 0°C air to obtain air of a specified temperature and is given by:

$$h_a = c_{pa} \times (T - T_o)$$

where,

h_a = enthalpy of the dry air (kJ/kg dry air)

c_{pa} = specific heat of air (kJ/kg-K) ≈ 1.005 kJ/kg-K

T = dry-bulb temperature (°C)

T_o = reference temperature (0°C)

Specific Heat of Water Vapour (c_{pv}): As per the experiments, the specific heat of saturated and superheated vapours slightly changes within a temperature range of -71 to 124°C. Hence, a specific heat value of 1.88 kJ/kg-K is used for calculations.

The **enthalpy of water vapour** at low temperatures is the sum of the energy required to vaporize liquid water at 0°C and the energy required to heat the resulting vapour from 0°C to the specified temperature.

$$h_v = h_g + c_{pv} \times (T - T_o)$$

where,

h_v = enthalpy of water vapour (kJ/kg water)

c_{pv} = specific heat of water vapour (kJ/kg water-K) ≈ 1.884 kJ/kg-K

T = temperature (°C)

T_o = reference temperature (0°C)

h_g = heat of vaporization of water at reference temperature of 0°C (kJ/kg air) ≈ 2501.4 kJ/kg

The **total enthalpy of an air-water vapour mixture** per kilogram of dry air is:

$$h = h_a + \omega h_v$$

where,

h = total enthalpy (kJ/kg air)

ω = humidity ratio (kg water/kg dry air)

Thus,

$$h = c_{pa} \times (T - T_0) + \omega[h_g + c_{pv} \times (T - T_0)]$$

$$= c_{pa} \times T + \omega(2501.4 + 1.884 \times T)$$

$$= (1.005 + 1.884\omega)T + 2501.4\omega \; = c_{pam}T + 2501.4\omega$$

where,

c_{pam} **is the specific heat of moist air (or humid heat)** in kJ/kg dry air-K

Humid heat is the heat required to raise the temperature of 1 kg of dry air and water vapour by 1 Kelvin. The specific heat of dry air is 1.005 kJ/kg and 1.884 kJ/kg for water. Hence, the humid heat of an air-vapor mixture.

$$c_{pam} = 1.005 + 1.884\omega$$

Dew-Point Temperature (T_{dp}): It is the temperature at which the liquid droplets just appear when the moist air is cooled continuously. At this temperature, the saturation pressure of water is equal to the partial pressure of the water vapour existing in the air-water vapour mixture. In other words, it is the saturation temperature corresponding to the humidity ratio of an existing mixture.

Wet-Bulb Temperature (T_{wb}): It is the temperature recorded by the mercury thermometer which bulb is wrapped by the moistened wick using the distilled water and kept with an air velocity of 4.5 m/s. It is the saturation temperature attained essentially by a constant enthalpy evaporation process.

Adiabatic Saturation

Adiabatic saturation refers to a process in thermodynamics and psychrometry where air is brought to a state of maximum humidity by evaporating water into it without any heat exchange with the surroundings. It is an important concept in the study of psychrometics, which deals with the properties of moist air. During adiabatic saturation, dry air at a specific temperature and humidity ratio comes into contact with water in a manner that allows water to evaporate into the air without any heat transfer occurring. As a result, the air becomes saturated with moisture, reaching its maximum humidity level at that particular temperature.

The process is illustrated in Figure 17.1. The partially saturated moist air enters an insulated chamber. The insulated chamber prevent the heat transfer between the chamber and ambient atmosphere. This air is brought into contact with water, usually through a wetted surface or spray and comes out saturated. The air absorbs moisture from the water, and the process occurs at constant enthalpy (adiabatic process). The dry-bulb temperature decreases due to evaporation while the specific humidity of the air increases. The wet-bulb temperature remains constant during the adiabatic saturation process. A thermal balance is maintained in the moist air mixture by shifting sensible heat to latent heat.

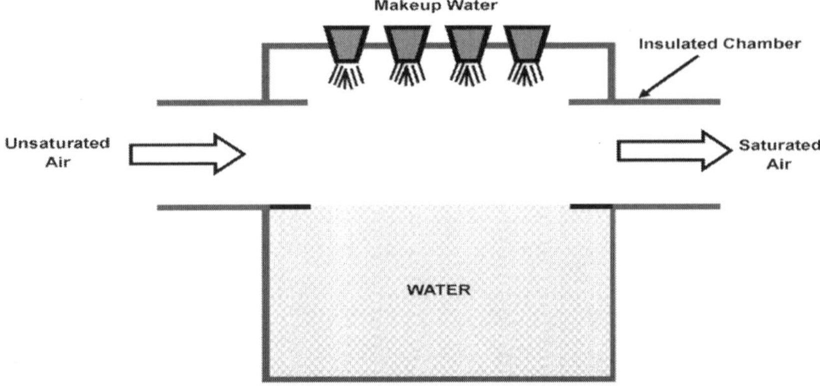

Figure 17.1. Adiabatic Saturation of Air.

One practical example of adiabatic saturation is the use of a wet-bulb thermometer. This instrument consists of a thermometer with a wetted wick or cloth covering its bulb. When the wet-bulb thermometer is exposed to moving air, water evaporates from the wick, causing the temperature to drop. The temperature indicated by the wet-bulb thermometer represents the adiabatic saturation temperature and can be used to determine various psychrometric properties.

Diffusion of Water Vapour

Diffusion refers to the process by which particles or molecules move from an area of high concentration to an area of low concentration. The diffusion of water vapour follows the same principle. When there is a difference in water vapour concentration between two regions, water vapour molecules will naturally move from the region of higher concentration to the region of lower concentration until the concentration becomes equalized. The movement of water vapour occurs due to the differences in partial pressure, which is a measure of the concentration of water vapour molecules in a gas mixture. The diffusion of water vapour will take place independently of the pressures of other gases in either region. The diffusion depends exclusively on the relationship between the partial pressures and the medium through which water vapour is diffusing. For example, diffusion through a porous material may be slower compared to diffusion through air. Understanding water vapour diffusion is crucial in the design and construction of agricultural building walls and ceilings to prevent moisture-related issues such as condensation, mould growth, and material degradation.

Moisture Holding Capacity of Air

The amount of water vapour necessary to saturate a given mass of dry air depends on the temperature of the air. The relationship between temperature and the maximum amount of water vapour the air can hold is described by the concept of relative humidity. Table 17.2 presents the maximum mass of water vapour per 1 kg of dry air at various temperatures, assuming 100 per cent relative humidity, or the mass of water vapour necessary to saturate 1 kg of dry air at various temperatures. As the air temperature increases by 10°C the water vapour holding capacity of the air nearly doubles. It means that the warmer air can hold more moisture than cooler air.

Table 17.2. Moisture Holding Capacity of Air at different Temperatures

Temperature (°C)	*Humidity Ratio (g/kg of dry air)*
0	3.8
5	5.5
10	7.7
15	10.7
20	14.8
25	20.1
30	27.3
35	33.0

These values represent the maximum amount of water vapour that can be held at saturation. The actual amount of water vapour present in the air may be lower if the relative humidity is less than 100 per cent.

17.3 Psychrometric Chart

A psychrometric chart is a graphical representation of the thermodynamic properties of moist air. It is used in the field of heating, ventilation, and air conditioning engineering to analyse and calculate various parameters related to environmental control problems for agricultural buildings. Figure 17.2 presents a skeleton psychrometric chart.

The chart displays the dry-bulb temperature on the x-axis and the humidity ratio on the y-axis. The wet-bulb and dew point temperatures are plotted on a curve that swings upward to the right. The wet-bulb temperature lines and enthalpy lines coincide with each other and are drawn obliquely. The relative humidity curves swing upward to the right. The curve at the left end of the chart represents 100 per cent relative humidity and is known as the saturation curve. The specific volume lines are obliquely drawn and have a different slope from the wet-bulb temperature lines. The chart is typically drawn for standard atmospheric pressure at sea level of 101.325 kPa (or 760 mm of Hg). It must be remembered that the psychrometric charts are based on a unit mass of dry air of 1 kg. The units are in quantity per kg of dry air.

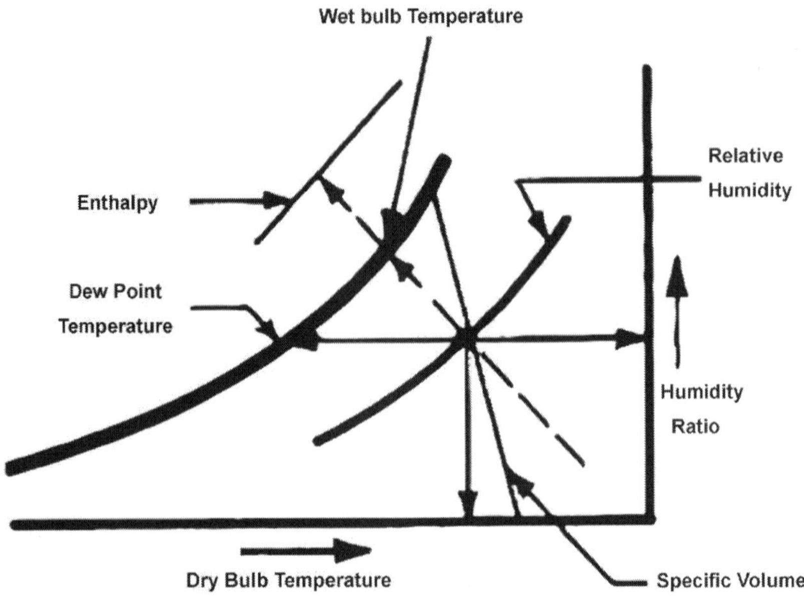

Figure 17.2. Skeleton Psychrometric Chart.

17.4 Psychrometric Processes

Psychrometric processes refer to the changes in the properties of moist air as it undergoes various transformations, such as heating, cooling, humidification, or dehumidification. These processes may be analysed with the help of equations or psychrometric chart.

17.4.1 Sensible Heating and Cooling of Moist Air

Sensible heating and cooling of moist air refers to the process of changing the temperature of air while keeping its moisture content constant. During cold season, it is common practice to provide supplemental artificial heat to livestock and poultry buildings for housing young animals and chicks. When heating moist air, the sensible heat is added to increase the air temperature while keeping the amount of moisture constant. This can be achieved using various heating methods, such as electric heaters, gas furnaces, or heat pumps. As the air temperature rises, the relative humidity decreases because the warmer air can hold more moisture before reaching saturation. However, the absolute amount of moisture in the air remains the same.

Cooling moist air involves removing sensible heat from the air to decrease its temperature while maintaining the same moisture content. Common cooling methods include air conditioning systems, evaporative coolers, or refrigeration systems. When the air is cooled, the relative humidity increases because cooler air has a reduced capacity to hold moisture. The straight and horizontal line on the chart showed the sensible cooling/heating process. During this process, the humidity ratio remains constant (Figures 17.3 and 17.4).

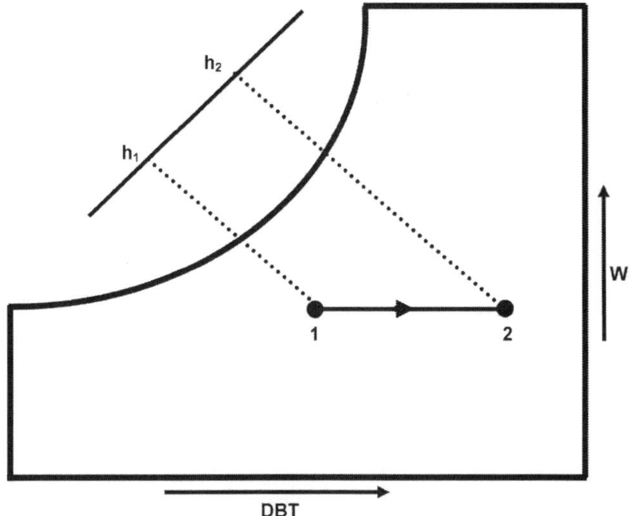

Figure 17.3. Sensible Heating Process.

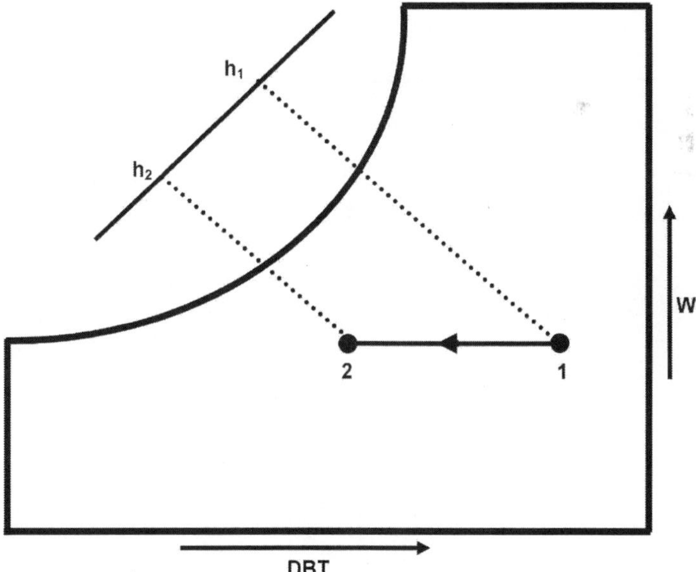

Figure 17.4. Sensible Cooling Process.

The heat added or removed from the air is determined by the following relationship:

$$q = m_a(h_2 - h_1)$$

where,

m_a is the mass of air in kg, h_1 is the initial enthalpy of air in kJ/kg, and h_2 is the final enthalpy of air in kJ/kg.

Example 17.1

In winter, a pig farrowing house receives heat at 20°C using a fan with the heater at 75 m^3/min. The ambient condition of air is 8°C DBT and 5°C WBT. What will be the hourly heat requirement for warming the pig house?

Solution

Using a psychrometric chart, locate intersection point '1' corresponding to 8°C DBT and 5°C WBT. Move right in the direction along the constant humidity ratio line up to the intersection with the vertical 20°C dry-bulb line. Read the enthalpy corresponding to points '1' and '2' and the specific volume at point '1' (since fan is located at point '1').

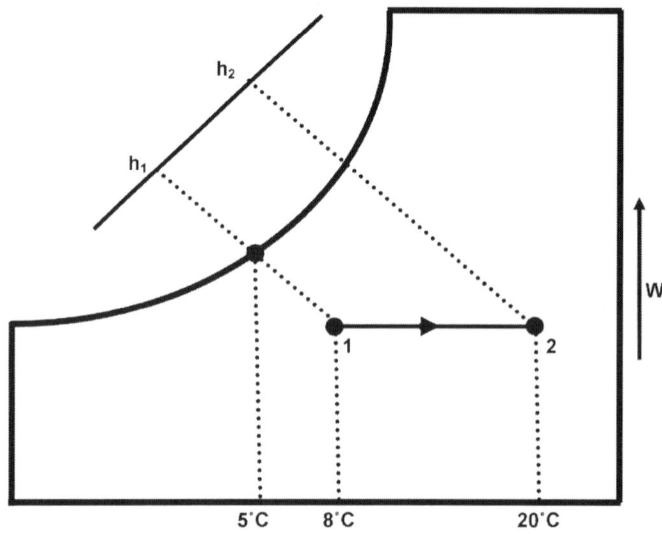

Enthalpy at point '1' (h_1) = 18.5 kJ/kg dry air

Enthalpy at point '2' (h_2) = 31.0 kJ/kg dry air

Specific volume at point '1' (v_1) = 0.802 m^3/kg dry air

Volumetric flow rate (Q) at point '1' = 75 m^3/min

Thus, mass flow rate at point '1' $(\dot{m}_a) = \frac{Q}{v_1} = \frac{75 \times 60}{0.802} \approx 5611$ kg/h

Hourly heat requirement = $\dot{m}_a(h_2 - h_1) = 5611 \times (31 - 18.5) \approx 70138$ kJ/h

Example 17.2

The air of 35°C DBT and 40 per cent RH was cooled to 21°C DBT. If the airflow rate is 100m³/min, calculate the sensible heat required to remove per hour.

Solution

Using a psychrometric chart, locate intersection point '1' corresponding to 35°C dry-bulb and 40 per cent relative humidity. Move left in the direction along the constant humidity ratio line up to the intersection with the vertical 21°C dry-bulb line. Read the enthalpy corresponding to points '1' and '2' and specific volume at point '2' (since fan is located at point '2').

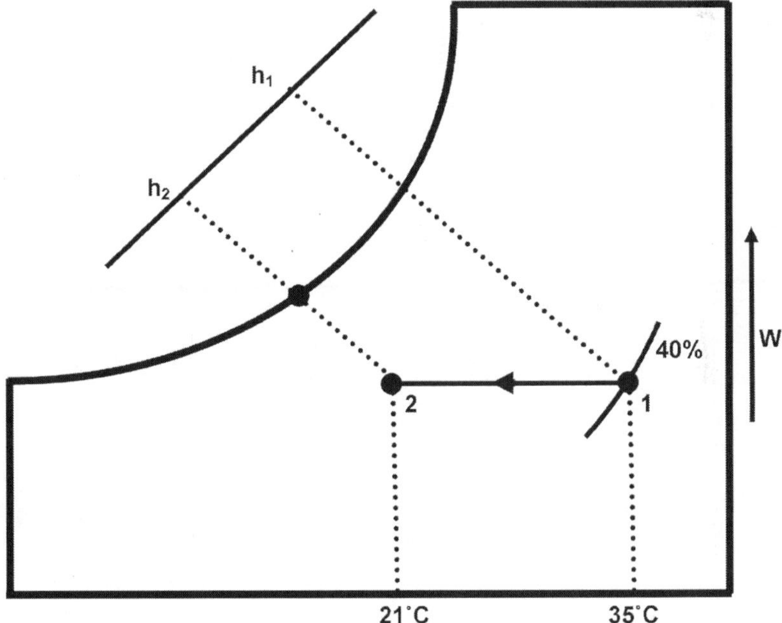

Enthalpy at point '1' (h_1) = 72.5 kJ/kg dry air

Enthalpy at point '2' (h_2) = 57.0 kJ/kg dry air

Specific volume at point '2' (v_2) = 0.852 m³/kg dry air

Volumetric flow rate (Q) at point '2' = 100 m³/min

Thus, mass flow rate at point '2' $(\dot{m}_a) = \frac{Q}{v_2} = \frac{100 \times 60}{0.852} \approx 7042$ kg/h

Sensible heat removed = $\dot{m}_a(h_1 - h_2) = 7042 \times (72.5 - 57) = 1{,}09{,}151$ kJ/h

17.4.2 Heating and Humidification of Moist Air

The heating and humidification process is commonly employed in winter air conditioning to warm and humidify the air. It increases the dry-bulb temperature and specific humidity of the air. The final relative humidity can be lower or higher compared to the entering air. It is the most common process that occurs with environmental control of agricultural buildings during cold climatic conditions.

During hot weather conditions, removal of sensible heat from livestock buildings is the critical problem, and water evaporation can be helpful. This is opposite of cold weather conditions. Sufficient ventilation air must be moved through the livestock buildings to carry out the sensible heat with a minimum temperature difference between inside and outside.

By understanding the properties of moist air and utilizing psychrometric charts, engineers and designers can analyse and optimize heating and humidification processes to achieve desired temperature and humidity conditions for maintaining comfortable and healthy indoor environments. Figure 17.5 illustrates the heating and humidification process on a skeleton psychrometric chart.

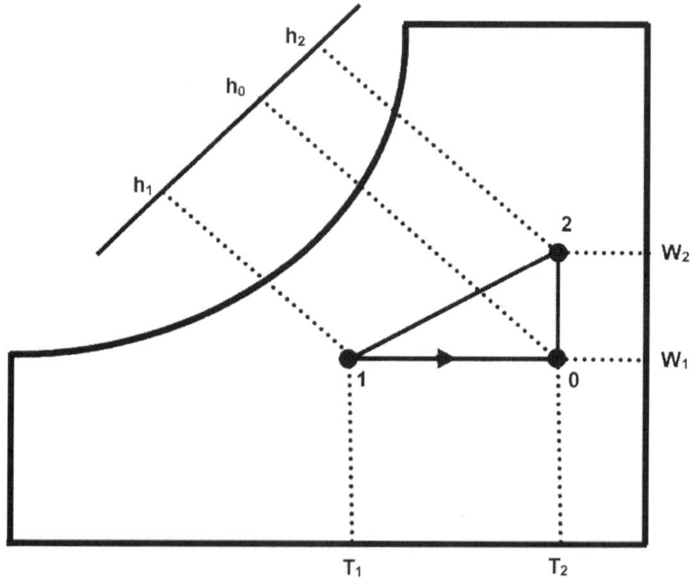

Figure 17.5. Heating and Humidification Process.

Example 17.3

Moist air at 4°C dry-bulb and 70 per cent relative humidity is brought into a growing house for hogs. The air is removed with a 400 m³/min exhaust fan at 18°C dry-bulb and 15°C wet-bulb. Calculate the amount of sensible and latent heat added to the building per hour?

Solution

Using a psychrometric chart, locate intersection point '1' corresponding to 4°C DBT and 70 per cent RH, and point '2' corresponding to 18°C dry-bulb and 15°C wet-bulb. Draw straight lines from points '1' and '2' to find out the intersection point '3'. Read enthalpy at points '1', '2' and '3' and specific volume at point '2' (since fan is located at point '2').

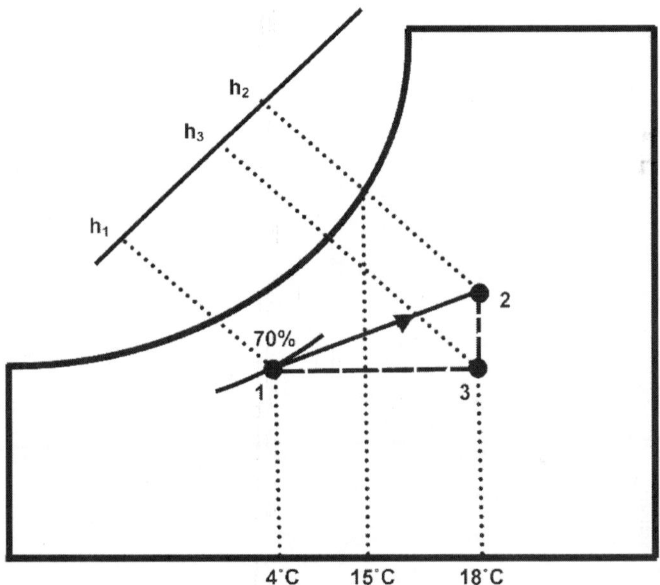

Enthalpy at point '1' (h_1) = 12.8 kJ/kg dry air

Enthalpy at point '2' (h_2) = 42.0 kJ/kg dry air

Enthalpy at point '3' (h_3) = 27.0 kJ/kg dry air

Specific volume at point '2' (v_2) = 0.836 m³/kg dry air

Volumetric flow rate (Q) at point '2' = 400 m³/min

Thus, mass flow rate at point '2' $(\dot{m}_a) = \frac{Q}{v_2} = \frac{400 \times 60}{0.836} \approx 28{,}708$ kg/h

Total heat added = $\dot{m}_a(h_2 - h_1) = 28708 \times (42 - 12.8) \approx 8{,}38{,}274$ kJ/h

Sensible heat added = $\dot{m}_a(h_3 - h_1) = 28708\times (27 - 12.8) \approx 4{,}07{,}654$ kJ/h

Latent heat added = $\dot{m}_a(h_2 - h_3) = 28708\times (42 - 27) = 4{,}30{,}620$ kJ/h

Example 17.4

In summer, a poultry house receives air of 30°C DBT and 40 per cent RH. The exhaust fans removed air at 60 m³/s at 35°C DBT and 27°C WBT. Calculate the ratio of sensible heat to total heat removed and water removal per hour.

Solution

Using a psychrometric chart, locate intersection point '1' corresponding to 30°C DBT and 40 per cent RH, and point '2' corresponding to 35°C dry-bulb and 27°C wet-bulb. Draw straight lines from points '1' and '2' to find out the intersection point '3'. Read enthalpy at points '1', '2' and '3', humidity ratio at points '1' and '2', and specific volume at point '2' (since fan is located at point '2').

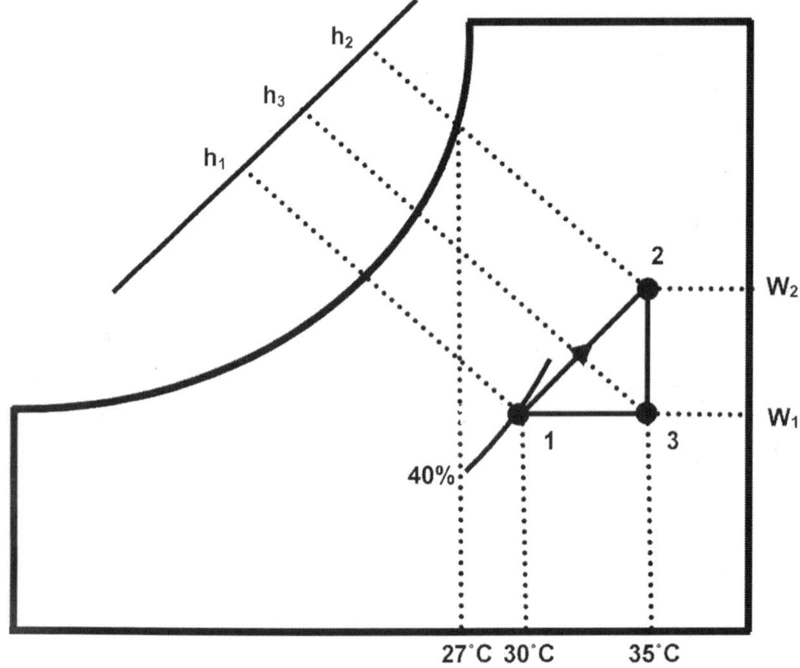

Enthalpy at point '1' (h_1) = 59.0 kJ/kg dry air

Enthalpy at point '2' (h_2) = 85.0 kJ/kg dry air

Enthalpy at point '3' (h_3) = 63.0 kJ/kg dry air

Humidity ratio at point '1' (W_1) = 0.0108 kg/kg dry air

Humidity ratio at point '2' (W_2) = 0.0192 kg/kg dry air

Specific volume at point '2' (v_2) = 0.90 m³/kg dry air

Volumetric flow rate (Q) at point '2' = 60 m³/s

Thus, mass flow rate at point '2' $(\dot{m}_a) = \frac{Q}{v_2} = \frac{60 \times 3600}{0.90} = 2{,}40{,}000$ kg/h

Total heat removed = $\dot{m}_a(h_2 - h_1) = 240000 \times (85 - 59) = 62{,}40{,}000$ kJ/h

Sensible heat removed = $\dot{m}_a(h_3 - h_1) = 240000 \times (63 - 59) = 9{,}60{,}000$ kJ/h

Latent heat removed = $\dot{m}_a(h_2 - h_3) = 240000 \times (85 - 63) = 52{,}80{,}000$ kJ/h

Ratio of sensible to the total heat removed = $\frac{9{,}60{,}000}{62{,}40{,}000} = 0.153$

It means that 15.3 per cent of the total heat was removed as sensible heat.

Mass of water evaporated = $\dot{m}_a(W_2 - W_1) = 240000 \times (0.0192 - 0.0108)$

$= 2016$ kg/h

17.4.3 Adiabatic Mixing of Two Steady Air Streams

Adiabatic mixing is a process where two steady air streams with different properties are combined without an exchange of either sensible or latent heat with the surroundings. In this process, the total energy content of the mixed air remains constant. The resulting mixture will have a new dry-bulb temperature, wet-bulb temperature, specific humidity, and relative humidity. The psychrometric chart used to assess the state of the mixed air by locating the conditions of two air streams on the chart. The two points 1 and 2 are joined by a straight line which is then divided in inverse proportion to the weights of the individual air quantities. Figure 17.6 shows schematically the process of adiabatic mixing of two steady air streams.

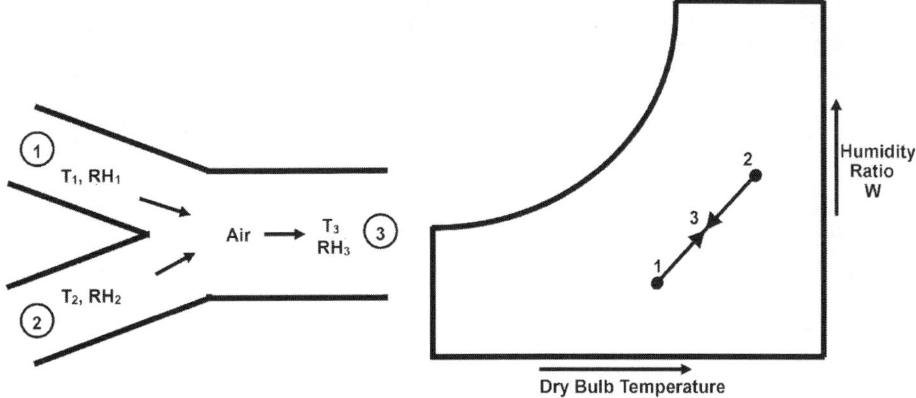

Figure 17.6. Adiabatic Mixing of Two Air Streams.

The enthalpy h_3 and humidity ratio W_3 of the mixed stream at state point 3 may also be readily calculated with the help of the following equations:

Mass balance: $m_1 + m_2 = m_3$

Enthalpy balance: $m_1h_1 + m_2h_2 = m_3h_3$

Water vapour balance: $m_1W_1 + m_2W_2 = m_3W_3$

On solving, we get

$$\frac{m_1}{m_2} = \frac{h_3 - h_2}{h_1 - h_3} = \frac{W_3 - W_2}{W_1 - W_3}$$

$$\text{Thus, } h_3 = \frac{m_1h_1 + m_2h_2}{m_1 + m_2}$$

$$W_3 = \frac{m_1W_1 + m_2W_2}{m_1 + m_2}$$

Example 17.5

A stream of air at 45°C and 10 per cent relative humidity merges with a stream at 30°C and 90 per cent relative humidity. The temperature of the mixture is 36°C.

1. What is the relative humidity of the mixture?
2. If the higher temperature air stream has a rate of 200 m³ per minute, what is the rate of the lower temperature stream?

Use psychrometric chart.

Solution

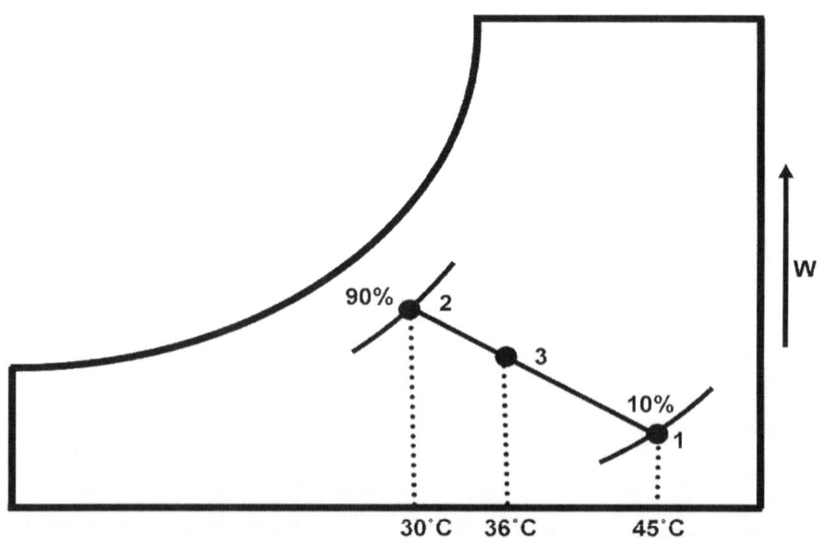

Using a psychrometric chart, locate intersection point '1' corresponding to 45°C dry-bulb temperature and 10 per cent relative humidity. Also, locate intersection point '2' corresponding to 30°C DBT and 90 per cent RH. Draw a straight line joining points '1' and '2'. Draw a vertical line through 36°C till it meets the straight line at point '3'. Point '3' indicates the state of mixture. Read relative humidity value at point '3'; specific volume at points '1' and '2'; and enthalpy at points '1', '2' and '3'.

1. Relative humidity of the mixture *i.e.* relative humidity at point '3' = 47 per cent
2. Specific volume at point '1' (v_1) = 0.91 m³/kg dry air

   Specific volume at point '2' (v_2) = 0.898 m³/kg dry air

   Volumetric flow rate at point '1' = 200 m³/min

   Thus, mass flow rate at point '1' (m_1) = 200/0.91 = 219.8 kg/min

   Enthalpy at point '1' (h_1) = 61.5 kJ/kg dry air

   Enthalpy at point '2' (h_2) = 93.0 kJ/kg dry air

   Enthalpy at point '3' (h_3) = 81.0 kJ/kg dry air

   Energy balance

   $$m_1h_1 + m_2h_2 = m_3h_3 = (m_1 + m_2)h_3$$

   $$219.8 \times 61.5 + m_2 \times 93 = (219.8 + m_2) \times 81$$

   or, m_2 = 357.2 kg/min

   Thus, volumetric flow rate at point '2' = mass flow rate × specific volume

   = 357.2 × 0.898 = 320.7 m³/min

17.4.4 Evaporative Cooling

Evaporative cooling is a process where heat is removed from the air through the evaporation of water. In this process, sensible heat is converted to latent heat, and there is no heat exchange with the surroundings. This process is commonly used for the cooling of agricultural buildings under certain hot weather conditions. This process can be represented as a diagonal line on the psychrometric chart (Figure 17.7).

The process of evaporative cooling involves passing dry air over a water source, typically in the form of a wetted surface or through a water spray. As the dry air comes into contact with the water, it absorbs moisture through evaporation, resulting in a decrease in the dry-bulb temperature and an increase in the specific humidity of the air. The wet-bulb temperature remains constant during the adiabatic saturation process. The effectiveness of evaporative cooling depends on factors such as the temperature and humidity of the incoming air, the water temperature, the surface area of the wetted surface, and the airflow rate. These factors influence the amount of heat that can be removed through evaporation and the resulting cooling effect. Evaporative coolers are commonly used in arid or dry climates where the humidity is relatively low. They are energy-efficient compared to traditional air conditioning systems since they do not require the use of refrigerants or compressors.

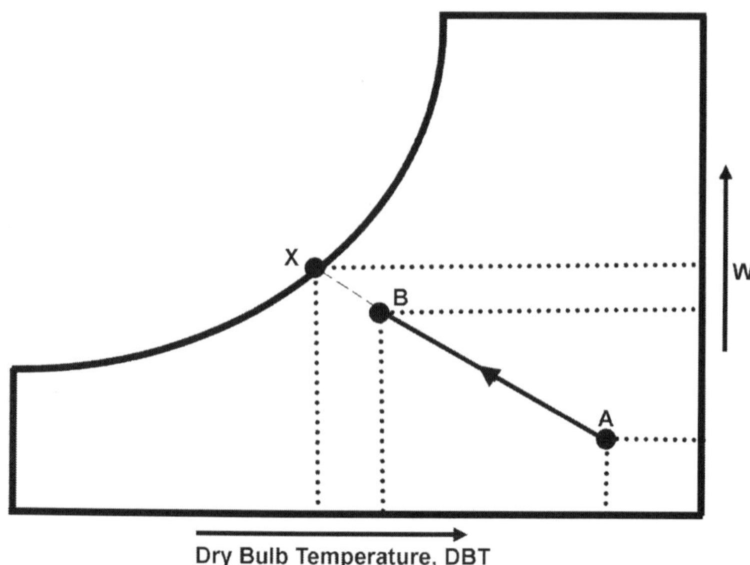

Figure 17.7. Evaporating Cooling Process on a Psychrometric Chart.

Example 17.6

The air is drawn from ambient of 40°C DBT and 40 per cent RH using an evaporative cooler at 8 m^3/s. If the effectiveness of the cooler is 80 per cent, then calculate the exit air temperature from the cooler. Also, calculate the ratio of sensible heat and latent heat of the cooler.

Solution

Using a psychrometric chart, locate intersection point '1' corresponding to 40°C DBT and 40 per cent RH. Move along the constant enthalpy line and read the wet-bulb temperature (T_{wb}).

Wet-bulb temperature corresponding to point '1' = 27.8°C

Effectiveness = 80 per cent = 0.8

$$\text{Exit dry-bulb temperature} = T_{db} - \text{effectiveness} \times (T_{db} - T_{wb})$$
$$= 40 - 0.8(40 - 27.8) \approx 30.2^{\circ}\text{C}$$

Locate intersection point '2' corresponding to 30.2°C dry-bulb temperature. Draw straight lines from points '1' and '2' to find out the intersection point '3'. Read enthalpy at points '1', '2' and '3' and specific volume at point '1'.

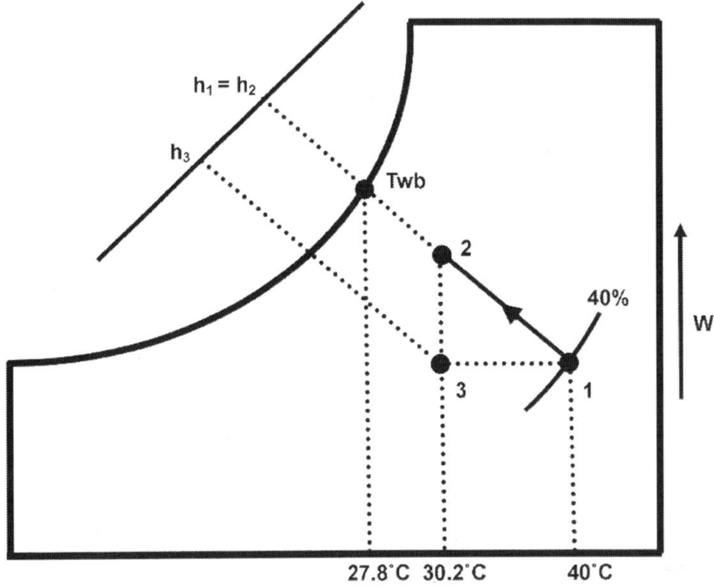

Enthalpy at point '1' (h_1) = 89.0 kJ/kg dry air

Enthalpy at point '2' (h_2) = 89.0 kJ/kg dry air

Enthalpy at point '3' (h_3) = 78.0 kJ/kg dry air

Specific volume at point '1' (v_1) = 0.915 m³/kg dry air

Volumetric flow rate (Q) at point '1' = 8 m³/s

Thus, mass flow rate at point '1' $(\dot{m}_a) = \dfrac{Q}{v_2} = \dfrac{8 \times 3600}{0.915} \approx 31{,}475$ kg/h

Sensible heat = $\dot{m}_a(h_1 - h_3) = 31475 \times (89 - 78) = 3{,}46{,}225$ kJ/h

Latent heat = Sensible heat = 3,46,225 kJ/h (being an adiabatic process)

Important Points to Remember

1. The ambient air temperature as recorded by ordinary thermometer is called dry-bulb temperature.
2. Specific humidity is also called as humidity ratio.
3. Specific humidity is defined as mass of water vapour contained in air-vapour mixture per kg of dry air.
4. Removal of moisture from the air, without change in its dry-bulb temperature is called dehumidification.

5. The addition of moisture to the air, without change in its dry-bulb temperature is called humidification.
6. For unsaturated air, the dew-point temperature is less than wet-bulb temperature.
7. Moist air is a mixture of dry air and water vapour.
8. Humidity is the mass of water vapour present in 1 kg of dry air.
9. The difference between the dry-bulb temperature and dew-point temperature of air is called dew-point depression.
10. The specific volume lines are inclined straight lines and uniformly spaced.
11. Relative humidity lines in psychrometry are curved lines and follow the saturation curve.
12. The heating of air without any change in its specific humidity is known as sensible heating.
13. Heat addition to the air without changing its humidity ratio is termed as sensible heat.
14. When wet-bulb temperature and dry-bulb temperature are same, relative humidity of air is 100 per cent.
15. Temperature at which condensation of water present in the air starts is dew-point.
16. Sling psycrometer/hygrometer is used for measurement of relative humidity.
17. The adiabatic process takes place in constant enthalpy of the working medium.
18. Psychrometric chart can be used if at least two properties of air are known.
19. When relative humidity is 100 per cent, dry-bulb temperature is equal to wet-bulb temperature of air.
20. When dry-bulb temperature is more than wet-bulb temperature, air is unsaturated.
21. Heat change in air at constant dry-bulb temperature is latent heat.
22. Psychrometer is used for measuring dry-bulb and wet-bulb temperatures.
23. For saturated air, dew-point temperature is equal to the wet-bulb temperature.
24. In a psychrometric chart, adiabatic cooling process is indicated by a straight line having constant enthalpy.
25. When relative humidity is 100 per cent, wet-bulb depression is zero.
26. During sensible cooling of air, the dry-bulb temperature decreases.

27. Heating and humidification process is generally used in winter air conditioning to warm and humidify the air.
28. On the psychrometric chart, sensible heating/cooling is indicated by straight horizontal line having constant humidity ratio.
29. The difference between the dry-bulb temperature and wet-bulb temperature of air is called wet-bulb depression.
30. The evaporative cooling is limited by wet-bulb temperature of entering air stream.

Exercises/Questions

1. Define the following psychrometric terms:

   (a) Dry-bulb temperature, (b) Wet-bulb temperature, (c) Degree of saturation, (d) Humidity ratio, (e) Dew-point temperature, (f) Relative humidity, (g) Specific volume, and (h) Enthalpy.

2. Draw a neat skeleton diagram of psychrometric chart and show all the lines.
3. What is psychrometry? Discuss various psychrometric processes with the help of a neatly illustrated psychrometric chart.
4. With the help of psychrometric chart, explain the following process:
   (a) Sensible heating and sensible cooling processes
   (b) Heating and humidification process
   (c) Adiabatic mixing process
   (d) Evaporative cooling
5. What is evaporative cooling? List its scope of applications in farm buildings.
6. An air/vapour mixture at atmospheric pressure has a dry-bulb temperature of 35°C and a wet- bulb temperature of 21°C. How much water is evaporated adiabatically until the dry-bulb temperature is 25°C.

   [**Answers**: 4 g/kg dry air]

7. A quantity of air having a volume of 300 m^3 at 10°C dry-bulb temperature and 7°C wet- bulb temperature is heated at 20°C dry-bulb temperature. Estimate the amount of heat added, final relative humidity and wet-bulb temperature. The air pressure is 1.01325 bar.

   [**Answers**: 4260 kJ, 32 per cent, 11.5°C]

8. Outside air with a dry-bulb temperature of 30°C and 30 per cent relative humidity is brought into a vegetable storage chamber through an evaporative cooler and adiabatically cooled to 70 per cent relative

humidity. There is a gain of 3000 kJ/hour as sensible heat inside the room and the temperature is maintained at 28°C. How much air flow is required? What is the relative humidity of outgoing air? What is the mass of water evaporated per hour?

[**Answers**: 401 m^3/h, 48 per cent, 1.57 kg/h]

Chapter 18

Heat and Vapour Transfer

Farm structures play a pivotal role in modern farming practices, providing shelter and controlled environments for crops and livestock. The application of basic principles of heat and vapour transfer within these structures helps in optimizing agricultural productivity, ensuring crop health, and maintaining livestock well-being.

Heat transfer in farm structures is an important aspect of agricultural operations as it directly affects the comfort and productivity of livestock, crop growth, and overall energy efficiency. Three primary modes of heat transfer are conduction for solids, convection for fluids, and radiation for distant objects by electromagnetic radiation.

Moisture condensation in walls is a common issue in buildings and structures, posing potential problems such as mould growth, structural damage, and reduced insulation efficiency. Condensation occurs when warm, moisture-laden air comes into contact with a cooler surface, causing water vapour to transform into liquid water.

In this chapter, the principles of heat and vapour transfer through the farm structures are discussed.

18.1 Conduction

Conduction is the transfer of heat through direct contact between objects or materials. In farm structures, conduction occurs when heat flows through the building materials, such as walls, floors, and roofs. Thermal conductivity of the materials used decides the rate of conduction. Insulation materials with low thermal conductivity, such as foam insulation or double-pane windows, can reduce heat transfer by conduction.

Heat transfer by conduction is the process of transferring heat between two objects or regions that are in direct contact with each other. In this mode of heat transfer, the heat energy flows from the region of higher temperature to the region of lower temperature, primarily through molecular collisions.

The rate of heat conduction is influenced by the thermal conductivity of the materials, temperature gradient, cross-sectional area, and thickness of the material.

Thermal Conductivity

Thermal conductivity is a property of the material and represents its ability to conduct heat. Materials with high thermal conductivity, such as metals, conduct heat more effectively than materials with low thermal conductivity, such as insulators. To enhance or inhibit heat conduction, materials can be selected based on their thermal properties. For example, in building insulation, materials with low thermal conductivity are used to minimize heat transfer through walls and roofs. Conversely, materials with high thermal conductivity, like copper or aluminium, are used in heat sinks to efficiently dissipate heat from electronic devices.

Thermal conductivity is typically denoted by the symbol "k" and is measured in units of watts per meter-kelvin (W/m-K). It represents the amount of heat (in watts) that can be conducted through a one-meter-thick material with a one-degree temperature difference between its two surfaces. Different materials have varying thermal conductivities due to differences in their atomic or molecular structures. The thermal conductivity is also influenced by other factors such as temperature, density, size of fibres or particles, degree of bond between particles and their arrangement within the material, and moisture content. Additionally, some materials, such as ceramics, can have anisotropic thermal conductivity, meaning it varies with direction. The thermal conductivity values of common materials are given in Table 18.1.

Table 18.1. Thermal Conductivity of some Common Materials at Normal Pressure and Temperature

Material	*Thermal conductivity (W/m-K)*
Construction materials	
Concrete	1.20
Brick (masonry)	0.65
Brick (first clay)	0.75 – 1.75
Earth	0.138
Glass (window)	0.75
Plaster	0.75 – 0.95
Wood	0.28

Material	*Thermal conductivity (W/m-K)*
Insulating materials	
Asbestos sheet	0.17
Cork, felt	0.05 – 0.10
Cardboard	0.07
Glass wool	0.03
Saw dust	0.07
Fibreboard	0.052
Polystyrene foam	0.036
Polyurethane foam	0.026
Wood shavings	0.09
Metals	
Aluminium	225
Brass	107
Copper	385
Cast iron	55 – 65
Steel	20 – 45
Miscellaneous	
Air	0.024
Ash	0.12
Ice	2.25
Water	0.55 – 0.70

Thermal Conductance

The thermal conductance is the quantity of heat that will pass in unit time, through unit area of a specified thickness of material, under unit temperature difference. It can be written as follows:

$$C = \frac{k}{x}$$

where,

C = conductance of the material in W/m^2-°C,

k = thermal conductivity of material in W/m-°C, and

x = thickness of material in meter.

Thermal conductance is sometimes used instead of thermal conductivity. It gives the heat flow for the entire thickness of a material, and is more general than thermal conductivity.

Fourier's Law

Fourier's law of heat conduction is a fundamental principle in the field of thermodynamics that describes the rate at which heat is transferred through a material. It was formulated by the French mathematician and physicist Jean-Baptiste Joseph Fourier.

According to Fourier's law, the heat flux (Q) through a material is directly proportional to the negative gradient of the temperature (dT/dx) in the direction of heat flow, and it is also proportional to the cross-sectional area (A) perpendicular to the heat flow. Mathematically, it can be expressed as:

$$Q = -kA\frac{dT}{dx} = \frac{kA(T_1 - T_2)}{x}$$

where,

Q = heat flux in Watts (J/s),

k = thermal conductivity of the material in W/m-°C,

(dT/dx) = temperature gradient (change in temperature per unit distance), and

A = cross-sectional area of flow path in m^2

x = thickness of the wall, m

T_1 and T_2 = temperature of the wall surface *i.e.* inside and outside, °C

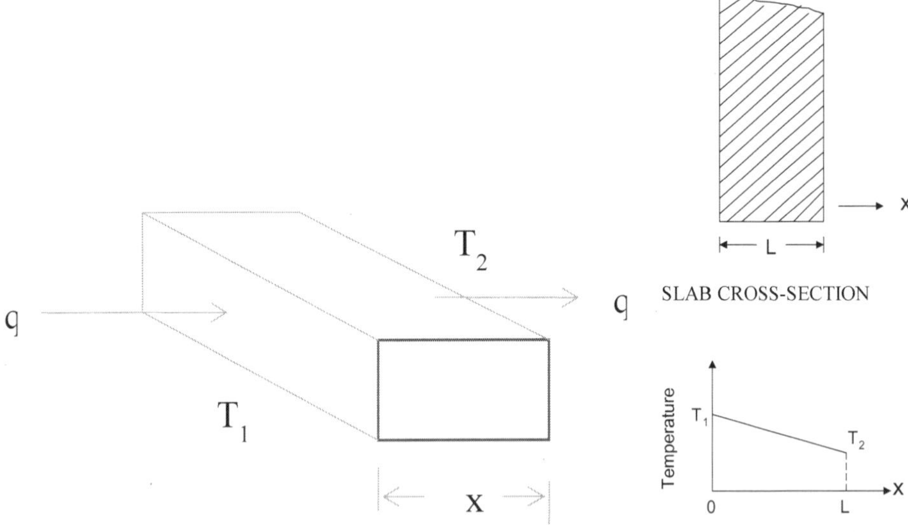

Figure 18.1. Heat Conduction through a Plane Wall.

Figure 18.2. Steady State Temperature Distribution Across a Plane Wall.

The equation explains that heat flux is directly proportional to the temperature gradient. The heat flows from higher to lower temperatures. The negative sign indicates that the heat flows in the opposite direction of the temperature increase.

Fourier's law is applicable to conductive heat transfer, which occurs primarily in solids and stationary fluids. It assumes that the material is homogeneous and isotropic (meaning its properties are the same in all directions), and that the heat transfer occurs in a steady state with no internal heat generation or phase change.

Example 18.1

Determine the heat flow through 1 m² of a wall of 20 cm thickness. The temperature is 2°C on one side and 25°C on the other. The thermal conductivity of the wall is 0.1 W/m-K.

Solution

Thickness of wall (x) = 20 cm = 0.2 m

Area of the wall (A) = 1 m²

Temperature gradient (ΔT) = 25 – 2 = 23°C

Thermal conductivity of the wall = 0.1 W/m-K

$$q = kA\frac{\Delta T}{x}$$

or, $q = 0.1 \times 1 \times \frac{23}{0.2} = 11.5$ W

Heat Transfer through Composite Wall

A composite wall consists of multiple layers of different materials with varying thermal conductivities. The temperature gradient across the wall drives the heat transfer from one side to the other.

The overall heat transfer through a composite wall can be calculated using the thermal resistance concept. The thermal resistance (R) of each layer is given by the ratio of the thickness to the thermal conductivity (R = x/k). For a composite wall, the total thermal resistance is the sum of the individual resistances of each layer.

When, there are three surfaces in contact with each other, then

$$\frac{Q}{A} = \frac{T_1 - T_4}{\frac{x_1}{k_1} + \frac{x_2}{k_2} + \frac{x_3}{k_3}} = \frac{\Delta T}{R} = \frac{Thermal\,potential}{Overall\ thermal\ resis\tan ce}$$

or, $Q = UA\Delta T$

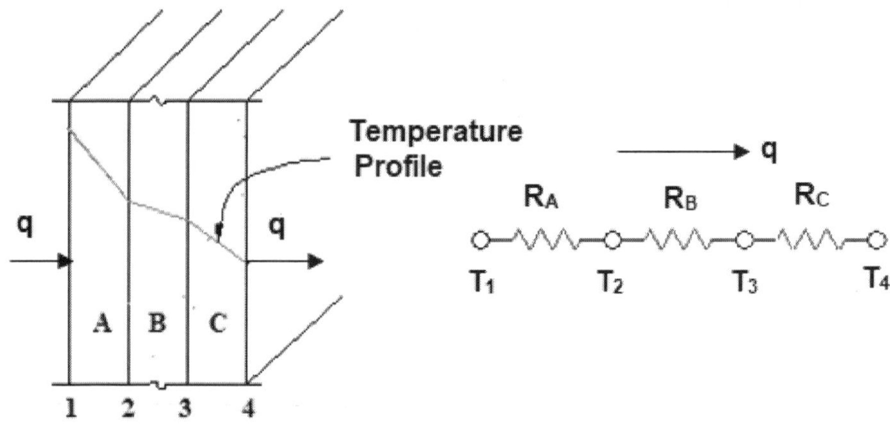

Figure 18.3. Steady State Conduction through a Composite Wall.

where,

$$R = \frac{1}{U} = \frac{x_1}{k_1} + \frac{x_2}{k_2} + \frac{x_3}{k_3} = R_A + R_B + R_C$$

Q = heat transfer rate in Watts,

T_1 = temperature on one side of the wall in degrees Celsius or Kelvin,

T_2 = temperature on the other side of the wall in degrees Celsius or Kelvin,

R_A, R_B.., R_n = thermal resistances of the individual layers,

R = overall thermal resistance per unit area of the wall,

A = cross-sectional area of the layer in square meters, and

U = overall heat transfer coefficient in W/m²-K.

By rearranging the equation, one can also calculate the heat transfer per unit area (q) by dividing Q by the surface area (A).

This equation assumes steady-state heat transfer and uniform properties within each layer of the composite wall.

Temperature at the Interface

The temperature at any interface of a material or wall may be calculated if certain information is available. In the steady state, the same quantity of heat per unit time must pass through each layer.

$$\therefore \quad Q = \frac{k_1 A_1 \Delta T_1}{x_1} = \frac{k_2 A_2 \Delta T_2}{x_2} = \frac{k_3 A_3 \Delta T_3}{x_3} = ...$$

$$\text{or, } Q = \frac{T_1 - T_2}{R_1} = \frac{T_2 - T_3}{R_2} = \frac{T_3 - T_4}{R_3} = \frac{T_1 - T_4}{R} = \frac{T_1 - T_x}{R_x}$$

Thus, $T_x = T_1 - \frac{R_x}{R}(T_1 - T_4)$

The above equation may be used to determine temperature at any interface within the wall or materials.

The sum of the temperature differences over each layer is equal to the difference in temperature of the two outside surfaces of the complete system, *i.e.*

$\Delta T_1 + \Delta T_2 + \Delta T_3 + \ldots\ldots = \Delta T$

where,

$\Delta T_1' = T_1 - T_2$

$\Delta T_2 = T_2 - T_3$

$\Delta T_3 = T_3 - T_4$

Heat Flow between Surface and Surroundings

The moving fluid over a stationary surface forms a stagnant boundary film with no relative velocity. The heat in this layer is transferred by conduction and convection processes. Although convection is the primary process due to the low fluid thermal conductivity. The rate of convective heat transfers between solid boundaries like walls or ceilings and adjacent fluids is determined by the Newton-Rikhman law.

$Q = h_s A(T_a - T_s)$

where,

h_s = surface heat transfer coefficient in W/m²-K,

T_a = temperature of the fluid in °C, and

T_s = temperature of the surface in °C.

The surface heat transfer coefficient (h_s) in a composite wall is determined by the presence of another layer at the surface. The factor ($1/h_s A$) represents the thermal resistance of the film in the heat transfer process. Heat passes through the surface, composite slab elements, and then surface film. The equation for overall heat flow is

$Q = UA(T_i - T_o)$

where,

Q = overall heat flow in watts,

U = overall heat transfer coefficient from air to air in W/m²-K,

A = area normal to direction of heat flow,

T_i = inside air temperature in °C, and

T_o = outside air temperature in °C.

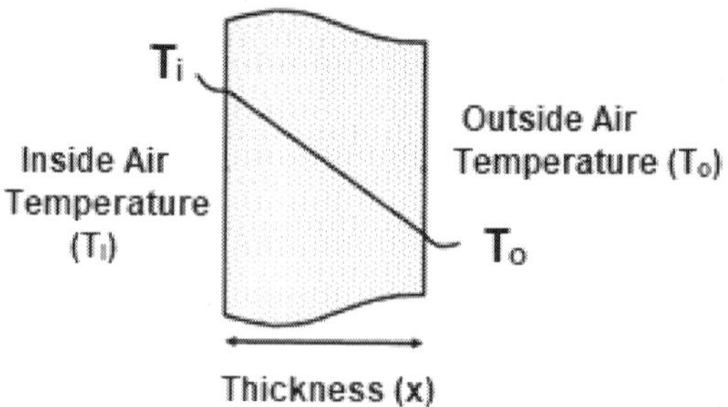

Figure 18.4. Temperature Gradient from Air to Air through a Single Solid Material.

The overall coefficient of heat transmission (U) may be calculated as follows:

$$\frac{1}{U}=\frac{1}{h_i}+\frac{x_1}{k_1}+\frac{x_2}{k_2}+\frac{x_3}{k_3}+.........+\frac{1}{h_o}$$

where,

h_i and h_o are the surface coefficients on either side of the composite slab in W/m^2-K.

Example 18.2

Heat is flowing through a vertical wall made of concrete. The thickness and area of concrete layer are 100 mm and 100 m^2 respectively. The convective heat transfer coefficients for outside and inside layers of air are respectively 100 W m^{-2} K^{-1} and 10 W m^{-2} K^{-1}. The thermal conductivity of concrete layer is 0.76 W m^{-1} K^{-1}. Determine the heat flow through a concrete layer when the temperature difference between inside and outside surfaces is 10°C.

Solution

Thermal conductivity of concrete (*k*) = 0.76 W/m-K

Thickness of concrete layer (x) = 0.1 m

Area (A) = 100 m^2

Convective heat transfer coefficient of air (h_i) = 10 W/m^2-K

Convective heat transfer coefficient of air (h_o) = 100 W/m^2-K

Temperature difference between inside and outside (T_i-T_o) = 10°C

Heat flow, $Q = UA(T_i - T_o)$

Overall heat transfer coefficient (U)

$$\frac{1}{U}=\frac{1}{h_i}+\frac{x}{k}+\frac{1}{h_o}$$

$$\text{or, } \frac{1}{U} = \frac{1}{10} + \frac{0.1}{0.76} + \frac{1}{100}$$

or, U ≈ 4.14 W/m²-K

$$Q = UA(T_i - T_o)$$

$$= 4.14 \times 100 \times 10 = 4140 \text{ W}$$

Example 18.3

A cold store has a wall made of 11 cm brick, 7.5 cm concrete, and 10 cm cork. The desired temperature of a cold store is -18°C, whereas the outside temperature of a brick wall is 20°C. Calculate 1) inside and outside wall temperatures 2) Interface temperatures at brick, concrete, and cork. The thermal conductivities for brick, concrete, and cork are 0.69, 0.76, and 0.043 W/m-K, respectively. The inside and outside heat transfer coefficients are 10 W/m² K.

Solution

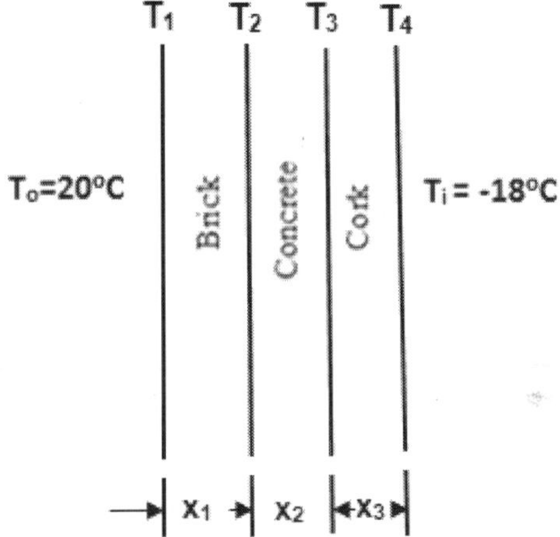

$$\text{Overall resistance, } R = \frac{1}{U} = \frac{1}{h_o} + \frac{x_1}{k_1} + \frac{x_2}{k_2} + \frac{x_3}{k_3} + \frac{1}{h_i}$$

$$= \frac{1}{10} + \frac{0.11}{0.69} + \frac{0.075}{0.76} + \frac{0.10}{0.043} + \frac{1}{10}$$

$$= 0.1 + 0.159 + 0.098 + 2.32 + 0.1 = 2.777 \text{ m}^2 \text{ K W}^{-1}$$

Temperature at a distance 'x'

$$T_x = T_o - \frac{R_x}{R}(T_o - T_i)$$

Outside Wall Temperature (T_1)

Place x as the outside surface

$$R_x = \frac{1}{h_o} = \frac{1}{10} = 0.1 \text{ m}^2\text{-K/W}$$

$$T_x = 20 - \frac{0.1}{2.777}(20-(-18)) = 18.63°\text{C}$$

or, T_1 = 18.63°C, *i.e.* outside wall temperature

Inside Wall Temperature (T_4)

Place x as the inside surface

$$R_x = \frac{1}{h_o} + \frac{x_1}{k_1} + \frac{x_2}{k_2} + \frac{x_3}{k_3}$$

$$= \frac{1}{10} + \frac{0.11}{0.69} + \frac{0.075}{0.76} + \frac{0.10}{0.043} = 2.677 \text{ m}^2\text{-K/W}$$

$$T_x = 20 - \frac{2.677}{2.777}(20-(-18)) = -16.63°\text{C}$$

or, T_4 = - 16.63°C, *i.e.* inside wall temperature

Interface Temperature between Brick and Concrete (T_2)

Place x between brick and concrete

$$R_x = \frac{1}{h_o} + \frac{x_1}{k_1}$$

$$= \frac{1}{10} + \frac{0.11}{0.69} = 0.259 \text{ m}^2\text{-K/W}$$

$$T_x = 20 - \frac{0.259}{2.777}(20-(-18)) = 16.45°\text{C}$$

or, T_2 = 16.45°C, *i.e.* interface temperature between brick and concrete

Interface Temperature between Concrete and Cork (T_3)

Place x between brick and concrete

$$R_x = \frac{1}{h_o} + \frac{x_1}{k_1} + \frac{x_2}{k_2}$$

$$= \frac{1}{10} + \frac{0.11}{0.69} + \frac{0.075}{0.76} = 0.357 \text{ m}^2\text{-K/W}$$

$$T_x = 20 - \frac{0.357}{2.777}(20 - (-18)) = 15.11^{\circ}\text{C}$$

or, $T_3 = 15.11^{\circ}$C, *i.e.* interface temperature between concrete and cork

18.2 Convection

Heat transfer by convection is a process in which heat is transferred from one place to another through the movement of a fluid, such as a liquid or a gas. In farm structures, convection plays a role in maintaining proper ventilation and air circulation. Good airflow is crucial to remove excess heat, moisture, and odours, as well as to distribute heat evenly during colder periods. Ventilation systems, fans, and natural air movement through openings like windows and vents help facilitate convective heat transfer.

There are two types of convection: natural convection and forced convection.

1. **Natural Convection**: Natural convection occurs when the fluid motion is caused by density differences in the fluid due to temperature variations. When a fluid is heated, it becomes less dense and rises, creating a convection current. As the fluid moves away from the heat source, it cools down, becomes denser, and sinks back. This cyclic process forms a convection loop, transferring heat from the hotter region to the cooler region. An everyday example of natural convection is the rising of hot air and the sinking of cold air in a room.
2. **Forced Convection**: Forced convection involves the use of an external force, such as a fan or a pump, to create fluid motion and enhance heat transfer. The forced motion of the fluid increases the rate of heat transfer compared to natural convection. In forced convection, the fluid is forced to flow over a surface or through a conduit, carrying the heat away with it. Common examples of forced convection include the use of fans in HVAC systems to distribute heated or cooled air and the circulation of coolant in car engines.

The rate of heat transfer by convection depends on the fluid properties, temperature difference, surface area, flow velocity, and shape and geometry of the surface.

Newton's Law of Cooling/Heating

This equation relates the rate of heat transfer (Q) by convection to the temperature difference (ΔT) between the solid surface and the fluid and the convective heat transfer coefficient (h_c):

$$Q = h_c A \Delta T$$

where,

Q = rate of heat transfer by convection in watts,

A = surface area in contact with the fluid in m^2,

ΔT = temperature difference between the surface and the fluid, and

h_c = convection heat transfer coefficient, $W/m^2\ ^oC$.

Convective heat transfer depends on fluid properties, flow conditions, and surface characteristics.

18.3 Radiation

Radiation is the transfer of heat through electromagnetic waves. In farm structures, radiation occurs when objects at different temperatures emit or absorb radiant energy. Unlike conduction and convection, which require a medium such as solids, liquids, or gases, radiation can occur in a vacuum as well. This mode of heat transfer does not rely on direct contact or movement of particles. For example, the sun radiates heat in the form of infrared radiation, which can be absorbed by building materials, livestock, or crops. Radiant barriers, reflective surfaces, and shading devices can be used to control the amount of radiant heat entering or leaving a structure.

In radiation, heat is emitted in the form of electromagnetic waves, primarily in the infrared region of the electromagnetic spectrum. These waves carry energy and can travel through empty space until they encounter a surface or are absorbed by an object.

Stefan-Boltzmann Law

This law states that the total energy radiated by an object is proportional to the fourth power of its absolute temperature. Mathematically, it is expressed as

$$Q = \varepsilon \sigma A T^4$$

where,

Q = rate of heat transfer in Watt

T = absolute temperature in K

σ = Stefan-Boltzmann constant (approx. $5.67 \times 10^{-8}\ W\ m^{-2}\ K^{-4}$)

ε = emissivity of the particular body and its value varies between 0 and 1, and

A = surface area in m^2

Emissivity: Every object has an emissivity value, which indicates how effectively it can emit and absorb thermal radiation. Emissivity ranges from 0 to 1, where 0 represents a perfect reflector (does not emit or absorb radiation) and 1 represents a perfect blackbody (absorbs and emits all radiation). Different materials and surfaces have different emissivity values. Approximate emissivity values of some common building materials is given in Table 18.2.

Table 18.2. Approximate Emissivity Values of some Common Building Materials

Material	Emissivity
Aluminium (polished)	0.04 - 0.05
Asphalt	0.90 - 0.95
Brick	0.90 - 0.95
Concrete	0.90 - 0.95
Glass	0.80 - 0.95
Paint (white enamel)	0.91
Paint (aluminium)	0.3 – 0.5
Plaster	0.90 - 0.95
Steel (galvanized)	0.15 - 0.20
Wood (unfinished)	0.85 - 0.95

Absorption and Reflection: When thermal radiation encounters a surface, it can be absorbed, reflected, or transmitted depending on the properties of the material and the angle of incidence. Darker and rougher surfaces tend to have higher absorption rates, while lighter and smoother surfaces reflect more radiation.

Net Radiant Heat Transfer between Two Surfaces

The net radiant heat transfer between two surfaces can be calculated using the Stefan-Boltzmann Law and the view factor between the surfaces. The Stefan-Boltzmann Law states that the radiant heat transfer rate between two surfaces is proportional to the fourth power of the temperature difference between them. The equation is as follows:

$$q = AF\sigma\left(T_1^4 - T_2^4\right)$$

where,

$$\frac{1}{F} = \frac{1}{\varepsilon_1} + \frac{1}{\varepsilon_2} - 1$$

ε_1 is the emissivity of the surface at temperature $T_{1,}$ and

ε_2 is the emissivity of the surface at temperature T_2

$$\text{View factor (F)} = \frac{1}{\dfrac{1}{\varepsilon_1} + \dfrac{1}{\varepsilon_2} - 1}$$

The view factor (F) between the two surfaces accounts for the geometry and orientation of the surfaces. It represents the fraction of radiation leaving one surface that reaches the other surface. The view factor ranges from 0 to 1.

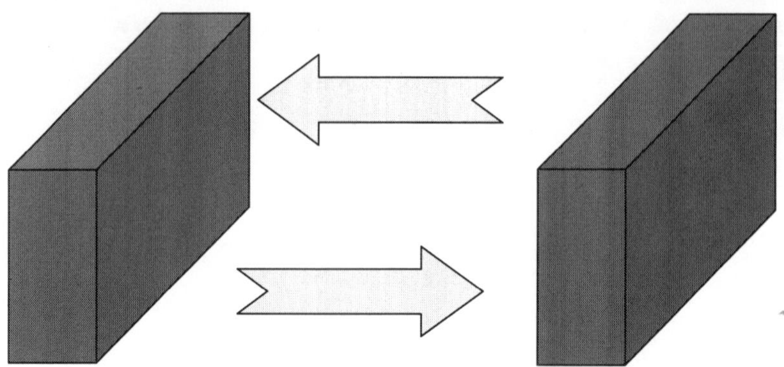

Figure 18.5. Radiation Heat Transfer between Two Surfaces

Radiation to a Small Body from its Surroundings

For a relatively small body in surroundings that are at a uniform temperature, the net heat exchange is given by the equation:

$$q = \sigma A \varepsilon \left(T_1^4 - T_2^4\right)$$

where,

ε is the emissivity of the body, T_1 is the absolute temperature of the body, and T_2 is the absolute temperature of the surroundings.

Radiation heat transfer from surface with area A_1, emissivity ε_1 and temperature T_1 to large enclosure at temperature T_2 is (surface 1 is relatively small compared to surface 2):

$$q_{12} = \sigma A_1 \varepsilon_1 \left(T_1^4 - T_2^4\right)$$

When two perfectly black surfaces 1 and 2 are exchanging radiation between them, the net heat transfer from 1 to 2 is given by:

$$q = \sigma A_1 \left(T_1^4 - T_2^4\right)$$

Net Radiant Heat Transfer between Two Gray Bodies

The earlier equation does not account for non-blackbodies. An expression that accounts for non-blackbodies and shape factor must be used. The expression would be:

$$q = AF\sigma(T_1^4 - T_2^4)$$

$$\frac{1}{F} = \frac{1}{F_{12}} + \frac{A_1}{A_2}\left(\frac{1}{\varepsilon_2} - 1\right) + \left(\frac{1}{\varepsilon_1} - 1\right)$$

where,

'F' factor accounts for both shape and emissivity.

Net Rate of Radiation Heat Transfer between the Two Cylinders

$$q = \frac{A_1 \sigma \left(T_1^4 - T_2^4\right)}{\frac{1}{\varepsilon_1} + \frac{1-\varepsilon_2}{\varepsilon_2}\left(\frac{d_1}{d_2}\right)}$$

where,

d_1 and d_2 are diameter of cylinders

Net Rate of Radiation Heat Transfer between the Two Spheres

$$q = \frac{A_1 \sigma \left(T_1^4 - T_2^4\right)}{\frac{1}{\varepsilon_1} + \frac{1-\varepsilon_2}{\varepsilon_2}\left(\frac{r_1^2}{r_2^2}\right)}$$

where,

r_1 and r_2 are radius of spheres

18.4 Total Building Heat Transmission

The overall coefficient of heat transmission (U) for each of the exposed areas (walls, ceilings, *etc.*) of a building shall be used to determine a weighted heat transfer coefficient (U_{av}). It is weighted according to the area that each portion of the building exposed surface contributing to the total. The weighted heat transfer coefficient is actually the total heat loss per degree temperature difference divided by the total area. The weighted overall coefficient of heat transmission (U_{av}) may be calculated as follows:

$$U_{av} = \frac{Q_t}{A_t}$$

where,

$Q_t = Q_1 + Q_2 + Q_3 + Q_4 + \ldots$, and

$A_t = A_1 + A_2 + A_3 + A_4 + \ldots$

Subscripts represent the different exposed surfaces of the building (walls, ceiling, doors, windows, *etc.*). Q_t is the total heat loss per °C temperature difference in W/°C, and A_t is the total exposed surface area in m². Q_1, Q_2, Q_3, Q_4, *etc.* can be calculated with the help of following relationships:

$Q_1 = U_1A_1$, $Q_2 = U_2A_2$, $Q_3 = U_3A_3$, $Q_4 = U_4A_4$, *etc.*

Exposure Factor (E.F.)

The conduction heat loss per animal or bird can be determined from the total building heat loss by dividing by the number of animals or birds in the building. The conduction heat loss per animal or bird may be written as

$$\text{E.F.} = \frac{A_t U_{av}}{N}$$

where,

E.F. = exposure factor in W/°C,

A_t = exposed area of walls, ceiling *etc.* of the building in m²,

U_{av} = average overall coefficient of heat transfer for the building in W/m²-K, and

N = number of animals or birds in the building.

The exposure factor represents the conductive heat loss from a building per animal or bird per degree temperature difference between the inside and outside. Thus, for a given building, the better it is insulated the smaller the exposure factor. Also, the greater the number of animals or birds in the building, the smaller the exposure factor.

Example 18.4

A 10 m × 50 m × 2.5 m poultry building houses 5500 laying hens, and the average overall coefficient of heat transfer for the building is 0.8 W/m²-K. What is the exposure factor?

Solution

Area of side walls = 2× (10×2.5+50×2.5)

= 300 m²

Area of ceiling = 10×50 = 500 m²

Total area of exposed surfaces (A_t) = 300 + 500 = 800 m²

Average overall coefficient of heat transfer for the building (U_{av}) = 0.8 W/m²-K

Number of laying hens (N) = 5500

$$\text{Exposure Factor} = \frac{A_t U_{av}}{N} = \frac{800 \times 0.8}{5500}$$

= 0.116 W/K (Hen)

18.5 Moisture Condensation in Walls

Moisture transfer through solid walls is a fundamental phenomenon that plays a crucial role in the overall performance and durability of buildings. Water vapour molecules move from an area of high vapour pressure to an area of low vapour pressure through the walls and ceiling. This leads to a potential issue of condensation within the wall, which may result in moisture accumulation. Understanding and managing moisture transfer is essential for maintaining a healthy and comfortable indoor environment, preventing structural damage, and mitigating the risk of mould growth and other moisture-related problems.

Controlling condensation in walls is essential to prevent moisture-related problems. Proper insulation is crucial in preventing condensation. Insulation helps maintain a consistent temperature on the wall surface, reducing the chances of moisture condensing. Ensure that the walls are adequately insulated, especially in colder climates. Condensation can also be controlled by installing a vapour barrier. A vapour barrier is a material that prevents moisture from entering the wall cavity. It is typically installed on the warm side of the wall. The vapour barrier helps prevent warm, moist air from reaching the colder wall surface where it can condense. Aluminium foil or plastic film materials are recommended vapour barriers. Sufficient ventilation of air may also help in the control of condensation.

Water vapour transmission is based on Fick's law. It is analogous to the Fourier law of heat conduction. Thus, for diffusion of water through wall, we have

$$\frac{W}{A\theta} = -\mu \frac{dp}{dx}$$

where,

W = weight of vapour transmitted in grams,

A = area of cross-section of flow path in m^2,

θ = time of transmission in hours,

μ = water vapour permeability in g-m/m^2-hr-kPa, and

dp/dx = vapour pressure gradient in the x-direction of flow.

By integrating the above equation from x = 0 to x = L and from p_1 to p_2, and rearranging, it becomes

$$\dot{W} = \frac{W}{\theta} = \bar{\mu} A \left(\frac{p_2 - p_1}{L} \right)$$

where,

$\dot{W}$ = vapour transfer rate in g/hour,

p_1 and p_2 = vapour pressure at each end of flow path in kPa,

$\bar{\mu}$ = average permeability in g-m/m^2-hr-kPa, and

L = thickness of material in meters.

For convenience in evaluating combined materials, permeability can be expressed as a coefficient of transmission (M) as

$$\dot{W} = MA(p_2 - p_1) = \frac{A(p_2 - p_1)}{D}$$

where,

M = permeance in g/m^2-hr-kPa = $\frac{\bar{\mu}}{L}$, and

D = vapour flow resistance in m^2-hr-kPa/g.

Resistance to vapour flow provided by the material is the reciprocal of the permeance. The overall permeance of a wall may be found in a way similar to the calculation of the overall coefficient of thermal conductance.

$$\frac{1}{M} = \frac{1}{M_1} + \frac{1}{M_2} + \ldots\ldots + \frac{1}{M_n}$$

The equation for overall vapour resistance becomes

$$D = D_1 + D_2 + \ldots\ldots + D_n$$

Permeance, like conductance, relates to any given material or combination of materials. Permeability, like conductivity, relates to a property of the substance, and is numerically equal to the permeance for the unit thickness.

For the prediction of condensation in the wall, a vapour pressure gradient through a wall may be determined, in conjunction with thermal calculations. The following calculations shall be carried out for the prediction of possible condensation in the wall.

1. Estimate temperatures at the surfaces and the interfaces within the wall with the help of following equation:

   $$T_x = T_o - \frac{R_x}{R}(T_o - T_i)$$

   where, T_x is the temperature at any point, R is overall resistance and R_x is the thermal resistance up to that point measured from outside of the wall.

2. From psychrometric/steam tables, find out saturation pressure at each temperature.

3. Estimate actual vapour pressures at the interfaces with the help of the following equation:

   $$p_x = p_o - \frac{D_x}{D}(p_o - p_i)$$

4. Plot the saturation pressures and actual vapour pressures throughout the wall.
5. If the actual vapour pressure at any section of the wall exceeds the saturation pressure, the condensation will occur, otherwise, there will be no condensation in the wall.

Example 18.5

A wood frame wall is exposed to inside design conditions of 20°C and 50 per cent RH and outdoor condition of 0°C and 80 per cent RH. The wall consists of 9.5 mm of gypsum board on the inside, 9.5 mm plywood on the outside with 100 mm of mineral wool between. Plot the saturation pressure and vapour pressure through the wall and determine if condensation would occur. The data is as follows:

Thermal resistance of 9.5 mm gypsum board	= 0.06 m²-K/W
Thermal resistance of 9.5 mm plywood	= 0.08 m²-K/W
Thermal resistance of inside surface	= 0.108 m²-K/W
Thermal resistance of outside surface	= 0.03 m²-K/W
Thermal conductivity of mineral wool	= 0.05 W/m-K
Permeance of gypsum board	= 10.34 kg/m²-h-Pa
Permeance of plywood	= 0.1448 kg/m²-h-Pa
Permeability of mineral wool	= 0.6094 kg-m/m²-h-Pa

Neglect the vapour resistance of the inside and outside surface.

Solution

$$\text{Overall resistance, R} = \frac{1}{U} = \frac{1}{h_i} + \frac{x_1}{k_1} + \frac{x_2}{k_2} + \frac{x_3}{k_3} + \frac{1}{h_o}$$

$$= R_1 + R_2 + R_3 + R_4 + R_5 = 0.108 + 0.06 + \frac{0.100}{0.05} + 0.08 + 0.03$$

$= 2.278$ m²-K/W

Temperature at a distance 'x'

$$T_x = T_i - \frac{R_x}{R}(T_i - T_o)$$

$$T_1 = 20 - \frac{0.108}{2..278}(20 - 0) = 19.05°\text{C}$$

$$T_2 = 20 - \frac{0.168}{2.278}(20 - 0) = 18.53°\text{C}$$

$$T_3 = 20 - \frac{2.168}{2.278}(20 - 0) = 0.97°C$$

$$T_4 = 20 - \frac{2.248}{2.278}(20 - 0) = 0.27°C$$

From psychrometric tables/steam tables, the saturation vapour pressures at each temperature can be obtained.

Temperature (°C)	*Saturation Vapour Pressure (kPa)*
$T_i = 20.0$	$P_{is} = 2.34$
$T_1 = 19.05$	$P_{1s} = 2.20$
$T_2 = 18.53$	$P_{2s} = 2.14$
$T_3 = 0.97$	$P_{3s} = 0.66$
$T_4 = 0.27$	$P_{4s} = 0.62$
$T_o = 0.0$	$P_{0s} = 0.61$

Actual Vapour Pressure Curve

The actual vapour pressure curve is determined by resistance to vapour transmission through the wall. The overall permeability of the wall is

$$\frac{1}{M} = \frac{1}{{}_1M_2} + \frac{1}{{}_2M_3} + .\frac{1}{{}_3M_4}$$

$$\text{or, } D = D_1 + D_2 + D_3 = \frac{1}{10.34} + \frac{0.100}{0.6094} + .\frac{1}{0.1448}$$

$$= 0.097 + 0.164 + 6.906 = 7.167 \text{ m}^2\text{-h-Pa/kg}$$

where, D is the vapour resistance

$$p_x = p_i - \frac{D_x}{D}(p_i - p_o)$$

$$p_i = 0.50(p_{is}) = 0.50 \times 2.34 = 1.17 \text{ kPa} = p_1$$

$$p_o = 0.80(p_{os}) = 0.80 \times 0.61 = 0.49 \text{ kPa} = p_4$$

$$p_2 = p_i - \frac{D_x}{D}(p_i - p_o) = 1.17 - \frac{0.097}{7.167}(1.17 - 0.49) = 1.16 \text{ kPa}$$

$$p_3 = p_i - \frac{D_x}{D}(p_i - p_o) = 1.17 - \frac{0.261}{7.167}(1.17 - 0.49) = 1.14 \text{ kPa}$$

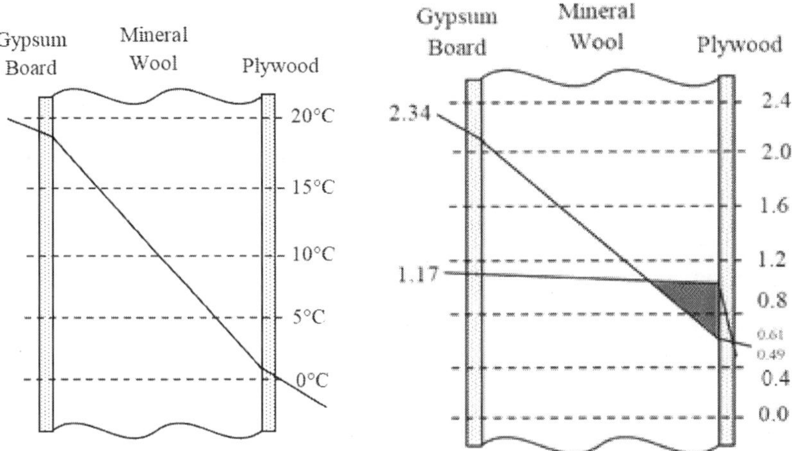

Plot the saturation vapour pressures and actual vapour pressures throughout the wall as shown in the figure. Perusal of figure shows that the saturation vapour pressure gradient line falls below the actual vapour pressure gradient line (shaded area). At this point of intersection and beyond towards the cold side of the wall, condensation will take place. Thus, materials with sufficient resistance to vapour transmission must be placed on the warm side of the wall. Aluminium foil or plastic film materials may be used as vapour barriers to prevent condensation in the wall.

Important Points to Remember

1. Heat transfer generally recognizes three distinct modes such as conduction, convection and radiation.
2. Conduction is the transfer of heat through direct contact between objects or materials.
3. Heat transfer in solids takes place by conduction.
4. Heat transfer is constant when temperature remains constant with time.
5. Units of overall heat transfer coefficient are W/m^2-K.
6. Heat transfer by conduction follows Fourier's law.
7. Unit of thermal conductivity is W/m K.
8. The coefficient of thermal conductivity is defined as the quantity of heat transfer per unit time per unit area per one degree temperature drop per unit length.
9. In the Fourier's law, $kA/\Delta x$ is called thermal conductance of the wall.
10. In the Fourier's law, $\Delta x/kA$ is called thermal resistance of the wall.
11. Convection is a process of heat transfer in which heat is transferred from one place to another through the movement of a fluid, such as a liquid or a gas.

12. Radiation is the transfer of heat through electromagnetic waves.
13. Every object has an emissivity value, which indicates how effectively it can emit and absorb thermal radiation.
14. The values of emissivity ranges from 0 to 1.
15. As the number of birds or animals in a building increase, exposure factor decreases.
16. Insulation can help in reducing vapour condensation.
17. Moisture transfer through solid wall varies directly as permeability of water vapour.
18. Moisture transfer through solid wall varies directly as vapour pressure.
19. Moisture transfer through solid wall varies inversely as temperature.
20. If the actual vapour pressure at any section of the wall exceeds the saturation vapour pressure, the condensation will occur.

Exercises/Questions

1. Write a short note on various modes of heat transfer.
2. State and explain the Fourier's equation for conduction heat transfer.
3. Define thermal conductivity, thermal resistance and thermal conductance.
4. Derive a mathematical relationship to estimate heat transfer through a flat composite wall having three layers of varying thickness and thermal conductivities.
5. Derive an expression for the determination of the temperature at any interface within the wall.
6. Explain Newton's law of cooling/heating.
7. Write a short note on radiation heat transfer.
8. Explain Stefan's Boltzman law as applied to radiation heat transfer.
9. Explain exposure factor and its importance.
10. State and explain Fick's law of mass transfer.
11. Explain the process of moisture transfer through solid wall.
12. How does condensation take place and how can it be prevented?
13. Define the following:

    (a) permeability (b) premeance (c) vapour resistance (d) emissivity

14. A cold storage is made of composite wall consisting of 25 mm concrete, 100 mm brick, 100 mm thermo-cole, and 5 mm ply-wood having thermal conductivities of 0.76, 0.69, 0.024, and 0.2 W m^{-1} K^{-1} respectively. The inside and outside wall temperatures are 2°C and

36°C respectively. Estimate the total thermal resistance of the wall and the rate of heat-gain by the cold storage.

[**Answers**: R = 4.37 m^2-K/W, Q = 7.78 W/m^2]

15. Determine the inside and outside wall temperatures of a cold store maintained at –15°C. The walls are of 15 cm cork with 2 cm plaster on each side, and outside temperature is 30°C. Inside and outside heat transfer coefficients are 10 W/m^2 K. Thermal conductivity of the cork is 1 W/m K and that of cement plaster 10 W/m K.

[**Answers**: Inside wall temperature = -2.3°C,
Outside wall temperature = 17.3°C]

16. Determine the exposure factor for a stall barn for dairy cattle 12 m × 50 m is designed for 75 cows. The average overall coefficient of heat transfer for the shed is 1.2 W/m^2-K and the ceiling height is 2.5 m.

[**Answer**: 14.6 W/K (cow)]

17. Find the required average overall coefficient heat transfer, U, for an animal housing unit 10 m × 20 m × 2.5 m high housing 50 animals. The expose factor should be about 11 W/K.

[**Answer**: 1.57 W/m^2-K]

Chapter 19

Reactions of Livestock to Environmental Factors

Animals, like humans, are highly influenced by their environment. They have evolved over millions of years to adapt and respond to various environmental factors, including climate, temperature, food availability, habitat conditions, and the presence of predators or other organisms. These reactions and responses are crucial for their survival, reproduction, and overall well-being.

The environment plays a crucial role in the health and productivity of livestock. A suitable and well-managed environment can enhance animal well-being, while an unfavourable or stressful environment can have negative effects. The comfort of animal inside their housing is the main objective of environment control.

The reactions of animals to environmental factors can be arbitrarily divided into two classes *i.e.* productive and physiological. Physiological responses involve the internal adjustments and processes that animals undergo to maintain homeostasis and optimize their physiological functions in response to environmental changes. The classification of physiological reactions is artificial but useful for practicing engineers to study the factors affecting the productivity of animals. The study of heat and moisture dissipation is essential for the design of heating, cooling, and ventilation systems. Other physiological reactions like body temperature, pulse rate, and respiration rate are useful indicators of thermoneutrality but less directly applicable to shelter design.

19.1 Homeostasis

Homeostasis is the process by which living organisms maintain a stable internal environment despite changes in the external environment. Many internal factors are maintained constant. These include body temperature, mature body weight, blood pressure, respiration, body fluid, hormones, and movement. Animals, including humans, have evolved various mechanisms to regulate and maintain homeostasis. This includes behaviours like seeking shade or shelter, adjusting their metabolic rate, and altering their blood flow.

19.1.1 Homeothermy

Homeothermy, an excellent example of homeostasis refers to the ability of animals to maintain a relatively constant body temperature regardless of the external temperature. It is also known as "warm-bloodedness." Homeothermy allows animals to be active and function optimally in a wide range of environmental conditions. There are two main types of homeothermic animals: endotherms and ectotherms.

1. **Endotherms**: These animals generate their own body heat internally through metabolic processes. They have a high metabolic rate and rely on internal heat production to maintain their body temperature. Mammals and birds are examples of endotherms. They have various physiological adaptations that help them regulate their body temperature, such as fur or feathers for insulation, sweat glands, and complex thermoregulatory systems.
2. **Ectotherms**: These animals derive their body heat from external sources, primarily the environment. Ectotherms have a lower metabolic rate compared to endotherms. They rely on behavioural adaptations to regulate their body temperature, such as basking in the sun or seeking shade. Reptiles, amphibians, and most fish are ectothermic. While they cannot maintain a constant internal body temperature, they can adjust their body temperature within a certain range to remain active.

Internal temperature regulation in mammals and birds is regulated through various feedback mechanisms. Each species has a thermoneutral range of optimal environmental temperature. This range minimises the action by regulating mechanisms and ensuring a balance between body temperature and metabolic activity. The skin of these homeotherm animals has peripheral receptors that send the signal to the brain for regulation of temperature.

Cold Stress: When the environment cools below a certain temperature, the peripheral senses are transmitted to the posterior hypothalamus (a part of the brain). This specific part of the brain activates protective actions such as increased appetite, decreased blood flow, and changes in hair or feather posture. Further decreases in temperature initiate shivering by muscle contractions that increase the rate of heat production.

Heat Stress: When the external temperature exceeds optimal levels, the body activates protective mechanisms. The body of animals experiences the dilation of surface blood vessels, decreased appetite, increased moisture evaporation, respiration, and profuse sweating. The animal starts seeking shade or cooler environments to maintain a stable temperature. These involuntary actions are mainly controlled by the anterior hypothalamus (a specific part of the brain).

Homeotherms may be divided into several classes, according to body temperatures, which range from 36°C for elephants to 43°C for small birds. Typical body temperatures are

Humans	37.0°C	Sheep	39.0 °C
Horse	38.0 °C	Swine	39.0 °C
Cattle	38.5 °C	Goat	40.0 °C
Cats and dogs	38.6 °C	Chicken	41.7 °C

19.1.2 Metabolism

The overall exchange of matter and energy by an animal is called total metabolism. The metabolic response and comfort level of animals can vary based on several factors, including their genetic makeup, environment, nutrition, and management practices.

Dairy cows have a relatively high metabolic rate due to their milk production. They require a balanced diet that meets their energy, protein, vitamin, and mineral needs to support milk synthesis and maintain body condition. Poultry (Chickens) have a rapid metabolic rate and require a diet that provides adequate energy, protein, vitamins, and minerals for growth, egg production (for laying hens), and maintenance. Their nutritional requirements are typically met through balanced formulated feeds. Swine (Pigs) have a relatively efficient metabolism and convert feed into growth relatively well. Their nutritional needs for energy, protein, vitamins, and minerals depend on their growth stage, with different requirements for piglets, growing pigs, and breeding stock.

19.1.3 Thermoneutrality

Thermoneutrality refers to the condition in which an organism's body temperature remains within a range where it does not need to expend extra energy to either cool down or warm up. In other words, it is the range of ambient temperatures in which an organism can maintain its body temperature without shivering or sweating. Within this range the thermal regulation is minimal *i.e.* neither heat nor cold is felt. In these terms, thermoneutrality defines a '**comfort zone'**. Animal thermoneutrality as related to environmental temperature is shown schematically in Figure 19.1.

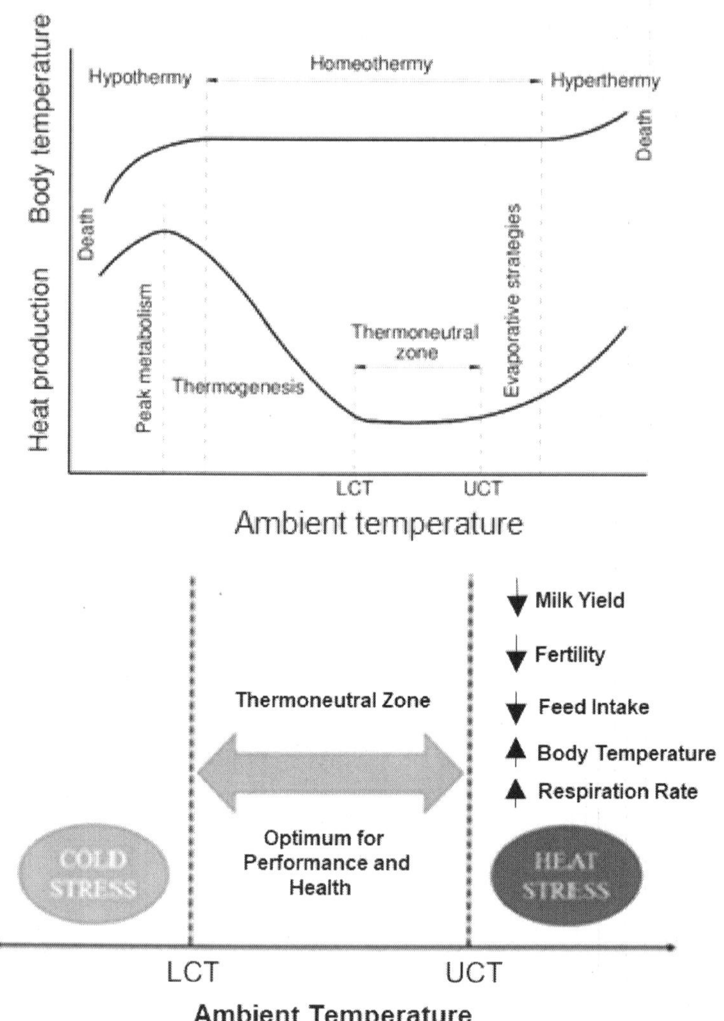

Figure 19.1. Effect of Environmental Temperature on Heat Production and Body Temperature of Homeotherms (LCT: Lower Critical Temperature; UCT: Upper Critical Temperature).

Different organisms have different thermoneutral zones, which are influenced by their physiology and habitat. For humans, the thermoneutral zone is typically between 20 to 27 degrees Celsius, although individual preferences may vary. Within this range, the body can maintain a stable core temperature without activating mechanisms like shivering or sweating. For dairy cows, the thermoneutral zone typically ranges from 5°C to 25°C. Beef cattle generally have a slightly wider thermoneutral zone, typically ranging from 0°C to 30°C. The thermoneutral zone for sheep and goats is similar to that of beef cattle, ranging from 0°C to 30°C. The thermoneutral zone for pigs varies with age and weight. For growing-finishing pigs, it typically ranges from 18°C to 22°C. Sows and piglets have slightly different thermoneutral zones. For chickens, the thermoneutral zone is generally between 18°C to 24°C, although broilers (meat chickens) can tolerate slightly higher temperatures. The thermoneutral zone for horses is similar to that of dairy cows, typically ranging from 5°C to 25°C.

It's important to note that thermoneutrality is specific to each species and can vary based on factors like age, health, acclimation, and activity level. Understanding the thermoneutral zone of different organisms is crucial for creating suitable environments for their well-being and also plays a role in areas such as building design, climate control, and animal husbandry.

The overall objective in animal housing systems design is to create environments in which the animals approach thermoneutrality, but with appropriate balance between the productive value of thermoneutrality and the cost of systems to maintain it.

19.2 Heat Production and Dissipation

Livestock generate heat through metabolic processes such as digestion, metabolism, and physical activity. When ambient temperatures are high, they experience heat stress and produce more metabolic heat. This additional heat production contributes to the overall heat load that the animal must dissipate.

Livestock dissipate metabolic heat primarily through several mechanisms, including radiation, conduction, convection, and evaporation. These processes help them regulate their body temperature and prevent overheating, which is particularly important in warm environments. Livestock animals have various ways of managing their body temperature to prevent overheating and maintain homeostasis.

Animals loose most of the sensible heat from the exterior surfaces and most of the latent heat from interior surfaces (lungs and the respiratory tract). Heat is produced throughout the body by the metabolizing tissues, and blood acts as a major carrier of heat to the exterior body surfaces and interior lung surfaces. Some conduction heat transfer also takes place since the deep body temperature is generally a few degrees above the exterior surface temperature. Inspired air is heated to body temperature, thus picking up some sensible heat. The expired air is also nearly saturated with water vapour so carries latent heat from the body. Warm-blooded animals like cattle, swine, sheep, and poultry emit heat and moisture. The rate of expulsion of moisture and heat varies based on their environmental conditions.

An animal's body temperature can be summed up with the simple formula:

$$Q_M - W = Q_e \pm Q_r \pm Q_{cond} \pm Q_c$$

where,

Q_M = metabolic heat produced in the body,

W = useful rate of working,

$Q_M - W$ = heat to be dissipated to the atmosphere,

Q_e = heat lost by evaporation,

Q_r = heat lost or gained by radiation,

Q_{cond} = heat lost or gained by conduction, and

Q_c = heat lost or gained by convection.

In order to maintain a constant temperature, heat production has to equal to heat loss. This concept is complicated by several factors, including environmental temperature and humidity.

Heat stress results when,

Heat Produced + Environmental Heat > Heat Loss

19.2.1 Sensible Heat (Non-evaporative Losses)

1. Radiation

Livestock animals can radiate heat to the surrounding environment. If the ambient temperature is lower than their body temperature, they will emit heat energy in the form of electromagnetic waves to the cooler surroundings. However, if the surroundings are warmer than the body temperature, radiation may not be as effective.

The rate of radiation transfer depends on the temperature and emissivity of the radiating surfaces, and some geometrical considerations. The net heat transfer rate by radiation per unit area of the animal is

$$Q_r = \sigma \varepsilon F_1 (T_1^4 - T_2^4)$$

where,

Q_r = heat transfer rate per unit area by radiation in W/m²,

σ = Stefan-Boltzmann constant (5.67×10^{-8} W/m²-K⁴),

ε = emissivity of animal surface (dimensionless),

F_1 = shape factor (dimensionless).

T_1 = temperature of animal surface in K, and

T_2 = temperature of the surrounding surface in K.

2. Conduction

Heat conduction occurs when animals come into direct contact with cooler surfaces. Livestock may seek out shaded areas or lie on the ground to transfer heat from their bodies to the cooler surface through conduction, facilitating heat dissipation. The average heat transfer per unit area of animal surface can be estimated by

$$Q_{cond} = K(T_s - T_f)\frac{p}{100}$$

where,

Q_{cond} = heat transfer rate per unit area of animal in W/m^2,

K = conductive heat transfer coefficient in W/m^2 °C,

T_s = animal surface temperature in °C,

T_f = floor temperature in °C, and

p = percentage of surface area touching the floor

3. Convection

Convection involves the transfer of heat through the movement of air currents. Livestock animals can increase convective heat loss by seeking out areas with air movement, such as open spaces, breezy spots, or well-ventilated barns. As air flows across their bodies, it carries away heat, contributing to cooling. The general equation for convection heat transfer is

$$Q_c = h_m (T_s - T_2)$$

where,

Q_c = heat transfer rate per unit area by convection in W/m^2,

h_m = convection heat transfer coefficient in W/m^2 °C,

T_s = surface temperature in °C, and

T_2 = air temperature in °C

The value of convective heat transfer coefficient depends on physical properties of air, velocity of air, besides other factors.

The greater the surface area, the greater will be the heat transfer by conduction, convection, and radiation. Empirical relationships have been developed by the research scientists for different species of animals. The generalized body surface area equation is

$$A = CW^K$$

where,

A = area in m^2,

W = body weight in kg,

C and K = constants

The values of C and K depend upon the animal and its breed. The exponent K was found to vary from 0.32 to 0.72 for individual animals.

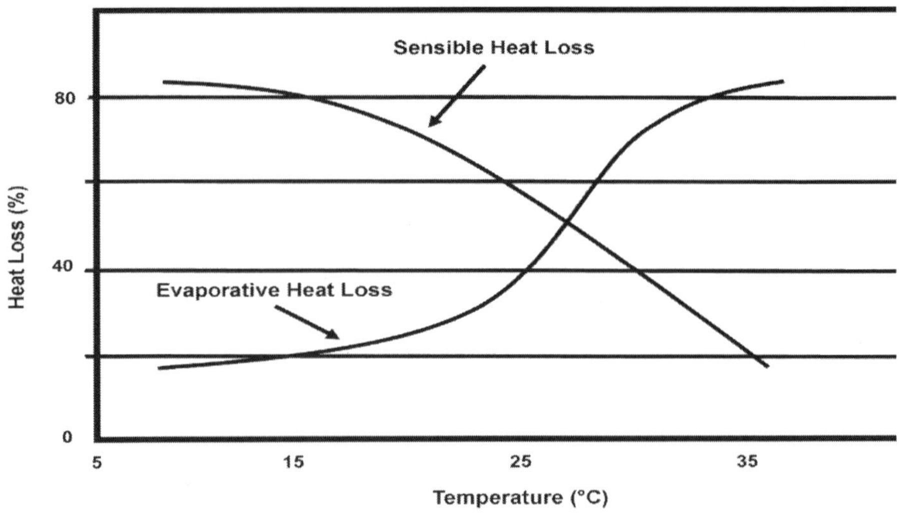

Figure 19.2. Effect of Temperature on the Method of Heat Loss from Animals.

19.2.2 Latent Heat (Evaporative Losses)

Evaporative cooling is crucial for livestock, especially in hot environments. They rely on evaporation to dissipate heat. Some species, such as cattle and horses, have limited sweat glands, primarily located in their noses and lower limbs. However, they rely more on respiratory evaporative cooling through panting. Panting involves rapid, shallow breathing, which allows moisture in the respiratory system to evaporate, taking heat away from the body.

The proportion of metabolic heat that is dissipated from an animal by evaporation increases with rising environmental temperatures (Figure 19.2) and decreasing temperature gradient between the animal surface and the ambient air. As temperature rises, total and sensible heat decrease and latent heat increases. Total heat is the sum of sensible and latent heat. Heat dissipation shall be 100 per cent evaporative in the case of animals with the same dry bulb temperature and skin temperature. 100 per cent evaporative means that only latent heat will be dissipated. Air velocity is one of the important factors affecting the evaporative heat transfer. It has been found that increase in air velocities over animals reduced skin and body temperatures when subjected to high environmental temperature. This increased cooling is accounted for by both convection and evaporation.

Cows regulate their body temperature by dissipating heat through respiration. As ambient temperature increases, cows try to cool themselves by panting and increasing their respiration rate. The animal releases more moisture through evaporation to dissipate more heat. It regulates its body temperature, and that results in a reduction in sensible heat production with increasing temperature.

The heat loss of evaporation is given by

$$Q_e = C_d (P_s - P_v) h_{fg}$$

where,

Q_e = heat loss by evaporation in W/m^2,

C_d = diffusion coefficient in kg of water evaporated per unit surface area and pressure difference per hour,

P_s = saturation vapour pressure corresponding to skin temperature,

P_v = vapour pressure of surrounding air, and

h_{fg} = latent heat of vaporization = 2450 kJ/kg

The heat loss by evaporation is always positive.

19.3 Reactions of Animals

Animals exhibit a wide range of reactions to thermal and other environmental factors. These reactions are often adaptations that help them survive and thrive in their specific habitats. Temperature is the most significant parameter.

19.3.1 Temperature

Dairy Cattle

Temperature can have a significant impact on the productivity of dairy cattle. Dairy cattle are highly sensitive to heat stress, and elevated temperatures can lead to reduced milk production, impaired reproductive performance, and overall decreased profitability for dairy farmers. Heat stress occurs when the animal's body temperature rises above its normal range, leading to various physiological and behavioural responses. Heat-stressed cattle may exhibit signs such as increased respiration rate, excessive panting, sweating, drooling, and reduced feed intake.

Heat stress can significantly reduce milk production in dairy cattle. High temperatures can negatively impact the cow's appetite, leading to reduced feed intake, and subsequently affecting milk yield. Additionally, heat stress can interfere with the cow's hormonal balance, affecting the synthesis and release of milk. Heat stress can also lead to changes in the composition of milk. The fat, protein, and lactose content may be affected, which can impact the quality and market value of the milk produced by the cows. While dairy cattle are more commonly associated with heat stress, extreme cold temperatures can also impact them. Cold stress can lead to increased energy expenditure as cows attempt to maintain their body temperature. Providing shelter, bedding, and windbreaks can help minimize the effects of cold stress.

Research findings: The effect of temperature was studied on two breeds, Holstein and Jersey, for their milk production. Both the breeds produced a normal level between 5 and 25°C. This range of temperature is the thermoneutral zone for both breeds. The milk production of these cows starts to decline above 25°C. At 35°C, the milk production was down to 50 per cent and 62 per cent of normal milk production for Holstein and Jersey, respectively. The results indicate that the heavier animals, the Holsteins, were more affected by the high temperature than the lighter Jerseys. Below 5°C, the cold affects the lighter animals (Jersey) more than the heavy animals (Holstein). In north Indian condition, livestock begins to suffer from mild heat stress when thermal heat index (THI) reaches higher than 72, moderate heat stress occurs at 80 and severe stress is observed after it reaches 90 (Upadhyay *et al.*, 2009). Thermal stress lowers feed intake and reduces animal productivity in terms of milk yield, body weight and reproductive performance. Heat stress reduces libido, fertility and embryonic survival in animals. The poor reproductive performance in buffaloes especially during summer months is due to inefficiency in maintaining the thermoregulation under high environmental temperature and relative humidity (RH) as those have dark skin and sparse coat of body hair which absorb more heat along with poor heat dissipation mechanism due to less number of sweat glands (Marai and Haeeb, 2010).

Poultry

Poultry birds are homeothermic animals, which means they maintain a relatively constant internal body temperature regardless of the external temperature. However, extreme temperatures, whether too hot or too cold, can disrupt their normal physiological processes and negatively affect their productivity.

Poultry birds are particularly susceptible to heat stress, especially in hot climates or during heat waves. High temperatures can cause physiological and metabolic changes in birds, leading to reduced productivity. They may exhibit increased respiration rate, panting, and wing spreading in an attempt to dissipate heat. Birds experiencing heat stress tend to eat less, leading to a decrease in nutrient intake and subsequent weight loss. Reduced feed consumption directly affects growth rate and egg production. Heat stress can impair the growth and development of poultry birds, especially young chicks. It slows down their metabolic processes, resulting in slower weight gain. High temperatures can disrupt the reproductive cycle of laying hens, causing a decrease in egg production. It may also lead to smaller egg sizes, poor eggshell quality, and increased mortality rates. Prolonged exposure to high temperatures can result in heat exhaustion, heatstroke, or even death, particularly if appropriate measures to alleviate heat stress are not implemented.

Extremely low temperatures can also impact poultry bird productivity, although it generally has a lesser effect compared to heat stress. They may huddle together, fluff up their feathers to trap air for insulation, and decrease their activity levels. To maintain their body temperature, poultry birds require more energy in cold environments. This increased energy demand may result in reduced growth, decreased egg production, or poor feed efficiency. Cold stress can cause feather loss in poultry birds, reducing their ability to maintain body heat. Featherless areas are more susceptible to frostbite, leading to further health issues. Birds may become less active in cold temperatures, resulting in reduced feed consumption and slower growth. Cold stress can also lead to decreased egg production due to disruptions in the reproductive cycle of laying hens.

Research findings: A study conducted on Rhode Island Red hens found that the optimal range for eggs production per day is 10 to 18°C, or possibly 16°C. At 18°C, the egg size and shell thickness begin to decrease. By 32°C, the production is severely reduced.

Swine

Temperature can have a significant impact on the productivity of swine, as pigs are susceptible to heat stress. Heat stress occurs when pigs are exposed to high temperatures, particularly when combined with high humidity. Heat stress can lead to increased respiration rate, reduced feed intake, and decreased weight gain. Pigs may also exhibit signs of discomfort, such as seeking shade or wallowing in mud to cool down. As temperatures increase, pigs may reduce their appetite, leading to a decrease in feed consumption. This can negatively impact their growth and overall productivity. Pigs experiencing heat stress may have slower growth rates. High temperatures can divert energy away from growth and towards cooling the body, resulting in reduced weight gain. This can affect the efficiency of swine production and delay market readiness. Heat stress can also affect the reproductive performance of swine. Sows and boars may experience reduced fertility, increased embryo mortality, and decreased litter size. High temperatures can disrupt normal reproductive processes, leading to lower conception rates and compromised litter quality. Heat stress increases the risk of mortality in swine, especially in younger pigs. Piglets are particularly vulnerable to high temperatures due to their limited ability to regulate body temperature. Heat stress can lead to dehydration, respiratory distress, and even death if proper cooling measures are not in place. In cold environments, pigs may experience cold stress, especially if they lack proper shelter and bedding. Cold stress can result in increased energy expenditure, shivering, reduced growth rate, and increased susceptibility to diseases.

Research findings: As per study conducted on pigs of 23-91 kg, the weight gain was reduced with the time when kept on temperature above and below 16°C. The temperature also affected their feed efficiency. The carcass quality of pigs was found best on rearing them temperature from 16 to 21°C.

19.3.2 Humidity

Dairy Cattle

Humidity can have a significant impact on the productivity and well-being of dairy cattle. High humidity levels, especially in combination with high temperatures, can lead to heat stress in dairy cows. Heat stress occurs when the cow's body temperature rises above the normal range, and it can have detrimental effects on milk production, reproductive performance, and overall health. The animals dissipate heat by evaporation. The high humidity hinders the evaporation that affects their production. The evaporation depends on the vapour pressure difference between the atmosphere and animal skin. In the case of the small pressure difference, the animal feels uncomfortable.

Heat stress due to high humidity can significantly reduce milk production in dairy cows. This is primarily due to the cow's energy being diverted towards dissipating heat to maintain a stable body temperature, rather than being used for milk synthesis. Additionally, heat-stressed cows may exhibit a decline in milk fat and protein content.

Poultry

Poultry birds, such as chickens, are sensitive to environmental conditions, including temperature and humidity. High humidity combined with high temperatures can lead to heat stress in poultry birds. Heat stress occurs when birds are unable to dissipate excess body heat effectively through evaporative cooling. It can result in increased respiration rate, reduced feed intake, decreased egg production, poor egg quality, decreased growth rate, and even increased mortality. High humidity can make birds feel uncomfortable, resulting in decreased feed intake. Reduced feed intake can lead to slower growth rates, lower body weight, and reduced egg production. Extremely low humidity levels can cause dehydration and respiratory issues in poultry. It can lead to dry mucus membranes, difficulty in breathing, and increased susceptibility to respiratory diseases.

Swine

Pigs are highly sensitive to heat stress, and high humidity levels can exacerbate the effects of heat stress on their bodies. Pigs are unable to sweat, so they rely on evaporation to dissipate heat from their bodies. High humidity hampers evaporative cooling through sweating or panting, making it difficult for pigs to regulate their body temperature. As a result, pigs may experience heat stress, which can lead to decreased feed intake, reduced weight gain, and poor reproductive performance. Humidity can also affect the appetite and digestion of swine. Pigs may reduce their feed intake in high-humidity conditions, as they may feel uncomfortable and have a decreased desire to eat. Reduced feed intake can result in slower growth rates and lower weight gains. Extremely low humidity can lead to dry skin and respiratory discomfort for pigs, as it can cause excessive moisture loss from their bodies.

19.3.3 Light

Dairy Cattle

Exposure to appropriate light-dark cycles helps maintain the cow's overall health and productivity. Light exposure has been found to have a direct impact on milk production in dairy cattle. Increasing the daily hours of light exposure beyond the natural photoperiod (length of daylight exposure) can stimulate milk production. This can be achieved by providing additional artificial lighting in the barns during the darker seasons or by extending the natural daylight using artificial lighting.

Poultry

Providing an appropriate photoperiod can enhance their feeding behaviour, metabolism, and nutrient utilization. Adequate light exposure encourages regular feeding patterns, leading to improved growth rates and optimal weight gain. Light has a direct impact on the reproductive performance of poultry birds, especially in terms of egg production and fertility. Manipulating the duration and intensity of light exposure can influence the onset and rate of sexual maturity, as well as the timing and consistency of egg-laying. Maintaining specific lighting programs can optimize egg production and improve hatchability rates. Poultry birds require exposure to natural or artificial sunlight to synthesize vitamin D. Vitamin D plays a crucial role in calcium metabolism and skeletal development. Insufficient levels of vitamin D can lead to skeletal disorders such as rickets and poor eggshell quality.

Swine

Adequate light levels and consistent lighting schedules promote better feed intake and digestion, leading to improved growth rates and higher weight gain in pigs. Lighting conditions can impact the feed consumption of swine. A well-lit environment allows pigs to see their surroundings and locate feed and water sources more easily. This promotes regular feeding and adequate nutrient intake, which is crucial for optimal growth and production.

19.3.4 Air Movement

Dairy Cattle

Air movement can have both positive and negative effects on the productivity of dairy cattle. Air movement helps dissipate heat, especially in hot and humid environments, by promoting evaporative cooling from the cow's skin and respiratory system. Cows prefer a comfortable environment, and air movement can contribute to their well-being. Adequate airflow helps to reduce the accumulation of moisture, condensation, and drafts, which can cause discomfort and stress for the cows. When cows are comfortable, they exhibit better behaviour, such as increased feed intake, improved rumination, and higher milk production.

While air movement is crucial for maintaining good ventilation and air quality, it is essential to manage it appropriately during the winter months. Excessive drafts and cold air can increase the risk of cold stress among cows, leading to reduced feed intake and decreased productivity. Therefore, the design and management of ventilation systems should consider the need for fresh air exchange while avoiding direct drafts on the cows.

Poultry

Poultry birds are sensitive to high temperatures, and excessive heat stress can reduce their productivity. Adequate air movement can help dissipate heat and prevent overheating, ensuring that birds remain comfortable and productive. Adequate air movement ensures a sufficient supply of oxygen for the birds. Poultry birds have high oxygen demands due to their rapid metabolic rates. Insufficient oxygen levels can lead to reduced growth, poor feed conversion, and decreased productivity. Proper air movement facilitates the exchange of oxygen and carbon dioxide, ensuring an ample supply of fresh air for the birds. It is important to note that while air movement is beneficial, excessive drafts or turbulent air can be detrimental to poultry birds.

Swine

Air movement can have both positive and negative effects on the productivity of swine. Adequate air movement helps in temperature control within swine facilities. Pigs are sensitive to heat stress, and high temperatures can significantly impact their productivity. Proper ventilation and air circulation help dissipate heat, reducing the risk of heat stress and improving overall productivity. However, it's important to note that excessive air movement can also have negative consequences. High-velocity drafts can cause chilling and stress in swine, leading to reduced productivity and health issues.

19.3.5 Solar Radiation

Dairy Cattle

Solar radiation can have both positive and negative effects on the productivity of dairy cattle. Solar radiation is essential for the synthesis of vitamin D in dairy cattle. Vitamin D plays a crucial role in calcium metabolism, bone health, and overall immune function. Adequate exposure to sunlight allows cattle to naturally produce vitamin D, supporting their overall productivity and well-being. Solar radiation contributes to increased ambient temperatures, which can lead to heat stress in dairy cattle. Heat stress negatively affects milk production, reproduction, and overall animal health. Cows under heat stress may reduce their feed intake, experience decreased milk yield, and have impaired reproductive performance.

Poultry

Solar radiation can have both positive and negative effects on the productivity of poultry birds, depending on various factors such as intensity, duration, and management practices. Poultry birds are sensitive to high temperatures, and solar radiation contributes to heat stress. Excessive heat can lead to reduced feed intake, decreased egg production, poor growth rates, and increased mortality. Day length, influenced by solar radiation, affects the reproductive behaviour of poultry birds. Artificial lighting is often used to manipulate the photoperiod and stimulate reproductive activities, such as egg production and broiler growth. Longer day lengths can enhance egg production, while shorter day lengths may stimulate broiler growth.

Swine

Solar radiation can have both positive and negative effects on the productivity of swine. Swine are susceptible to heat stress, especially in hot climates. Solar radiation contributes to the ambient temperature, and excessive heat can negatively affect swine productivity. Heat stress can lead to reduced feed intake, decreased growth rate, lower fertility, and increased susceptibility to diseases. To mitigate the negative effects of solar radiation, proper ventilation, cooling systems, and shade should be provided in swine housing. Solar radiation varies with seasons, and this can impact swine productivity. In colder seasons with reduced sunlight, swine may experience slower growth rates and decreased appetite.

19.4 Strategies for Heat Stress Management

Heat stress can be a significant issue for livestock and poultry, particularly in hot climates or during heatwaves. To overcome the impact of heat stress on livestock and poultry, following technologies/strategies can be used.

19.4.1 Livestock

1. Ensure that there are shaded areas available for the animals to rest and escape direct sunlight during the hottest parts of the day.
2. Make sure that clean and cool water is readily available at all times. Consider installing water misters/foggers or sprinklers to help cool the animals down. Provide insulation on top of the shed with water sprinklers.
3. Ensure good airflow in barns or shelters. This can be achieved through the use of fans, open windows, or other ventilation systems to reduce heat build-up.
4. Modify feeding schedules to avoid feeding during the hottest times of the day. Provide proper ration including concentrates during morning and evening to increase the feed intake. Additionally, consider adjusting the composition of feed to include more electrolytes and minerals that can help animals cope with heat stress.

5. Install sprinkler systems in outdoor areas to provide periodic cooling for the animals. This can also help reduce dust and improve air quality.
6. Use methods such as evaporative cooling pads or misting systems to help lower the ambient temperature in barns or shelters.
7. Lower stocking density to reduce the amount of heat generated within the herd or flock. This allows individual animals more space to dissipate heat.
8. Use bedding material that helps absorb moisture and provides insulation against hot surfaces.
9. Regularly monitor animals for signs of heat stress, such as excessive panting, drooling, or lethargy. Take immediate action if any signs are observed.
10. For species with wool or heavy coats, consider shearing or clipping to help animals dissipate heat more effectively.
11. Encourage indoor feeding with green fodder instead of outdoor grazing to reduce the heat stress and increase the feed intake.
12. Provide water bath to lactating animals to reduce the sun stroke.

By implementing these strategies, livestock producers can help mitigate the effects of heat stress and ensure the welfare and productivity of their animals, even during hot weather conditions.

19.4.2 Poultry

Provide shaded areas within poultry houses or outdoor areas where birds can rest and cool down. Shade cloth or natural vegetation can be used to create shade.

1. Ensure proper ventilation in poultry houses to facilitate air exchange and reduce heat build-up. Fans, vents, and air inlets can help improve airflow.
2. Ensure constant access to cool and clean drinking water to keep the body temperature under control. Consider adding electrolytes or vitamins to the water to help replenish lost nutrients due to heat stress.
3. Install misting/fogger systems or sprinklers to periodically wet the birds and cool the air. Provide insulation on top of the roof with water sprinklers. This helps lower the ambient temperature and reduces heat stress.
4. Minimize dust inside poultry houses, as it can exacerbate heat stress. Regularly clean and maintain the house, and consider using dust control measures such as wetting down surfaces or using dust suppressants.

5. Avoid overcrowding by reducing stocking density, which allows birds more space to dissipate heat and move freely.
6. Install cooling pads or evaporative cooling systems in the house to lower the temperature. These systems use water evaporation to cool the air. Alternatively, provide side curtains and sprinkle water on the curtains.
7. Feed the birds during early morning (before 5 AM) and late evening (after 5 PM) to increase feed intake and avoid production of metabolic heat during peak hours of high ambient temperature. Provide easily digestible feeds and consider supplementing with electrolytes or feed additives that support heat tolerance.
8. For birds with heavy feathers, consider shearing or clipping to help them dissipate heat more effectively.
9. Ensure sufficient roosting space to allow birds to perch comfortably, as this helps them stay off the hot ground and improves airflow around their bodies.

Implementing these strategies can help poultry producers effectively manage heat stress and maintain the health and welfare of their birds, ultimately supporting optimal performance and productivity.

Important Points to Remember

1. The reactions of animals to environmental factors can be arbitrarily divided into two classes *i.e.* productive and physiological.
2. Homeostasis is the process by which living organisms maintain a stable internal environment despite changes in the external environment.
3. Homeothermy refers to the ability of animals to maintain a relatively constant body temperature regardless of the external temperature.
4. Endotherms animals generate their own body heat internally through metabolic processes.
5. Ectotherms animals derive their body heat from external sources, primarily the environment.
6. Ectotherms have a lower metabolic rate compared to endotherms.
7. The overall exchange of matter and energy by an animal is called total metabolism.
8. Thermoneutrality refers to the condition in which an organism's body temperature remains within a range where it does not need to expend extra energy to either cool down or warm up.
9. For humans, the thermoneutral zone is typically between 20 to 27 degrees Celsius.

10. Warm-blooded animals, such as cattle, swine, sheep, and poultry, give off heat and moisture during normal living processes.
11. Animals loose most of the sensible heat from the exterior surfaces and most of the latent heat from interior surfaces (lungs and the respiratory tract).
12. The non-sweating animals adjust to high ambient temperature by increasing rate of respiration.
13. Heat produced by dairy animals varies with body size and the environmental conditions.
14. The greater the surface area of the animal, the greater will be the heat transfer by conduction, convection, and radiation.
15. High air speeds accelerate elimination of the body heat in farm animals.

Exercises/Questions

1. Discuss the effect of environmental temperature on heat production and body temperature of homeotherms with the help of neat diagram.
2. Discuss the effect of temperature on the method of heat loss from animals.
3. Discuss the reactions of animals (dairy cattle, poultry and swine) to environmental temperature.
4. Discuss the reactions of animals (dairy cattle, poultry and swine) to relative humidity.
5. Discuss the reactions of animals (dairy cattle, poultry and swine) to light.
6. Discuss the reactions of animals (dairy cattle, poultry and swine) to air movement.
7. List down the different strategies for heat stress management in livestock.
8. List down the different strategies for heat stress management in poultry.
9. Define the following terms:
   - (a) Homeostasis
   - (b) Homeothermy
   - (c) Endotherms
   - (d) Ectotherms
   - (e) Cold stress
   - (f) Heat stress
   - (g) Metabolism
   - (h) Thermoneutrality

Chapter 20

Ventilation

Ventilation refers to the process of exchanging or replacing air in an enclosed space to maintain a fresh and healthy indoor environment. It involves the circulation of outdoor air into a building or the removal of stale or polluted indoor air. Ventilation systems are designed to control temperature, humidity, remove contaminants, reduce odours, and replenish oxygen levels in a space. Agricultural buildings, such as barns, poultry houses, and greenhouses, require proper ventilation to maintain a healthy and productive environment for the animals or crops housed within them.

20.1 Importance of Ventilation

1. **Temperature Regulation**: Proper ventilation helps regulate the temperature inside agricultural buildings, preventing overheating during hot seasons and minimizing heat loss in colder months. It allows for the removal of excess heat and provides fresh air, creating a comfortable environment for livestock or plants.
2. **Humidity Control**: Excessive humidity can lead to condensation, mold growth, and a range of health issues for animals or plants. Ventilation systems aid in maintaining optimal humidity levels by removing moisture-laden air and replacing it with drier air. This helps prevent the growth of pathogens, respiratory problems, and heat stress.
3. **Air Quality and Disease Control**: Adequate ventilation helps remove airborne contaminants, such as dust, ammonia, and volatile organic compounds (VOCs), which can negatively impact animal health and productivity. It also helps dilute and disperse odours, reducing

nuisance for neighbouring areas. Additionally, proper airflow can reduce the transmission and spread of diseases among livestock or crops.

20.2 Determinant Effects of Inadequate Ventilation in Animal Housing

To ensure the health and welfare of animals, proper ventilation systems should be in place in animal housing facilities. These systems should provide adequate air exchange, control humidity levels, and remove harmful gases and airborne pathogens from the environment. Regular monitoring of air quality and temperature, along with appropriate management practices, are essential to maintaining optimal ventilation in animal housing. Some potential consequences of inadequate ventilation are:

1. **Poor Air Quality**: Inadequate ventilation leads to a build-up of harmful gases, such as ammonia, carbon dioxide, and dust particles. These pollutants can irritate the respiratory system of animals, leading to respiratory distress, coughing, and other respiratory problems.
2. **Heat Stress**: Insufficient airflow and ventilation can contribute to high temperatures and heat stress in animals, especially in hot climates or during summer months. Heat stress can lead to reduced feed intake, decreased productivity, and even death in severe cases.
3. **Increased Humidity**: Insufficient ventilation can result in high humidity levels within animal housing. High humidity promotes the growth of bacteria, fungi, and molds, which can cause respiratory diseases, skin infections, and other health issues in animals.
4. **Spread of Diseases**: Inadequate ventilation can contribute to the spread of airborne diseases among animals. Insufficient air exchange allows pathogens to accumulate and persist in the environment, increasing the risk of contagious diseases and compromising the overall health of the animal population.
5. **Poor Ammonia Control**: In animal housing, ammonia is released from urine and faeces. Inadequate ventilation hampers the removal of ammonia from the air, leading to its accumulation. High levels of ammonia can irritate the respiratory tract, causing respiratory problems and making animals more susceptible to respiratory infections.
6. **Reduced Productivity**: Animals housed in poorly ventilated environments may experience reduced productivity, including decreased growth rates, lower reproductive performance, and compromised immune function. Inadequate ventilation can stress animals and make them more susceptible to diseases, resulting in reduced overall productivity.

7. **Increased Mortality**: In extreme cases, inadequate ventilation can lead to increased mortality rates in animals. Poor air quality, high temperatures, and the spread of diseases can all contribute to a higher risk of animal deaths within the housing facility.

20.3 Heat and Moisture Balance

The heat and moisture balance equations for livestock buildings are essential for understanding and managing the thermal environment within these structures, which directly impacts the health, welfare, and productivity of the animals housed within them. These equations typically take into account various factors such as heat production from animals, ventilation rates, thermal properties of the building materials, ambient weather conditions, and moisture generation from animals and their waste.

Heat Balance Equation

The heat balance equation accounts for all heat gains and losses within the building. It can be expressed as:

$$Q = Q_{in} - Q_{out} + Q_{met} + Q_{res}$$

where,

Q = net heat gain or loss within the building (W),

Q_{in} = heat gained from sources like solar radiation, equipment, and animals (W),

Q_{out} = heat lost through ventilation, conduction, and radiation (W),

Q_{met} = metabolic heat produced by the animals (W), and

Q_{res} = heat released or absorbed due to phase changes (*e.g.*, water evaporation) in W.

Moisture Balance Equation

The moisture balance equation accounts for the moisture generated within the building from animal respiration, urination, and evaporation from manure and bedding, as well as moisture removal through ventilation and other processes.

$$M_{total} = M_{generation} - M_{removal}$$

where,

M_{total} = total moisture input or output to the building (kg/s),

$M_{generation}$ = moisture generation rate within the building (kg/s), and

$M_{removal}$ = moisture removal rate through ventilation and other means (kg/s).

20.4 Ventilation and Thermal Loads

Air exchange increases a building thermal load in two ways.

1. Incoming air must be cooled or heated from the outdoor air temperature to indoor air temperature. The rate or energy consumption due to this sensible heating or cooling is given by

$$Q_s = V_a \rho C_p \Delta T \qquad (1)$$

where,

Q_s = sensible heat load in watts,

V_a = volumetric air flow rate in m^3/s,

ρ = density of air in kg/m^3,

C_p = specific heat of air in J/kg-K, and

ΔT = indoor-outdoor temperature difference.

2. Air exchange modifies the moisture content of air in building. The rate of energy consumption with these latent loads is given by

$$Q_L = V_a \rho h_{fg} \Delta W \qquad (2)$$

where,

Q_L = latent heat load in watts,

h_{fg} = latent heat of vapour in J/kg-K (about 2.34×10^6), and

ΔW = difference in indoor and outdoor humidity ratios.

20.5 Estimating Air Flow Rate

The generalized equation for estimating air flow rate to maintain a constant difference in concentrations of the air constituent is

$$\dot{m} = \frac{\dot{C}}{C_i - C_o} \qquad (3)$$

where,

$\dot{m}$ = air flow rate required to maintain a constant difference in concentrations in kg/hour,

C_i and C_o = concentration inside and outside of an enclosure, and

$\dot{C}$ = rate at which the constituent being generated.

The concentration may refer to any one of the several air constituents such as heat, moisture, odour, gas, and dust suspended in the air.

20.6 Control of Temperature, Humidity and other Constituents

The equations of air flow rates required for controlling temperature, humidity and other constituents are given in Table 20.1.

Table 20.1. Air Flow Rate Equations based upon Concentrating Rates and Concentrations of Air Constituents

Constituent	*Inside and Outside Concentration (C) (per kg of dry air)*	*Air Flow Rate ($\dot{m}$) (kg/h)*	*Eqn.*
Sensible heat in moist air (Q_s)	$q_s = C_p T$	$\dot{m}_s = \frac{\dot{Q}_s}{q_{si} - q_{so}} = \frac{\dot{Q}_s}{1.005(T_i - T_o)}$	(4)
Latent heat in moist air (Q_L)	$q_L = Wh_{fg}$	$\dot{m}_L = \frac{\dot{Q}_L}{q_{Li} - q_{Lo}} = \frac{\dot{Q}_L}{(Wh_{fg})_i - (Wh_{fg})_o}$	(5)
Water vapour in moist air (W)	$W = \mu W_s$	$\dot{m}_w = \frac{\dot{W}}{W_i - W_o} = \frac{\dot{W}}{(\mu W_s)_i - (\mu W_s)_o}$	(6)
Air constituents: suspended dust, gases, odours (B)	b, concentration	$\dot{m}_B = \frac{\dot{B}}{b_i - b_o}$	(7)

The air flow rate required for controlling temperature is determined by the equation (4). Equation (5) or (6) may be used for the determination of air flow rate for the control of humidity inside the building. Equations (5) and (6) are the same, and therefore yield the same result. In the similar manner, the equations for air flow rates for other constituents of air can be written. The amount of sensible heat, latent heat and water vapour of the inside and outside air can be determined with the help of psychrometric relationships or psychrometric chart. The equations (1) and (4), and (2) and (5) are same. The volumetric air flow rate can be determined by dividing mass air flow rate by the density of air.

In the analysis and design of ventilating systems for livestock and poultry, the interrelationship between the ventilation rate and the heat and moisture removed in the ventilating air is important. In some cases, the desired minimum ventilation rate has been determined for moisture control, and available heat must equal the heat removed in the ventilating air. In other cases, the available heat is fixed and the ventilation rate must be balanced accordingly.

Example 20.1

A 30 m×11 m×2.5 m cattle barn is used to house 50 cows. It is desired to maintain an inside temperature of 20°C and 70 per cent RH when the outside temperature is 2°C and 80 per cent RH. Each cow produces 600 kcal sensible heat and 200 g of moisture per hour. Estimate the rate of airflow by ventilation necessary to remove the moisture.

Solution

Plot the points corresponding of inside (20°C and 70 per cent RH) and outside (2°C and 80 per cent RH) environmental conditions on a psychrometric chart. Read the inside and outside humidity ratios and specific volume at inside conditions corresponding to these points.

Inside humidity ratio (W_i) = 0.0105 kg/kg of dry air

Outside humidity ratio (W_o) = 0.0034 kg/kg of dry air

Specific volume (v) = 0.845 m^3/kg of dry air

Moisture produced by 50 cows = (200/1000) ×50 = 10 kg/h

Rate of airflow ($\dot{m}_w$)

$$\dot{m}_w = \frac{\dot{W}}{W_i - W_o} = \frac{\text{moisture produced by 50 cows}}{\textit{moisture} \text{ that can be carried by 1 kg of air}}$$

$$\dot{m}_w = \frac{10}{0.0105 - 0.0034} = 1408.5 \text{ kg/h}$$

Volumetric rate of airflow (V_a) = $\dot{m}_w \times v$

= 1408.5 × 0.845 = 1183.1 m^3/h = 0.33 m^3/s

20.7 Types of Ventilation Systems

Ventilation system can either be natural or mechanical (forced). Proper ventilation is essential for maintaining a healthy and comfortable indoor environment. While natural ventilation relies on air movement caused by wind and temperature differences, sometimes it is insufficient to meet the ventilation needs of a space. In such cases, forced ventilation becomes necessary. Forced ventilation involves the use of mechanical devices, such as fans and blowers, to circulate and exchange air within a building. A well-designed forced ventilation system can greatly contribute to the well-being and satisfaction of occupants while ensuring optimal indoor air quality. Closed buildings require mechanical air movement, while open ones harness the forces of nature to circulate air.

20.8 Natural Ventilation

Natural ventilation relies on natural forces like wind and buoyancy to circulate air. It usually involves the use of windows, doors, or vents strategically placed to allow fresh air in and stale air out. Natural ventilation is cost-effective, energy-efficient, and suitable for buildings with fewer ventilation requirements.

Wind and thermal forces, acting alone or together may be utilized for the movement of ventilation air through livestock, storage, or plant growth structures. Natural ventilation can significantly reduce reliance on mechanical cooling and ventilation systems, leading to substantial energy savings. One air change per minute or more is required during hot weather to remove the metabolic heat produced in livestock and poultry buildings. Natural ventilation systems in open-type buildings are not controllable during cold weather to the extent necessary to conserve available body heat and maintain inside environmental air temperatures more than 5 to 7°C above outside climatic temperatures. Naturally ventilated buildings are mainly designed to protect the animals from cold drafts, rain, and snow. These buildings are significantly less costly to build and operate than environmentally controlled buildings.

Principles

Natural ventilation operates on the basic principles of pressure differentials and airflow. The driving forces behind natural ventilation are wind and buoyancy. Wind-induced ventilation relies on the external pressure differences created by wind flow around a building. Buoyancy-driven ventilation, on the other hand, depends on the temperature differences between indoor and outdoor air, which cause warm air to rise and be replaced by cooler air entering from lower openings.

20.8.1 Benefits of Natural Ventilation

Natural ventilation offers several benefits to building occupants and the environment. Some key advantages include:

1. **Improved Indoor Air Quality**: By allowing a constant flow of fresh air, natural ventilation helps to remove indoor pollutants, odours, and excessive moisture. This promotes a healthier indoor environment, reducing the risk of respiratory problems and improving overall well-being.
2. **Energy Efficiency**: Natural ventilation can significantly reduce reliance on mechanical cooling and ventilation systems, leading to substantial energy savings. By utilizing natural airflow, buildings can minimize the need for artificial cooling and the associated energy consumption.
3. **Thermal Comfort**: Properly designed natural ventilation systems can help regulate indoor temperatures, ensuring thermal comfort for occupants. By harnessing cooler outdoor air during the evening or early morning and expelling warm air during the day, natural ventilation can mitigate the need for air conditioning and maintain comfortable conditions.
4. **Sustainable Design**: Incorporating natural ventilation into building design aligns with sustainable practices by reducing greenhouse gas emissions associated with mechanical ventilation. It also decreases

reliance on non-renewable energy sources, contributing to a more sustainable and environmentally friendly built environment.

Natural ventilation has several benefits. It can provide fresh air, reduce reliance on mechanical systems, improve indoor air quality, and contribute to energy savings. However, it also has limitations. The effectiveness of natural ventilation is highly dependent on external factors like wind availability and direction, making it inconsistent in certain locations and climates. Additionally, natural ventilation may not provide sufficient airflow in densely populated or sealed buildings.

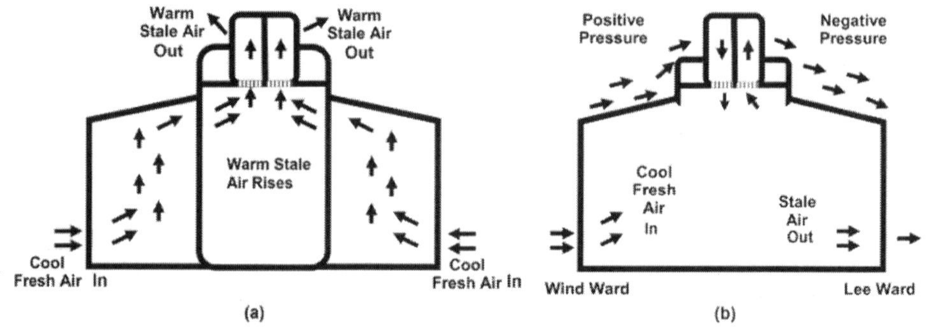

Figure 20.1. Ventilation Systems (a) Stack Driven, (b) Wind Driven.

20.8.2 Natural Air Exchange

Incoming air must be cooled or heated from the outdoor air temperature to indoor air temperature. The quantity of ventilation air exchange necessary for open livestock buildings can be calculated if the amount of heat produced and the desirable temperature difference (inside to outside) are known.

Amount of heat produced,

$$q = V_{air} C_p \rho (T_i - T_o)$$

Thus, quantity of air exchange,

$$V_{air} = \frac{q}{C_p \rho (T_i - T_o)} = \frac{q}{1280(T_i - T_o)}$$

where,

V_{air} = air exchange in m^3/s,

q = heat removed in watts,

C_p = specific heat of air at constant pressure (1004 J/kg-K),

ρ = density of standard air (1.2743 kg/m^3), and

$(T_1 - T_0)$ = average indoor–outdoor temperature difference in K.

Example 20.2

Determine the ventilation air exchange necessary to remove heat produced from 50 dairy cows. Each cow produces 1000 W of heat. The inlet and outlet temperatures of the building are 21°C and 15°C respectively.

Solution

Heat produce by each cow = 1000 W

Number of dairy cows = 50

Total amount of heat produced (q) = 1000×50 = 50000 W

Inlet temperature (T_i) = 21°C

Outlet temperature (T_o) = 15°C

Quantity of air exchange required (V_{air}) $= \dfrac{q}{1280(T_i - T_o)}$

$$= \frac{50000}{1280(21-15)} = 6.5 \text{ m}^3/\text{s}$$

20.8.3 Ventilation by Wind

Ventilation by wind, also known as natural ventilation or wind-driven ventilation (Figure 20.1b), is a process by which air is exchanged within an enclosed space through the action of wind. It is a passive method of ventilating buildings or other structures without the use of mechanical systems such as fans or air conditioning. When wind blows against a building, it creates a pressure difference between the windward (facing the wind) and leeward (opposite side) sides of the structure. This pressure difference can drive the movement of air through openings such as windows, doors, vents, or gaps in the building envelope. The air enters the building through openings on the windward side, flows through the interior, and exits on the leeward side, carrying away stale air, pollutants, and excess heat. The rate of ventilation in building depends upon the wind speed and its direction, seasonal and daily variation in velocity and direction, presence of buildings and trees, and the size, location and arrangement of outlet and inlet openings. Wind velocities are usually lower in summer than in winter, and vary in direction between summer and winter seasons.

The air exchange caused by wind forces is calculated by:

$$V_{air} = EAV$$

where,

V_{air} = air flow in m³/s,

A = free area of inlet openings in m²,

V = wind velocity in m/s, and

E = effectiveness of openings.

The effectiveness of an opening depends chiefly on its orientation with respect to wind direction and its location in the building. 'E' may be assumed as 0.5 to 0.6 for perpendicular to the opening winds and 0.25 to 0.35 for diagonal winds.

Example 20.3

What inlet opening area would be necessary to provide 6.5 m^3/s air exchange through the building? Assume a 0.5 effectiveness of opening and a wind velocity of 4 m/s.

Solution

Air flow (V_{air}) = 6.5 m^3/s

Wind velocity (V) = 4 m/s

Effectiveness of openings (E) = 0.5

Air exchange caused by wind forces (V_{air}) = EAV

$$\text{Thus, inlet opening area, A} = \frac{V_{air}}{EV} = \frac{6.5}{0.5 \times 4} = 3.25 \text{ m}^2$$

20.8.4 Stack Ventilation

Stack or flue ventilation, also known as chimney ventilation or natural draft ventilation (Figure 20.1a), is a passive cooling and ventilation system that utilizes the principle of buoyancy to induce airflow in a building. It takes advantage of the temperature difference between the interior and exterior of a building to create a natural flow of air. The stack ventilation system typically consists of a vertical shaft or chimney-like structure called a stack. The stack is designed to be taller than the surrounding structures to ensure effective ventilation. The stack can be an independent feature of the building or integrated into the architectural design.

The process of stack ventilation relies on the principle that warm air rises while cool air sinks. In a stack ventilation system, the stack acts as a vertical pathway for the warm air to escape from the building. This warm air rises due to its lower density compared to the surrounding cooler air. As the warm air exits the top of the stack, it creates a negative pressure inside the building, which draws in fresh air from the outside through other openings, such as windows or vents, to replace the expelled air.

To enhance the effectiveness of stack ventilation, certain design considerations are important. These include the height and size of the stack, the location of openings for fresh air intake, and the arrangement of internal spaces to allow for proper airflow. The stack should be positioned to take advantage of prevailing winds and solar exposure to maximize the airflow.

Stack ventilation is considered a sustainable and energy-efficient method of cooling and ventilating buildings, as it relies on natural forces rather than mechanical systems. It can be particularly useful in climates with mild to moderate temperatures and adequate wind flow. However, in hot and humid climates, stack ventilation alone may not provide sufficient cooling, and additional strategies like shading, insulation, or mechanical ventilation might be necessary.

Overall, stack ventilation offers a passive and environmentally friendly solution for providing airflow and maintaining comfort inside buildings. It is often used in combination with other ventilation strategies to create a well-ventilated and energy-efficient indoor environment.

Thermal forces can be complementary to wind forces in providing air exchange through naturally ventilated structures. During times when there is effectively no wind force, thermal forces must be relied on entirely.

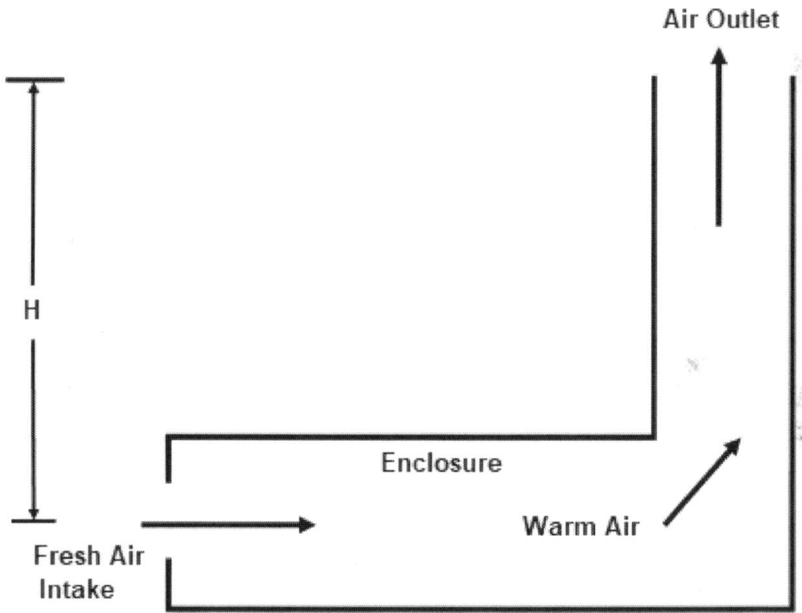

Figure 20.2. Stack Ventilation System.

The thermal (buoyancy) forces due to the difference in air density at different temperatures can cause air flow due to stack effect (Figure 20.2) as:

$$V_a = A\theta\sqrt{\frac{2gh(T_i - T_o)}{T_i}}$$

where,

V_a = air flow in m³/s,

A = free area of inlets or outlets (the smaller area) in m²,

g = acceleration due to gravity, 9.81 m/s²,

h = height from inlets to outlets in m,

θ = ratio of flow with heat and frictional losses to theoretical flow, usually 0.3 to 0.5,

T_i = inside temperature in K, and

T_o = outside temperature in K.

This equation is based on the concept that the potential energy of the buoyancy force is converted to kinetic energy when the air moves through the stack, and that the perfect gas law is applicable.

Example 20.4

What inlet opening area would be necessary to provide 6.5 m³/s air exchange through the building? Assume a ridge opening at a height of 8 meter. The inside and outside temperatures of the building are 21°C and 15°C respectively.

Solution

Air flow (V_a) = 6.5 m³/s

Acceleration due to gravity (g) = 9.81 m/s²

Height from inlets to outlets (h) = 8 m

Ratio of flow with heat and frictional losses to theoretical flow (θ) = 0.3 to 0.5

Inside temperature (T_i) = 21 + 273 = 294 K

Outside temperature (T_o) = 15 + 273 = 288 K

$$V_a = A\theta\sqrt{\frac{2gh(T_i - T_o)}{T_i}}$$

$$\text{or, } A = \frac{V_a}{\theta\sqrt{\frac{2gh(T_i - T_o)}{T_i}}} = \frac{6.5}{\theta\sqrt{\frac{2\times9.81\times8\times(294-288)}{294}}}$$

$$= \frac{3.65}{\theta}$$

$$\text{For } \theta = 0.3, \text{ Area (A)} = \frac{3.65}{0.3} = 12.2 \text{ m}^2, \text{ and}$$

$$\text{For } \theta = 0.5, \text{ Area (A)} = \frac{3.65}{0.5} = 7.3 \text{ m}^2$$

20.8.5 Natural Ventilation Rules

To effectively use natural ventilation in a building, there are several rules and considerations that should be taken into account. These guidelines can help optimize airflow, maintain indoor air quality, and enhance occupant comfort. Here are some key rules to consider:

1. **Building Orientation**: Orient the building to take advantage of prevailing wind directions. Positioning windows, vents, and openings on windward and leeward sides of the building can facilitate natural airflow.
2. **Window Placement and Size**: Place windows strategically to promote cross-ventilation and create a flow of air. Windows should be large enough to allow sufficient airflow but not so large that they create excessive drafts.
3. **Stack Effect**: Utilize the stack effect, which relies on temperature differences, by incorporating high and low openings in the building. Warm air tends to rise, creating a natural upward airflow, while cool air enters through lower openings.
4. **Airflow Paths**: Inlet openings should not be obstructed by buildings, trees, signboards, *etc.* Establish clear airflow paths within the building to enable the movement of air. Direct short circuits between openings on two sides at a high level may clear the air at that level without producing any appreciable ventilation at the lower level of occupancy.
5. **Opening Size and Location**: Determine the appropriate size and location of openings (*e.g.*, windows, vents) based on factors such as building orientation, prevailing wind directions, and the desired airflow rate. Outlets in regions where downward air currents or high pressures exist are less effective than outlets in other regions. In order that the inlet openings be fully effective, there must be an approximately equal (or larger) outlet area.
6. **Ventilation Rates**: Calculate the required ventilation rates based on the size of the space, the occupancy level, and the intended use of the area. This ensures an adequate supply of fresh air and helps maintain indoor air quality.
7. **Environmental Conditions**: Consider local climate conditions, including temperature, humidity, and wind patterns, when designing natural ventilation systems. This helps optimize the effectiveness of airflow and ensures comfort for occupants. Different climate zones require tailored natural ventilation strategies. In hot climates, for example, it is crucial to incorporate shading devices and consider night time ventilation to take advantage of cooler temperatures. In colder climates, strategies may focus on minimizing cold drafts while maintaining adequate air exchange.

8. **Adjustable Openings**: Provide mechanisms to adjust the size of openings, such as windows or vents, to control the amount of airflow according to occupant needs and environmental conditions.

20.9 Mechanical (Forced) Ventilation

Mechanical ventilation involves the use of fans, blowers, and ductwork to actively move air in and out of the building. It provides greater control over airflow rates, allowing for precise regulation of temperature, humidity, and air quality. Mechanical ventilation systems can be classified into two main types:

1. Exhaust Ventilation

Exhaust ventilation systems (Figure 20.3) use fans to remove stale air and contaminants by creating a negative pressure inside the building, which pulls fresh air in through strategically placed inlets. These systems are commonly used in poultry houses and barns, effectively removing odours, excess heat, and humidity. This system is simple and quite reliable in that the ventilation air moves across the building and provides good air distribution.

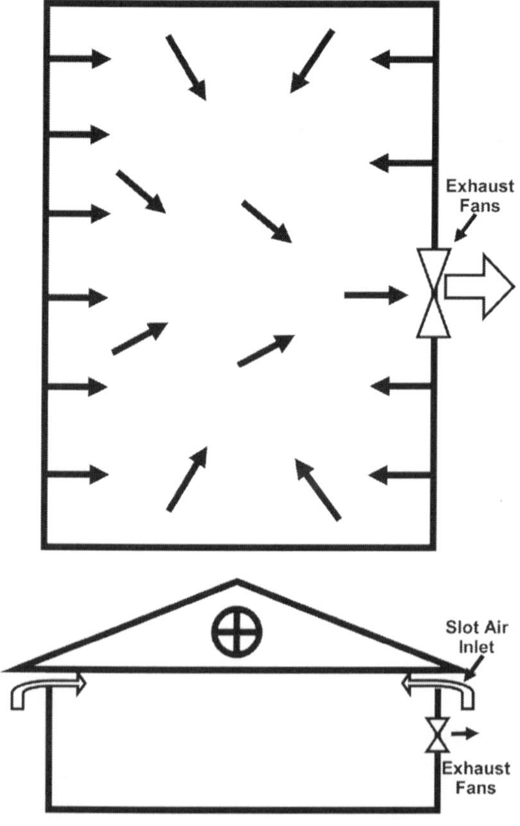

Figure 20.3. Exhaust Ventilation System.

2. Positive Pressure Ventilation

Positive pressure ventilation systems (Figure 20.4) use fans to introduce fresh air into the building at a higher pressure, creating a positive pressure environment that forces stale air out through exhaust vents or openings. Pressure systems are often used for summer ventilation in large-animal housing, because high-velocity fresh air around the animals' aids in cooling them. This system has poor air distribution of incoming fresh air unless a system of ducts is used. During cold weather, it causes cold drafts in the vicinity of the fan inlet. To avoid cold draft, duct distribution systems can be installed, but they are expensive and require additional power. Pressure fans without the duct network would be acceptable during hot weather. Pressurized systems are seldom used in cold climates during winter, because moisture can be forced through small openings into wall cavities and attics where it will wet the insulation. Also, frost may form at air leaks around and under doors, causing them to freeze shut.

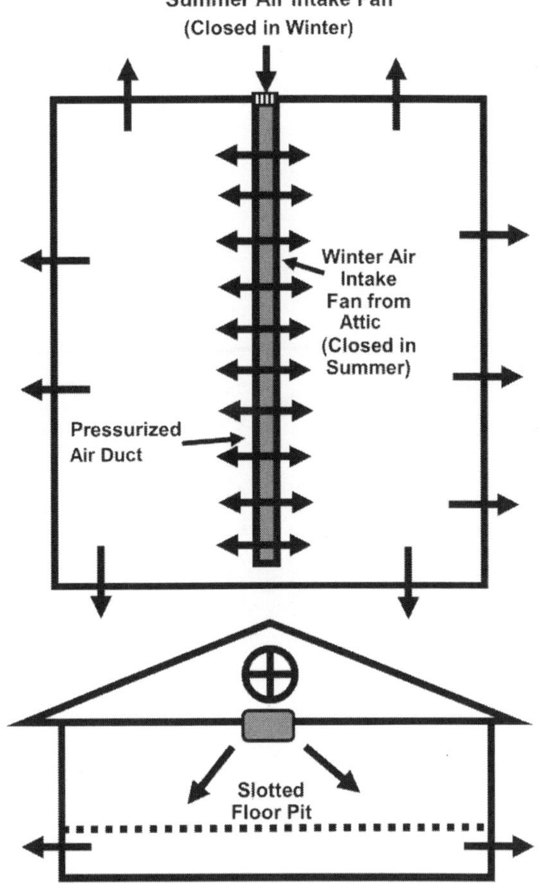

Figure 20.4. Pressure Ventilation System.

20.9.1 Need for Forced Ventilation

1. **Inadequate Natural Ventilation**: Natural ventilation depends on external factors like wind and temperature differences. In some situations, these factors may not provide sufficient air movement, especially in tightly sealed or overcrowded spaces.
2. **Contaminant Removal**: Certain spaces generate pollutants or contaminants that cannot be adequately eliminated through natural ventilation alone. Examples include fumes from industrial processes, odours, or harmful gases.
3. **Temperature and Humidity Control**: Forced ventilation can help regulate indoor temperature and humidity levels, ensuring a comfortable environment for occupants. It can help remove excess heat and moisture generated by occupants, equipment, or processes.
4. **Airborne Pathogens**: In environments where the risk of airborne diseases is high, forced ventilation plays a crucial role in diluting and removing pathogens, reducing the chances of transmission.

For successful ventilation, the points exhaust and natural inlets must be so arranged that fresh air is drawn across all parts of the space. Typical bad and good arrangements are shown in Figure 20.5.

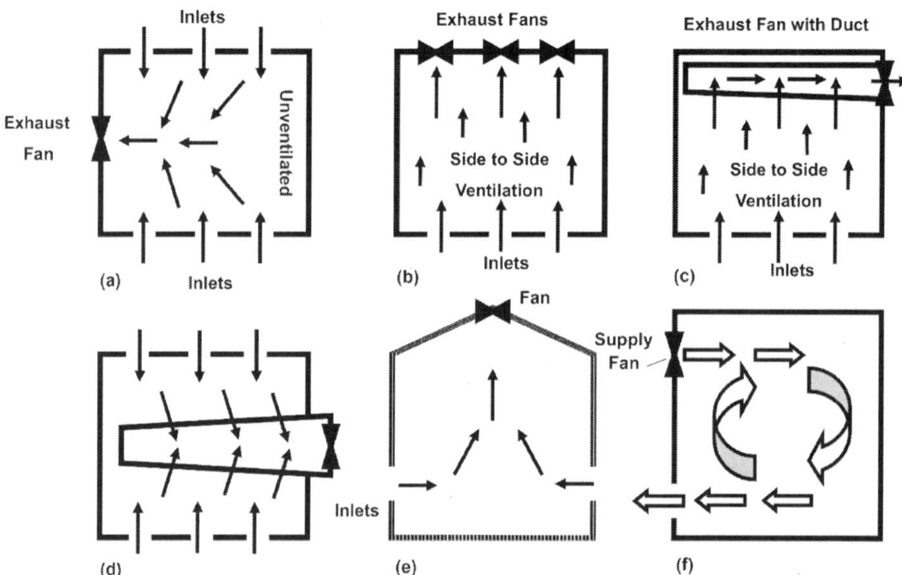

Figure 20.5. (a) Bad Arrangement Causing Short Circuiting, (b) Good Distribution by Side-to-Side Ventilation, (c) Side-to-Side Ventilation with Extraction Duct, (d) Central Extraction Duct, (e) Roof Extraction Unit, (f) Plain Supply System with Fan Blowing Direct from Atmosphere.

20.9.2 Air Inlets and Outlets

Air Inlets

The size of air inlets for ventilation can vary depending on several factors, including the purpose of ventilation, the size of the space being ventilated, and the air exchange rate required. For effective ventilation, it is important to distribute the air evenly throughout the space. Consider the placement and number of air inlets to achieve good air circulation and minimize stagnant areas. Locate air inlets at ceiling level so that cold incoming air has a chance to mix with warmer air before it reaches the animals. Using a baffle to restrict the inlet opening increases air velocity and aids in the mixing of air. The exhaust fans in one building should not blow foul air into the fresh air inlets of another building. Thus, a distance of at least 10 meter shall be provided between two buildings.

Inlet-opening size must match fan capacity to maintain proper air velocity. Usually, this means the inlet is sized for the highest (hot weather) ventilation rate and then closed down with a baffle for lower ventilation rates. To determine air inlet area, an average effective air velocity of 3 m/s is assumed for the required ventilation air exchange through the inlet.

Air Outlets

The size of outlets for ventilation is also an important consideration when designing an effective ventilation system. The size of the outlets depends on factors such as the airflow requirements, the type of space, and the air exchange rate desired. It is important to distribute the outlets evenly throughout the space to ensure proper air circulation and prevent dead zones. Consider the placement and number of outlets to achieve uniform airflow and minimize air recirculation.

Air in the exhaust-type ventilation systems is uniformly drawn from all directions. Air velocity decreases rapidly with increasing distance from the inlet side of the fan. Air is being moved by an exhaust fan at inside environmental temperature and thus does not cause a cold draft. Figure 20.6 shows a fan outlet which is drawing air from all directions of the hemisphere.

According to the continuity equation, the flow through point '1' must equal that through point '2'.

Thus, $\frac{\pi}{4} D^2 V_1 = 2\pi r^2 V_2$

or, $V_2 = \frac{D^2 V_1}{8r^2}$

V_1 = average velocity of air through the outlet,

V_2 = velocity of air at some radial distance from the centre of the outlet,

D = full diameter of outlet, and

r = radial distance from centre of outlet.

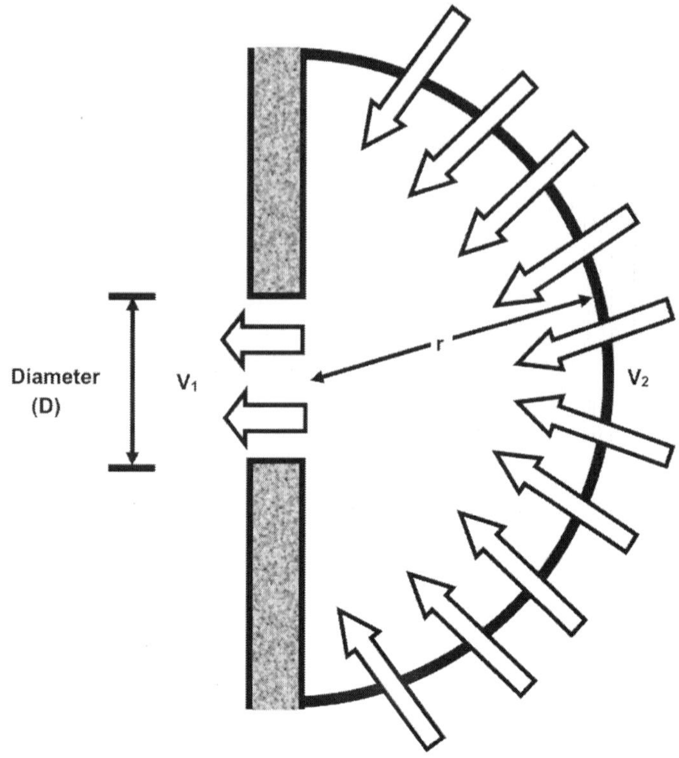

Figure 20.6. Exhaust Fan Outlet.

Example 20.5

One-meter diameter fan having an exhaust air flow capacity of 8 cubic meter per second is installed in an animal shelter. What will be the average exit air velocity? What will be the velocity at an equal distance of 2.5 meter from the centre of the fan?

Solution

Flow rate = 8 m^3/s

Diameter of fan (D) = 1 m

Radial distance from the centre of outlet (r) = 2.5 m

Cross-sectional area of fan = $\frac{\pi}{4}D^2 = \frac{\pi}{4}(1)^2 = 0.785 \text{ m}^2$

Thus, average exit air velocity (V_1) = $\frac{\text{flow rate}}{\text{cross-sectional area of fan}} = \frac{8}{0.785}$

$$= 10.2 \text{ m/s}$$

Velocity of air (V_2) at an equal distance of 2.5 meter from the centre of the fan

$$V_2 = \frac{D^2 V_1}{8r^2}$$

$$= \frac{1^2 \times 10.2}{8 \times (2.5)^2} = 0.204 \text{ m/s}$$

20.9.3 High Volume Low Speed Fan

A high volume low speed (HVLS) fan is a large ceiling fan designed to move significant amounts of air at low rotational speeds. Unlike traditional fans that operate at high speeds, HVLS fans create a gentle breeze by moving a large volume of air at a slow pace, typically between 50 to 150 revolutions per minute (RPM). The slower, steadier fan speed and longer fan props push large volumes of air from the ceiling and down to the floor level. The air moves horizontally until reaching a column or wall, which then projects airflow upward and sideways. These fans are commonly used in large spaces such as warehouses, industrial facilities, gymnasiums, and commercial buildings to improve air circulation, enhance ventilation, and maintain comfortable indoor temperatures. HVLS fans are particularly beneficial for preventing heat stress in livestock during periods of high temperatures. By promoting air circulation and evaporative cooling, HVLS fans can help animals dissipate excess body heat more efficiently, reducing the risk of heat-related illnesses and maintaining optimal performance and health. This is particularly important for species like dairy cows, which are sensitive to heat stress and may experience decreased productivity in high temperatures.

Figure 20.7. High Volume Low Speed Fan.

The design of HVLS fans allows them to efficiently circulate air throughout a space, helping to alleviate issues such as stagnant air, temperature stratification, and moisture build-up. By promoting air movement on a larger scale, HVLS fans can also contribute to energy savings by reducing the need for mechanical cooling systems and improving the overall effectiveness of HVAC systems. In addition to their practical benefits, HVLS fans are known for their quiet operation and low maintenance requirements. They are often equipped with aerodynamically designed blades and energy-efficient motors to maximize airflow while minimizing noise and power consumption. However, HVLS fans typically have a higher upfront cost compared to traditional fans. This cost may be offset by energy savings and long-term benefits. A high-volume low-speed fan is greater than 2.1 meter in diameter.

Advantages of High Volume Low Speed Fans

1. HVLS fans move a large volume of air at low speeds, promoting better air circulation throughout large spaces. This helps to prevent stagnant air pockets and reduces temperature differentials within the space.
2. Compared to traditional high-speed fans or air conditioning systems, HVLS fans consume less energy while providing effective air movement. This can lead to significant energy savings.
3. HVLS fans help to improve ventilation by circulating fresh air and exhausting stale air. This can contribute to better indoor air quality by reducing the build-up of pollutants, odours, and moisture.
4. By improving air circulation and ventilation, HVLS fans can reduce the workload on HVAC systems. This may extend the lifespan of HVAC equipment and decrease maintenance costs.
5. The gentle breeze created by HVLS fans provides a cooling effect, making occupants feel more comfortable even in large spaces with high ceilings. This can enhance productivity and well-being of human beings or animals.

Important Points to Remember

1. Ventilation refers to the process of exchanging or replacing air in an enclosed space to maintain a fresh and healthy indoor environment.
2. Inadequate ventilation leads to a build-up of harmful gases, such as ammonia, carbon dioxide, and dust particles in an animal housing.
3. Air exchange due to ventilation increases a building thermal load in two ways *viz.* sensible heating or cooling load, and latent heat load (due to change of moisture content).
4. Ventilation system can either be natural or mechanical (forced).
5. Natural ventilation relies on air movement caused by wind and temperature differences (buoyancy).

6. Forced ventilation involves the use of mechanical devices, such as fans and blowers, to circulate and exchange air within a building.
7. Closed buildings require mechanical air movement, while open ones usually harness the forces of nature to circulate air.
8. Natural ventilation is cost-effective, energy-efficient, and suitable for buildings with fewer ventilation requirements.
9. The driving forces behind natural ventilation are wind and buoyancy.
10. The effectiveness of an opening in the wall depends chiefly on its orientation with respect to wind direction and its location in the building.
11. The process of stack ventilation relies on the principle that warm air rises while cool air sinks.
12. Windows should be placed strategically to promote cross-ventilation and create a flow of air.
13. Inlet openings should not be obstructed by buildings, trees, signboards, *etc.*
14. Mechanical ventilation systems can be classified as exhaust and pressure ventilation systems.
15. Exhaust ventilation systems creates a negative pressure inside the building, which pulls fresh air in through strategically placed inlets.
16. In exhaust ventilation, air distribution within the building is uniform.
17. Exhaust ventilation systems are commonly used in poultry houses and barns for effectively removing odours, excess heat, and humidity.
18. Pressure systems are often used for summer ventilation in large-animal housing, because high-velocity fresh air around the animals' aids in cooling them. This system has poor air distribution of incoming fresh air.
19. Uneven air distribution inside a room occurs during pressure ventilation.
20. Air should be evenly distributed throughout the space for effective ventilation.
21. Ventilation is required for controlling temperature, moisture and proper air composition.
22. During pressure ventilation, positive pressure is created inside the enclosure.
23. Vacuum is created with in the building during exhaust ventilation.
24. Even in a warm environment, air drafts should be avoided in barns.

Exercises/Questions

1. Why is ventilation required in a farm building? List the factors affecting ventilation.
2. What are the determinant effects of inadequate ventilation in animal housing?
3. Write an equation for heat and moisture balance in a building.
4. What are the benefits of natural ventilation?
5. Write a short note on stack ventilation.
6. Discuss the rules and considerations for natural ventilation.
7. Explain different types of mechanical ventilation systems.
8. Under what circumstances, forced ventilation is recommended?
9. Write a short note on air inlets and outlets for ventilation system in an animal housing.
10. Differentiate between

    (a) natural and forced ventilation, (b) pressure and exhaust ventilation

11. Calculate the ventilation air exchange required to remove 30000 W of heat from a shed housing 30 dairy cows with a temperature difference of 5°C between inside and outside environment.

    [**Answer**: 4.7 m^3/s]

12. What inlet opening area would be necessary to provide the 5 m^3/s air exchange through the building? Assume a 0.3 effectiveness of opening and a wind velocity of 5 m/s.

    [**Answer**: 3.3 m^2]

13. What inlet opening area would be necessary to provide the 5 m^3/s air exchange? Assume a ridge opening at a height of 7 m. The inside and outside temperatures of the building are 15°C and 5°C respectively. Assume the value of θ as 0.3.

    [**Answer**: 7.74 m^2]

14. One-meter diameter fan having an exhaust air flow capacity of 5 m^3/s is installed in an animal shelter. What will be the average exit air velocity? What will be the velocity at an equal distance of 2 m from the centre of the fan?

    [**Answers**: 6.37 m/s, 0.19 m/s]

15. Explain the working of high volume low speed fans. Also enlist the advantages of high volume low speed fans.

Chapter 21

Livestock Waste Management

Animal housing facilities generate a significant amount of waste, including manure, bedding material, and feed residues. Gases, dust particles, and odorous vapours are also produced in these facilities. If not managed properly, these by-products can lead to environmental pollution, unpleasant odours, and the potential spread of diseases. Moreover, the enclosed nature of these environments intensifies the impact, making it crucial to adopt integrated waste management strategies. Effective waste management not only promotes environmental conservation but also ensures the health and well-being of both animals and humans involved in these operations.

21.1 Gases and Odors

Gaseous pollutants in animal buildings can pose significant challenges to both animal welfare and the health of those working in such environments. Among the commonly monitored gases are ammonia (NH_3), hydrogen sulphide (H_2S), methane (CH_4), and carbon dioxide (CO_2). Ammonia, a by-product of microbial decomposition of animal waste, can cause respiratory distress and compromise the immune system in animals. Hydrogen sulphide, emanating from manure decomposition, is a highly toxic gas that, in elevated concentrations, can lead to respiratory issues and pose serious health risks. Methane, produced during enteric fermentation in ruminants, contributes to greenhouse gas emissions, impacting the environment. Carbon dioxide, a by-product of respiration, requires proper ventilation control to prevent its accumulation, which can lead to respiratory stress in animals. Effective waste management practices, well-designed ventilation systems, and regular monitoring are essential to mitigate the adverse effects of these gaseous pollutants in animal buildings, ensuring the well-being of both livestock and caretakers.

Permissible limits for gases in animal housing environments can vary depending on the specific type of animal, its age, health status, and the purpose of the facility. For example, recommended ammonia levels typically range from 5 to 25 parts per million (ppm), while hydrogen sulphide may have permissible exposure limits of around 10 ppm over an 8-hour workday. Methane and carbon dioxide are often monitored for ventilation purposes, and their levels can vary based on the specific needs of the animals.

Carbon Dioxide

Elevated levels of carbon dioxide (CO_2) in the environment can have profound effects on animal health and performance. Increased CO_2 concentrations can lead to respiratory distress in animals. The normal atmospheric concentration of CO_2 is approximately 0.03 per cent, but it increases significantly due to animal respiration and manure decomposition, leading to a depletion of oxygen in the building atmosphere. While the gas is non-toxic on its own, concentrations of up to 10 per cent can be endured by both humans and animals, provided that exposure does not exceed one hour. CO_2 concentration of 10 per cent causes violent panting. Above 10 per cent level, even with sufficient oxygen, the narcotic effect persists due to the greater affinity of red blood cells for carbon dioxide than for oxygen.

Ammonia

Ammonia, a compound composed of nitrogen and hydrogen, poses significant risks to animal health and performance, particularly in agricultural and confined settings. Commonly found in animal waste, such as urine and manure, ammonia can accumulate in high concentrations in poorly ventilated areas like barns or pens. Inhalation of elevated ammonia levels can lead to respiratory distress in animals, causing irritation to the eyes, nose, and respiratory tract. In the case of poultry, high ammonia levels have been linked to reduced egg production and compromised egg quality.

The gas is highly soluble in water. Due to this, solid floors or units with litter tend to produce more pronounced ammonia odours compared to liquid-manure systems. Ammonia odours can be discerned at concentrations as low as 1 ppm. It causes eye irritation at 25-30 ppm. Chickens may experience eye inflammation at 50 ppm (0.005 per cent). Animals exposed to concentrations of up to 200 ppm may experience a loss of appetite, increased salivation, and sneezing. While there is no significant impact on feed efficiency, there is a notable decrease in the rate of gain. A concentration of 5000 ppm is considered hazardous for humans.

Hydrogen Sulphide

Hydrogen sulphide (H_2S), a colourless and highly toxic gas, poses severe threats to animal health and performance. Livestock housed in environments where manure decomposition occurs, such as in confined animal feeding operations, may be exposed to elevated levels of hydrogen sulphide (rotten egg odour). Hydrogen sulphide is considered as most poisonous gas in liquid-manure storage systems. It is produced by anaerobic decomposition of manure. The gas is soluble in water and can be controlled somewhat through the dilution of manure. In a mixture with oxygen, hydrogen sulphide has the potential to burn and undergo violent explosions. Agitation and the pumping of liquid wastes from storage can result in the release of high concentrations of hydrogen sulphide.

Even at low concentrations, hydrogen sulphide can cause severe irritation to the eyes and respiratory tract in humans within an hour. Animals exposed to levels of 20 ppm may exhibit fear of light, nervousness, and loss of appetite. Concentrations of 50 – 200 ppm can lead to symptoms such as vomiting, nausea, and diarrhoea. Higher concentrations of 800-1000 ppm during liquid manure agitation are sufficient to cause death to animals and birds. Adequate ventilation can help maintain levels well below 20 ppm, except during the agitation of liquid manure.

Methane

Methane does not directly impact animal health, but its release into the atmosphere contributes significantly to global warming. Methane is highly flammable and posing the greatest hazard as an explosive gas when combined with air at concentrations between 5-15 per cent. Methane tends to rise due to its lighter-than-air properties and accumulates near the upper regions of confined spaces, such as enclosed manure storage pits. While ruminant animals release lesser quantities of methane through exhalation, the majority emanates from the decomposition of manure under anaerobic conditions. In elevated concentrations, methane can also pose a risk of asphyxiation. The methane gas can be dissipated fairy rapidly with normal ventilation.

Aerial Dust

Aerial dust, often prevalent in agricultural and confined animal environments, can have detrimental effects on animal health and performance. Microscopic particles suspended in the air, including organic matter, dried faeces, and feed particles, contribute to the formation of airborne dust. Prolonged exposure to high levels of dust can irritate the respiratory systems of animals, leading to conditions such as coughing, nasal discharge, and laboured breathing. Dust particles may carry allergens and pathogens, increasing the risk of respiratory infections. In addition to compromising respiratory health, the inhalation of dust can result in reduced feed intake and hinder overall animal performance. Dust-related stress can manifest in behaviours such as restlessness and decreased weight gain. Proper ventilation, dust control measures, and regular cleaning practices are essential for minimizing airborne dust in animal facilities, safeguarding respiratory health, and optimizing the well-being and productivity of animals in various agricultural and captive settings.

Control of Gases, Odor and Other Constituents

The primary and crucial factor in managing gases and odours within a livestock building is maintaining continuous ventilation to expel these substances and introduce fresh external air. The problems with odours and gases are minimal in solid floor housing. When incorporating underfloor storage pits into the livestock housing system, it is crucial to incorporate exhaust ventilation for the pits. Without proper ventilation through the pits, the levels of carbon dioxide and hydrogen sulphide within the pits may rise to toxic levels.

Ventilation rates designed for temperature and humidity control in confined livestock production structures are generally sufficient to uphold gas levels that are not harmful to animals, except during activities such as liquid manure agitation or pumping from storage. Special precautions and measures are necessary during these tasks, such as relocating animals from the building, employing additional ventilation through extra fans or open doors and windows, or a combination of both. If comprehensive data on gas generation rates are available, the use of an equation (7) of Chapter 20 can help determine the necessary ventilation rates.

It's crucial for animal care professionals and facility managers to regularly monitor gas concentrations, implement proper ventilation systems, and adhere to recommended guidelines to maintain a safe and healthy environment for both animals and humans.

21.2 Animal Waste Production and Characteristics

Animal waste production and characteristics vary depending on the species, diet, housing conditions, and management practices. Different animals produce varying amounts and types of waste. For example, ruminants like cattle produce large quantities of solid manure, while poultry generate a mix of faeces and bedding material. The composition of animal waste is influenced by the type and quality of the animal's diet. Diets rich in fibre can result in more fibrous and solid waste. The design and conditions of animal housing, such as whether animals are kept indoors or outdoors, can impact waste management.

Manure is organic matter, mostly derived from animal faeces and urine, but normally also containing plant material (often straw), hair, feathers, and other debris. Plant material has been used as bedding for animals and has absorbed the faeces and urine. Examples of livestock/animals wastes/manure include dairy shed effluent (containing urine, dung, wash water, residual milk, and waste feed), dairy manure, poultry litter (a mix of manure, water, spilled feed, feathers, and bedding material), renderings, and other wastes from livestock finishing operations. Manure containing up to 15 per cent solids (85 per cent moisture content) can be handled as liquid, but may need special equipment for pumping.

Livestock Waste Production

Livestock manure includes dung and urine, which are valuable sources of organic matter and nutrients for soil fertility. The amount of wastes an animal can produce depends on several factors such as the type of animal/bird, breed, age, diet and management practices. The livestock waste production rate of principal farm animals (ASAE 1982) is given in Table 21.1. These are rough estimates and the actual amounts can vary based on specific conditions. The specific conditions may be diet, animal age, and management practices. The raw waste values give the weight of faeces and urine, only. The production rate given in table is per 1000 kg of live weight, and thus involve more than one animal.

Table 21.1. Manure Production (ASAE 1982)

Animal/Bird	*Raw Waste (kg/day/1000 kg live weight)*	*Faeces/ Urine Ratio*
Dairy		
Cow	82.0	2.2
Heifer	85.0	1.2
Beef		
Yearling (182-318 kg)	90.0	1.8
Feeder (> 318 kg)	60.0	2.4
Swine		
Feeder	65.0	1.2
Breeder	50.0	
Sheep	40.0	1.0
Horse	45.0	4.0
Poultry		
Layer	53.0	
Broiler	71.0	

Composition of Raw Manure

Raw manure is a mixture of solid and liquid waste produced by animals, and its composition can vary depending on the type of animal, diet, age, and management practices. However, in general, the composition of raw manure includes organic matter, nutrients, water, and other substances. Manure has a high water content, which varies depending on factors such as animal species, diet, and environmental conditions. The moisture content can range from 70 per cent to 90 per cent or more. Manure contains essential nutrients for plant growth, including nitrogen (N), phosphorus (P), and potassium (K). These nutrients are present in varying proportions, depending on the type of animal and its diet. Other micronutrients such as sulphur, calcium, magnesium, and trace elements may also be present. The contents of N, P, and K in the manure may vary from 1-3 per cent, 0.5-2 per cent, and 1-2 per cent, respectively. The organic matter content can range from 20 per cent to 80 per cent or more, depending on the type of animal and its diet. As a rough estimate, the density of raw manure might fall in the range of 1005 to 1050 kg/m^3.

Cow manure typically contains around 10-15 per cent total solids, but this can vary. Volatile solids are the organic matter in manure that can be broken down by microbial activity. This includes materials like carbohydrates, proteins, and fats. The percentage of volatile solids in raw manure is typically a subset of the total solids. For example, in cow manure, the volatile solids content might range from approximately 70 per cent to 80 per cent of the total solids. It's important to note that these values are general estimates, and the actual composition of manure can be influenced by numerous factors.

21.3 Livestock Waste Collection and Handling

Animal or poultry wastes may be handled as solids or liquids. When managed as solids, the solids content shall be 20 per cent or more. The solids content can be enhanced by adding absorbent material such as straw, wood shavings, *etc.* and also through ventilation air. If managed as liquids, the waste production must be diluted with additional water to achieve pumpability with mechanical equipment. Manure that has up to 5 per cent solids can be handled as liquid. Conventional materials handling equipment cannot manage raw manure. Typically, the raw manure is allowed to undergo separation into solid and liquid components, and each component is then handled separately.

Adding Bedding to Raw Manure

Adding bedding to raw manure is a common practice in animal husbandry, and it serves several purposes in managing manure. Bedding materials are typically added to provide comfort to animals, absorb moisture, and create a more conducive environment. Bedding can make it easier to handle and manage as solid manure, especially when it comes to composting or incorporating manure into soil. Common bedding materials include straw, hay, wood shavings, sawdust, and similar materials.

The amount of bedding required to increase the solids content to a specific level for handling as a solid can be determined from the of following relationship:

$$\text{Amount of bedding to be added } (W_b) = \frac{W_m(MC_i - MC_f)}{(MC_f - MC_b)}$$

where,

W_b = weight of bedding material to be added to the manure in kg,

W_m = initial weight of manure in kg,

MC_i = initial moisture content of manure in decimal,

MC_f = final or desired moisture content of manure in decimal, and

MC_b = initial moisture content of bedding material in decimal.

Example 21.1

How much straw at 12 per cent moisture content must be added to 100 kg of raw dairy manure (87.5 per cent moisture content) to make it possible to handle as a solid. The final moisture content of the mixture shall be 80 per cent.

Solution

$$\text{Amount of bedding to be added } (W_b) = \frac{W_m(MC_i - MC_f)}{(MC_f - MC_b)}$$

$$= \frac{100(0.875 - 0.8)}{(0.8 - 0.12)}$$

$$= 11.03 \text{ kg}$$

Thus, 11.03 kg of straw is required for each 100 kg of manure.

Note: This example can also be solved with the help of mass balance equation.

Adding Water to Manure

Moisture content influences the consistency of manure. Adding water may make the manure easier to pump, stir, transport, and apply. The amount of water required to dilute manure to a specific level for handling as a liquid can be determined from the following relationship:

$$\text{Amount of water to be added } (W_w) = W_m\left[\frac{(1 - MC_i)}{(1 - MC_f)} - 1\right]$$

where,

W_w = weight of water to be added to the manure in kg,

W_m = initial weight of manure in kg,

MC_i = initial moisture content of manure in decimal, and

MC_f = final or desired moisture content of manure in decimal.

Example 21.2

How much water must be added to 100 kg of raw dairy manure (87.5 per cent moisture content) to make it pumpable? Assume 5 per cent solids for pumping.

Solution

Initial moisture content of manure (MC_i) = 87.5 per cent = 0.875

Final moisture content (MC_f) = 100 – 5 = 95 per cent = 0.95

Weight of raw manure (W_m) = 100 kg

Amount of water to be added (W_w) $= W_m\left[\frac{(1-MC_i)}{(1-MC_f)}-1\right]$

$$= 100\left[\frac{(1-0.875)}{(1-0.95)}-1\right]$$

$$= 150 \text{ kg}$$

Thus, 150 kg of water is required to make the manure pumpable.

Note: This example can also be solved with the help of mass balance equation.

Collection and Handling Methods

Effective animal waste collection and handling are crucial aspects of livestock management to ensure environmental sustainability, animal well-being, and compliance with regulations. The methods used may vary based on the type of animals, and farm size. The collection of manure from the building can be done either manual or mechanical. Manual methods such as shovelling or scraping may be used in smaller operations to remove solid waste from stalls, pens, or housing areas. For larger facilities, skid loaders, automated or tractor-drawn scrapers are often used to collect manure from the floors and alleys. For liquid manure, flushing or scraping may be used to move liquid waste to collection points. Gutters and drainage systems are designed to collect urine and other liquids separately from solid manure. This separation can help in managing the waste more efficiently.

Mechanical scrapers (Figure 21.1) with blades are integral components in modern livestock facilities, contributing to effective waste management and ensuring optimal living conditions for animals. These systems are specifically designed to efficiently collect and remove manure from solid floors, under slotted floors, or in gutters or channels to convey wastes to storage treatment or spreading equipment. The scraper blades are engineered to effectively scrape and push manure along the barn floor, directing it towards designated collection points. Many mechanical scrapers are automated and run on a predetermined schedule, ensuring consistent waste removal without manual intervention. This automation enhances efficiency and reduces the labour required for daily maintenance. They are suitable for use in dairy barns, beef feedlots, and other livestock confinement systems. The scraper blades are typically constructed from durable materials such as steel or high-density plastics, ensuring longevity and resistance to wear and tear in the challenging barn environment. Front-end loaders on farm tractors may be used to remove solid manure from open lot surfaces, storages, and building floors. The flooring in livestock housing may either be slotted or solid.

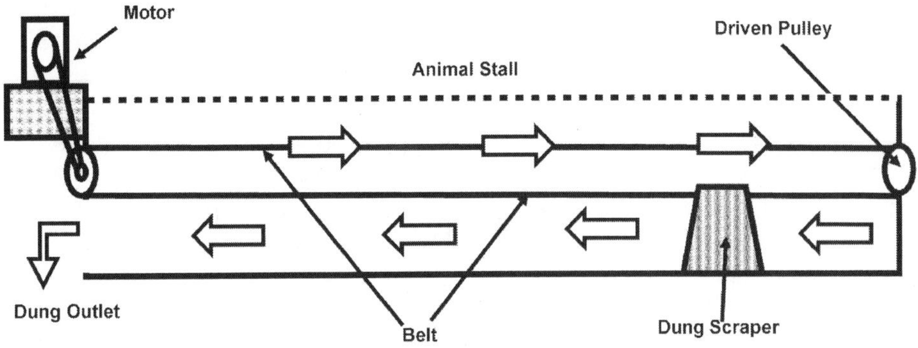

Figure 21.1. Manure Cleaning Scraper System.

Slotted Floors

Slotted floors are commonly used in confinement systems of livestock production, especially in barns and buildings where animals are housed. These floors have openings or slots that allow manure and urine to pass through, facilitating waste management and improve hygiene. Slotted floors are made of wood, plastic, concrete, metal, extruded aluminium, and fiberglass materials with openings or slots that are designed to be wide enough for waste to pass through but narrow enough to provide a comfortable and stable surface for the animals to stand or lie on. The size of slots for slotted floors in animal housing can vary depending on the type of animals being housed, their size, age, and specific management practices. The slot size should provide good support for the animals' feet to prevent injuries and lameness. The floor should have enough strength to support the weight of animals. Narrow slots hamper the fall-through of manure, whereas excessively wide slots may result in broken legs or damage to the animals' feet. For pigs' slot sizes typically range from 1.9 to 3.8 cm. Smaller slots may be suitable for piglets to prevent them from getting stuck, while larger slots may be used for larger pigs. For poultry in broiler production, slots are often smaller to provide good support for the birds' feet. For adult sheep and goats, slot sizes typically range from 2.5 to 5 cm. For lambs and kids, smaller slot sizes may be necessary to prevent them from slipping through or getting stuck. A slot size of around 2.5 cm is often recommended for young animals. The separation of animals from their waste contributes to a more hygienic environment, reducing the risk of diseases and improving overall animal health. Slats of various materials for slotted floors is shown in Figure 21.2.

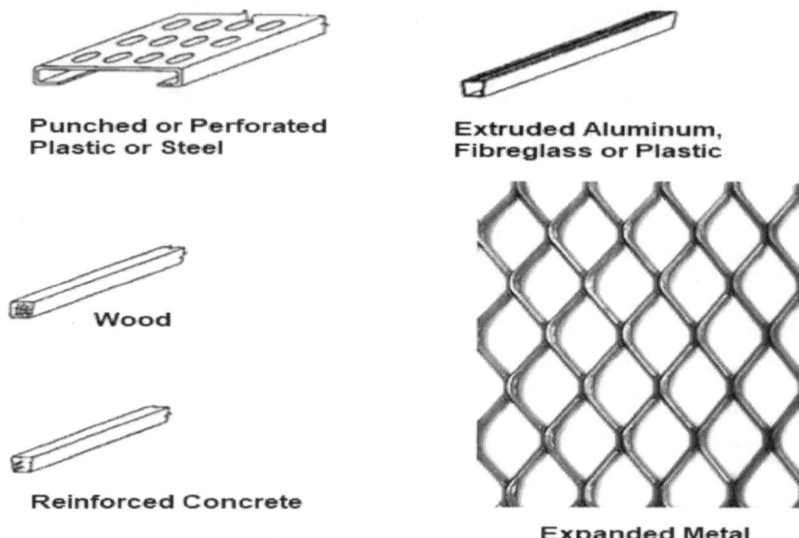

Figure 21.2. Slats of Various Materials for Slotted Floors.

Solid-Concrete Floors

Collecting livestock wastes on solid concrete floors involves the use of strategies and systems to manage manure and urine in facilities where animals are housed. Solid concrete floors provide a durable and easily cleanable surface. Mechanical scrapers or skid loaders equipped with manure blades are commonly used to scrape and collect manure from solid concrete floors. Concrete floors can be designed with gutters or channels along the edges to facilitate the collection of liquid waste. The gutters direct urine and runoff to a collection point or a drainage system, preventing pooling and promoting efficient waste removal. Water flushing systems can be implemented to clean and remove manure from solid concrete floors. The solid concrete floors are designed with 4 to 8 per cent slope to facilitate drainage towards collection points. Insufficient slope hinders the effective movement of manure, while excessive slopes lead to animal footing problems. Efficient removal of manure from animal shelters can be achieved through a well-designed flushing system. However, a significant drawback of this system is the elevated volume of waste material due to the excess water used. A flushing velocity ranging from 0.9 to 1.0 m/s is needed for dislodging fresh manure from a wetted surface. To effectively transport the manure, the velocity must exceed 0.6 m/s.

21.4 Waste Storage

Proper storage of animal manure is crucial part of waste handling management. The location of manure storage should be carefully chosen to avoid water bodies and minimize the risk of pollution. Various containers, such as lagoons or solid structures, can be employed, depending on factors like the type and volume of manure. Covering the storage helps control odours and prevents rainwater from entering, reducing the likelihood of nutrient runoff. Adequate aeration is essential to minimize the build-up of anaerobic conditions and the release of harmful gases. Regular management practices, including stirring and removing solids, ensure a homogeneous mixture and prevent storage capacity issues. Choosing the appropriate animal manure storage system involves considering factors like farm size, type of manure (solid, liquid, or semi-solid), animal housing system, climate, and available resources.

1. **Lagoons**: Common for liquid manure, lagoons are large, open-air ponds designed to hold and treat the waste. They facilitate natural microbial breakdown but require careful management to prevent issues like odour and nutrient runoff.
2. **Pits and Ponds**: In-ground pits or ponds are used for both liquid and solid manure storage. Impermeable liners prevent leaching into the soil, and covers help control odours and minimize environmental impact. By storing manure below ground, farmers can utilize space more efficiently and, when properly managed, reduce the potential for water pollution.
3. **Above-Ground Tanks**: Tanks provide containment for liquid manure and allow for controlled application. They are often equipped with agitators to maintain a consistent mixture and reduce solids settling.
4. **Solid Manure Storage Structures**: Solid-manure storage is a crucial component, particularly for farms with livestock such as cattle, poultry, or swine. Solid manure can be stored in covered structures like sheds or bunkers. These structures allow for the controlled decomposition of organic matter, reducing odours and providing an opportunity to transform manure into valuable compost. Regular turning and aeration help manage decomposition. Proper solid-manure storage not only prevents nutrient runoff but also facilitates the strategic application of the nutrient-rich material to fields as an organic fertilizer.
5. **Composting Systems**: Some farms opt for composting as a storage and treatment method. Properly managed composting systems convert manure into a valuable, stabilized product with reduced odour and pathogen content.
6. **Stacking or Piling**: Dry or semi-dry manure can be stacked or piled in designated areas. Careful management, such as covering to prevent nutrient loss, is essential to control odours and minimize environmental impact.

Storage Volume

The storage volume required for solid and liquid storages depends upon the number of animals, production of manure per day, storage time, and amount of dilution by spilled and cleaning water. Storage pits are usually made from masonry or reinforced concrete. The maximum recommended depth for a pit is 1.5 meter. The storage volume of a pit for liquid manure may be calculated from the following formula:

$$V = \frac{N \times P \times t}{\rho_m} + V_c$$

where,

V = storage volume (m^3),

N = number of animals,

P = daily manure production per animal (kg/day),

t = storage time (days),

ρ_m = density of manure (kg/m^3), and

V_c = cleaning and spilled water, and initial-fill water (m^3).

When there is high solids content in the raw manure as with poultry manure or when litter, bedding, *etc.* are added with the manure, the manure is stored as a solid. The storage space requirement for solid manure is then calculated by adding the volume of litter in the above formula. The storage capacity for solid manure then becomes:

$$V_s = V + 0.5V_b$$

where,

V_s = storage capacity for solid manure (m^3), and

V_b = volume of the litter (m^3).

Volume of the litter is multiplied by a factor 0.5 to take into account loss of volume due to intermixing of materials.

Important Points to Remember

1. The commonly monitored gases in a confined animal housing are ammonia (NH_3), hydrogen sulphide (H_2S), methane (CH_4), and carbon dioxide (CO_2).
2. Ventilation rates designed for temperature and humidity control in confined livestock production structures are generally sufficient to uphold gas levels that are not harmful to animals, except during activities such as liquid manure agitation or pumping from storage.
3. Density of raw manure varies in the range of 1005 to 1050 kg/m^3.
4. The contents of N, P, and K in the manure may vary from 1-3 per cent, 0.5-2 per cent, and 1-2 per cent, respectively.

5. Animal or poultry wastes may be handled as solids or liquids. When managed as solids, the solids content shall be 20 per cent or more, and the manure that has up to 5 per cent solids can be handled as liquid.
6. Slotted floors are made of wood, plastic, concrete, metal, extruded aluminium, and fiberglass materials with openings or slots that are designed to be wide enough for waste to pass through but narrow enough to provide a comfortable and stable surface for the animals to stand or lie on.
7. The solid concrete floors are designed with 4 to 8 per cent slope to facilitate drainage towards collection points.

Exercises/Questions

1. What are the impact of carbon dioxide, ammonia, hydrogen sulphide, methane, and aerial dust on animals when exposed to various concentrations?
2. Discuss briefly about animal waste production and its characteristics.
3. Describe different livestock waste collection and handling methods.
4. Write short notes on the following:
   (a) Control of gases, odour and other constituents in a livestock building
   (b) Waste storage
   (c) Estimation of volume for storage of manure
5. How much straw will be needed to raise the solids content of raw manure (12.7 per cent solids content) to a final content of 20 per cent if the straw has an initial moisture content of 11 per cent ?

   [**Answer**: 10.6 kg of straw per 100 kg of raw manure]
6. How much water should be required to be added to raw manure (12.7 per cent solids content) to make it suitable for use in sprinkler system? The solids content of the mixture should not be more than 4 per cent.

   [**Answer:** 217.5 kg of water per 100 kg of raw manure]

Chapter 22

Greenhouses

A greenhouse is a framed or an inflated structure covered with a transparent or translucent material in which crops could be grown under partial or fully controlled environment to get optimum plant growth and productivity. Greenhouse protects the plants from adverse climatic conditions such as wind, cold, rain, excessive radiation and extreme temperature. It also controls the inside environment of the structure which affects plant growth.

Greenhouse technology has the potential of providing a quantum jump of horticultural produce in the country besides production of quality produce, as it is able to provide favourable environment conditions to the plants. Design of greenhouse depends upon environmental parameters such as ambient temperature, relative humidity, wind velocity, and snow; latitude and longitude of place, *etc.* The proper design of greenhouse shall be able to provide conducive environment to the plants at minimum input cost. This chapter is designed to provide the information on site selection and layout, orientation, utilities, climate control systems, and maintenance of greenhouse so that an appropriate and efficient greenhouse can be constructed as per the requirement.

Greenhouse Effect: The sunlight admitted to the greenhouse is absorbed by the crops, floor, and other objects. These objects in turn emit long wave thermal radiation in the infrared region for which the glazing material has low transparency. As a result, the solar energy remains trapped in the greenhouse, thus raising its temperature. This phenomenon is called the 'greenhouse effect'. Greenhouse effect, the term for the role the atmosphere plays in insulating and warming the earth's surface. Without the greenhouse effect, the earth would be a frozen planet with an average temperature of about –18°C.

22.1 Advantages and Disadvantages of Greenhouse Cultivation

Greenhouse offers several advantages and disadvantages, and understanding them is crucial for effective greenhouse management.

Advantages of Greenhouse

1. Precise control over temperature, humidity, and light creates an optimal environment for plant growth. This control enables the cultivation of a wide variety of crops and facilitates year-round production.
2. Significant increase in crop productivity inside greenhouse compared to open field conditions.
3. The production of superior or export quality produce can be achieved by growing it in a controlled environment.
4. The crop can be grown in the offseason as per market requirements. The horticultural crops fetch more price in the offseason.
5. Greenhouse systems allow for efficient water use through controlled irrigation methods. Water can be recycled and reused within the greenhouse, minimizing wastage.
6. Greenhouses can be automated for irrigation, input application, and environmental control using computers and artificial intelligence techniques.
7. The enclosed structure of a greenhouse provides a barrier against many pests and diseases, reducing the need for chemical pesticides.
8. Greenhouse structures are ideally suited for production of genetically engineered and micro-propagated varieties and hybrids of crops.
9. Various intensive agriculture growing mediums like peat mass, vermiculate, rice hulls, and compost can be effectively utilized in greenhouses.
10. Green house provides self-employment opportunities for educated rural youth.
11. Modern techniques of hydroponic (soil less culture), aeroponics and nutrient film techniques are possible only under greenhouse cultivation.
12. The entrapped heat inside the green house can be utilized for drying and related agro-operations of harvested produce when crops are not grown.

Disadvantages of Greenhouse

1. Building and equipping a greenhouse are expensive.
2. Maintaining the ideal conditions inside a greenhouse often requires the use of energy for heating, cooling, and lighting. This can contribute to

high operational costs and environmental impact, particularly if the energy source is not sustainable.

3. While greenhouses can provide protection against some diseases, they can also create an environment conducive to certain pathogens. Proper sanitation and disease management practices are crucial to prevent outbreaks.
4. Greenhouses heavily rely on technology for environmental control. Malfunctions or power outages can disrupt the carefully regulated conditions, potentially impacting crop health and yield.
5. In enclosed environments, natural pollinators may have limited access, requiring growers to implement manual pollination methods for certain crops.

22.2 Classification of Greenhouse

Greenhouse structures classified on the basis of shape, utility, covering material, construction and cost are briefly described below:

1. Greenhouse Type Based on Shape

Greenhouses are classified based on their shape or style. The uniqueness of their cross section is the key factor in this classification (Figure 22.1). Common types include:

(a) **Lean to Type Greenhouse**: A lean-to design is a greenhouse that is built against an existing building. One or more sides of the existing structure are used for its construction. The selection of sides depends on the best direction for adequate sun exposure.

(b) **Gable Even Span Type Greenhouse**: The even-span structure is a full-size and standard structure. The roof slope of this greenhouse are of equal pitch and width with straight sidewalls. The gable roof is common in these greenhouse due to better space utilization and advantages in framing.

(c) **Gable Uneven Span Type Greenhouse**: This greenhouse is designed for hilly terrain. The uneven width of roofs allows this green house to adapt to the side slopes of the hill.

(d) **Ridge and Furrow Type (or gutter connected)**: This type of greenhouse design utilizes multiple single span greenhouses connected along the eave. The eaves serve as a furrow or gutter to carry rain and snow away. The advantages include significant crop growing space inside the greenhouse and lower heating costs per unit area covered.

(e) **Saw Tooth Type**: The roof of this greenhouse is of a right-angled triangular shape. The provision of natural ventilation and the shape of the roof make this greenhouse different than the Ridge and Furrow-type greenhouse. The roof of this greenhouse develops a specific path for the natural ventilation.

(f) Quonset Greenhouse: This greenhouse uses pipe arches or trusses supported by pipe purlin. The use of pipes makes it more efficient than a gable building in terms of the utilisation of materials in construction. However, some applications may have unused space due to side wall curvature. Polyethylene is typically used as the covering material.

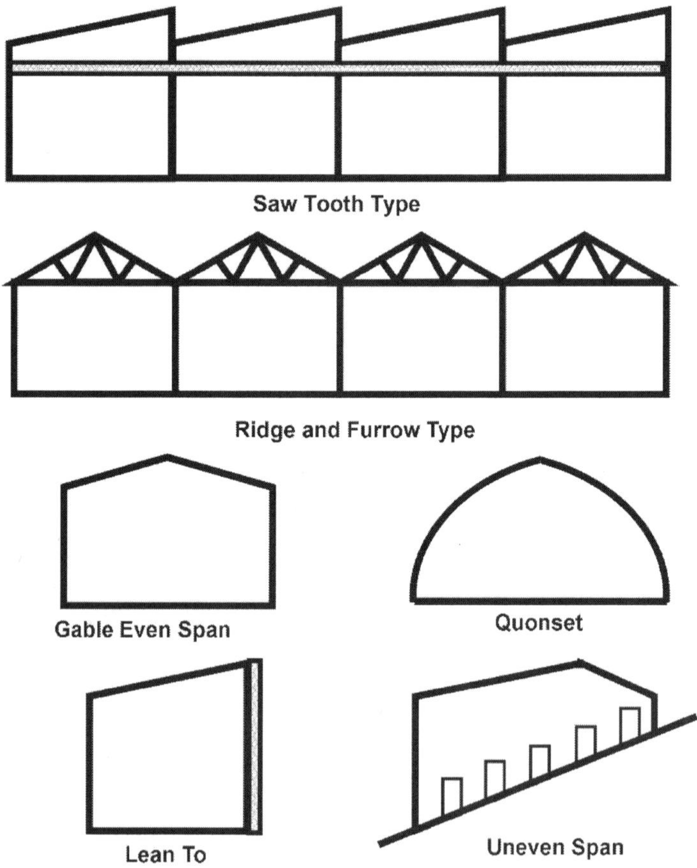

Figure 22.1. Different Shapes of the Greenhouse.

2. Greenhouse Type Based on the Basis of Utility

Greenhouses are classified based on their functions or utilities. A greenhouse can be equipped with a costly artificial cooling or heating system. These systems are useful and cost effective at the location that have extreme weather condition like extreme cold or heat.

(a) **Active Heating Greenhouse:** The air temperature inside the greenhouse decreases during the night of winter. The plants grown in the green house may experience frost bites. Therefore, to save the plants, extra heat can be supplied in the greenhouse to raise the air temperature. Equipment like unit heaters, central heaters, radiant heaters, or solar heating systems can be used to heat the air.

(b) **Active Cooling Greenhouse:** The greenhouse effect increases the temperature of green house that need to be reduced for effective crop growth. The temperature can be reduced in the greenhouse either by evaporative cooling pad system or by fogging.

3. Greenhouse Type Based on Construction Material

The structural material used for their construction primarily determines the type of construction in greenhouses. Although the covering materials also influence the type of greenhouse. The greenhouse can be divided into three main categories: wooden, pipe, and truss.

(a) **Wooden Framed Structure**: Wooden framed structure has the side posts and columns constructed of wood. These low-cost structures are typically recommended for greenhouses with a span of less than 6 meters.

(b) **Pipe Framed Structure**: Pipes are used as side posts, columns, cross ties, and purlins. The pipes of different sizes can be used as per the requirements of strength in different components of a green house. The truss is not required in the pipe greenhouse (Figure 22.2).

(c) **Truss Framed Structure**: Truss frames are commonly used in greenhouses with a 15m or greater span. These frames are ideal for pre-fabrication (Figure 22.2).

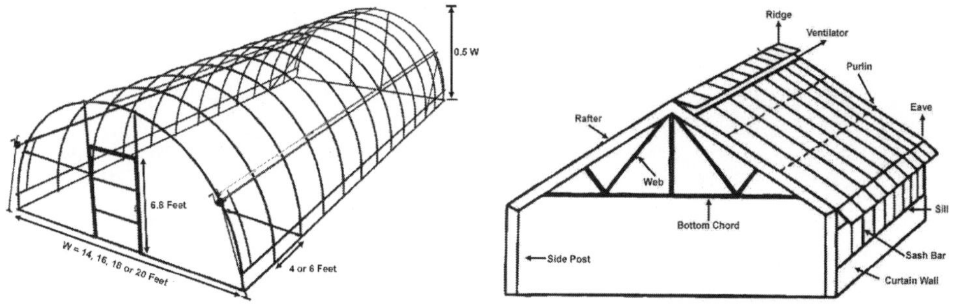

Pipe Framed Structure **Truss Framed Structure**

Figure 22.2. Greenhouse Based on Construction Material.

4. Greenhouse Type Based on Covering Material

Covering materials are very important part of the greenhouse structures. They control the solar radiation and influence the greenhouse effect that alter the air temperature within the house. Based on the type of covering materials, the greenhouses are classified as glass, plastic film and rigid panel greenhouses.

(a) **Glass Glazing**: Glass as covering material has the advantage of greater interior light intensity. These greenhouses have higher air infiltration rate which leads to lower interior humidity and better disease prevention. Lean-to type, even span, ridge and furrow type of designs are used for construction of glass greenhouse.

(b) **Fiber Glass Reinforced Plastic (FRP) Glazing**: Rigid panels of polyvinyl chloride, acrylic, or polycarbonate and fibre glass-reinforced plastic are used in quonset or ridge and furrow type frames in greenhouses. Panels can be plain or corrugated in shape. These panels are resistant to breakage and uniform in light intensity, with a lifespan of up to 20 years. However, these panels can collect dust and harbour algae that reduce light transmission. These panels are also not fire-resistant.

(c) **Plastic Film**: Flexible plastic films like polyethylene, polyester, polycarbonate, fiberglass, acrylic, and PVC are used as covering materials in greenhouses. These coverings are popular due to their affordability and lower heating costs compared to glass greenhouses. However, these plastic films are not fire-resistant and have a short lifespan. Even UV stabilized LDPE film has 2 to 3-years life. Quonset design as well as gutter-connected design is suitable for using this covering material.

Figure 22.3. Natural Ventilation Providing Crosswise Ventilation.

5. Based on the Cost of Construction Involved

The greenhouse can be classified as high cost, medium cost and low cost depending upon the cost per unit area.

(a) **High Cost Green House**: High cost greenhouse has facilities like auto control mechanism of heating, cooling and humidification system; drip irrigation system, *etc.*

(b) **Medium Cost Green House**: It has a cooling and heating arrangement, and may have a double layer or UV stabilized plastic film.

(c) **Low Cost Green House**: Low cost greenhouse does not have any control system.

Figure 22.4. Open Roof Ventilation Design.

22.3 Components of Greenhouse

The key components (Figure 22.5) of a greenhouse are:

Frame: The greenhouse frame should be able to withstand wind, snow, and crop loads with minimum obstruction to sunlight. The materials used for the construction of greenhouses may be wood, aluminium, steel or reinforced concrete. Wood shall be protected against decay. Wood frames include post beam and rafter systems, posts and trusses, glued laminated arches and rigid frames. Steel and aluminium are used for posts, beams, purlins, girts, trusses, and arches in greenhouses. These metals shall be protected from corrosion due to ground contact.

Covering Material: Light transmittance of glazing materials is an important factor. It should be high as light is responsible for plant growth. Most glazing materials transmit direct light, but some, like fibreglass reinforced plastics, have a high percentage of diffused light. The various glazing material to cover the greenhouse structures are glass, acrylic structural sheet, polycarbonate structured sheet, fibreglass reinforced plastic, polyethylene, and polyvinyl chloride (PVC).

Doors and Windows: Entry doors and windows are essential for ventilation and access. They can be manually or automatically operated to regulate temperature and humidity.

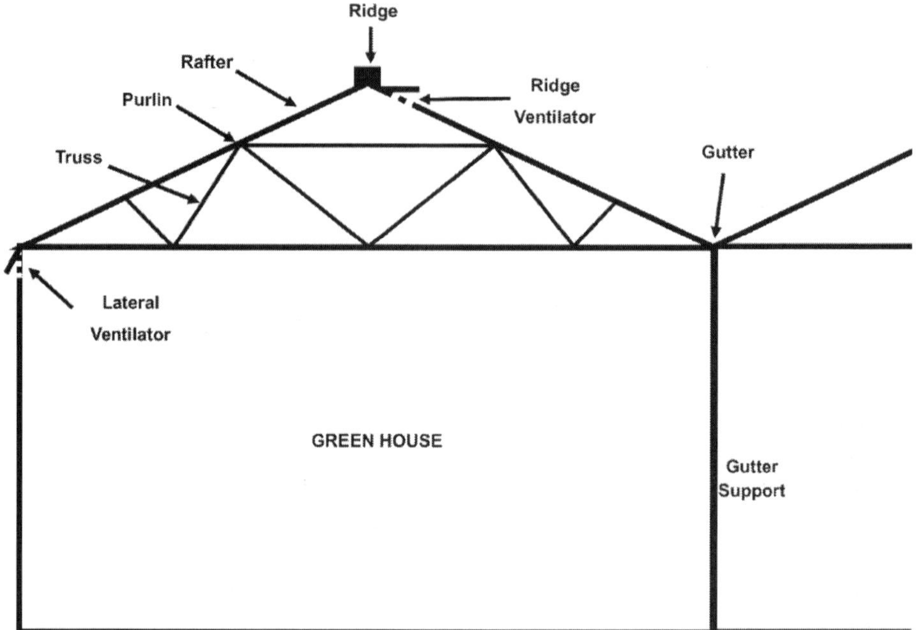

Figure 22.5. Components of a Typical Greenhouse.

Ventilation Systems: Louvered side vents, roof vents, or exhaust fans are installed to facilitate air circulation and regulate the internal temperature. Automated systems may include sensors that trigger ventilation based on pre-set conditions.

Heating Systems: Greenhouses often require supplemental heating, especially in colder climates. Heating systems can include gas heaters, electric heaters, or hot water heating systems.

Cooling Systems: Cooling systems, such as shade cloths, misting systems, or evaporative cooling, help prevent overheating during warm periods.

Control/Monitoring Equipment: Equipment for control and monitoring of ventilation, heating, cooling, watering, CO_2 generation, lighting system are provided in the greenhouse structure. Computerised environmental control system helps towards greater saving of energy and also helps in the progress of growth and plant management.

Irrigation Systems: Drip irrigation systems provide a controlled and efficient way to deliver water directly to the plants' root zones, reducing water wastage. While, misting systems help maintain proper humidity levels and can be used for cooling during hot periods.

22.4 Plant Response to Greenhouse Environment

The yield of a crop is influenced not solely by its genetic makeup but also by the microclimate around it. The elements comprising the microclimate of a crop include light, temperature, air composition, and the nature of the root medium. In open fields, the only feasible manipulations involve altering the nature of the root medium through practices like tillage, irrigation, and fertilizer application. While, greenhouse, with their enclosed boundaries, affords the ability to control one or more components of the microclimate.

Light

Light is essential for photosynthesis process in plants. The plants convert light energy into chemical energy to fuel their growth. In a greenhouse, the quality and quantity of light significantly impact photosynthetic rates. Different wavelengths of light affect plant growth differently. Visible light (400 to 700 nm) is essential for plant development as the conversion of radiant to chemical energy proceeds only in visible light. This range of wavelength is called photosynthetically active radiation (PAR). Wavelengths of the visible spectrum are selectively absorbed by plants through chlorophyll and various pigments, mainly phytochrome, carotenoids, and flavones. Ultraviolet light (290 to 400 nm) is generally detrimental to plant growth. Glass screens are transparent to UV light and light below the 325 nm range. Far red light (700 to 750 nm) affects plants, besides causing photosynthesis. Infrared rays of longer wavelengths are not useful in photosynthesis. Photosynthesis activity is higher in the blue and red bands. The blue light retards growth and makes plants hard and dark. Red light, on the other hand, softens the growth and produces tall plants with long internodes. Blue light is associated with vegetative growth, while red light is crucial for flowering. Greenhouses may use specific light spectra from artificial sources to tailor the light environment for various stages of plant development. Excessive light, especially in combination with high temperatures, can lead to light stress in plants. Greenhouses employ shading systems to protect plants from intense sunlight and prevent damage. Greenhouse operators often supplement natural light with artificial lighting, especially during periods of low natural light, to enhance photosynthesis and promote optimal plant growth. The reaction for the photosynthesis is:

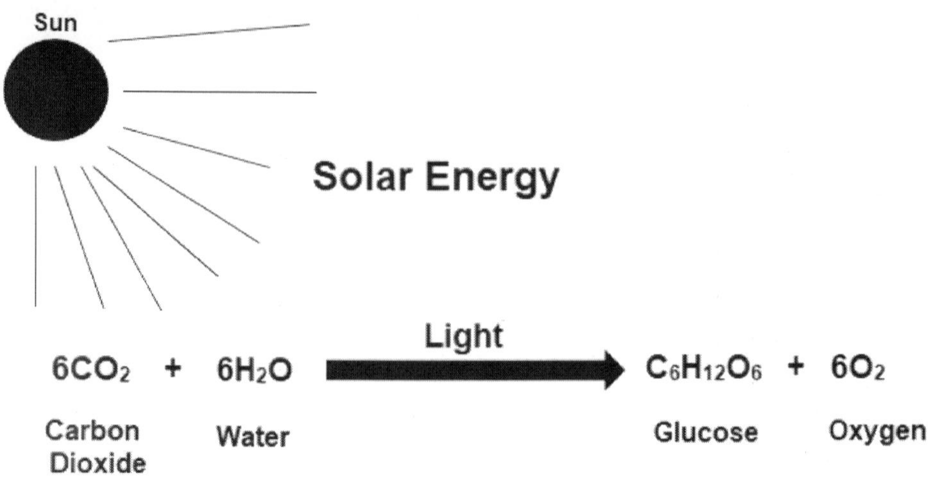

Temperature

In a greenhouse, temperature plays a pivotal role in shaping the growth and development of plants with different species having specific temperature requirements. Temperature profoundly influences crucial plant processes such as photosynthesis, respiration, and transpiration. The rate of photosynthesis generally increases with temperature up to an optimal range, beyond which efficiency may decline. Respiration rates also rise with higher temperatures, impacting the delicate balance between carbon gain and loss. Transpiration, the process through which plants lose water, is influenced by temperature. Extreme temperatures, whether excessively high or low, can induce stress in plants, leading to wilted foliage, reduced growth, and tissue damage. Greenhouses incorporate temperature control mechanisms, including shading, ventilation, heating and cooling systems, to mitigate such stress and maintain an optimal growing environment. As a thumb rule, the day temperature of the greenhouse shall be 3-6 °C higher on cloudy days and 8 °C higher on clear days compared to the night temperature to grow the crops.

Relative Humidity

Relative humidity affects the rate of transpiration, the process by which plants release water vapour through their leaves. Higher relative humidity reduces the rate of transpiration, leading to potentially elevated humidity levels within the plant canopy. This can influence nutrient uptake, water balance, and overall plant hydration. High humidity levels in a greenhouse can create conditions conducive to the development of fungal diseases, such as powdery mildew and botrytis. Proper humidity control is crucial to minimize disease susceptibility and maintain plant health. Extremely low humidity can lead to high transpiration, water stress and reduce photosynthetic rates. For most crops, the acceptable range of relative humidity is between 50 to 80 per cent. However, for plant propagation work, relative humidity up to 90 per cent may be desirable.

Greenhouses often employ humidity control systems, like humidification or dehumidification to manage these stress factors.

Air Velocity

Air velocity, or the movement of air within a greenhouse, influences many factors that affects plant growth and overall greenhouse climate such as transpiration, evaporation, and carbon dioxide availability. Air movement helps regulate the transpiration process in plants. Adequate air velocity around the plant canopy can enhance the removal of water vapour from the leaves, promoting cooling. This is especially important in preventing heat stress during periods of high temperature in the greenhouse. Air movement facilitates the exchange of gases, including carbon dioxide (CO_2) and oxygen, around the plant. Improved gas exchange supports photosynthesis, enhancing overall plant growth and productivity. It also helps prevent the build-up of stale air, ensuring a fresh and oxygen-rich environment. Proper air circulation helps reduce humidity levels around the plant, minimizing the conditions favourable for the development of fungal diseases. Improved ventilation can also discourage the growth of pathogens by reducing leaf wetness, which is crucial for disease prevention. Some plants rely on the movement of air for pollination. Adequate air circulation within the greenhouse ensures the dispersal of pollen, facilitating pollination and supporting the development of fruits and seeds. The amount of fresh air to supply carbon dioxide for photosynthesis in a greenhouse has been estimated to 20 to 30 air changes per hour. Typically, air velocities ranging from 0.1 to 0.25 m/s promote effective carbon dioxide absorption across the surface of a leaf. However, when the air speed reaches 0.5 m/s, there is a decrease in carbon dioxide uptake. Growth inhibition occurs at air speeds of 1.0 m/s, and any speeds exceeding 4.5 m/s can lead to physical damage to plants. Greenhouse operators often use fans and ventilation systems to control air movement and create the optimal environment for plant growth.

Carbon Dioxide

Managing carbon dioxide levels in a greenhouse is a critical aspect of modern horticulture. By optimizing CO_2 concentrations, greenhouse operators can enhance photosynthesis, improve water-use efficiency, and ultimately promote robust plant growth and higher yields. Carbon dioxide is a critical component for photosynthesis, the process by which plants convert light energy into chemical energy. Outside atmosphere contains about 0.03 per cent (or 340 ppm) of CO_2. Most of the plant species grow well at this level. However, increasing the concentration of CO_2 in the greenhouse environment can enhance photosynthetic rates, leading to improved plant growth and productivity. If the level of CO_2 is less than ambient levels, CO_2 may retard the plant growth. Plants regulate the opening and closing of stomata, small pores on the leaf surface, to control water loss and facilitate gas exchange. Elevated CO_2 levels often lead to increased stomatal resistance, reducing water loss through transpiration. This improved water-use efficiency is especially advantageous in arid or water-limited conditions. At low concentration of CO_2, the stomatal resistance has been shown to decrease by 50 per cent, resulting in more evaporative demands. During the day time, when plants engage in photosynthesis using natural light, greenhouse plants reduce the concentration of CO_2 to levels below 200 ppm. In this scenario, the introduction of outside air through infiltration or ventilation becomes necessary to elevate carbon dioxide levels and maintain the desired ambient CO_2 concentrations. However, in colder climates, relying on ventilation to sustain ambient CO_2 levels can be economically impractical due to the associated need to heat the incoming air to preserve optimal growing temperatures. In such regions, the practice of enriching the greenhouse with additional CO_2 is adopted. The specific CO_2 requirements for a particular crop vary and must be carefully calibrated in conjunction with other factors in greenhouse production, including light exposure, temperature, nutrient levels, cultivar selection, and the stage of maturity. Typically, most crops exhibit favourable responses to CO_2 concentrations ranging from 1000 to 1200 ppm. Commercial greenhouses often use CO_2 supplementation systems to maintain elevated levels, particularly in enclosed environments where natural CO_2 levels can be limiting.

22.5 Site Selection and Layout

Site Selection

The functional and environmental operation of a greenhouse structure may be affected by the site selected for its construction. The factors which shall be considered while selecting a site for the construction of greenhouse are:

1. The area should be nearly level with the ideal gradient being 1 in 100 to 1 in 200.
2. The area should be well drained. Adequate provision should be made to direct surface water away from the greenhouse.
3. Adequate and good quality water should be available near the site.
4. The soil should be good for raising of the crops/plants.
5. The location should be shadow free. A location with maximum sunlight exposure should be chosen.
6. The greenhouse shall not be located in the vicinity of an industry and vehicular pollution. Pollution from the industries/vehicular traffic may cause injury to greenhouse crops and reduction in light levels.
7. The site shall have a south-facing slope (maximum 3 per cent) for protection against the northern wind and for optimal winter light.
8. Site shall be well connected to market. It shall facilitate in convenient supply of inputs and the sale of its produce.

9. Site shall be located near the motorable road. Proper transport facilities reduce the cost of transportation of raw and finished materials.
10. Availability of adequate labour force near the site.
11. Site shall be near to existing electric supply for the operation of greenhouse.
12. Site shall have proper security arrangement.
13. There shall be provision for future expansion.

Orientation

Orientation of the greenhouse is mainly dependent on the availability of light level and prevailing winds direction. The light levels in the greenhouse should be adequate and uniform for crop growth. While the winds should not adversely affect either the structure or the operation of the facility. In free standing/single span greenhouse, more sunlight in winter is available in an east-west oriented greenhouse as compared to north-south oriented greenhouse. It also helps in reducing heating cost during winter. In multi-span greenhouses, the gutters should be oriented in north-south to avoid continuous shading of certain portions of the greenhouse due to structural members. Gutter shading the same area during each day may result in uneven growth in some plants. In areas above 40° north latitude, the ridge in either single span greenhouse structures or gutter-connected range shall be east-west to transmit maximum winter sunlight to the plants. If a wind direction at the location is known, then in a naturally ventilated greenhouse, the ventilators should open on the leeward side.

Site Layout

Following factors should be taken in view while deciding the general layout of the entire facility:

1. A head house for office, utilities, employee areas, storage, and dispatch shall be established. It shall be located north of the production area to avoid shading. The head house should have an area of 7–14 m^2 per 100 m^2 of greenhouse area. The windbreaks should be at least 30 meter away to the side of the prevailing winter winds to reduce energy consumption.
2. The way to sale section and storage/receiving section shall not be same.
3. Avoid movement of customer into the production area to protect crops from disease and interruption to the ongoing work in greenhouse.
4. Provision of parking shall be provided.

Interior Layout

Crops can be grown on the floor or on benches. Benches are used for pot plant production, whereas floors are used for bedding plants. Benches are made of metal, wood, or plastic and have a solid or mesh bottom. They are placed 0.5 to 1.0 meters above the floor, with bench widths ranging from 0.9 meters for one side to 2.0 meters for both sides, with a minimum aisle width of 0.5 meters.

Arrangement of benches: The arrangement of benches in a greenhouse depends on its dimensions, walkways, and doors. Longitudinal arrangements allow for continuous water lines, heat pipes, and plant support systems. Peninsular arrangements, with one centre aisle and perpendicular side aisles, provide more growing space. Movable benches, which can move about 0.5 m sideways, allow for all areas except one or two work aisles. Connections for water, heat, and electrical systems must be flexible. Overall, one main aisle of 1–1.5 m and a side aisle of 0.5m are recommended inside the greenhouse.

22.6 Utilities

Water

Requirement of water depends on area under production, crop and climate. The water requirement for different crops is presented in Table 22.1. However, general rule is to make a provision of 20 litres per square meter per watering. In summers, watering shall be done more frequently. Good quality water free from impurities *etc.* shall be used for irrigation of greenhouse crops. The pH of irrigation water shall be between 6.5 to 7.5. Method of irrigation depends to some extent on the quantity of water available.

Table 22.1. Water Requirement for different Crops (IS 14462: 1997)

Crop	*Water Required (litres/square meter)*
Bedding plants	20.0
Bench crops	16.0
Chrysanthenum	41.0
Pot plants	20.0
Roses, tomatoes	29.0

Electricity

An assured electrical supply for uninterrupted operations of motors, fans, lights, and other equipment is required for environment control and mechanization needs of the green house. An auxiliary generating system should be available to cope with the failure of regular electrical supply. The electrical system should be located inside or adjacent to the head house. The underground electric wiring system should be provided to improve appearance, avoid damage, and reduce hazards. Electric wires should be buried at least 0.5-meter-deep to avoid damage from surface traffic. The size and number of motors, fans, lights and other electric appliances shall determine the total electric load required for the greenhouse. Natural ventilated greenhouse/poly house may not need any power supply.

22.7 Management of Greenhouse Environment

Environmental control in a greenhouse involves managing and regulating various factors to create an optimal environment for plant growth. The key environmental factors to control in a greenhouse include temperature, humidity, light, carbon dioxide (CO_2) levels, and air circulation.

22.7.1 Temperature

Temperature control in a greenhouse is crucial for creating an optimal environment for plant growth. Maintaining the right temperature range helps ensure that plants thrive and produce healthy yields. In order to create better growing conditions, the greenhouse is required to be cooled whenever the outside temperature goes beyond 30°C and heating is required where the winter temperature is very low. Both cooling and heating system in the greenhouse are required at places where the climate is extreme cold and hot.

22.7.1.1 Cooling

The regular ventilation system is not able to provide sufficient cooling in the greenhouse during summer seasons. Thus, additional cooling systems such as evaporative cooling, shade cloth or paint may be necessary. The most convenient cooling system for the greenhouse is evaporative cooling. The evaporative cooling systems are most effective in hot and dry climate when the humidity of the outside air is low. The degree of cooling obtained from an evaporative system is related to wet bulb depression (difference between dry bulb and wet bulb temperatures) of the ambient air. The temperature drops increases with the increase in wet bulb depression. The systems based on the principles of evaporative cooling are fan and pad system, fog cooling system, and mist cooling system.

Evaporative Cooling

1. **Fan and Pad Cooling System:** Fan and pad systems (Figure 22.6) are efficient, economical, and adaptable methods for cooling greenhouses. They consist of exhaust fans, a pad, a water tank, and a pump. The pump circulates water through fibrous pads. As water evaporates from pads, heat energy is absorbed by the water, lowering the air temperature in the greenhouse. The air then moves across the greenhouse to exhaust fans, picking up heat from solar radiation, plants, and soil. This results in a temperature gradient across the greenhouse, with the pad side being the coolest and the fan side being the warmest.

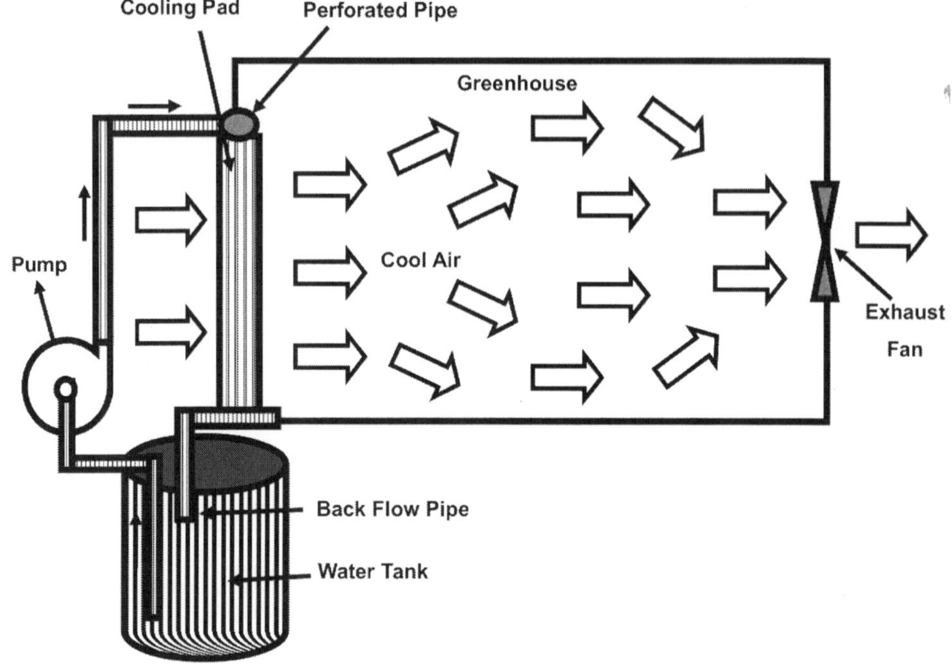

Figure 22.6. Fan and Pad Cooling System.

2. **Fog Cooling System:** A high-pressure pumping system generates fog containing small water droplets using the nozzles. The mean size of the droplet or fog is less than 10 microns. The suspended, tiny droplets evaporate in the air. Therefore, the plants inside the greenhouse stay dry during the operation of the fog cooling system. This system is beneficial for seed germination and propagation as it eliminates the need for a mist system.
3. **Mist Cooling System:** Mist cooled greenhouses use water spray through nozzle to create a fine mist that cools the air as it evaporates. Most of the mist evaporates before reaching the plants, but some settles on foliage, reducing leaf temperatures. This system is most effective during the warmest part of the day, achieving a temperature difference of 5-10°C.

Comparison between Fan Pad and Fog Cooling System

The droplet size is smaller in the fog cooling system than the fan pad system. Therefore, complete evaporation is achieved in fog cooling system.

The fan-and pad system can lower the temperature of incoming air by about 80 per cent of the difference between the dry and wet bulb temperatures while the fog cooling system can lower the temperature by nearly 100 per cent difference.

Greenhouse Cooling Efficiency

The cooling efficiency of the system can be calculated using the following equation:

$$\text{Cooling efficiency } (\eta) = \frac{T_o - T_i}{T_o - T_{wb}} \times 100$$

where,

T_o, T_i and T_{wb} are temperature of the outside air, temperature of air exiting the cooling pad and wet-bulb temperature of the outside air, respectively.

Shading

Greenhouse shading is a cooling technique that reduces solar radiation reaching plants grown in the greenhouse. It reduces the air temperature and leaf surface temperature inside the greenhouse. This can reduce the growth and development rates of plants by limiting photosynthesis due to lowered light intensity. Roll-up screens made of wood or aluminium, vinyl netting, and paint can be used as shading materials. Traditional methods involve applying semi-permanent shading compounds to the outer glazing material of the greenhouse. The glasshouses and polyethylene greenhouses use a white liquid. Mesh screens of different density are available in the market, providing shading from 30 to 92 per cent.

22.7.1.2 Heating

Heat is required to be added in the greenhouse to maintain desirable temperature during winter.

The desired temperature for the optimum growth of plants, the location and type of greenhouse, exposed area of greenhouse are the major factors that decide the heat requirement of green house. The capacity of heating system shall match the heating requirement of greenhouse. The conventional and non-renewable fuels like diesel, coal, wood *etc.* are available for burning in boiler. The solar energy that renewable source of energy can be used of heating purpose. Although, the choice of source of energy depends upon its availability, cost and plant requirement.

Heating Systems Based on Conventional Fuels

Heating systems based on conventional fuels are unit space heater, hot water system and steam heating system.

1. **Unit Space Heater:** Unit space heaters are either floor mounted or supported. These heaters are normally fuelled with natural or bottled gas or fuel oil and use fans for heat distribution. This system requires

a relatively moderate capital investment. It is simple to install, and provides for easy expansion of facilities.

2. **Hot Water System:** The pipes of the hot water are provided under the crop benches; perimeter or fins of overhead fan can be used to deliver heat inside the greenhouse. Boiler, valves and some other necessary accessories are required in hot water system. They are simpler to install and normally requires less maintenance than a steam system. There is slower heating and cooling of pipes, but temperatures are normally more uniform. This type of system is mainly used in smaller ranges.
3. **Steam Heating System:** This system consists of boiler, valves and pipes. The steam heating system is known for its rapid heating. It may be used for the pasteurization of the soil. The system requires high initial cost, although cost can be justified by its long life.
4. **Unit Radiant Heaters:** The heating by radiation is the oldest method. The radiation can be absorbed by any object without physical contact between them. It also heats the surrounding air of crop inside the greenhouse.

Heating Systems Based on Non-Conventional Energy Sources

Solar or biogas based heating systems have been developed for greenhouse.

1. **Solar Based Heating System:** Solar heating is a popular alternative to fossil fuels. The solar heating system consists of a collector, heat storage facility, exchanger, backup heater, and control set. This system can be treated as partial or complete alternative option to fossil fuels. Flat plate collectors are commonly used. It shall be installed perpendicular to the sun at solar noon to achieve best efficiency. It can fulfil the 20 to 50 per cent of the heat requirement depending on the location.
2. **Biogas Based Heating System:** Biogas can be used for heating of greenhouse. The combustion of biogas will give off carbon dioxide. The flue gas thus produced could directly be admitted into the greenhouse with advantage. The availability of the additional CO_2 shall improve the growth of greenhouse crops.

22.7.2 Humidity

Humidity control in a greenhouse is essential for creating an environment that promotes healthy plant growth and helps prevent the development of diseases. The greenhouse structures need either humidification or dehumidification to maintain the desired level of relative humidity.

Humidification: In summers, relative humidity can be increased through fan-pad evaporative and fogging systems. These systems also lower the temperature inside the greenhouse. In winters, relative humidity levels can fall below acceptable limits that require an additional supply of sensible heat to raise the temperature. In that situation, humidifiers shall be used to maintain relative humidity level.

Dehumidification: Dehumidification of greenhouse air can be achieved through ventilation. Ventilation could be practiced whenever ambient relative humidity is lower than inside relative humidity. In rainy seasons, when ambient relative humidity is high, ventilation will not lower the relative humidity of greenhouse air. In that situation, other methods of dehumidification such as chemical dehumidification or use of refrigeration systems (cooling coils) *etc.* may be used.

22.7.3 Light

Controlling light in a greenhouse is crucial for optimizing plant growth and productivity. Different plants have varying light requirements, and providing the right amount of light at the right time is essential for photosynthesis and overall development. Insufficient solar radiation can be supplemented with artificial light to increase photosynthetic active radiation (PAR) and maintain active growth. The positive effect of increased PAR on growth is more relevant at low PAR levels. Techniques to improve light availability include painting greenhouse components white and using light-reflecting materials. High-intensity discharge lamps (HID) or LEDs are the most reliable and effective method for increasing light availability in winter. To prevent overheating and protect light-sensitive crops, install shade cloths or movable screens to regulate the amount of sunlight reaching the plants.

22.7.4 Carbon Dioxide

Controlling carbon dioxide (CO_2) levels in a greenhouse is important for promoting plant growth, photosynthesis, and overall productivity. The optimal CO_2 level for plant growth varies among different crops. Regular monitoring and adjustment based on the specific needs of the cultivated plants are essential for successful greenhouse management. As the sun rises in the morning to provide light, the plants begin to produce food energy (photosynthesis). The level of CO_2 drops in the greenhouse as it is used by the plants. Maximise ventilation rates after 1-2 hours of sunrise to utilise the CO_2 present in greenhouses. During the night, the plant respires and increases the level of carbon dioxide. Later on, the ventilation replenishes CO_2 in the greenhouse. Electric lighting combined with CO_2 injection is used to increase the yield of vegetable and flower crops. CO_2 sources include liquid or bottled gas, dry ice, and the combustion of sulphur-free fuels. Automated CO_2 injection systems can be used in commercial greenhouses. CO_2 enrichment is not necessary as long as crops grow and develop to the optimum level.

22.7.5 Air Circulation

Air circulation (ventilation) in a greenhouse is vital for maintaining uniform environmental conditions, preventing the build-up of heat, controlling humidity, replenishing carbon dioxide supply and promoting the overall health of plants. Air temperatures above 35°C are generally not suited to greenhouse. Ventilation can reduce the temperature of the greenhouse to 35°C in spring and autumn. There are two types of ventilation, *i.e.*, natural and mechanical ventilation. Mechanical ventilation is also known as forced ventilation. The roof vents with side inlets (louvers) are used for natural ventilation. In natural ventilation, air gets lighter as it gets heated, and thus rises. This is called chimney effect. The warm air moves out and cool air in. The natural ventilation requires inlet and outlet openings. Winds also cause additional natural ventilation in the greenhouses. Natural ventilation is quite effective in small greenhouse (less than 6-meter-wide) during spring and winter seasons. Forced ventilation is crucial for precise control of greenhouse air temperature, humidity, and carbon dioxide levels. The components of forced ventilation are inlet openings and exhaust fans. The exhaust fans operate on electricity. All fans should be equipped with louvers and fan guards for safety. Fans are mounted on one end of the house and air inlets on the opposite end. More number of smaller fans may be selected as against a few large diameter fans. Ventilation requirements for greenhouses vary based on the weather and season. In tropical conditions, two air changes per minute are recommended for green houses. However, the small greenhouses require more air exchange. In winter, 20–30 per cent of one air volume exchange per minute is sufficient for cool air mixing without damaging the plants.

22.8 Passive Solar Greenhouse

A passive solar greenhouse maximizes solar energy while minimizing thermal exchange to maintain desirable winter temperatures. Consider the following points while design of passive greenhouse:

1. Orientation of greenhouse shall be in east-west directions.
2. Provide thermal insulation to the north side wall of greenhouse.
3. The north side of the greenhouse shall be reflective surface to ensure maximum availability of sunlight on the crops.
4. The greenhouse admitting sunlight should use two layers of plastics with a stagnant air layer in between. It shall result in less thermal exchange between the greenhouse and surrounding environment as air acts as an insulation layer.
5. The thermal mass in the form of pebbles or water should be kept in the greenhouse for storage of heat and subsequent use during a period of heating.
6. An insulated and reflected night curtain should be used inside the greenhouse to reduce heat losses.

22.9 Crop Selection and Management

Crop selection and management in a greenhouse are critical factors that directly impact the success and productivity of the greenhouse operation. By carefully selecting crops based on market demand, and the environmental suitability, and by implementing effective crop management practices, one can optimize the productivity and sustainability of greenhouse operation. The general guidelines for effective crop selection and management are:

Crop Selection

1. **Market Demand**: Choose crops that have a high demand in the local market. Consider market trends, consumer preferences, and potential profitability.
2. **Climate Adaptability**: Select crops that are well-suited to the local climate and can thrive in the controlled environment of the greenhouse.
3. **Crop Rotation**: Implement crop rotation to prevent soil-borne diseases and maintain soil fertility. Plan rotations based on the specific needs and growth cycles of different crops.
4. **Diversity**: Diversify your crop selection to reduce risks associated with market fluctuations, pest outbreaks, or crop failures.
5. **High-Value Crops**: Consider growing high-value crops that may fetch a premium price in the market, increasing the economic viability of the greenhouse.
6. **Organic or Speciality Crops**: Explore opportunities to grow organic or specialty crops that cater to niche markets and may command higher prices.

Crop Management

1. **Optimal Plant Spacing**: Space plants appropriately to allow for proper air circulation, light penetration, and ease of management. Avoid overcrowding.
2. **Irrigation Management**: Implement efficient irrigation practices based on the water needs of each crop. Use drip irrigation or other water-saving methods to prevent overwatering.
3. **Nutrient Management**: Develop a nutrient management plan to provide plants with the necessary macro and micronutrients. Regularly monitor nutrient levels and adjust as needed.
4. **Temperature Control**: Maintain optimal temperature ranges for different crops. Utilize heating and cooling systems as required to create a conducive environment.
5. **Humidity Control**: Manage humidity levels to prevent fungal diseases. Use proper ventilation, dehumidifiers, or humidification systems as needed.

6. **Pruning and Training**: Implement pruning and training techniques to shape plants, improve light distribution, and increase airflow. This is especially important for vining or tall crops.
7. **Pest and Disease Monitoring**: Regularly monitor crops for signs of pests and diseases. Implement integrated pest management (IPM) strategies, including biological controls, to minimize the use of chemical interventions.
8. **Harvest Timing**: Harvest crops at the optimal stage of maturity for the best quality and yield. Plan harvesting schedules to meet market demands.
9. **Record Keeping**: Maintain detailed records of crop management activities, including planting dates, irrigation schedules, pest control measures, and harvest yields.
10. **Workers Training**: Train greenhouse staff on proper crop management practices, including planting techniques, irrigation, pest identification, and harvesting methods.

22.10 Maintenance of Greenhouse

A robust preventive maintenance programme can significantly reduce corrective maintenance actions. It saves money on expenses incurred during equipment downtime, which wastes labour hours. The maintenance requirement of different systems of greenhouse are described below:

Ventilation System

The efficiency of a ventilation fan can be significantly reduced by 30-50 per cent due to dust build-up or malfunctioning shutters. Inadequate ventilation can lead to stagnant air, insufficient cooling, high heating costs, heavy condensation in winter, reduced equipment life and reliability, and higher repair bills. The following periodic maintenance schedule be followed for the maintenance of ventilation system:

1. Clean the fan blades, fan housing, and shutters.
2. Lubricate the fan bearings, motor, and shutters.
3. Check fan wheel for proper rotation. Proper direction of rotation is generally marked on the fan housing.
4. Check the fan belts for proper tension to prevent slippage. If the belts are cracking, splitting, or fraying, replace immediately.
5. Inspect the electrical supply cords to fans.
6. Don't allow vegetation near to the fans of greenhouse.
7. Avoid placement of any obstructions in the air flow path of fan. The distance between the obstruction and fan shall be more than the length of one fan blade.

8. Check and close the openings for air infiltration inside the greenhouse.
9. Calibrate the sensors like thermostats and humidistat for verification of prescribed environmental conditions.

Evaporative Cooling System

Evaporative cooling efficiency depends on factors such as compacting cooling pads, improper fan operation, open greenhouse doors, and insufficient water supply to cooling pads. The lower cooling efficiency increases the greenhouse air temperature. The following periodic maintenance schedule be followed for the maintenance of evaporative cooling system:

1. Check pads and supporting frames for a tight fit.
2. Prevent air infiltration through open doors, cracks, or openings.
3. Repair holes in pads or structures.
4. Replace the pads when the void area is 10 per cent or greater of the total pad area.
5. Clean or replace the pads, if they are covered with algae. Presence of algae on the pads increase resistance to air flow, resulting in reduction of cooling efficiency.
6. Fans should be properly maintained to perform at the rated capacities so that the proper volume of air is drawn through the cooling pads at the required static pressure resistance.
7. The water supply rate to the cooling pads should be checked at the start of each cooling season.
8. Fungicides are being utilized in the water supply system to prevent the growth and accumulation of algae in cooling pads.

Heating System

Unit space heaters, hot water heaters, and steam heaters are the three primary types of heating systems commonly utilized. The following periodic maintenance schedule be followed for the maintenance of heating system:

1. Check nozzle of all burners regularly.
2. Clean the blades and oil of the motor of fan as per the recommendation of manufacturer.
3. Ensure all flues are checked for leaks and the stacking of flues shall be at least four feet above the house ridge or any nearby building.
4. Inspect all pipes in a hot water or steam system for any potential leaks.
5. Proper orientation of unit heaters ensures optimal heat and air distribution at plant level. Check belt-driven fans for good condition of belts and pulleys.

6. Maintain clean pipes and fins in hot water systems and steam systems. It prevents dust accumulation, which increases the efficiency of the unit heater.
7. Ensure all boiler components are in optimal working condition well in advance of the heating season.

Maintenance of Shade Cloth

Shade cloths are used to regulate greenhouse temperature by diffusing the sunlight before it enters the greenhouse. Shade cloths also regulate light intensity in the greenhouse. Shade cloth must be attached properly to the structure, and be able to apply or remove as per the requirement.

22.11 Energy Conservation

Greenhouse energy conservation is crucial for reducing operational costs, promoting sustainability, and minimizing environmental impact. Implementing energy-efficient practices in greenhouse operations can lead to significant savings and contribute to a more sustainable and resilient agricultural system. Energy-efficient heating and cooling systems, wall insulation, passive solar design, energy-efficient electrical equipment, natural ventilation, thermal curtains, smart environmental controls, *etc.* are some of the energy conservation strategies.

Important Points to Remember

1. A greenhouse is a framed or an inflated structure covered with a transparent or translucent material in which crops could be grown under partial or fully controlled environment to get optimum plant growth and productivity.
2. The productivity of the crop in the greenhouse is increased considerably as compared to productivity under open field conditions.
3. Off-season vegetables can be produced in greenhouses.
4. Wooden framed structures are low cost and generally recommended for the greenhouses with span less than 6 m.
5. Truss frames are used for the greenhouses whose span is greater than or equal to 15 m.
6. Plastics as covering material are cheap and the cost of heating is also less when compared to glass greenhouses.
7. A gable greenhouse has sloping, flat roofs connected to vertical sidewalls.
8. Saw tooth type green house is category of greenhouse based upon its shape.

9. Roof slope is an important parameter affecting light transmission in greenhouse.
10. High cost greenhouse has facilities like auto control mechanism of heating, cooling and humidification system; drip irrigation system, *etc.*
11. In photosynthesis, plants produce the foodstuffs using water and carbon dioxide.
12. Light transmittance of glazing materials should be high as light is responsible for plant growth.
13. Light is essential for photosynthesis, the process by which plants convert light energy into chemical energy to fuel their growth.
14. Visible light (400 to 700 nm) is essential for plant development as the conversion of radiant to chemical energy proceeds only in visible light.
15. Photosynthetic Active Radiation (PAR) is defined as the photons of radiation in the 400 to 700 nm waveband.
16. Light intensity affects photosynthesis and influences overall plant growth.
17. The day temperature of greenhouse shall be 3-6°C higher in cloudy day and 8°C higher in clear days than the night temperature.
18. Extremely low humidity in the greenhouse can lead to high transpiration, water stress and reduce photosynthetic rates.
19. Increasing the concentration of CO_2 in the greenhouse environment can enhance photosynthetic rates, leading to improved plant growth and productivity.
20. The location of the greenhouse should be shadow free.
21. In free standing/single span greenhouse, more sunlight in winter is available in an east-west oriented greenhouse as compared to north-south oriented greenhouse.
22. In multi-span greenhouses, the gutters should be oriented in north-south to avoid continuous shading of certain portions of the greenhouse due to structural members.
23. The head house should be located to the north of the production area to avoid shading due to it.
24. As a general rule, water requirement in the greenhouse is 20 litres per square meter per watering.
25. The pH of irrigation water shall range between 6.5 to 7.5.
26. The evaporative cooling systems are most effective in hot and dry climate when the humidity of the outside air is low.
27. The temperature drop in the evaporative cooling system increases with the increase in wet bulb depression.

28. Complete evaporation of the water does not take place in fan and pad system due to bigger droplet size, whereas in the fog cooling system, complete evaporation takes place because of the minute size of the water droplets.
29. Shade cloths regulate light intensity in the greenhouse.
30. Mesh screens of different density can be used to provide shading from 30 to 92 per cent.
31. Maximize ventilation rates whenever possible starting 1 or 2 hours after sunrise when the overnight build-up of CO_2 has been depleted due to photosynthesis.
32. CO_2 in a green house is more at 12.00 midnight than at 12.00 noon.
33. When the solar radiation is insufficient, it may be complemented with artificial light.
34. The greenhouse air temperature can be brought below 35°C during spring and autumn seasons through ventilation.
35. UV light from the sun causes the plastic film used as greenhouse covering material to darken, thereby lowering the transmission of light, also making it brittle, which leads to its breakage due to wind.

Exercises/Questions

1. Define greenhouse. What is the status and prospects of protected cultivation in India?
2. List various advantages and disadvantages of greenhouse cultivation.
3. Discuss the classification criteria of green houses. Name various green houses on the basis of these criteria.
4. Name various covering (glazing) materials. What factors are considered in selecting the covering materials? Discuss the advantages and disadvantages of plastic films/sheets as covering materials.
5. Discuss the various components of the green houses.
6. Discuss response of plants to greenhouse environment.
7. What factors would you consider while deciding to locate a green house? In which direction a greenhouse should be oriented and why?
8. Describe briefly about the water and electricity requirement for a greenhouse.
9. Discuss the need and methods of cooling a greenhouse.
10. Describe the evaporative cooling phenomenon.
11. Describe the various methods used in a greenhouse for heating, cooling, and light control.

12. What do you understand by ventilation of a green house? Differentiate between active and passive ventilation.
13. What is the role of CO_2 in the green houses? How does it vary during day/night? Describe how CO_2 can be managed in green houses?
14. Describe the maintenance requirements of ventilation system, evaporative cooling system, heating system, and shade cloth.
15. Write short notes on:
    (a) Greenhouse effect
    (b) Fogging system for cooling
    (c) CO_2 levels in greenhouse
    (d) Interior layout in greenhouse
    (e) Passive solar greenhouse
    (f) Energy conservation in greenhouse
16. Differentiate between the following:
    (a) Glass house and poly house
    (b) Fogging and misting system
    (c) Green house and growth chamber
    (d) Thermal screen and shade screen

Appendices

Appendix A: Conversion Factors

Length	1 inch	= 2.54 cm
	1 ft	= 12 inch = 30.48 cm = 0.3048 m
Area	1 in^2	= 6.4516 cm^2
	1 ft^2	= 0.0929 m^2
Volume	1 ft^3	= 0.0283 m^3 = 28.317 litres
	1 gal English	= 0.004546 m^3 = 4.546 litres
	1 gal US	= 0.003785 m^3 = 3.785 litres
	1 litre	= 0.001 m^3 = 1000 cc
Mass	1 lb	= 0.4536 kg = 453.6 g
	1 kg	= 2.2046 lb = 100 dag
	1 tonne	= 10 quintal = 1000 kg
Density	1 lb/ft^3	= 16.0185 kg/m^3
Pressure	1 lb/in^2	= 6895 Pa =6895 N/m^2
	1 torr	= 133.3 Pa
	1 atm	= 1.013 x 10^5 Pa = 760 mm Hg = 1.0132 bar
	1 Pa	= 1 N/m^2 = 1 kg/m s^2 = 10^{-5} bar
	1 bar	= 10^5 Pa = 10^5 N/m^2 = 1.0197 kgf/cm^2
Force	1 Newton	= 1 kg m/s^2 = 10^5 dyne = $\frac{1}{9.80665}$ kgf
	1 dyne	= 1 g cm/s^2 = 10^{-5} kg m/s^2
Viscosity (dynamic)	1 cP	= 0.001 N s/m^2 = 0.001 Pa s
	1 lb/ft sec	= 1.49 N s/m^2 = 1.49 kg/m s
	1 kg/m sec	= 1 Pa-s = 1 N s/m^2 = 10 dyne-s/cm^2 = 10 poise
	1 g/cm s (poise)	= 10^{-1} kg/m s = 100 centipoise
Viscosity (kinematic)	1 m^2/s	= 10^4 cm^2/s (or stokes) = 38750 ft^2/h
Energy	1 J	= 1 W-s = 10^7 ergs = 0.239 cal
	1 kJ	= 10^3 Nm = 101.972 kgf m
	1 Btu	= 1055 J = 252.16 cal
	1 kcal	= 4.184 kJ = 3.9657 Btu
	1 kWh	= 856.66 kcal = 3.6 MJ

Power	1 W	= 1 J/s = 14.34 cal/min
	1 horsepower	= 0.746 kW = 75 kgf m/s = 10.54 kcal/min = 596.8 kcal/h
	1 ton refrigeration	= 3.519 kW = 50 kcal/min = 200 Btu/min
Temperature	T°C	= (T+273.16) K = (1.8T+32)°F = (1.8T+459.69)°R
Temperature units	(°F)	= 5/9 (°C) = 5/9 (K)
Specific heat	1 Btu/lb°F	= 4.186 J/g°C = 1 cal/g°C
Heat-transfer coefficient	1 Btu/ft² h °F	= 5.678 J m² s °C = 5.678 W/m² °C
	1 kcal/m² h °C	= 0.205 Btu/ft² h °F
Thermal conductivity	1 Btu/ft h°F	= 1.731 J/m s °C = 1.731 W/m °C
	1 kcal/m h °C	= 0.6719 Btu/ft h °F

Appendix B: Weight per meter of equal angles (Refer Fig. 3.4)

Designation	*Size (mm)*	*Weight per meter (kg)*
ISA 2020	20×20×3	0.9
ISA 2020	20×20×4	1.1
ISA 2525	25×25×3	1.1
ISA 2525	25×25×4	1.4
ISA 2525	25×25×5	1.8
ISA 3030	30×30×3	1.4
ISA 3030	30×30×4	1.8
ISA 3030	30×30×5	2.2
ISA 3535	35×35×3	1.6
ISA 3535	35×35×4	2.1
ISA 3535	35×35×5	2.6
ISA 3535	35×35×6	3.0
ISA 4040	40×40×3	1.8
ISA 4040	40×40×4	2.4
ISA 4040	40×40×5	3.0
ISA 4040	40×40×6	3.5
ISA 4545	45×45×3	2.1
ISA 4545	45×45×4	2.7
ISA 4545	45×45×5	3.4
ISA 4545	45×45×6	4.0
ISA 5050	50×50×3	2.3
ISA 5050	50×50×4	3.0
ISA 5050	50×50×5	3.8
ISA 5050	50×50×6	4.5
ISA 5555	55×55×5	4.1
ISA 5555	55×55×6	4.9
ISA 5555	55×55×8	6.4
ISA 5555	55×55×10	7.9
ISA 6060	60×60×5	4.5
ISA 6060	60×60×6	5.4
ISA 6060	60×60×8	7.0
ISA 6060	60×60×10	8.6
ISA 6565	65×65×5	4.9

Designation	*Size (mm)*	*Weight per meter (kg)*
ISA 6565	65×65×6	5.8
ISA 6565	65×65×8	7.7
ISA 6565	65×65×10	9.4
ISA 7070	70×70×5	5.3
ISA 7070	70×70×6	6.3
ISA 7070	70×70×8	8.3
ISA 7070	70×70×10	10.2
ISA 7575	75×75×5	5.7
ISA 7575	75×75×6	6.8
ISA 7575	75×75×8	8.9
ISA 7575	75×75×10	11.0
ISA 8080	80×80×6	7.3
ISA 8080	80×80×8	9.6
ISA 8080	80×80×10	11.8
ISA 8080	80×80×12	14.0
ISA 9090	90×90×6	8.2
ISA 9090	90×90×8	10.8
ISA 9090	90×90×10	13.4
ISA 9090	90×90×12	15.8
ISA 100100	100×100×6	9.2
ISA 100100	100×100×8	12.1
ISA 100100	100×100×10	14.9
ISA 100100	100×100×12	17.7
ISA 200200	200×200×12	36.9
ISA 200200	200×200×16	48.5
ISA 200200	200×200×20	60.0
ISA 200200	200×200×25	73.9

Appendix C: Weight per meter of unequal angles (Refer Fig. 3.5)

Designation	*Size (mm)*	*Weight per meter (kg)*
ISA 3020	30×20×3	1.1
ISA 3020	30×20×4	1.4
ISA 3020	30×20×5	1.8
ISA 4025	40×25×3	1.5
ISA 4025	40×25×4	1.9
ISA 4025	40×25×5	2.4
ISA 4025	40×25×6	2.8
ISA 4530	45×30×3	1.7
ISA 4530	45×30×4	2.2
ISA 4530	45×30×5	2.8
ISA 4530	45×30×6	3.3
ISA 5030	50×30×3	1.8
ISA 5030	50×30×4	2.4
ISA 5030	50×30×5	3.0
ISA 5030	50×30×6	3.5
ISA 7550	75×50×5	4.7
ISA 7550	75×50×6	5.6
ISA 7550	75×50×8	7.4
ISA 7550	75×50×10	9.0
ISA 8050	80×50×5	4.9
ISA 8050	80×50×6	5.9
ISA 8050	80×50×8	7.7
ISA 8050	80×50×10	9.4
ISA 9060	90×60×6	6.8
ISA 9060	90×60×8	8.9
ISA 9060	90×60×10	11.0
ISA 9060	90×60×12	13.0
ISA 10065	100×65×6	7.5
ISA 10065	100×65×8	9.9
ISA 10065	100×65×10	12.2
ISA 10075	100×75×6	8.0
ISA 10075	100×75×8	10.5
ISA 10075	100×75×10	13.0
ISA 10075	100×75×12	15.4

Appendix D: Weight per meter of rolled steel channels (Refer Fig. 3.6)

Designation	*Size (mm)*	*Flange Thickness (mm)*	*Web Thickness (mm)*	*Weight per meter (kg)*
ISJC 100	100×45	5.1	3.0	5.8
ISJC 125	125×50	6.6	3.0	7.9
ISJC 150	150×55	6.9	3.6	9.9
ISJC 175	175×60	6.9	3.6	11.2
ISJC 200	200×70	7.1	4.1	13.9
ISLC 75	75×40	6.0	3.7	5.7
ISLC 100	100×50	6.4	4.0	7.9
ISLC 125	125×65	6.6	4.4	10.7
ISLC 150	150×75	7.8	4.8	14.4
ISLC 175	175×75	9.5	5.1	17.6
ISLC 200	200×75	10.8	5.5	20.6
ISLC 225	225×90	10.2	5.8	24.0
ISLC 250	250×100	10.7	6.1	28.0
ISLC 300	300×100	11.6	6.7	33.1
ISLC 350	350×100	12.5	7.4	38.9
ISLC 400	400×100	14.0	8.0	45.8
ISMC 75	75×40	7.3	4.4	6.8
ISMC 100	100×50	7.5	4.7	9.2
ISMC 125	125×65	8.1	5.0	12.8
ISMC 150	150×75	9.0	5.4	16.4
ISMC 175	175×75	10.2	5.7	19.2
ISMC 200	200×75	11.4	6.1	22.2
ISMC 225	225×80	12.4	6.4	26.0
ISMC 250	250×80	14.1	7.1	30.4
ISMC 300	300×90	13.6	7.6	35.9
ISMC 350	350×100	13.5	8.1	42.2
ISMC 400	400×100	15.3	8.6	49.5

ISJC – Indian Standard Junior Channels; ISLC – Indian Standard Light Weight Channels; ISMC – Indian Standard Medium Weight Channels

Appendix E: Weight per meter of rolled steel beams (Refer Fig. 3.8)

Designation	*Size (mm)*	*Flange Thickness (mm)*	*Web Thickness (mm)*	*Weight per meter (kg)*
ISJB 150	150×50	4.6	3.0	7.1
ISJB 175	175×50	4.8	3.2	8.1
ISJB 200	200×60	5.0	3.4	9.9
ISJB 225	225×80	5.0	3.7	12.8
ISLB 75	75×50	5.0	3.7	6.1
ISLB 100	100×50	6.4	4.0	8.0
ISLB 125	125×75	6.5	4.4	11.9
ISLB 150	150×80	6.8	4.8	14.2
ISLB 175	175×90	6.9	5.1	16.7
ISLB 200	200×100	7.3	5.4	19.8
ISLB 225	225×100	8.6	5.8	23.5
ISLB 250	250×125	8.2	6.1	27.9
ISLB 275	275×140	8.8	6.4	33.0
ISLB 300	300×150	9.4	6.7	37.7
ISMB 100	100×75	7.2	4.0	11.5
ISMB 125	125×75	7.6	4.4	13.0
ISMB 150	150×80	7.6	4.8	14.9
ISMB 175	175×90	8.6	5.5	19.3
ISMB 200	200×100	10.8	5.7	25.4
ISMB 225	225×110	11.8	6.5	31.2
ISMB 250	250×125	12.5	6.9	37.3
ISMB 300	300×140	12.4	7.5	44.2
ISWB 150	150×100	7.0	5.4	17.0
ISWB 175	175×125	7.4	5.8	22.1
ISWB 200	200×140	9.3	6.1	28.8
ISWB 225	225×150	9.9	6.4	33.9
ISWB 250	250×200	9.0	6.7	40.9
ISWB 300	300×200	10.0	7.4	48.1

ISJB – Indian Standard Junior Beams; ISLB – Indian Standard Light Weight Beams; ISMB – Indian Standard Medium Weight Beams; ISWB – Indian Standard Wide Flange Beams

Appendix F: Weight per meter of rolled steel tee bars (Refer Fig. 3.9)

Designation	*Size (mm)*	*Flange Thickness (mm)*	*Web Thickness (mm)*	*Weight per meter (kg)*
ISNT 20	20×20	3.0	3.0	0.9
ISNT 30	30×30	3.0	3.0	1.4
ISNT 40	40×40	6.0	6.0	3.5
ISNT 50	50×50	6.0	6.0	4.5
ISNT 60	60×60	6.0	6.0	5.4
ISNT 80	80×80	8.0	8.0	9.6
ISNT 100	100×100	10.0	10.0	15.0
ISNT 150	150×150	10.0	10.0	22.8
ISHT 75	75×150	9.0	8.4	15.3
ISHT 100	100×200	9.0	7.8	20.0
ISHT 125	125×250	9.7	8.8	27.4
ISHT 150	150×250	10.6	7.6	29.4
ISST 100	100×50	10.0	5.8	8.1
ISST 150	150×75	11.6	8.0	15.7
ISST 200	200×165	12.5	8.0	28.4
ISST 250	250×180	14.1	9.2	37.5
ISLT 50	50×50	6.4	4.0	4.0
ISLT 75	75×80	6.8	4.8	7.1
ISLT 100	100×100	10.8	5.7	12.7
ISJT 75	75×50	4.6	3.0	3.5
ISJT 87.5	87.5×50	4.8	3.2	4.0
ISJT 100	100×60	5.0	3.4	5.0
ISJT 112.5	112.5×80	5.0	3.7	6.4

Appendix G: Weight, sectional area, perimeter and length per ton of square and round bars

Diameter or Width Across Flat (mm)	*Weight per meter (kg)*		*Sectional Area (cm^2)*		*Perimeter (cm)*		*Length in meter per ton*	
	Square	*Round*	*Square*	*Round*	*Square*	*Round*	*Square*	*Round*
5	0.20	0.15	0.25	0.20	2.0	1.57	5000	6667
6	0.28	0.22	0.36	0.28	2.4	1.88	3571	4545
8	0.50	0.39	0.64	0.50	3.2	2.51	2000	2564
10	0.78	0.62	1.00	0.79	4.0	3.14	1282	1613
12	1.13	0.89	1.44	1.13	4.8	3.77	885	1124
16	2.01	1.58	2.56	2.01	6.4	5.03	498	633
20	3.14	2.47	4.00	3.14	8.0	6.28	319	405
22	3.80	2.98	4.84	3.80	8.8	6.91	263	336
25	4.91	3.85	6.25	4.91	10.0	7.85	204	260
28	6.15	4.83	7.84	6.16	11.2	8.80	163	207
32	8.04	6.31	10.24	8.04	12.8	10.05	124	159
36	10.17	7.99	12.96	10.18	14.4	11.31	98	125
40	12.56	9.86	16.00	12.57	16.0	12.57	80	101
45	15.90	12.49	20.25	15.90	18.0	14.14	63	80
50	19.62	15.41	25.00	19.64	20.0	15.71	51	65
56	24.62	19.34	31.36	24.63	22.4	17.59	41	52
63	31.16	24.47	39.69	31.17	25.2	19.79	32	41
71	39.57	31.08	50.41	39.59	28.4	22.30	25	32
80	50.24	39.46	64.00	50.26	32.0	25.13	20	25
90	63.58	49.94	81.00	63.62	36.0	28.27	16	20
100	78.50	61.66	100.00	78.54	40.0	31.42	13	16

Appendix H: Weight per metre of flat mild steel in kg per meter

Width in mm	*Thickness in mm*									
	5	*5.5*	*6*	*7*	*8*	*10*	*11*	*12*	*14*	*16*
12	0.5	0.5	0.6	0.7	0.8	0.9	1.0	1.1	1.3	1.5
16	0.6	0.7	0.8	0.9	1.0	1.3	1.4	1.5	1.8	2.0
20	0.8	0.9	0.9	1.1	1.3	1.6	1.7	1.9	2.2	2.5
25	1.0	1.1	1.2	1.4	1.6	2.0	2.2	2.4	2.7	3.1
32	1.3	1.4	1.5	1.8	2.0	2.5	2.8	3.0	3.5	4.0
40	1.6	1.7	1.9	2.2	2.5	3.1	3.5	3.8	4.4	5.0
50	2.0	2.2	2.4	2.7	3.1	3.9	4.3	4.7	5.5	6.3
63	2.5	2.7	3.0	3.5	4.0	4.9	5.4	5.9	6.9	7.9
80	3.1	3.5	3.8	4.4	5.0	6.3	6.9	7.5	8.8	10.0
100	3.9	4.3	4.7	5.5	6.3	7.8	8.6	9.4	11.0	12.6
125	4.9	5.4	5.9	6.9	7.8	9.8	10.8	11.8	13.7	15.7
160	6.3	6.9	7.5	8.8	10.0	12.6	13.8	15.1	17.6	20.1
200	7.8	8.6	9.4	11.0	12.6	15.7	17.3	18.8	22.0	25.1
250	9.8	10.8	11.8	13.7	15.7	19.6	21.6	23.6	27.5	31.4

Appendix I: Task or outturn per day per skilled labour

Sl.No.	Particulars of Items	Quantity	Per day
1.	Brick work in lime or cement mortar in foundation	1.25 cum	Per mason
2.	-do- in super structure	1.00 cum	Per mason
3.	Brick work in mud mortar in foundation and plinth	1.50 cum	Per mason
4.	-do- in super structure	1.25 cum	Per mason
5.	Half brick wall in partition	5.00 sqm	Per mason
6.	Coursed rubber stone masonry in lime or cement mortar including dressing	0.80 cum	Per mason
7.	Random rubble stone masonry in lime or cement mortar	1.00 cum	Per mason
8.	Ashlar masonry in lime or cement mortar	0.40 cum	Per mason
9.	Stone arch work	0.40 cum	Per mason
10.	Lime concrete in foundation or floor	8.50 cum	Per mason
11.	Lime concrete in roof terracing	6.00 cum	Per mason
12.	Cement concrete 1: 2 : 4	5.00 cum	Per mason
13.	R.C.C. work	3.00 cum	Per mason
14.	12 mm plastering with cement or lime mortar	10.00 sqm	Per mason
15.	Pointing with cement or lime mortar	15.00 sqm	
16.	White washing or colour washing three coats	70.00 sqm	Per white washer
17.	White washing or colour washing one coat	200.00 sqm	Per white washer
18.	Painting or varnishing door or windows one coat	25.00 sqm	Per white washer
19.	Painting large surface one coat	35.00 sqm	Per painter
20.	Distempering one coat	35.00 sqm	Per painter
21.	2.5 cm. C.C. floor	7.50 sqm	Per mason
22.	Timber framing sal or teak wood	0.07 cum	Per carpenter
23.	-do- Country wood	0.15 cum	Per carpenter

Sl.No.	*Particulars of Items*	*Quantity*	*Per day*
24.	Door or window shutters panel or glazed	0.70 sqm	Per carpenter
25.	-do- battened	0.80 sqm	Per carpenter
26.	Single Allahabad tiling or Mangalore tiling	6.00 sqm	Per tile layer
27.	Double Allahabad tiling	4.00 sqm	Per tile layer
28.	Breaking of brick ballast- 40 mm	0.75 cum	Per labourer breaker
29.	Breaking or brick ballast- 25 mm	0.55 cum	Per labourer breaker
30.	Breaking or stone ballast- 40 mm	0.40 cum	Per labourer breaker
31.	Breaking or stone ballast- 25 mm	0.25 cum	Per labourer breaker
32.	Earth work in excavation in ordinary soil	3.00 cum	Per beldar mazdoor
33.	Earth work in excavation in hard soil	2.00 cum	Per beldar mazdoor
34.	Number of bricks laid by a mason in brick work up to a height of 3 meter	600 bricks	Per mason

Appendix J: Sizes and dimensions of various works

Sl.No.	*Particulars*	*Size/Dimension (approx.)*
1.	Sizes of doors	1.20×2.10 m
		1.10×2.00 m
		1.00×1.90 m
		0.90×1.80 m
		0.75×1.80 m
2.	Sizes of windows	1.00×1.50 m
		0.90×1.50 m
3.	Sizes of *chowkhats* for doors and windows	12×8 cm
		10×10 cm
		10×8 cm
		8×8 cm
		7×7 cm
4.	Thickness of DPC	2.0 cm/2.5 cm
5.	Thickness of plaster	13 mm/10 mm
6.	Sizes of ventilators	1.20×0.60 m
		1.00×0.60 m
		0.90×0.50 m
7.	Thickness of doors and windows shutters	3.0 cm/4.0 cm/4.5 cm
8.	Thickness of RCC work in slab	7.5 cm/10 cm/12 cm
9.	Height of plinth	45 cm/60 cm/75 cm/90 cm
10.	Insulation layer in roof	2.5 cm/4.0 cm
11.	Height of building	3.0 m/3.3 m/3.6 m/4.2 m
12.	Sizes of rooms	3.0×3.0 m
		3.0×3.6 m
		3.6×3.6 m
		3.6×4.2 m
		3.6×4.8 m

Appendix K: BIS standards for storage of food grains, storage structures and equipment

Sl.No.	Standard	Code
1.	General requirements for silos for grain storage: Part 1 constructional requirements (Reaffirmed In : 2020)	IS 5503 (Part 1) : 1969
2.	General Requirements for Silos for Grain Storage, Part II Grain Handling equipment and accessories	IS 5503 (Part 2) : 2020
3.	Specification for outdoor aluminum food grain storage bins	IS.631.1979
4.	Improvement of Existing Structures Used or Intended to be Used for Food Grain Storage — Code of Practice	IS.609.202
5.	Terminology for Food grains (Third Revision)	IS 2813 : 2018
6.	Methods of analysis for food grains: Part 1 refractions (Third Revision)	IS 4333 (Part 1) : 2018
7.	Methods of analysis for food grains: Part 2 determination of moisture content (Second Revision) (Reaffirmed In : 2022)	IS 4333 (Part 2) : 2017ISO 712 : 2009
8.	Methods of analysis for food grains: Part 3 determination of bulk density, called mass per hectolitre (Second Revision) (Reaffirmed In :2023)	IS 4333 (Part 3) : 2018ISO 7971-3 : 2019
9.	Methods of analysis for food grains: Part 4 determination of the mass of 1000 grains (Second Revision) (Reaffirmed In : 2022)	IS 4333 (Part 4) : 2017ISO 520 : 2010
10.	Methods of analysis for food grains: Part 5 determination of uric acid (Reaffirmed In : 2020)	IS 4333 (Part 5) : 1970
11.	Specification for steel bins for grain storage (Reaffirmed In : 2020)	IS 5606 : 1970
12.	Storage Management Code Part 1 Terminology (First Revision)	IS 6151 (Part 1) : 2020
13.	Storage management code: Part 2 general care in handling and storage of agricultural produce and inputs (Reaffirmed In : 2020)	IS 6151 (Part 2) : 1971
14.	Storage management code: Part 3 specific care in handling and storage of agricultural produce and inputs (Reaffirmed In : 2020)	IS 6151 (Part 3) : 1976
15.	Methods of analysis for detection of insect and rodent contamination in grains and milled products	IS 6261 : 1971

Sl.No.	Standard	Code
16.	Method for determination of angle of repose of grains (Reaffirmed In : 2020)	IS 6663 : 1972Reaffirmed In : 2020
17.	Specification for steel bins for domestic storage: Part 1 *gharelu kothi* (Reaffirmed In : 2020)	IS 7147 (Part 1) : 1973
18.	Code of practice for fumigation of agricultural produce: Part 1 methyl bromide (Reaffirmed In : 2020)	IS 7247 (Part 1) : 1974
19.	Fumigation of Agricultural Produce ï¿½ Code of Practice Part 3 Aluminum Phosphide (Phosphine) (First Revision) (Reaffirmed In : 2020)	IS 7247 (Part 3) : 1974
20.	Code of practice for fumigation of agricultural produce: Part 4 ethylene dichloride and carbon tetrachloride mixture (Reaffirmed In : 2020)	IS 7247 (Part 4) : 1975
21.	Fumigation of Agricultural Produce Code of Practice Part 5 General Requirements (First Revision) (Reaffirmed In : 2022)	IS 7247 (Part 5) : 1985
22.	Method for testing suitability of bins for safe storage of food grains (Reaffirmed In : 2020)	IS 7715 : 1975
23.	Method for testing efficacy of fumigation for disinfestation of grains in domestic bins (Reaffirmed In : 2020)	IS 7716 : 1975
24.	Method for determination of ergot in food grains (Reaffirmed In : 2020)	IS 8184 : 1976
25.	Code of practice for construction of polyethylene embedded earthen bins for bulk storage of food grains (Reaffirmed In : 2020)	IS 8453 : 1977
26.	Determination of coefficient of friction of food grains (Reaffirmed In : 2020)	IS 8972 : 1978
27.	Method for determination of thermal conductivity of food grains (Reaffirmed In : 2020)	IS 9027 : 1978
28.	Method for determination of thermal diffusivity of foodgrains (Reaffirmed In : 2020)	IS 10698 : 1983
29.	Method for determination of specific heat of food grains (Reaffirmed In : 2020)	IS 10699 : 1983
30.	Method of test for quality characteristics of pulses (Reaffirmed In : 2020)	IS 10768 : 1984
31.	Test methods for determination of storability (Safe Storage Life) of food grains (Reaffirmed In : 2020)	IS 11396 : 1985

Sl.No.	*Standard*	*Code*
32.	Method for measurement of carbon dioxide in the inter-granular atmosphere (Reaffirmed In : 2020)	IS 11454 : 1985
33.	Storage of cereals and pulses: Part 1 general recommendations for the keeping of cereals (First Revision) (Reaffirmed In : 2020)	IS 11816 (Part 1) : 2010ISO 6322Part 1:1996
34.	Storage of cereals and pulses: Part 2 practical recommendations (First Revision) (Reaffirmed In : 2020)	IS 11816 (Part 2) : 2009
35.	Storage of cereals and pulses: Part 3 control of attack by pests (First Revision)	IS 11816 (Part 3) : 2018
36.	Storage of foodgrains - Storage losses by insects - Methods for estimation (Reaffirmed In : 2020)	IS 12529 : 1988
37.	Food grain storage godowns - Code of practice (Reaffirmed In : 2020)	IS 16144 : 2014

Appendix L: Thermodynamic properties of moist air at standard atmospheric pressure (101.325 kPa)

Temperature (°C)	*Saturation Pressure (kPa)*	*Humidity Ratio at Saturation (kgw/kgda)*	*Specific Volume (m^3/kg dry air)*		*Specific Enthalpy (kJ/kg dry air)*	
			Dry Air	*Saturated Mixture*	*Dry Air*	*Saturated Mixture*
0	0.6112	0.003789	0.7734	0.7781	0.000	9.473
1	0.6571	0.004076	0.7762	0.7813	1.006	11.204
2	0.7060	0.004381	0.7791	0.7845	2.012	12.982
3	0.7581	0.004707	0.7819	0.7878	3.018	14.811
4	0.8135	0.005054	0.7848	0.7911	4.024	16.696
5	0.8725	0.005424	0.7876	0.7944	5.030	18.639
6	0.9353	0.005818	0.7904	0.7978	6.036	20.644
7	1.0020	0.006237	0.7933	0.8012	7.041	22.713
8	1.0729	0.006683	0.7961	0.8046	8.047	24.852
9	1.1481	0.007157	0.7990	0.8081	9.053	27.064
10	1.2280	0.007661	0.8018	0.8116	10.059	29.352
11	1.3128	0.008197	0.8046	0.8152	11.065	31.724
12	1.4026	0.008766	0.8075	0.8188	12.071	34.179
13	1.4979	0.009370	0.8103	0.8225	13.077	36.726
14	1.5987	0.010012	0.8132	0.8262	14.084	39.370
15	1.7055	0.010692	0.8160	0.8300	15.090	42.113
16	1.8185	0.011413	0.8188	0.8338	16.096	44.963
17	1.9380	0.012178	0.8217	0.8377	17.102	47.926
18	2.0643	0.012989	0.8245	0.8414	18.108	51.008
19	2.1979	0.013848	0.8274	0.8457	19.114	54.216
20	2.3389	0.014758	0.8302	0.8498	20.121	57.555
21	2.4878	0.015721	0.8330	0.8540	21.127	61.035
22	2.6448	0.016741	0.8359	0.8583	22.133	64.660
23	2.8105	0.017821	0.8387	0.8627	23.140	68.440
24	2.9852	0.018963	0.8416	0.8671	24.146	72.385
25	3.1693	0.020170	0.8444	0.8717	25.153	76.500
26	3.3633	0.021448	0.8472	0.8764	26.159	80.798
27	3.5674	0.022798	0.8501	0.8811	27.165	85.285
28	3.7823	0.024226	0.8529	0.8860	28.172	89.976
29	4.0084	0.025735	0.8558	0.8910	29.179	94.878

Temperature (°C)	*Saturation Pressure (kPa)*	*Humidity Ratio at Saturation (kgw/kgda)*	*Specific Volume (m^3/kg dry air)*		*Specific Enthalpy (kJ/kg dry air)*	
			Dry Air	*Saturated Mixture*	*Dry Air*	*Saturated Mixture*
30	4.2462	0.027329	0.8586	0.8962	30.185	100.006
31	4.4961	0.029014	0.8614	0.9015	31.192	105.369
32	4.7586	0.030793	0.8643	0.9069	32.198	110.979
33	5.0345	0.032674	0.8671	0.9125	33.205	116.857
34	5.3242	0.034660	0.8700	0.9183	34.212	123.011
35	5.6280	0.036756	0.8728	0.9242	35.219	129.455
36	5.9468	0.038971	0.8756	0.9303	36.226	136.209
37	6.2812	0.041309	0.8785	0.9366	37.233	143.290
38	6.6315	0.043778	0.8813	0.9431	38.239	150.713
39	6.9988	0.046386	0.8842	0.9498	39.246	158.504
40	7.3838	0.049141	0.8870	0.9568	40.253	166.683
41	7.7866	0.052049	0.8898	0.9640	41.261	175.265
42	8.2081	0.055119	0.8927	0.9714	42.268	184.275
43	8.6495	0.058365	0.8955	0.9792	43.275	193.749
44	9.1110	0.061791	0.8983	0.9872	44.282	203.699
45	9.5935	0.065411	0.9012	0.9955	45.289	214.164

Bibliography

ASAE (1982). *Agricultural Engineers Yearbook of Standards*. 1982. Data D384, American Society of Agricultural Engineers, St. Joseph, Michigan.

ASAE (1983-1984). *Agricultural Engineers Yearbook of Standards*. American Society of Agricultural Engineers, St. Joseph, Michigan.

ASHRAE. (2009). *ASHRAE Handbook of Fundamentals*. Atlanta, Georgia, USA.

Adhikari, K. and Fedler, C. B. (2020). *Water sustainability using pond-in-pond wastewater treatment system: Case studies*. Journal of Water Process Engineering, 36, 101281.

Aggarwal, R.K., Ohri, M.L. and Dhall, P.K. (1986). *Concrete Technology*. India Publishing House, Delhi.

Anarase, D. and Garg, M.K. (2017). *Development of expert system for the design of food grain storage bin and facility*. International J. of Agricultural Sciences, **9(1)**:3594-3598.

Anarase, D. and Garg, M.K. (2021). *Development of decision support system for designing the layout of grain storage warehouse*. J. Agric. Engng., 58(4): 352 – 367.

Arora, S C and Domkundwar, S. (1991). *A Course in Heat & Mass Transfer*. Dhanpat Rai and Sons. New Delhi.

Arora, S C and Domkundwar, S. (1992). *A Course in Refrigeration and Air Conditioning*. Dhanpat Rai and Sons. New Delhi.

Bal, S.K., Prasad, J.V.N.S and Singh, V.K. (2022). *Heat Wave 2022 Causes, impacts and way forward for Indian Agriculture*. Technical Bulletin No. ICAR/CRIDA/TB/01/2022, ICAR-Central Research Institute for Dryland Agriculture, Hyderabad, Telangana, India.

Ballaney, P L (2002). *Refrigeration and Air Conditioning*. 13th ed. Khanna Publishers, Delhi.

Banga, K. S., Kotwaliwale, N., Mohapatra, D., and Giri, S. K. (2018). *Techniques for insect detection in stored food grains: An overview*. Food Control, 94, 167-176.

Banga, K. S., Kumar, S., Kotwaliwale, N., and Mohapatra, D. (2020). *Major insects of stored food grains*. International Journal of Chemical Studies, 8(1), 2380-2384.

Banga, K. S., Kumar, S., and Khan, K. A. (2019). *Waste-reduction techniques in fresh produce*. Processing of fruits and vegetables: From farm to fork, 241-278.

Bansal, N. K. and Minke, G. (1988). *Climatic zones and rural housing in India. Part 1 of the Indo-German project on passive space conditioning*. Germany: Germany: N. p., 1988. Web. (199004).

Barre, H.J., Sammet, L.L. and Nelson, G.L. (1988). *Environmental and Functional Engineering of Agricultural Buildings*. An AVI book, published by Van Nostrand Reinhold Company, New York.

Bhadauria, P., Keshava, Mamgai, P., Murai, A. and Jadoun, Y.S. (2016). *Management of Heat Stress in Poultry Production System*. Technical Bulletin, ICAR-ATARI, Zone-1, PAU Campus, Ludhiana.

Bhatnagar, A.P. (1976). *Farm Buildings in Punjab*. PAU Ludhiana publication.

Bhoosreddy, Sanjay (2014). *Poultry Farm Manual*. A Reference Guide for Central & State Poultry Farms. Department of Animal Husbandry, Dairying and Fisheries, Ministry of Agriculture and Fisheries Welfare, Government of India, Krishi Bhawan, New Delhi.

Bhutani, R. (2016). *Quality Surveying and Valuation (Civil Estimating and Costing)*. Ishan Publications, Jalandhar City.

Birewar, B.R. (1985). *Recent Developments in Storage Structures of Agriculture Durable and Semi-Perishable*. CIAE Publication on Storage of Agriculture Durables & Semi-Perishable.

Breedt, Hentie (2005). *Manual on housing for pigs*. Institute for Agricultural Engineering, 141 Cresswell Road, Silverton, Pretoria, South Africa.

Buffington, D.E., Collazo-Arocho, A. and Canton, G.H. (1981). *Black globe-humidity comfort index (BGHI) as comfort equation for dairy cows*. Transactions of the ASAE 24(4):711-714.

Bureau of India Standards. (2005). *National Building Code of India (2005th ed.)*. Bureau of India Standards.

Bureau of Indian Standard. (2016). *National Building Code of India 2016 part 1* (3rd ed., Vol. 2). India: Bureau of Indian Standard.

Chakraverty, A. (1988). *Post-harvest Technology of Cereals, Pulses and Oilseeds*. 3rd edition. Oxford and IBH Publishing Co. Pvt. Ltd., New Delhi.

Chandra, P. (1992). *Instruction manual for establishing a 4 m × 20 m plastic greenhouse*. Division of Agricultural Engineering, IARI, New Delhi, 14 pp.

Choomthi, A. (2008). *Principles of fish farm establishment: Site selection, pond/cage design and development. In Handbook on Community-based Aquaculture for*

Remote Rural Areas of Southeast Asia (pp. 16-25). Secretariat, Southeast Asian Fisheries Development Center.

Christensen, C.M. and Kaufmann, H.H. 1969. *Grain Storage, The Role of Fungi in Quality Loss*. University of Minnesota Press, Minneapolis.

Danner, G. R., & Merrill, P. (2006). *Disinfectants, disinfection and biosecurity in aquaculture*. Aquaculture biosecurity: prevention, control, and eradication of aquatic animal disease, 91-128.

Dhamsaniya, N.K. and Dabhi, M.N. (2018). *Agricultural Process Engineering (Numerical Problems)*. Agrotech Publishing Academy, Udaipur.

Domkundwar, S. (1981). *A Course in Heat and Mass Transfer*. Dhanpat Rai and Sons, Delhi.

Ebeneezar, S., Linga Prabu, D., & Sajina, K. A. (2023). *Feeds and feed management in mariculture*. In: Winter School on Mariculture Technologies for Income Multiplication, Employment, Livelihood and Empowerment. ICAR-Central Marine Fisheries Research Institute, Kochi, pp. 242-255.

Esmay, M.L. and Dixon, J.E. (1986). *Environmental Control for Agricultural Buildings*. The AVI Publishing Company, Inc. Westport, Connecticut.

Flor, O., Palacios, H., Suárez, F., Salazar, K., Reyes, L., González, M., & Jiménez, K. (2022). *New Sensing Technologies for Grain Moisture*. Agriculture, 12(3), 386

Food and Agriculture Organization (1960). *Agricultural Machinery Workshops: Design, Equipment and Management*, FAO Agricultural Development Paper No. 66, Rome.

Food and Agriculture Organization (2011). Rural Structures in the Tropics-Design and Development. FAO, Rome.

Garg, M.K. (2021). *Processing and Food Engineering (Solved Examples)*. Jain Brothers, New Delhi.

Garg, M.K. and Dhingra, D.P. (1986). *Development of thermal environmental model of an enclosure*. J. Agric. Engng., **23**(4):359-67.

Garg, M.K. and Dhingra, D.P. (1987). *Validation and application of thermal environmental model*. J. Agric. Engng., **24**(2):206-13.

Garg, M.K. and Chandra, P. (2005). *Question-Bank on Agricultural and Food Engineering*. 2nd ed. Reliance Publishers, Delhi.

Garg, M.K. and Chandra, P. (2012). *Processing and Food Engineering (Objective questions and answers)*. Jain Brothers, New Delhi.

Goo, J., Kwak, Y., Kim, J., Kang, J., Shin, H., Jo, S. K. and Huh, J. H. (2024). *Development of early design tool for aquaculture buildings using building performance simulation: A case study of an indoor fish farm*. Developments in the Built Environment, 100363.

Gopi, Satheesh (2010). *Basic Civil Engineering*. Pearson, Delhi.

Hall, D.W. 1980. *Handling and Storage of Food Grains in Tropical and Subtropical Area*. Oxford & IBH Publishing Co. Pvt. Ltd., New Delhi.

Hodgson, H.J. (1979). *Role of the Dairy Cow in World Food Production*. Journal of Dairy Science, 62 (2).

IS:278 (2009). *Indian Standard Code of Galvanized Steel Barbed Wire for Fencing – Specification (Fourth revision)*. Bureau of Indian Standards, New Delhi.

IS:607 (1971). *Indian Standard Code of Practice for Construction of Bagged Foodgrain Storage*. Bureau of Indian Standards, New Delhi. (Withdrawn in 2013).

IS:712 (1984). *Indian Standard Code of Specification for Building Limes (Third revision)*. Bureau of Indian Standards, New Delhi.

IS:1080 (1985). *Indian Standard Code of Practice for Design and Construction of Shallow Foundations in Soils other than Raft, Ring and Shell (Second revision)*. Bureau of Indian Standards, New Delhi.

IS:1172 (1993). *Indian Standard Code of Basic Requirements for Water Supply, Drainage and Sanitation. (Fourth revision)*. Bureau of Indian Standards, New Delhi.

IS:1624 (1986). *Indian Standard Code of Galvanized Steel Chain Link Fence Fabric - Specification (Second revision)*. Bureau of Indian Standards, New Delhi.

IS:2470 (1985). *Indian Standard Code of Methods of Field Testing of Building Lime (Second revision)*. Bureau of Indian Standards, New Delhi.

IS:2721 (2003). *Indian Standard Code of Galvanized Steel Chain Link Fence Fabric - Specification (Second revision)*. Bureau of Indian Standards, New Delhi.

IS:2732 (1985). *Indian Standard Code of Practice for Poultry Housing (Second revision)*. Bureau of Indian Standards, New Delhi.

IS:2733 (1985). *Indian Standard Code of Practice for Sheep and Goats Housing (First revision)*. Bureau of Indian Standards, New Delhi.

IS:3916 (1966). *Indian Standard Code of Practice for Pig Housing*. Bureau of Indian Standards, New Delhi.

IS:6313 (2022). *Indian Standard Code of Practice for Anti-Termite Measures in Buildings – Part 3 (Fourth revision)*. Bureau of Indian Standards, New Delhi.

IS:7147 (1973). *Indian Standard Specification for Steel Bins for Domestic Storage. Part – I Gharelu Kothi*. Bureau of Indian Standards, New Delhi.

IS:8453 (1977). *Indian Standard Code of Practice for Construction of Polyethylene Embedded Earthen Bins for Bulk Storage of Grains*. Bureau of Indian Standards, New Delhi.

IS:10500 (2012). *Indian Standard Code for Drinking Water - Specification (Second revision)*. Bureau of Indian Standards, New Delhi.

IS:11786 (1986). *Indian Standard Recommendation for Cattle Housing for an Average Farmer*. Bureau of Indian Standards, New Delhi.

IS:11799 (2005). *Indian Standard Recommendations for Cattle Housing in Rural Areas (First revision)*. Bureau of Indian Standards, New Delhi.

IS:11942 (1986). *Indian Standard Recommendations for Gaushala and other organized milk producers*. Bureau of Indian Standards, New Delhi.

IS:14462 (1997). *Recommendations for Layout, Design and Construction of Greenhouse Structures*. Bureau of Indian Standards, New Delhi.

Jha, S. N., Vishwakarma, R. K., Ahmad Tauqueer, Rai, Anil and Dixit, A. K. (2015). *Report on assessment of quantitative harvest and post-harvest losses of major crops and commodities in India*. ICAR-All India Coordinated Research Project on Post-Harvest Technology, ICAR-Central Institute for Post-Harvest Engineering and Technology, Ludhiana.

Kargwal, Raveena, Garg, M. K., Singh, V.K., Garg, Ruby and Kumar, Nitin (2020). *Principles of modified atmosphere packaging for shelf life extension of fruits and vegetables: An overview of storage conditions*. International Journal of Chemical Studies, 8(3): 2245-52.

Kepenyes, J. (1984). *Recirculating systems and re-use of water in aquaculture*. In ADCP Inter-Regional Training Course in Inland Aquaculture Engineering, Budapest (Hungary), 6 Jun 1983. FAO.

Kohli, D.D. and Kohli, R.C. (2011). *A Text Book of Estimating and Costing (Civil)*. S. Chand & Company Ltd., New Delhi.

Kooner, D.S., Murty, V.V.N. and Sundaresan, D. (1969). *Dairy Cattle Housing*. Farm Bulletin No. 23, Punjab Agricultural University, Ludhiana.

Kumar, D.S. (2004). *Heat and Mass Transfer*. S.K. Kataria & Sons, Nai Sarak, Delhi.

Kumar, S., Mohapatra, D., Kotwaliwale, N., and Singh, K. K. (2017). *Vacuum hermetic fumigation: A review*. Journal of stored products research, 71, 47-56.

Kumar, Sunil, Somavat, Pavel and Garg, M.K. (2017). *RVA study of stored wheat under hermetic storage*. Environment and Ecology, **35(1A)**: 313-317.

Lindley, J.A. and Whitaker, J.H. (1996). *Agricultural Buildings and Structures*. American Society of Agricultural Engineers (ASAE).

Lindmark, M., Audzijonyte, A., Blanchard, J. L., & Gårdmark, A. (2022). *Temperature impacts on fish physiology and resource abundance lead to faster growth but smaller fish sizes and yields under warming*. Global change biology, 28(21), 6239-6253.

Mangaraj, S. (2013). *Modified Atmosphere Packaging of Fruits and Vegetables*. Jain Brothers, New Delhi.

Mangaraj, S., Swain, S. and Alagusundaram, K. (2021). *Agricultural Structures Engineering*. Jain Brothers, New Delhi.

Marai, I.F.M. and Haeeb, A.A.M. (2010). *Buffalo's biological functions as affected by heat stress – a review*. Livestock Sci. 127:89–109.

Mathur, M.L. and Mehta, F.S. (1991). *Steam and Other Tables in S.I. Units*. Jain Brothers, New Delhi.

Michael, A M and T P Ojha (1985). *Principles of Agricultural Engineering*, Vol. I & II. Jain Brothers, New Delhi.

Midwest Plan Service (1977). *Farmstead Planning Handbook*. Ames, Iowa.

Midwest Plan Service (1983). *Structures and Environment Handbook*. MWPS-1, 11th edition, Midwest Plan Service, Iowa State University, Ames, Iowa.

Midwest Plan Service (1984). *Small Farms – Livestock Buildings and Equipment*. Ames, Iowa.

Midwest Plan Service (2013). *Dairy Freestall Housing and Equipment*. Ames, Iowa.

Mohapatra, D., Kumar, S., Kotwaliwale, N., and Singh, K. K. (2017). *Critical factors responsible for fungi growth in stored food grains and non-Chemical approaches for their control*. Industrial Crops and Products, 108, 162-182.

NFDB (2024). National Fisheries Development Board. https://nfdb.gov.in/PDF/Fish% 20&%20 Fisheries%20of%20India/1.Fish%20and%20 Fisheries%20of%20India.pdf

Nagpaul, P.K. and Joshi, B.K. (1987). *Housing Management for Dairy Livestock*. Livestock Adviser, 12(11).

Neubauer L.W. and Walker, H.B. (1961). *Farm Building Design*. Prentice Hall, Inc., New Jersey.

Nikhade, J.S., Rawat, S.N. and Victor, V.M. (2016). *Numerical Problems in Agricultural Engineering*. Jain Brothers, New Delhi.

Ojha, T.P. and Michael, A.M. (2016). *Principles of Agricultural Engineering*, Volume 1. 8th edition. Jain Brothers, New Delhi.

Pal, R.K., Padaria, R.N., Kumbhare, N.V. and Kumar, Pankaj (2009). *Pusa zero energy cool chamber*. Division of Post-Harvest Technology, India Agricultural Research Institute, New Delhi.

Panchdhari, A.C. (1993). *Water Supply and Sanitary Installations (Within Building)*. Wiley Eastern Limited, New Delhi.

Pandey, P.H. (2019). *Principles and Practices of Agricultural Structures and Environmental Control*. Kalyani Publishers, New Delhi.

Pandya, N.C. and Shah, C.S. (1979). *Logarithms, Steam and Other Tables*. Charotar Book Stall, Anand.

Patil, R.T., Jha, S.N. and Wanjari, O.D. (2009). *Agricultural Structures and Environment Control for Plants, Animals and Fisheries*. Central Institute of Post-Harvest Engineering and Technology, Ludhiana.

Prasad, Rajendra (1986). *Improved cook stoves*. Technical manual. Department of non-conventional energy sources. Government of India.

Prasad, Rajesh K. and Ojha, T.P. (2012). *Non-conventional Energy Sources*. Jain Brothers, New Delhi.

Proctor, D.L. 1994. *Grain Storage Techniques*. FAO, Rome.

Rai, G D (1992). *Non-conventional Energy Sources*. Khanna Publishers, Delhi.

Rangwala, S.C., Rangwala, K.S. and Rangwala, P.S. (1996). *Engineering Materials*. Charotar Publishing House, Anand.

Richey, C.B., Jacobson, P., and Hall, C.W. (1961). *Agricultural Engineers' Handbook*. McGraw Hill Book Company, New York.

Roy, K. S., & Khurdiya, D. S. (1982). *Keep vegetables fresh in summer (Vol. 27)*. Indian Horticulture.

Roy, S. K., & Khurdiya, D. S. (1985). *Zero energy cool chamber (Vol. 43)*. India Agricultural Research Institute: New Delhi, India. Research Bulletin.

Roy, S. K. (1989). *On-farm storage technology can save energy and raise farm income*. Amity Science, Technology & Innovation Foundation. http://ucce.ucdavis.edu/files/ datastore/234-2143.pdf.

Sahay, K.M. and Singh, K.K. (1996). *Unit Operations of Agricultural Processing*. Vikas Publishing House Pvt. Ltd., New Delhi.

Sahoo, P. R., Ananth, P. N., Nandi, S., Sundaray, J. K., Barik, N. K., & Jayasankar, P. (2017). *Early breeding and seed production of Indian major carps: Attributes of the innovation from an adaptive trial*. Current Agriculture Research Journal, 5(1), 59-65.

Sastry, N.S.R. and Thomas, C.K. (1976). *Farm Animal Management*. Vikas Publication, New Delhi.

Sastry, N.S.R. and Thomas, C.K. (1991). *Livestock Production Management*. 3rd ed. Kalyani Publishers, New Delhi.

Seilheimer, T. S., Wiermaa, E. and Jescovitch, L. N. (2021). *Fisheries, Hatcheries, and Aquaculture—What's the Difference?* Choices, 36(4), 1-9.

Sharma, P.C., Gopalkrishnan, S., Raman, N.V. and Kumar, G.V.S. 1978. *Small capacity ferrocement bins for food grain storage*. International conference on materials of construction for developing countries, Bangkok, Thailand.

Sharma, Vikesh and Dash, Madhusmita (2022). *Household energy use pattern in rural India: A path towards sustainable development*. Environmental Challenges, 6, 1 – 9, doi.org/ 10.1016/j.envc.2021.100404.

Singh, S.K., Singh, B.R., Samsher and Singh, Jaivir (2007). *Training manual on greenhouse practices*. Department of Agricultural Engineering & Food Technology, Sardar Vallabh Bhai Patel University of Agriculture & Technology, Merrut.

Somavat, Pavel, Huang, H., Kumar, Sunil, Garg, M.K., Danao, M.C., Singh, V., Rausch, K.D. and Paulsen, M.R. (2017). *Comparison of hermetic storage of wheat with traditional storage methods in India*. Applied Engineering in Agriculture,**33(1)**:121-130.

Subrot Panigrahi, S., Singh, C. B., Fielke, J., and Zare, D. (2020). *Modelling of heat and mass transfer within the grain storage ecosystem using numerical methods: A review*. Drying Technology, 38(13), 1677-1697.

Upadhyay, R.C., Ashutosh, Raina, V.S. and Singh, S.V. (2009). *Impact of climate change on reproductive functions of cattle and buffaloes*. In: Aggarwal PK (ed) Global climate change and Indian agriculture: case studies from the ICAR network project. ICAR Publication, New Delhi, pp 107–110.

Vasavada, A.A. (1954). *Farm Structures*. Yeravada Prison Press, Poona.

Vaugh, Mason (1962). *Farm Buildings for India*. Department of Agricultural Engineering, Allahabad Agriculture Institute, India.

Sooch, S.S. and Grewal, N.S. 2016. *Renewable Energy: Biogas – An ideal source*. Punjab Agricultural University, Ludhiana.

Verma, Deepak and Aggarwal, Anand (2018). *Water Supply and Waste Water Engineering*. Ishan Publications, Jalandhar City.

Tanveer, M., Balasubramanian, S., Sivakumar, M., Manimehalai, N., & Jagan, P. (2018). *A technical review on feeders in aquaculture. International Journal of Fisheries and Aquatic Studies* 6(4): 305-309.

Threlked, J.L. (1977). *Thermal Environmental Engineering*. Prentice Hill Inc., Englewood Cliffs, New Jersey.

Tom, A. P., Jayakumar, J. S., Biju, M., Somarajan, J. and Ibrahim, M. A. (2021). *Aquaculture wastewater treatment technologies and their sustainability: A review*. Energy Nexus, 4, 100022.

Tucker, C. (2005). *Pond Aeration*. SRAC publication, 3700.

Whitaker J.H. (1979). *Agricultural Buildings and Structures*. Reston, Virginia, Reston Publishing Co.

World Health Organization (2016). *Burning opportunity: clean household energy for health, sustainable development, and wellbeing of women and children*. Geneva, Switzerland.

Worley, J.W. (2000). *Fences for the Farm*. Biological and Agricultural Engineering Department. College of Agricultural and Environmental Sciences. Cooperative Extension Service. The University of Georgia.

Yap, W. G. (2001). *Developments in marine and brackish water fish culture in Southeast Asia*. In Responsible Aquaculture Development in Southeast Asia. Proceedings of the Seminar-Workshop on Aquaculture Development in Southeast Asia organized by the SEAFDEC Aquaculture Department, 12-14 October 1999, Iloilo City, Philippines (pp. 45-66). Aquaculture Department, Southeast Asian Fisheries Development Center.

https://dahd.nic.in Department of Animal Husbandry & Dairying, New Delhi.

https://dahd.nic.in/schemes/programmes/animal-husbandry-statistics

https://en.wikipedia.org/wiki/Deep_litter

https://vikaspedia.in/agriculture/poultry/breeds-of-poultry/housing-for-layers

Solar Power Fencing for Crop Protection from Monkey Menace and Domestic/ Wild Animals in Himachal Pradesh.

https://www.pericoli.com/en/temperature-humidity-index-what-you-need-to-know-about-it/

https://www.researchgate.net/publication/248626590_The_pros_and_cons_of_cages

https://hobbyfarms.com/jaguzafarm.com/

https://agricultureguruji.com/greenhouse-farming/Greenhouse farming in India (Beginner Guide 2021)

https://www.researchgate.net/figure/Recommended-temperature-and-relative-humidity-for-fruits-and-the-approximate-storage-life_tbl3_259841724

www.pericoli.com

Vikaspedia.in/agriculture/livestock/pig farming.1

Vikaspedia.in/agriculture/livestock/role-of-livestock-in-indian-economy

ecourseonline.iari.res.in.

Index

D

E

O

P

Q

R

S

Y

Z